Open Quantum Systems and Feynman Integrals

Fundamental Theories of Physics

A New International Book Series on the Fundamental Theories of Physics: Their Clarification, Development and Application

Open Quantum Systems and Feynman Integrals

by

Pavel Exner

Nuclear Centre, Charles University,
Prague, Czechoslovakia

D. Reidel Publishing Company

A MEMBER OF THE KLUWER ACADEMIC PUBLISHERS GROUP

Dordrecht / Boston / Lancaster

Library of Congress Cataloging in Publication Data

Exner, Pavel, 1946-
Open quantum systems and Feynman integrals.

(Fundamental theories of physics)
Bibliography: p.
Includes index.
1. Open systems (Physics) 2. Quantum theory. 3. Feynman integrals.
I. Title. II. Series.
QC174.85.06E96 1984 530.1'2 84-17853

ISBN-13: 978-94-010-8803-9 e-ISBN-13: 978-94-009-5207-2
DOI: 10.1007/978-94-009-5207-2

Published by D. Reidel Publishing Company,
P.O. Box 17, 3300 AA Dordrecht, Holland.

Sold and distributed in the U.S.A. and Canada
by Kluwer Academic Publishers,
190 Old Derby Street, Hingham, MA 02043, U.S.A.

In all other countries, sold and distributed
by Kluwer Academic Publishers Group,
P.O. Box 322, 3300 AH Dordrecht, Holand.

TO MY PARENTS

Who taught me to perceive
beauty and honour truth

Table of Contents

Preface

Every part of physics offers examples of non-stability phenomena, but probably nowhere are they so plentiful and worthy of study as in the realm of quantum theory. The present volume is devoted to this problem: we shall be concerned with open quantum systems, i.e. those that cannot be regarded as isolated from the rest of the physical universe. It is a natural framework in which non-stationary processes can be investigated.

There are two main approaches to the treatment of open systems in quantum theory. In both the system under consideration is viewed as part of a larger system, assumed to be isolated in a reasonable approximation. They are differentiated mainly by the way in which the state Hilbert space of the open system is related to that of the isolated system – either by orthogonal sum or by tensor product. Though often applicable simultaneously to the same physical situation, these approaches are complementary in a sense and are adapted to different purposes. Here we shall be concerned with the first approach, which is suitable primarily for a description of decay processes, absorption, etc. The second approach is used mostly for the treatment of various relaxation phenomena. It is comparably better examined at present; in particular, the reader may consult a monograph by E. B. Davies.

In the existing literature, unstable systems are seldom studied from first principles. From the practitioner's point of view, this is not necessary as we have a functioning method, developed in the early years of quantum theory, by which we can satisfactorily describe the observed decays and often predict their characteristics. However, in a theory pretending to be a complete and coherent description of the micro-world, such a method should admit derivation from the postulates. This task is fraught with peculiar difficulties, mainly because the mathematics here is somewhat more involved than in the 'stationary' quantum mechanics, and formal considerations sometimes yield misleading or seemingly paradoxical results. One of the aims of this volume is to prevent the permanent rediscovering of these 'paradoxes'. At the same time, a careful analysis of the problem can provide a deeper insight into some of the fundamental problems of quantum theory.

By the methods used, the book is addressed primarily to those engaged in mathematical physics, but it is hoped that the work will be accessible and useful to a wider audience among mathematicians, physicists and quantum chemists. The choice of scope reflects the author's opinion that mathematical physics is *not* a part of mathematics *or* physics but rather a bridge between the two – a bridge whose existence must be defended at any expense.

The book can be divided into two parts. The first three chapters contain what one might call a sketch of the general theory of unstable quantum systems. Here we examine the Hilbert-space kinematics of decays, the dynamical mechanism that governs most of the practically interesting decay processes, their symmetries, etc. Particular attention is paid to a possible influence of interaction with the environment to the decay laws. These considerations illustrate how the commonly used semigroup ansatz for time evolution of the unstable systems arises. We know that it is necessarily approximative, but this is nothing to be afraid of if we understand the nature of this approximation and if we are able to estimate its degree of accuracy.

In the second part, we treat the semigroup time evolution from the viewpoint of its infinitesimal generator, thus giving a more rigorous meaning to the phenomenological non-selfadjoint Hamiltonians widely used in some branches of quantum physics. For these operators, which we call pseudo-Hamiltonians, a sort of generalized quantum mechanics can be constructed. Various relevant results can be found in the mathematical literature; usually one has only to add the dissipativity condition which makes the natural physical interpretation possible. A particular interest concerns the class which appears most frequently in the applications; namely, the Schrödinger-type operators with complex-valued potentials. Notice that aside from its primary physical aim, the theory of pseudo-Hamiltonians also has some technical merit. For instance, it may sometimes be useful to regularize a mathematical construction for a given Schrödinger operator by introducing a suitable absorptive term to the potential instead of imposing some truncations or constraints.

Rather than a detailed exposition of all the known results on non-selfadjoint second-order differential operators, we present a new method of treating the Schrödinger pseudo-Hamiltonians. This method is based on extending to complex potentials the (rigorous version of) Feynman's path-integral solution to the Schrödinger equation. As a preliminary, we include an extensive and more or less self-contained chapter where the Feynman path integrals are examined. This problem has attracted attention for over thirty years, during which time many people have tried to set this appealing concept on a firm mathematical basis. In fact, a physicist need not master any such theory. It is sufficient for him to know that his considerations can lean on a mathematically sound formalism which might be consulted in the case of ambiguities. A comparison with the theory of distributions arises naturally, but the present problem is further complicated by the fact that it admits various mathematical approaches of which none, until now, could win a dominating role. Hence, it is likely that we shall still not have a satisfactory and commonly accepted theory of Feynman path integrals. Nevertheless, considerable progress has been achieved in the last decade, and some of these new results will be reported here.

Let me add a few words about the history of this volume. It began eleven years ago when our teacher Prof. V. Votruba, in order to terminate an ardent seminaral dispute, asked M. Havlíček and me to formulate the basic features common to all decaying quantum systems. As this was an interesting problem, we often returned to it, with the constant and valuable help of Prof. Votruba, after we had finished the required analysis. There were also other proposals that inspired some parts of the present study; I have to mention Prof. I. Úlehla who pleaded for a better justification of the complex-potential methods, and Dr G. I. Kolerov who brought to my attention the problems of functional integration. Most of the material has been discussed in a series of seminars held at the Joint Institute for Nuclear Research in Dubna, and also in Prague, Leipzig and other places. A small part of it was used in the course of lectures on Hilbert-space operators for quantum theory given at the Charles University in Prague.

In conclusion, let me acknowledge the help I have received from numerous colleagues. Beside those mentioned above, I would particularly like to thank: Profs N. N. Bogoliubov and V. A. Meshcheryakov, for the hospitality extended to me at the Laboratory of Theoretical Physics, JINR, and constant support; my teacher Dr M. Havlíček, and Dr J. Blank, for the years of delight and inspiration in common work; Profs A. Degasperis and G. Lassner, and Drs L. Dadashev, J. Dolejší, J. Hořejší, E.-M. Ilgenfritz, R. Kotecký, M. Lokajíček, M. I. Shirokov, J. Stará and J. Tolar, for helpful discussions; Profs B. M. Barbashov, J. Formánek, B. Geyer, D. Robaschik, B. Simon, O. G. Smolianov, A. Uhlmann, and the referees, for valuable comments; Profs G. Johnson, A. Truman and others who made available their results prior to publication; Prof. A. van der Merwe, and Dr D. J. Larner of D. Reidel, for useful advice in preparing the manuscript; Drs J. Hořejší, O. Navrátil and P. Seba, for help with proofreading; the staff of LTP, especially Mrs V. Ryzhova for copying and Mrs Z. Pavliukevicz for drawing the illustrations; and all others whose support made it possible to finish this book. Last, but not least, I am deeply obliged to my wife, Jana, for her understanding and encouragement.

Dubna, July 1983 P. E.

Prerequisites for Reading

A brief outline of the book was given in the Preface; a more detailed guide can be found in the opening to each chapter. General motivating considerations are presented at appropriate places in the text, especially in Sections 1.1 and 5.1. As mentioned above, we shall be concerned mainly with the non-stability phenomena in open systems, and therefore we shall use mostly the more specified term 'unstable quantum system' in the sense concretized in Section 1.2. Later we shall also speak occasionally about 'dissipative systems' as those whose time evolution is governed by a pseudo-Hamiltonian. As a rule, the terminology is introduced in the course of the exposition. We avoid formal definitions; a new term is distinguished typographically and can be identified in its following appearances by means of the subject index. The mathematical notation is given in the List of Symbols and Notations.

The mathematics used is standard for the most part, and should be accessible to students of higher undergraduate courses. The main mathematical prerequisite is linear functional analysis, particularly the theory of linear operators on Hilbert spaces. In order not to exceed a reasonable volume extent, I have declined the idea of including such a preliminary in the book. There are many excellent monographs and textbooks to consult, and we shall refer to them if necessary. Some assertions, which are important in the deductions, are presented in the Appendix. Of course, their choice is rather subjective.

From the physical side, acquaintance with the fundamentals of quantum theory is assumed. The emphasis is on a consistent and rigorous formulation rather than on new physical concepts. In particular, treatment of the repeated-measurements problem in Chapter 2 relies fully on the standard interpretation of quantum measurements. We use mainly a rationalized system of units, with the exception of those places where the values of c, $\hbar$ or other dimensional physical quantities are of importance.

To facilitate reference within a chapter we employ a consecutive dual-numbering system for theorems, propositions, lemmas, etc. (e.g. 1.1, ...), and also for formulae (e.g. (1.1), ...). The chapter number is only included when the item referred to is not contained in the chapter under consideration.

Each chapter is accompanied by notes which contain most of the bibliographical references, except those which mainly serve technical purposes. In general, the bibliography to the problems discussed here is very extensive, and in order to obtain a finite result in a finite time we are forced to omit many interesting results.

The system of bibliographical referencing is more or less standard. If an assertion is to be located more exactly in a book or a paper, reference has been made to the number of the theorem, section, etc., rather than to the page. This seemed to be more practical, since our world is a multiply-connected manifold, and often a quoted source is accessible in translated form only. The book also contains a list of selected problems. It may be regarded as an invitation; some open questions can be found in the text, and many more will appear when you begin to work actively in this field.

List of Symbols and Notations

"Praise the small disorder, easy to survey. You can preserve it for many days, while a perfect order spoils during few hours."

M. HORNÍČEK, *A Well-Hidden Violin*

$\langle A \rangle_\rho$	expectation of A in the state ρ
$(\Delta A)_\rho$	dispersion of A in the state ρ
AC^p	see Example 4.3.9
$AC_x[J^t; \mathbb{R}^d]$	path space, see Section 5.3
$B(H)$	bounded-energy states
$\mathrm{Bor}(\mathscr{H})$	Borel sets in $\mathscr{H}$
$\mathscr{B}(\mathscr{H})$	bounded operators
$\mathscr{B}_+(\mathscr{H})$	the positive cone in $\mathscr{B}(\mathscr{H})$
$\mathscr{B}(\mathscr{H}_1, \mathscr{H}_2)$	bounded operators from $\mathscr{H}_1$ to $\mathscr{H}_2$
c	velocity of light, in Sections 2.3 and 3.5
C^∞	infinitely differentiable functions
C_0^∞	C^∞-functions with a compact support
$C_x[J^t; \mathbb{R}^d]$	path space, see Section 5.3
$\|C\|_H$	H-norm of the operator C, see (A.14)
$\mathbb{C}$	complex numbers
$\mathbb{C}_F, \mathbb{C}_F^0$	see Section 5.2
$\det M$	determinant of M
$D(T)$	domain of the operator T
$\mathrm{D}\gamma$	formal Lebesgue measure on path spaces
$\mathrm{D}\phi_s(\gamma)$	formal Feynman measure, see Section 5.1
$E_A(\,.\,)$	spectral measure of the operator A
$E(t)$	measured decay law, in Chapter 2
E_u	projection on the subspace $\mathscr{H}_\mathrm{u}$, in Chapters 1 and 3
F_d	Fourier–Plancherel operator
$\mathscr{F}(\mathscr{H})$	B-algebra of Fresnelian-integrable functions, see Section 5.2
$\mathscr{F}_s^\alpha(\mathscr{H}_x)$	F-integrable functions according to the definition (α), in Chapter 5
$\mathscr{F}(\mathbb{R}^d)$	see Remark 6.1.2(a)
g	coupling constant, in Chapter 3 and Section 4.4

$G_t(x, y)$	propagator kernel, in Section 6.2
$G(\tau, \xi)$	reproducing kernel, in Section 5.3, 5.4
$\mathscr{G}(T)$	graph of the operator T
$\hbar$	Planck's constant
H	Hamiltonians, pseudo-Hamiltonians
H_0	free Hamiltonian, unperturbed Hamiltonian
$\mathscr{H}$	Hilbert spaces
$\mathscr{H}_{\mathrm{d}}$	orthogonal complement to $\mathscr{H}_{\mathrm{u}}$
$\mathscr{H}_{\mathrm{u}}$	state space of an unstable system, in Chapters 1 and 3
$\mathscr{H}_x^\sigma$	polygonal paths, see Section 5.4
$\mathscr{H}_{\mathrm{ac}}(H)$	absolutely continuous subspace of H, see also Section 4.5
$\mathscr{H}_{\mathrm{bound}}(H)$	subspace spanned by the bound states of H
$\mathscr{H}_{\mathrm{dec}}(H)$	decaying states, see Section 4.5
$\mathscr{H}_{\mathrm{nre}}(H)$	see Section 4.5
id	identical mapping
inf ess	essential infimum (with respect to a given measure)
I_s^{a}	analytic F-integral
I_s^{f}	Fresnelian F-integral (F-map)
I_s^{g}	Gaussian-regularized F-integral
I_s^{luc}	limiting uniform cylindrical F-integral
I_s^{p}	product F-integral (F-map)
I_s^{rc}	regular cylindrical F-integral (F-map)
I_s^{ri}	regular improperly-cylindrical F-integral (F-map)
I_s^{uc}	uniform cylindrical F-integral (F-map)
I_s^{ui}	uniform improperly-cylindrical F-integral (F-map)
I_s^{up}	uniform product F-integral (F-map)
$I_s^{\mathrm{up}}[\,.\,]$	uniform product operator-valued F-integral
J	conjugation
J^t	the interval $[0, t]$
$\mathscr{J}_1$	trace-class operators
$\mathscr{J}_2$	Hilbert–Schmidt operators
Ker T	kernel (null-space) of the operator T
liminf	limits inferior
limsup	limits superior
L^p	(equivalence classes of) functions integrable with the pth power
L^∞	(equivalence classes of) bounded measurable functions
$L^p + L^\infty$	see Section 4.3
$L^p + L^\infty_\epsilon$	see Section 4.3
L^2_{loc}	(equivalence classes of) locally square-integrable functions
$L_{\mu\nu}$	generators of a Lorentz-group representation
$L_{\mathscr{P}}$	Lie algebra of $\mathscr{P}$
M_{lin}	complex linear span of M
M^{T}	transpose of the matrix M
$M^\perp$	orthogonal complement to M

$M(H)$	finite-energy states
$\mathcal{M}(H)$	see (4.5.1)
$\mathcal{M}(\mathcal{H})$	finite Borel measures on $\mathcal{H}$, see Section 5.2
pr	see Proposition 1.2.1
P_k, P_μ	momentum operators
$P(t)$	primary decay law, in Chapter 2
$P_\rho(t)$	decay law, see Section 1.2
$P_{ac}(H)$	projection to $\mathcal{H}_{ac}(H)$
$P(\sigma), P^c(\sigma)$	projections on polygonal paths, see Section 5.4
$P_H(\{\lambda\})$	generalized spectral projection, see Section 4.3
$\mathcal{P}$	Poincaré group
$\mathcal{P}(J^t)$	partitions of J^t
$\mathcal{P}_x[J^t; \mathbb{R}^d]$	polygonal paths, see Section 5.4
Q, Q_k	position, coordinate operators
$Q(A)$	form domain of the operator A
Ran T	range of the operator T
$R(z, H)$	resolvent of the operator H
$R_u(z, H)$	reduced resolvent of H, see Section 3.1
$R_u^{II}(z, H)$	analytically continued reduced resolvent
$\mathbb{R}, \mathbb{R}^d$	real axis, d-dimensional Euclidean space
$\mathbb{R}_+, \mathbb{R}_-$	the intervals $[0, \infty)$ and $(-\infty, 0]$, respectively
s	inverse-mass parameter, in Chapter 5
supp	support
supp ess	essential supremum (with respect to a given measure)
s-lim	strong-operator limit
S	scattering operator
$S(\gamma)$	action along the path γ
$\mathcal{S}(\mathbb{R}^d)$	Schwartz space of rapidly decreasing functions
t_p	chamber-structure period, in Chapter 2
T_ρ	lifetime of the state ρ
T^*	adjoint operator to T
$\bar{T}$	closure of T
Tr B	trace of B
u, U	complex potentials
u-lim	operator-norm limit
U_t	unitary propagator
$U(t, s; E)$	propagator in the limit of continual observation
$U_{opt}(E)$	Feshbach optical potential, in Section 4.4
v, V	real potentials
V_t	reduced propagator
$V(t, s)$	contraction-valued propagator
w	Wiener measure
w_c	Wiener measure with dispersion c
w-lim	weak-operator limit

$W(\,.\,,.\,)$	chamber-structure function, in Chapter 2
x, y	vectors in $\mathbb{R}^d$
X, Y	Banach spaces
$\bar{z}$	complex conjugate of z
$\int^{\alpha} \mathrm{D}\phi_s(\gamma)$	F-integral according to the definition (α), see I_s^{α}
β	velocity of the particle divided by c, in Sections 2.3, 3.5
γ	paths, in Chapters 5 and 6
Γ	decay rate
$\Gamma(g)$	decay rate, in Chapter 3
Γ_x	path space, unspecified
δ_j	subinterval length in a partition
$\delta(\sigma)$	maximal subinterval length of the partition σ
Δ	Laplacian
Δ_D^M	Dirichlet Laplacian relative to the region M
Δ_S	Laplace–Beltrami operator
Δ_j	subinterval of a partition
Θ	antiunitary symmetries, in Sections 2.4, 3.4
Θ	Heaviside function
$\Lambda_{\mu\nu}$	Lorentz group transformation, in Section 3.5
$\mu * \nu$	convolution of μ, ν
$\mu \otimes \nu$	product-measure of μ, ν
$\|\mu\|$	total variation of the measure μ
ν_φ	measurement frequency, in Section 2.2
ρ	states, in Chapter 1
$\rho(T)$	resolvent set of the operator T
σ	partitions, see Section 5.4
$\sigma[V]$	energy support of V, see Section 1.5
$\sigma(H)$	spectrum of the operator H
$\sigma_{\mathrm{ac}}(H)$	absolutely continuous spectrum
$\sigma_{\mathrm{c}}(H)$	continuous spectrum of H
$\sigma_{\mathrm{disc}}(H)$	discrete spectrum of H
$\sigma_{\mathrm{ess}}(H)$	essential spectrum of H
$\sigma_{\mathrm{p}}(H)$	point spectrum of H
τ_j	points of a partition
φ, ψ	vectors in Hilbert spaces, pure states
χ_M	characteristic function of the set M
Ω	complex frequency matrix, in Section 6.2
$\Omega_{\pm}(H, H_0)$	wave operators, see Sections 3.2, 4.5
$\subset$	inclusions
$\oplus, \Sigma^{\oplus}$	orthogonal sums of Hilbert spaces and operators
$\oplus$	B-space direct sum of Banach spaces
$\circ$	see Problem 28
$\otimes$	tensor products
$\times$	Cartesian product

$\setminus$	set difference
$\restriction$	restriction of a mapping
$\circ$	composition of mappings
$(\cdot,\cdot), \langle\cdot,\cdot\rangle$	scalar products
$\Vert\cdot\Vert$	Hilbert-space norm, $\mathscr{B}(\mathscr{H})$-norm
$\Vert\cdot\Vert_p$	L^p-norm
$\Vert\cdot\Vert_\infty$	supremum norm
$\Vert\cdot\Vert_0$	norm on $\mathscr{F}(\mathscr{H})$, see Section 5.2
$\Vert\cdot\Vert_1$	trace norm on $\mathscr{T}_1$
$\Vert\cdot\Vert_2$	Hilbert–Schmidt norm
$\vert\cdot\vert$	modulus, absolute value of an operator
■	end of a proof

Chapter 1

Quantum Kinematics of Unstable Systems

"... while the experimentalist might collect all his data between breakfast and lunch in a small cluttered laboratory, his theoretical colleagues interpret those results in terms of isolated systems moving eternally in infinitely extended space. The validity of appropriate approximations of this type is the heart and soul of theoretical physics and has the same fundamental significance as the reproducibility of experimental data."

O. BRATELLI and D. W. ROBINSON

Let us first present a few historical facts and a brief survey of the problems related to unstable quantum systems. This chapter is devoted to a discussion of the kinematical properties (in the quantum-theoretical sense) of unstable systems, especially their time evolution. We begin with the basic notions and formulation of postulates which are expected to be obeyed be any unstable quantum system (Section 2). Since they represent just one of several approaches towards a description of open systems within the quantum theory, we also review here the other existing methods. We then discuss how the decaying system behaves over short intervals (i.e. soon after the instant of preparation), and dependence of this behaviour on the energy distribution of the initial state (Section 3). Since each unstable system can be viewed as a part of some larger isolated system, there arises the natural question: 'Can one reconstruct a complete description of the decay from the knowledge of the time evolution of the unstable system itself?' This is the so-called inverse decay problem treated in Section 4: we present there the basic criterion for existence and uniqueness of the solution as well as its simple properties. The most frequently used and practically successful way of describing the decay processes starts from the assumption that the time evolution of an unstable system is governed by an operator semigroup. However, the well-known difficulty of the below unbounded total energy then arises. In Section 5 we discuss this problem in detail, and at the same time present other properties of the energy spectrum together with some examples. Thus we have to show the sense in which the semigroup description may be used as an approximation to the true reduced evolution. With this purpose, we treat the so-called bounded-energy approximation (Section 6), which has a natural physical meaning. A discussion of the quantitative aspects of the problem is postponed until the next chapter.

1.1. Is There Anything Left to Study on Unstable Systems?

Some irreversible processes are of extreme physical importance: the approach to the thermodynamic equilibrium; the interaction of a microsystem with macroscopic measuring apparatus; or the spontaneous decays of some elementary particles, nuclei, etc. Here we shall be mainly concerned with the last case.

Let us recall first some historical facts. The investigation of unstable nuclei began when H. Bequerel discovered natural radioactivity and his followers proved the existence of the related transmutations of elements. According to the empirical law formulated by J. Elster and H. J. F. Geitel in 1899, the amount of radioactive material decreases exponentially with time, $N(t) = N(0)\exp(-\Gamma t)$, and each type of decay is characterized by the mean life Γ^{-1}. An elementary explanation says that the decay probability of each nucleus does not depend on the instant when it was prepared; this was first postulated by E. von Schweidler in 1905. One should realize it was a bold argument at that time because it assumed a non-deterministic description for the decay of an individual nucleus.

However, within classical (non-quantum) physical theories one can neither explain this *ad hoc* assumption nor determine the decay probability per unit time Γ. The first attempt to derive a decay law from a microphysical concept was undertaken by G. Gamow (1928) in his semiclassical theory of α-radioactivity. A little later, V. F. Weisskopf and E. P. Wigner (1930) formulated the first fully quantum theory of a decay process, namely damping of the radiation emitted from an excited atom. They considered the atom to be initially in a state ψ_1, which then passes to a state ψ_2, with lower energy while a photon is radiated. The central *assumption* of this theory was that the probability of finding the atom in the initial state ψ_1 decreases as $\exp(-\gamma t)$ with time. In such a case, the frequency distribution of radiation intensity assumes a Breit–Wigner (or Lorentzian) shape. Weisskopf and Wigner found the width γ of the distribution peak (i.e. of the spectral line) to be equal to the sum of probabilities per unit time of all energetically allowed transitions from the states ψ_1, ψ_2 (in particular, of the transition $\psi_1 \to \psi_2$, if ψ_2 is the ground state and ψ_1 the lowest excited state of the atom). These probabilities are calculated within the framework of perturbation theory and the results show a reasonable agreement with experimental data – cf. Heitler (1945, §V.18).

Half a century later, we made two more steps into the structure of matter. First, many elementary particles were discovered, then many of their properties were explained using even more elementary quarks. Nevertheless, the problem of non-stability remains. We disregard the quarks, which are not expected to be seen as single free objects; however, most of the elementary particles decay. In fact, of nearly twenty particles that were conventionally denoted as stable, there are actually only few stable ones (within the range of experimental errors: γ, ν_e, ν_μ, e, p). Furthermore, some recent results, both experimental (indications for a non-zero rest mass of $\bar{\nu}_e$) and theoretical (connected with attempts to unify all non-gravitational forces), suggest that even neutrinos and protons might decay.

What is important is that the scheme devised originally for radiation damping applies to a surprising number of processes of decay ranging from molecular to subnuclear level, mainly in cases where the interaction responsible for decay is reasonably weak in some sense. On the other hand, there are unstable systems which need a slightly more complicated description. As an example, consider the *neutral kaons* described conventionally as a superposition of two exponentially decaying states with different lifetimes, which are moreover non-orthogonal due to non-conservation of the CP-parity (Perkins, 1972, §4.12). Nevertheless, essentially the same method is applicable in this case if the original assumption concerning the exponential decay is replaced by a more general one, namely that the time evolution in the kaon state space is governed by a one-parameter family $\{V_t : t \geqslant 0\}$ of operators obeying the semigroup condition

$$V_t V_s = V_{t+s}, \quad t, s \geqslant 0. \tag{1.1}$$

There are other cases in which the introduction of condition (1.1) can be avoided but where it conveniently simplifies the formalism: remember, for instance, the hyperfine-structure measurements of atomic levels using the beam-foil technique (Martinson, 1974), where the unstable state is often an orthogonal superposition of two or more excited states, each of which decays exponentially (up to experimental errors).

The generalized Weisskopf–Wigner method, which starts from assumption (1.1) and expresses the generator of the semigroup $\{V_t\}$ by means of the perturbative technique (most frequently in the lowest order – via the so-called *Fermi golden rule*), represents a simple and effective tool for describing the decay processes; it well satisfies the practical requirements of all branches of quantum microphysics. On the other hand, the reliability of such a method is usually much higher if it leans on a theory starting from first principles in some well-defined sense. This is so in the various methods of 'stationary' quantum mechanics; in particular, those which express the shift of an isolated energy eigenvalue under the influence of a perturbation by means of a power series converging to a locally analytical function of an appropriate parameter (coupling constant). Surprisingly enough, the Weisskopf–Wigner method has a much weaker footing. The difficulties stem particularly from the very semigroup condition (1.1), which is known to be incompatible with the natural physical requirement that the total energy must be bounded from below. It means that the method cannot in principle pretend to give an exact description, although it is able to yield an excellent approximation to it.

The deviations are likely to be negligible from the practical point of view, but the problem has other aspects. The following is important: the deviations are not only unavoidable but are also dependent on the way in which the considered unstable system is prepared (as we shall see later, particularly in Sections 5 and 2.3). Thus the *question of identity* arises: one might ask, say, whether two pions produced in different experimental conditions are actually alike.

Another problem concerns the *dynamical mechanism of decays*. Typically the total Hamiltonian is expressed as a sum of two parts: the decaying states refer to eigenvalues of the unperturbed Hamiltonian but belong to the continuous spectrum of the total one. In this sense, therefore, it is clear that the decay is caused by switching on the perturbation; on the other hand, quantitative analysis of such cases relies on the perturbation theory of eigenvalues embedded in the continuous spectrum which is much less developed than the 'stationary' perturbation theory mentioned above.

There are other problems which should be solved within any theory which pretends to be a complete description of unstable quantum systems. In the first place, let us look at the *connection with the scattering theory*. Of course, there is a substantial difference between metastable objects and scattering resonances which is well known from non-relativistic potential scattering. If we are able to determine the shape of the resonance (from cross section or phase shift), then the times involved are too short to allow measurement of the decay plot, and vice versa. On the other hand, a common phenomenological framework for metastable objects and resonances is often useful (e.g. in the quark models) so this connection should be kept in mind and clarified further. Secondly, there is the problem of *relativistic invariance* concerning primarily the unstable elementary particles. We know that the stable elementary objects are classified by means of unitary irreducible representations of the Poincaré group $\mathscr{P}$. One is therefore tempted to associate with unstable elementary particles suitable irreducible non-unitary representations of $\mathscr{P}$. On the other hand, the postulate of relativistic invariance demands existence of a unitary (reducible) representation of $\mathscr{P}$ ascribed to the isolated system formed by the unstable particle itself and its decay products. The connection between these two representations is not *a priori* clear and has to be studied.

Yet another problem concerns specific features of unstable systems within the *quantum field theories*. This can be divided into two distinct parts. The first is connected with the observation that all the 'functioning' field theories (beside their other mathematical flaws) deal with the basic quanta of the fields as though they were stable objects. Thus it is certainly an approximation when one ascribes an external line with fixed energy in the appropriate Feynman graph to a muon when calculating the cross-section of its scattering on a proton. In spite of its negligible importance for a comparison with experimental data, this problem should disturb each theoretically inclined physicist. An early attempt to construct a quantum field theory of unstable particles was undertaken by Matthews and Salam (1958, 1959); however, it was soon evident that the obtained object belonged to the Borchers class of free fields – see Williams (1971). Thus the task remains, and seems to be difficult. The other part of the problem is practically more important; namely, the dynamical derivation of decay characteristics. Here again the situation is more complicated compared to quantum-mechanical systems, when one tries to go beyond the standard low-order perturbative calculations. Much attention has been paid to the solvable models of quantum fields – particularly to the Lee model.

A suitable starting point for studying these and other problems can be found in the *kinematical concept of unstable systems*, which consists of treating those properties which are expected to be common for all such systems being implied by the basic postulates of the quantum theory only. In this way, one can hope to obtain a framework for solving the dynamical problems of decays. We shall begin our discussion of the kinematical concept in the following section.

1.2. Basic Notions

"... mathematical rigor although not necessary and sometimes even harmful (if it obscures the physical content for instance) is nothing to be rejected if it can be obtained with relative ease, as in a well-formulated axiomatic theory."

J. M. JAUCH

The description of any microsystem in the standard quantum-theoretical framework requires specification of the state Hilbert space and the set of observables. These choices are, of course, mutually closely related. Further, one has to determine a representation of the Poincaré group $\mathscr{P}$ acting on the state space, and if there is some internal symmetry, then also representation of the corresponding symmetry group. This applies particularly to unstable systems, and we must single out their characteristic properties. It seems reasonable to ascribe to them the same observables as to the stable ones (four-momentum, spin, parity, etc.). For unstable elementary particles, this choice of observables is sometimes used to define their 'integrity'; though we shall not try to formulate such a requirement in a strict way here, it will be contained implicitly in the following considerations. Later we shall return to this problem (particularly in Sections 3.4 and 3.5).

First of all, we assume that the considered unstable system can be distinguished as a physical entity; this fact we express by ascribing to it a state Hilbert space $\mathscr{H}_\mathrm{u}$. Each unstable system must be regarded as non-isolated, i.e. as a part of some larger isolated system which typically consists of the unstable system itself together with its decay products. The corresponding state Hilbert space $\mathscr{H}$, therefore, contains $\mathscr{H}_\mathrm{u}$ as a proper subspace; the orthogonal complement will be denoted by $\mathscr{H}_\mathrm{d}$. Since the system related to $\mathscr{H}$ is assumed to be isolated, relativistic invariance implies existence of a unitary (strongly continuous) representation $U : \mathscr{P} \to \mathscr{B}(\mathscr{H})$ of the Poincaré group, which transforms observables of the system in a specific way. Stable elementary objects are connected with unitary irreducible representations of $\mathscr{P}$ characterized by mass and spin; representations corresponding to unstable objects are expected to be reducible.

The main interest concerns the representation of one-parameter subgroup of time translations, i.e. the strongly continuous group of unitary operators $U : \mathbb{R} \to \mathscr{B}(\mathscr{H})$, conventionally called **propagator** (or **evolution operator**). Due to Stone's theorem there is a self-adjoint operator H on $\mathscr{H}$, which generates $U : U_t = \exp(-iHt)$ for each $t \in \mathbb{R}$. The fundamental dynamical postulate asserts that H coincides with the total Hamiltonian of the system. A typical feature of unstable systems is that their state vectors do not stay within $\mathscr{H}_\mathrm{u}$ during the time evolution described by U, i.e. that $\mathscr{H}_u$ is not invariant under U_t. Summing the above discussion, we can formulate the following two postulates:

(u1) *The state spaces obey* $\mathscr{H}_u \subset \mathscr{H}$ *with the orthogonal complement* $\mathscr{H}_d \neq \{0\}$.

(u2) *For every* $t > 0$, $\mathscr{H}_u$ *fails to be invariant under* U_t.

Before proceeding further, we shall recall some mathematical notions. Let $\mathscr{G}$ be a subspace in $\mathscr{H}$ with an associated projection P, then to any $B \in \mathscr{B}(\mathscr{H})$ we define the operator $\mathrm{pr}\, B = PB \upharpoonright \mathscr{G}$; if necessary we index 'pr' by P, $\mathscr{G}$ or some other subscript specifying the considered subspace. There is an obvious relation between $\mathrm{pr}\, B$ and PBP, and often these operators are identified in the literature. Let us list some simple properties of 'pr':

PROPOSITION 1.2.1. (a) $\|\mathrm{pr}\, B\| \leqslant \|B\|$ *for any* $B \in \mathscr{B}(\mathscr{H})$;

(b) pr *is transitive: we have* $\mathrm{pr}_{\mathscr{M}}(\mathrm{pr}_{\mathscr{N}} B) = \mathrm{pr}_{\mathscr{M}} B$ *for* $\mathscr{M} \subset \mathscr{N} \subset \mathscr{H}$ *and* $B \in \mathscr{B}(\mathscr{H})$;

(c) pr *is linear*: $\mathrm{pr}(\alpha B + \beta C) = \alpha\, \mathrm{pr}\, B + \beta\, \mathrm{pr}\, C$ *for all* $\alpha, \beta \in \mathbb{C}$ *and* $B, C \in \mathscr{B}(\mathscr{H})$;

(d) $(\mathrm{pr}\, B)^* = \mathrm{pr}\, B^*$ *for any* $B \in \mathscr{B}(\mathscr{H})$;

(e) *if* $\mathscr{G}_i \subset \mathscr{H}_i$, $i = 1, 2$, *and* $\mathscr{H} = \mathscr{H}_1 \oplus \mathscr{H}_2$, *then* $\mathrm{pr}_1 B_1 \oplus \mathrm{pr}_2 B_2 = \mathrm{pr}(B_1 \oplus B_2)$ for all $B_i \in \mathscr{B}(\mathscr{H}_i)$;

(f) *if a sequence* $\{B_n\}_{n=1}^{\infty}$ *converges weakly (strongly, in the operator norm) to some* $B \in \mathscr{B}(\mathscr{H})$, *then* $\{\mathrm{pr}\, B_n\}_{n=1}^{\infty}$ *converges weakly (strongly, in the operator norm) to* $\mathrm{pr}\, B$;

(g) *if* $\mathscr{G}_i \subset \mathscr{H}_i$, $i = 1, 2$, *and* $\mathscr{H} = \mathscr{H}_1 \otimes \mathscr{H}_2$, *then* $\mathrm{pr}_1 B_1 \otimes \mathrm{pr}_2 B_2 = \mathrm{pr}(B_1 \otimes B_2)$ *for all* $B_i \in \mathscr{B}(\mathscr{H}_i)$.

Assertions (a)–(f) can be found in Sz.-Nagy and Foias (1970, §I.4); the last one following from the fact that if E_i are projections corresponding to the subspace $\mathscr{G}_i$ in $\mathscr{H}_i$, then $E_1 \otimes E_2$ projects on the subspace $\mathscr{G}_1 \otimes \mathscr{G}_2$ in $\mathscr{H}$ – cf. Blank and Exner (1980, Lemma 6.3.1). Assertions (e), (g) generalize easily to the case of an n-fold orthogonal sum (tensor product) for any natural number n. We shall also use 'pr' for unbounded operators; however, in these cases the domains must be treated carefully (eventually $\mathrm{pr}\, T$ for an unbounded T can make no sense at all). The situation is especially simple if $P\mathscr{H} \subset D(T)$ and PTP is bounded.

Another necessary prerequisite concerns the trace class operators. For a given $\mathscr{H}$ they form an ideal in $\mathscr{B}(\mathscr{H})$, usually denoted by $\mathscr{J}_1$. The following properties will be particularly useful:

PROPOSITION 1.2.2. (a) *Let* B, C *be bounded operators on* $\mathscr{H}$, $C \in \mathscr{J}_1$, *then* $\mathrm{Tr}\, |BC| \leqslant \|B\|\, \mathrm{Tr}\, |C|$ *and* $\mathrm{Tr}\, |C^*| = \mathrm{Tr}\, |C|$.

(b) *Let* $\{B_\alpha\}$ *be a one-parameter family* $\subset \mathscr{B}(\mathscr{H})$, $\text{w-lim}_{\alpha \to \beta} B_\alpha = B$, *then for an arbitrary* $C \in \mathscr{J}_1$ *we have* $\lim_{\alpha \to \beta} \mathrm{Tr}(B_\alpha C) = \mathrm{Tr}(BC)$.

Proof. (a) The polar decomposition theorem implies existence of partial isometries U, V, W such that $|BC| = U^*BC$, $C = V|C|$, $C^* = W|C^*|$. The operator $|C|$ is positive and compact so there is an orthonormal basis $\{\varphi_k\}$ in $\mathscr{H}$, $|C|\varphi_k = \mu_k\varphi_k$. Both sides of the first inequality make sense and

$$\mathrm{Tr}\,|BC| = \sum_k (\varphi_k, U^*BV|C|\varphi_k) \leqslant \|B\| \sum_k \mu_k = \|B\|\,\mathrm{Tr}\,|C|.$$

Further, we have $|C^*| = W^*|C|V^*$ so $\mathrm{Tr}\,|C^*| = \mathrm{Tr}\,V^*W^*|C| \leqslant \mathrm{Tr}\,|C|$ and $\mathrm{Tr}\,|C| = \mathrm{Tr}\,|C^{**}| \leqslant \mathrm{Tr}\,|C^*|$.

(b) It is sufficient to check validity of the assertion for an arbitrary sequence $\{\alpha_n\}$, $\alpha_n \to \beta$, i.e. for all sequences $\{B_n\} \subset \mathscr{B}(\mathscr{H})$ which converge weakly to B. In the same way as above, we obtain

$$|\mathrm{Tr}(B_nC) - \mathrm{Tr}(BC)| \leqslant \sum_k \mu_k |(\varphi_k, (B_n - B)V\varphi_k)| \leqslant \sum_k \mu_k \|B_n - B\|. \qquad (*)$$

Due to the assumption, the sequence $\{(B_n - B)V\varphi\}$ converges weakly for each $\varphi \in \mathscr{H}$, thus it is bounded in view of the Banach–Steinhaus theorem. Further, we apply a stronger version of the uniform boundedness principle (cf. Appendix (A.1)) to the sequence $\{p_n\}$ of convex functionals $p_n(\psi) = \|(B_n - B)\psi\|$; it implies the existence of a positive M such that $p_n(\psi) \leqslant M\|\psi\|$, i.e. $\|B_n - B\| \leqslant M$ for all n. Consequently, the series in (*) converge uniformly with respect to n, and the limit $n \to \infty$ can be interchanged with the sum. ■

Now we return to the above formulated postulates. For a unitary operator U on $\mathscr{H}$, the following three statements are equivalent:

(i) U is reduced by the subspace $\mathscr{H}_\mathrm{u}$, i.e. both $\mathscr{H}_\mathrm{u}$ and $\mathscr{H}_\mathrm{d}$ are invariant under U;
(ii) the projection E_u with the range $\mathscr{H}_\mathrm{u}$ commutes with U;
(iii) $V = \mathrm{pr_u}\,U$ is unitary, i.e. $V^*V = VV^* = I_\mathrm{u}$, where I_u is the unit operator on $\mathscr{H}_\mathrm{u}$.

Thus U_t commutes with E_u for no $t > 0$, and also for no $t < 0$, because U_t commutes with E_u iff the same is true for $U_t^* = U_{-t}$. We obtain

$$[U_t, E_\mathrm{u}] \neq 0 \quad \text{for all non-zero } t \in \mathbb{R}. \qquad (2.1)$$

It implies further that the Hamiltonian does not commute with E_u: $E_\mathrm{u}H \not\subset HE_\mathrm{u}$.

Time evolution in $\mathscr{H}_\mathrm{u}$ alone is characterized by the **reduced propagator** V : $V_t = \mathrm{pr}\,U_t$ (throughout this chapter, 'pr' always refers to $\mathscr{H}_\mathrm{u}$ unless otherwise stated). The properties of U_t together with Proposition 2.1 show that $\{V_t : t \in \mathbb{R}\}$ is a *strongly continuous one-parameter family of contractions* on $\mathscr{H}_\mathrm{u}$ which fulfils

$$V_t^* = V_{-t} \quad \text{for all } t \in \mathbb{R}. \qquad (2.2)$$

Let us remark that writing the propagators U, V as functions on $\mathbb{R}$ (or $\mathbb{R}_+ = [0, \infty)$) is mainly a matter of convenience: they are relevant to time evolution from the moment when the system is prepared to the next measurement. There are at least three typical cases:

(i) the system is prepared at a given moment, conventionally $t = 0$, and we are interested in its further evolution undisturbed by measurements (that is, we assume the next measurement to take place at some – in general, arbitrary – instant t);

(ii) the system is prepared in the far past (in an appropriate time scale) and we are interested in the whole following undisturbed evolution; this is the case of an unstable system born and decayed again during a scattering process;

(iii) the system suffers frequently repeated measurements or it is even observed continually; such situations will be discussed in Chapter 2.

With these facts in mind, we notice that V_t *cannot fulfil the group law*, i.e. the condition (1.1) for all real t, s, because owing to (2.2) and the equivalence of (i) and (iii) above it would contradict (u2).

Further, we introduce the **decay function** (or **decay operator**) by $P : P(t) = V_t^* V_t$. The function $P(\,.\,)$ is obviously strongly continuous and its values are positive contractions on $\mathscr{H}_u$. Let us assume that the system is at $t = 0$ in a state described by a density matrix (i.e. normalized positive trace class operator) ρ, $\operatorname{Ran}\rho \subset \mathscr{H}_u$, then we define the **decay law** for this state by $P_\rho : P_\rho(t) = \operatorname{Tr}\{\rho P(t)\}$, i.e.

$$P_\rho(t) = \operatorname{Tr}\{V_t \rho V_t^*\} = \operatorname{Tr}\{U_t \rho U_t^* E_u\}. \tag{2.3a}$$

In particular, if ρ is a pure state, i.e. one-dimensional projection corresponding to a unit vector $\psi \in \mathscr{H}_u$, then the decay law can be rewritten in the form

$$P_\psi(t) = \|V_t \psi\|^2 = \|E_u U_t \psi\|^2. \tag{2.3b}$$

The decay law defined in this way has a natural physical interpretation. Let us assume the *yes–no* experiment (Jauch, 1968) corresponding to the projection E_u: a positive result means that the system is found undecayed and vice versa. The state in the Schrödinger picture is given by $\rho_t = U_t \rho U_t^*$; the system is *ex definitio* undecayed at $t = 0$, $\operatorname{Tr}\{\rho E_u\} = \operatorname{Tr}\rho = 1$. Thus $P_\rho(t)$ denotes the probability that the system will be found undecayed in the next measurement performed at t. There are other definitions of the decay law (Sinha, 1972; Shirokov, 1975). Also, in the case of repeated measurements, the decay law must be understood in a different way (cf. Chapter 2).

Properties of the reduced propagator together with Proposition 2.2 give the following assertion:

PROPOSITION 1.2.3. *For each state* ρ, $\operatorname{Ran}\rho \subset \mathscr{H}_u$, *the decay law* $P_\rho(\,.\,)$ *is a continuous function which obeys* $0 \leqslant P_\rho(t) \leqslant P_\rho(0) = 1$.

Let us exhibit two simple examples:

EXAMPLE 1.2.4. Let $\mathscr{H}_u$ be a *one-dimensional* subspace in $\mathscr{H}$ spanned by a unit vector ψ, then $V_t = v(t)E_\psi$, where $v(t) = (\psi, U_t\psi)$ and $E_\psi \equiv E_u = (\psi, .\,)\psi$. The decay law is given by $P_\psi(t) = |v(t)|^2 = |(\psi, U_t\psi)|^2$. Let us denote by $E_H(\,.\,)$

the spectral measure of the Hamiltonian H and by $E_\lambda = E_H((-\infty, \lambda])$ the corresponding decomposition of unity. Then, owing to $U_t = \exp(-iHt)$, we have

$$P_\psi(t) = \left| \int_{\mathbb{R}} e^{-i\lambda t}\, d(\psi, E_\lambda \psi) \right|^2 . \tag{2.4}$$

In particular, if ψ and H are such that the energy distribution has a *Breit–Wigner shape*,

$$d(\psi, E_\lambda \psi) = (\Gamma/2\pi)[(\lambda - \lambda_0)^2 + \tfrac{1}{4}\Gamma^2]^{-1}\, d\lambda \quad \text{with some } \Gamma > 0,$$

then (2.4) gives a purely exponential decay law, $P_\psi(t) = \exp(-\Gamma t)$ for all $t \geqslant 0$.

EXAMPLE 1.2.5. The postulate (u2) does not exclude the possibility that $\mathscr{H}_d$ *is invariant under* U_t. Consider $\mathscr{H} = L^2(\mathbb{R})$ with $\mathscr{H}_u = L^2(\mathbb{R}_-)$, $\mathbb{R}_- = (-\infty, 0]$, and the unitary group $U : (U_t\varphi)(x) = \varphi(x - t)$. Clearly, $\mathscr{H}_d = L^2(\mathbb{R}_+)$ is invariant under U_t for each $t \geqslant 0$. It is easy to see that H acts as a multiple of the first derivative, $H\varphi = -i\varphi'$ in this case; it is self-adjoint on $AC[\mathbb{R}] = \{\psi \in L^2(\mathbb{R}) : \psi$ absolutely continuous, $\psi' \in L^2(\mathbb{R})\}$ (Akhiezer and Glazman, 1966, §IV.55). This illustrates that H does not commute with E_u: the vector $\varphi : \varphi(x) = \exp(-x^2)$ belongs to $D(H) = D(E_u H)$, but $E_u\varphi$ is discontinuous so $E_u\varphi \notin D(H)$; consequently, $E_u H \subset HE_u$ does not hold. For a vector $\psi \in \mathscr{H}_u$, the decay law is given by

$$P_\psi(t) = \int_{-\infty}^{-t} |\psi(y)|^2\, dy. \tag{2.5}$$

Another important quantity is the **lifetime**: for a system which is originally in a state ρ, we define it by the relation

$$T_\rho = \int_0^\infty P_\rho(t)\, dt. \tag{2.6}$$

It is closely connected with the *mean life* of the considered system; however, these concepts do not coincide fully. In order to illustrate this, let us first derive the correspondence formally. If the probability that the system would decay within (s, t) is given by $P_\rho(s) - P_\rho(t)$, then the mean life is given by $\langle t \rangle_\rho = -\int_0^\infty t\, dP_\rho(t)$, so integration *per partes* gives

$$\langle t \rangle_\rho = -\int_0^t t\dot{P}_\rho(t)\, dt = -\lim_{t \to \infty} tP_\rho(t) + \int_0^\infty P_\rho(t)\, dt.$$

The main difficulty with this deduction, ignoring the technicalities, stems from the fact that $P_\rho(\,.\,)$ is not necessarily monotonous as it would need to be to generate a probability measure. In fact, Example 5.15 below shows that an occurrence of local minima in the decay law is quite natural, though they are usually expected to be located outside the experimentally attainable region, and the difference mentioned plays no practical role. In the one-dimensional case considered in Example 2.4, an alternative expression for lifetime is given by relation (5.20).

The framework specified by the postulates (u1, u2) is very wide; in fact, every case, when a state of a quantum system 'leaves' some submanifold of the state space, belongs there. In particular, one often meets the situation when the 'small' Hilbert space is connected with a 'yes–no' experiment ascertaining that the value of some observable lies in a chosen set.

EXAMPLE 1.2.6. Let M be a Borel set $\subset \mathbb{R}^d$ such that both M and $\mathbb{R}^d \backslash M$ have a non-zero Lebesgue measure, and denote by E_M the projection referring to the subspace $L^2(M) \subset L^2(\mathbb{R}^d)$. Suppose that we have a quantum system with a configuration space $\mathbb{R}^d$, the dynamics of which is described by some H (e.g. a single structureless particle in $\mathbb{R}^3$, with H being a Schrödinger operator $L^2(\mathbb{R}^3)$). If the system is localized initially to M – i.e. if it is described at $t = 0$ by a density matrix ρ with $\operatorname{Ran} \rho \subset L^2(M)$ – then the relation (2.3a) with $E_u = E_M$ expresses the 'decay law' $P_\rho(t)$ which now means obviously the probability of again finding coordinates of the system at the instant t in the set M. Furthermore, T_ρ defined by (2.6) may be used to characterize the time spent by the system within M (this is sometimes called the **sojourn time**).

Of particular interest is the case in which the time evolution is given by the free Hamiltonian $H_0 = -\frac{1}{2}\Delta$, where Δ is the Laplacian on $L^2(\mathbb{R}^d)$. The corresponding propagator can be written explicitly (Reed and Simon, 1975, §IX.7):

$$(U_t\varphi)(x) = (2\pi i t)^{-d/2} \operatorname*{l.i.m.}_{r \to \infty} \int_{B_r} \exp\left\{\frac{i}{2t}|x - y|^2\right\} \varphi(y)\, dy \tag{2.7}$$

holds for each $\varphi \in L^2(\mathbb{R}^d)$ and all non-zero t, where $B_r = \{y \in \mathbb{R}^d : |y| \leqslant r\}$ and l.i.m. means the L^2-norm limit (if φ belongs simultaneously to $L(\mathbb{R}^d)$, one can simply write the integral over $\mathbb{R}^d$). If the set M is nice enough, e.g. an open region in $\mathbb{R}^d$ with a smooth boundary, then one can use the relation (2.7) to check validity of postulate (u2) in a straightforward way. To this purpose, it is only necessary to fix t and to choose a point x with $d(x, M) > 0$; then one can always find a function $\varphi \in L^2(\mathbb{R}^d)$ with the support in M such that $(U_t\varphi)(\,.\,)$ is continuous and non-zero at x.

The integral representation (2.7) yields an explicit expression for the non-decay probability (2.3). In particular, if $\varphi \in L(\mathbb{R}^d) \cap L^2(\mathbb{R}^d)$ and M is bounded, then Fubini's theorem, together with the relation $|x - z|^2 - |x - y|^2 = (2x - y - z) \cdot (y - z)$, give

$$P_\varphi(t) = \int_{\mathbb{R}^{2d}} \bar{\varphi}(y)\varphi(z) p_t(y, z)\, dy\, dz, \qquad t \neq 0, \tag{2.8a}$$

$$p_t(y, z) = (2\pi t)^{-d} \int_M \exp\left\{\frac{i}{2t}(y - z) \cdot (2x - y - z)\right\} dx. \tag{2.8b}$$

Before proceeding further, let us comment briefly on alternative ways of describing the time evolution of open quantum systems. The existing methods can be divided roughly into two groups. The first includes the cases when one treats (at least, in principle) the system under consideration as part of some larger *quantum* system that may be regarded as isolated in a reasonable approximation.

The approach formulated above clearly belongs to this group. There is another way to embed the state Hilbert space $\mathscr{H}_{\text{open}}$ into one of the 'large' system, namely, by means of the tensor product, $\mathscr{H} = \mathscr{H}_{\text{open}} \otimes \mathscr{H}_{\text{res}}$, where $\mathscr{H}_{\text{res}}$ refers to the complementary subsystem, usually called *reservoir* or *heat bath*. There are peculiar distinctions between these two cases. If the evolution on $\mathscr{H}$ is governed by a strongly continuous unitary group and starts from some initial state $\rho(0) = \rho_{\text{open}} \otimes \rho_{\text{res}}$ at $t = 0$, then the reduced states $\rho_{\text{open}}(\rho(t))$ (see Section 3.4) do not 'leave' the space $\mathscr{H}_{\text{open}}$, i.e. their traces remain to be equal one for each t. On the other hand, a pure state ρ_{open} may become mixed in this way, which is certainly impossible for the evolution on $\mathscr{H}_{\text{u}}$ governed by $\{V_t\}$. Nevertheless, the two approaches are closely related in many cases of physical interest. For instance, consider the problem of radiation damping. The isolated system here consists of a chosen atom and the electromagnetic field. In that case, we have two possible frameworks for a description of the same process. Either we choose the state space $\mathscr{H}_{\text{a}}$ of the atom as $\mathscr{H}_{\text{open}}$ and $\mathscr{F}$, the Fock space referring to the electromagnetic field, as $\mathscr{H}_{\text{res}}$, or alternately we can construct $\mathscr{H}_{\text{u}}$ from a given excited state and the vacuum vector of $\mathscr{F}$, and identify its orthogonal complement to $\mathscr{H}$ with $\mathscr{H}_{\text{d}}$ (see Section 3.1). It is important that neither of the two methods may be universally preferred. Loosely speaking, the first choice is appropriate when we are interested primarily in the approach to equilibrium (to the ground state in the above particular case), while the second should be chosen when the emphasis is on the decay aspect of the problem.

The second group contains the methods that describe the behaviour of open systems in some phenomenological way. Such an approach is clearly less satisfactory, but it is frequently the only method when the full problem involves a large number of degrees of freedom and its solution is practically impossible. Moreover, most of the solution is not even required as it contains a lot of redundant information. This shows that phenomenological methods are not only tools for handling experimental data, but they deserve serious consideration within the theory.

Of course, the origin and purpose of phenomenological methods are not all alike. As for the first point of view, it is desirable to have methods that can be justified on microphysical grounds, with the number of ad hoc hypotheses reduced to a minimum. In this connection, a special remark should be made on those descriptions of open quantum systems which are obtained as a result of quantization of some classical dissipative system. The point is the *physical* role of quantization. In our opinion, which is unlikely to be shared by everyone, it is neither more nor less than a set of prescriptions on how to construct possible quantum theories which would yield some initially chosen classical limit, i.e. behaviour of the system in the situation when the involved quantities of dimension of action are much larger than $\hbar$. If more than one candidate arises in this way, then only by experiment are we able to decide which is the 'correct' one. Hence a method that relies on quantization might generally be expected to yield a plausible quantum-theoretical model if there are physical reasons why the latter should lead back just to a given classical limit. There are some cases when this is so, e.g. in the 'frictional' models of heavy nuclei which can be classically described as drops of a viscous liquid. However, applicability of quantization to dissipative systems is rather limited and, in general, the 'microphysical' justification of a phenomenological description must be preferred.

According to the intended application, the phenomenological methods divide again roughly into two groups. If we are concerned mainly with the effects of decay, absorption, etc., we usually describe the time evolution by means of a suitable non-self-adjoint Hamiltonian. On the other hand, the 'ergodic' problems are often treated phenomenologically by diverse methods in which either the Hamiltonian is time-dependent, including eventually its kinetic part, or the corresponding Schrödinger-type equation is non-linear.

In this book, we are going to concentrate on various non-stability phenomena in open systems. We shall discuss in detail fundamentals of their quantum-theoretical treatment as well as the methods that employ the non-self-adjoint Hamiltonians. As the above brief survey suggests, we are forced to omit many interesting and important problems. In the notes we present some basic references which can be used as a guide to further reading.

1.3. Small-Time Behaviour

We have already mentioned that the exponential decay law, $P(t) = \exp(-\Gamma t)$, is experimentally confirmed for most unstable particles and nuclei in wide range of time. Defining the **initial decay rate** as the right derivative $-\dot{P}(0+)$, we have in this case $-\dot{P}(0+) = \Gamma > 0$. On the other hand, deriving formally the decay law (2.3b) we get $\dot{P}_\psi(t) = 2\,\mathrm{Im}(U_t^* E_u U_t \psi, H\psi)$ so $\dot{P}_\psi(0) = 2\,\mathrm{Im}(\psi, H\psi) = 0$. This seems to be a paradox, however, and one has to remember that

(i) the deduction was only formal;
(ii) we are able to measure the decay-law shape with some finite time resolution, not the initial decay rate directly.

We shall start with the first comment. The above deduction is naturally expected to be valid for $\psi \in D(H)$; a more careful analysis shows that the mean value of energy is decisive. A density matrix ρ is said to describe a **finite-energy state** if the integral

$$\langle H\rangle_\rho = \int_{\mathbb{R}} \lambda\, d\mu_\rho(\lambda), \qquad \mu_\rho(\lambda) = \mathrm{Tr}\,\{\rho E_\lambda\} \tag{3.1}$$

converges, where $\{E_\lambda\}$ is the decomposition of unity related to H (cf. Example 2.4), i.e. if $f_{\mathrm{id}} \in L(\mathbb{R}, \mu_\rho)$. The function $\mu_\rho(\,.\,)$ is non-decreasing because the same is true for $\lambda \mapsto E_\lambda$ and ρ is positive; further, Proposition 2.2(b) shows that it is right-continuous. Thus $\mu_\rho(\,.\,)$ generates a Lebesgue–Stieltjes measure (for the sake of simplicity, denoted again by μ_ρ); obviously

$$\mu_\rho(M) = \mathrm{Tr}\,\{\rho E_H(M)\} \quad \text{for each Borel } M \subset \mathbb{R}. \tag{3.2}$$

The set of all finite-energy states will be denoted by $M(H)$. For a pure state ψ, we have $\langle H\rangle_\psi = \int_{\mathbb{R}} \lambda \, d(\psi, E_\lambda \psi)$ so that the pure finite-energy states span $Q(H)$, the form domain of H.

Each density matrix ρ is a positive trace class operator so we can write the spectral decomposition

$$\rho = \sum_k w_k E_k, \qquad \sum_k w_k = 1, \tag{3.3}$$

where $E_k = (\varphi_k, .\,)\varphi_k$ and each positive eigenvalue w_k is counted with its (finite) multiplicity. Further, we introduce positive operators $H_\pm = \pm\int_{\mathbb{R}_\pm} \lambda \, dE_\lambda$, which allow us to write $H = H_+ - H_-$ and $|H| = H_+ + H_-$; a state ρ belongs to $M(H)$ iff $\langle |H| \rangle_\rho = \langle H_+\rangle_\rho + \langle H_-\rangle_\rho < \infty$. We denote $\mu_k(\lambda) = (\varphi_k, E_\lambda \varphi_k)$; further, $\langle H\rangle_k \equiv \langle H\rangle_{\varphi_k}$, and analogously for $\langle H_\pm\rangle_k$, $\langle |H| \rangle_k$. The relations (3.2), (3.3) give $\mu_\rho(M) = \Sigma_k\, w_k \mu_k(M)$ for each Borel $M \subset \mathbb{R}$, so that

$$\langle H\rangle_\rho = \sum_k w_k \langle H\rangle_k, \qquad \langle |H| \rangle_\rho = \sum_k w_k \langle |H| \rangle_k. \tag{3.4}$$

The last relation makes it possible to formulate an alternative definition for the *mixed* finite-energy states: $\rho \in M(H)$ iff

(i) each pure state contained in the mixture is a finite-energy state, i.e. $\varphi_k \in Q(H)$ for each eigenvector φ_k of ρ corresponding to a non-zero w_k;

(ii) the series (3.4) converge.

We can now formulate the result concerning the initial decay rate:

THEOREM 1.3.1. *If ρ is a finite-energy state, then $\dot{P}_\rho(0) = 0$.*

Proof. It is sufficient to check the existence of $\dot{P}_\rho(0)$ because, according to Proposition 2.3, the continuous function P_ρ is maximal at $t = 0$. The relations (3.3) and (2.3) give

$$P_\rho(t) = \sum_k w_k P_k(t), \quad P_k(t) = \| V_t \varphi_k \|^2. \tag{3.5}$$

Further, we introduce

$$p_\rho(t) = \sum_k w_k p_k(t), \quad p_k(t) = |(\varphi_k, V_t \varphi_k)|^2.$$

The Schwarz inequality implies $0 \leqslant p_\rho(t) \leqslant P_\rho(t) \leqslant 1$. In view of this relation and $p_\rho(0) = P_\rho(0) = 1$ it is enough to verify $\dot{p}_\rho(0) = 0$. Each φ_k corresponding to a non-zero w_k belongs to $Q(H)$, so we can interchange derivative with integral in

$$\frac{\mathrm{d}}{\mathrm{d}t} \int_{\mathbb{R}} \mathrm{e}^{-i\lambda t}\, \mathrm{d}\mu_k(\lambda)$$

thus obtaining

$$\frac{\mathrm{d}}{\mathrm{d}t} (\varphi_k, U_t \varphi_k) = -i \int_{\mathbb{R}} \lambda\, \mathrm{e}^{-i\lambda t}\, \mathrm{d}\mu_k(\lambda).$$

The last relation gives

$$\dot{p}_k(t) = 2\, \mathrm{Im} \left\{ (\varphi_k, U_t^* \varphi_k) \int_{\mathbb{R}} \lambda\, \mathrm{e}^{-i\lambda t}\, \mathrm{d}\mu_k(\lambda) \right\}.$$

These derivatives are bounded and continuous: $|\dot{p}_k(t)| \leqslant 2 \int_{\mathbb{R}} |\lambda|\, \mathrm{d}\mu_k(\lambda) = 2\langle |H| \rangle_k$; further, φ_k belongs to the domain of $\sqrt{|H|} = \sqrt{H_+} + \sqrt{H_-}$, so

$$\left| \int_{\mathbb{R}} \lambda (\mathrm{e}^{-i\lambda t} - \mathrm{e}^{-i\lambda s})\, \mathrm{d}\mu_k(\lambda) \right| \leqslant \sum_{j=\pm} \| \sqrt{H_j}\, \varphi_k \| \, \| (U_t - U_s) \sqrt{H_j}\, \varphi_k \|$$

and the strong continuity of U implies the continuity of $\dot{p}_k(\,.\,)$. Finally, the series $\Sigma_k w_k \dot{p}_k(t)$ is majorized by $2 \Sigma_k w_k \langle |H| \rangle_k = 2\langle |H| \rangle_\rho$ and therefore converges uniformly with respect to t. This means that $\dot{p}_\rho(\,.\,)$ is a continuous function given by the relation

$$\dot{p}_\rho(t) = 2 \sum_k w_k\, \mathrm{Im} \left\{ (\varphi_k, U_t^* \varphi_k) \int_{\mathbb{R}} \lambda\, \mathrm{e}^{-i\lambda t}\, \mathrm{d}\mu_k(\lambda) \right\};$$

in particular, $\dot{p}_\rho(0) = 2 \Sigma_k w_k\, \mathrm{Im}\langle H \rangle_k = 0$. ■

REMARK 1.3.2. The condition $\rho \in M(H)$ is sufficient for $\dot{P}_\rho(0) = 0$, but not necessary. Consider Example 2.5 with the vector $\psi = 2^{-1/2} \chi_{[-3,-1]}$, where χ_M denotes the characteristic function of the set M. The 'Hamiltonian' coincides with the conventional momentum operator, thus $H = F^{-1} Q F$, where $(Q\varphi)(x) = x\varphi(x)$ and F is the Fourier–Plancherel operator. This relation, together with $(F\varphi)(\lambda) = \pi^{-1/2}\, \mathrm{e}^{2i\lambda}\, \lambda^{-1} \sin \lambda$, give $\langle |H| \rangle_\psi = (2/\pi) \int_0^\infty \lambda^{-1} \sin^2 \lambda\, \mathrm{d}\lambda = \infty$. However, the decay law $P_\psi(t) = 1$ for $t \leqslant 1$ (and then decreases linearly for $1 \leqslant t \leqslant 3$) so $\dot{P}_\psi(0) = 0$.

We have mentioned that, strictly speaking, the initial decay rate is not measurable. The theorem we have proved could provide some indirect information about it if we were able to check the energy finiteness of a given state. This problem is closely related to the question of whether the infinite-energy states are physically realizable. Some authors answer it negatively (cf. Streater and Wightman (1964, chap. 1)); others, like Bogoliubov *et al.* (1969, chap. 2) require even that the 'physical' pure states must belong to the domain of Hamiltonian. However, one can object that the only actually measured quantities for any observable A are $\mathrm{Tr}\{\rho E_A(M)\}$, i.e. the probabilities that the value of A lies in the sets M (Jauch, 1968). Each real experiment consists of a finite number of such '*yes–no*' experiment; thus convergence of the mean values like (3.1) is a *matter of our extrapolation* and *cannot be tested experimentally*. This conclusion is not affected by a possible rescaling of the measuring apparatus corresponding to replacement of the observable A by some bounded function of A, say $\mathrm{arctg}\,A$; all the same we are not able to determine with sufficient accuracy the behaviour of $\mathrm{Tr}\{\rho E_{\mathrm{arctg}\,A}([-\pi/2, .\,])\}$ near $\lambda = \pm\frac{1}{2}\pi$. There is also other argument against the above-mentioned point of view: why should it especially be the mean values of energy that are finite for the 'true' states, and not also those of momentum, angular momentum, etc.?

Since the infiniteness of mean values has no measurable consequences, it seems reasonable to assume that the set of states is closed (cf. Blank and Exner (1980, §7.2.3)) and *to discuss realizability of states only if it is experimentally meaningful*, for instance, in the case of superselection rules. We shall return to this problem in Section 6 from a more general point of view. The considerations presented show at the same time that neither indirect determination of $\dot{P}_\rho(0)$ based on Theorem 3.1 is possible.

We shall now discuss the small-time behaviour from a somewhat different point of view. The semigroup condition (1.1) can be reformulated equivalently as $E_u U_t E_u U_s E_u = E_u U_{t+s} E_u$ for all $t, s \geqslant 0$, or further rewritten in the form

$$E_u U_t E_d U_s E_u = 0, \quad t, s \geqslant 0, \tag{3.6}$$

where E_d is the projection corresponding to the orthogonal complement $\mathscr{H}_d$ to $\mathscr{H}_u$ in $\mathscr{H}$. Thus there is a direct connection between validity of the semigroup condition and **regeneration** of the considered system: the relation (3.6) says that once the state vector has left the subspace $\mathscr{H}_u$ it can never return. As we shall see in Section 5, there is always some regeneration, because the semigroup condition cannot be exact due to the semiboundedness of the total energy. Below we prove an assertion concerning the regeneration rate at short times, which in combination with the mentioned result shows that the regeneration at the beginning of the decay process is not only non-zero, but even it must not be too slow. We shall start with the following auxiliary assertion:

LEMMA 1.3.3. *Let* $F : \mathbb{R}_+ \to \mathscr{B}(\mathscr{H})$ *obey* $F(0) = I$, *and assume that* $C(t) \equiv \text{s-lim}_{n\to\infty} F(t/n)^n$ *exists for all t and* $\text{w-lim}_{t\to 0+} C(t) = I$, *then* $C : \mathbb{R}_+ \to \mathscr{B}(\mathscr{H})$ *is a (strongly) continuous semigroup.*

Proof. First, assume t, s to be rationally related positive numbers so that there are natural n, m which obey $(t+s)/(m+n) = t/n = s/m$. Thus, for each natural k the following identity holds

$$F((t+s)/k(m+n))^{k(m+n)} = F(t/kn)^{kn} F(s/km)^{km}.$$

Further, we let $k \to \infty$; in view of the assumption, and of the strong sequential continuity of the operator product, we obtain $C(t+s) = C(t)C(s)$ for all $t, s > 0$.

The assumption $F(0) = I$ gives $C(0) = I$ so the last relation also holds for $t, s \geqslant 0$ and $\{C(t) : t \geqslant 0\}$ forms a semigroup $\subset \mathscr{B}(\mathscr{H})$; it is continuous due to (A.2). ∎

THEOREM 1.3.4 (Misra and Sinha). *Let $\{F(t) : t \geqslant 0\}$ be a weakly continuous contractive family in $\mathscr{B}(\mathscr{H})$ with $F(0) = I$. If there is a dense set $D \subset \mathscr{H}$ such that*

$$\|[F(t)F(s) - F(t+s)]\psi\| \leqslant C_\psi t^\alpha s^\alpha \tag{3.7a}$$

for all $t, s \geqslant 0$, $\psi \in D$ and some $\alpha > 1$, where C_ψ is a constant independent of t, s, then $\{F(t) : t \geqslant 0\}$ is a continuous contrative semigroup. In particular, if $\{V_t : t \geqslant 0\}$ is a reduced propagator and

$$\|E_{\mathrm{u}} U_t E_{\mathrm{d}} U_s E_{\mathrm{u}} \psi\| \leqslant C_\psi t^\alpha s^\alpha, \qquad t, s \geqslant 0 \tag{3.7b}$$

for all $\psi \in D$ and some $\alpha > 1$, then $\{V_t : t \geqslant 0\}$ is a continuous contractive semigroup, or equivalently, there is no regeneration, i.e. relation (3.6) *holds.*

Proof. Let $\{\delta_i\}_{i=1}^n$ be a set of positive numbers, then

$$\begin{aligned} &\left\| F\left(\sum_{i=1}^n \delta_i\right)\psi - F(\delta_1)\dots F(\delta_n)\psi \right\| \\ &\quad \leqslant \left\| F\left(\sum_{i=1}^n \delta_i\right)\psi - F(\delta_1)F\left(\sum_{i=2}^n \delta_i\right)\psi \right\| + \\ &\quad + \left\| F(\delta_1)\left[F\left(\sum_{i=2}^n \delta_i\right) - F(\delta_2)\dots F(\delta_n)\right]\psi \right\| \end{aligned}$$

so the contractivity together with (3.7a) imply

$$\begin{aligned} &\left\| F\left(\sum_{i=1}^n \delta_i\right)\psi - F(\delta_1)\dots F(\delta_n)\psi \right\| \\ &\quad \leqslant C_\psi \delta_1^\alpha \left(\sum_{i=2}^n \delta_i\right)^\alpha + \left\| F\left(\sum_{i=2}^n \delta_i\right)\psi - F(\delta_2)\dots F(\delta_n)\psi \right\|. \end{aligned}$$

Further, we estimate the r.h.s. in the same way and continue this procedure substituting at the end $\delta_i = t/n$ for all i. This gives

$$\|F(t)\psi - F(t/n)^n\psi\| \leqslant C_\psi\left(\frac{t}{n}\right)^{2\alpha} \sum_{j=1}^{n-1} j^\alpha \leqslant C_\psi\left(\frac{t}{n}\right)^{2\alpha} \int_1^n x^\alpha \,\mathrm{d}x \leqslant \frac{t^{2\alpha}}{\alpha+1} n^{1-\alpha}.$$

Thus, $F(t/n)^n\psi$ converges to $F(t)\psi$ as $n \to \infty$ for all $\psi \in D$. Further, D is dense and the considered family of operators is uniformly bounded with respect to n, $\|F(t/n)^n\| \leqslant 1$, so we obtain $\text{s-lim}_{n\to\infty} F(t/n)^n = F(t)$. Now the assertion follows from the above lemma with $F(t) = C(t)$. ∎

It is clear that the inequalities (3.7) represent an actual restriction for short times only, being satisfied automatically for all t, s sufficiently large. Let us notice further that the operator-valued function F in the theorem is not necessarily a reduced propagator, because the latter has to obey additional conditions as we shall see in the next section.

1.4. The Inverse-Decay Problem

Assume that the state space $\mathscr{H}_u$ of an unstable system and the corresponding reduced propagator V are known. There naturally arises the question: 'Is this information sufficient to provide the complete dynamical description?' In other words, we are interested in the conditions which must be imposed on $\{V_t : t \geqslant 0\}$ in order to ensure the existence of a Hilbert space $\mathscr{H}$, a continuous group $\{U_t : t \in \mathbb{R}\}$ of unitary operators on $\mathscr{H}$ and a projection E_u such that $\mathscr{H}_u = E_u \mathscr{H}$ and $V_t = \mathrm{pr}\, U_t$ for all $t \geqslant 0$. This is the so-called **inverse-decay problem**, which was formulated by Williams (1971), and independently by Horwitz *et al.* (1971).

Let us start with the problem of *existence and uniqueness*. This can be solved by means of the unitary-dilations theory – cf. Sz.-Nagy and Foias (1970, chap. I) and notes to this section. We shall present below the basic result that is relevant to our problem; but we shall formulate it in a more general way, with the group of reals replaced by an arbitrary topological group.

Let G be a group. An operator-valued function $V: G \to \mathscr{B}(\mathscr{H})$ is called **positive definite** (or of a **positive type**) if for any n natural and arbitrary $g_1, g_2, \ldots, g_n \in G$, $\varphi_1, \varphi_2, \ldots, \varphi_n \in \mathscr{H}$, the inequality

$$\sum_{i,j=1}^{n} (\varphi_i, V(g_i^{-1} g_j)\varphi_j) \geqslant 0 \tag{4.1a}$$

holds. In particular, if G is the one-parameter translation group, then the positive definiteness of $t \mapsto V_t$ means

$$\sum_{i,j=1}^{n} (\varphi_i, V_{t_j - t_i}\varphi_j) \geqslant 0 \tag{4.1b}$$

for all $\varphi_1, \varphi_2, \ldots, \varphi_n \in \mathscr{H}$ and $t_1, t_2, \ldots, t_n \in \mathbb{R}$. Since the Hilbert space is complex, the relation (4.1a) for $n = 2$ implies

$$V(g)^* = V(g^{-1}) \qquad \text{for all} \quad g \in G, \tag{4.2}$$

especially for $G = \mathbb{R}$ we get (2.2). Now the mentioned assertion reads:

THEOREM 1.4.1. (Naimark and Sz.-Nagy). (a) *Let U be a unitary representation of G on $\mathscr{H}$. For a subspace $\mathscr{H}_u \subset \mathscr{H}$, the function $V : V(g) = \mathrm{pr}\, U(g)$ is positive* *definite and* $V(e) = I_u$*, where e is the unit element of G. Moreover, if G is a topological group and* $U(\,.\,)$ *is continuous (strongly or weakly, which amounts to the same in view of unitarity), then* $V(\,.\,)$ *is also strongly continuous.*

(b) *Conversely, let* $V: G \to \mathscr{B}(\mathscr{H}_u)$ *be positive definite with* $V(e) = I_u$*, then there is a Hilbert space* $\mathscr{H} \supset \mathscr{H}_u$ *and a unitary representation* $U: G \to \mathscr{B}(\mathscr{H})$ *such that* $V(g) = \mathrm{pr}\, U(g)$ *for all* $g \in G$*. Furthermore, if the 'minimality condition'*

$$\mathscr{H} = \left[\bigcup_{g \in G} U(g)\mathscr{H}_u\right]_{\mathrm{lin}} \tag{4.3}$$

holds, then the triple $\{\mathscr{H}, U, E_u\}$ *is unique up to an isometric isomorphism. Finally, if G is a topological group and* $V(\,.\,)$ *is (weakly) continuous, then* $U(\,.\,)$ *is also (weakly, hence also strongly) continuous.*

If the triple $\{\mathscr{H}, U, E_u\}$ obeys (4.3), we call it **minimal unitary dilation** of V. Before we prove this theorem, let us state its consequence for the inverse-decay problem:

COROLLARY 1.4.2. *An operator-valued function* $V: \mathbb{R} \to \mathscr{B}(\mathscr{H}_u)$ *can represent the reduced propagator of some unstable system if:*

(a) *it is weakly continuous, positive definite and* $V_0 = I_u$,
(b) $V_t^* V_t \neq I_u$ *for all positive t.*

The condition (b) means not only non-unitarity of V_t but also that the subspace $\mathscr{H}_u$ must not be invariant under U_t: if $V_t^* V_t = I_u$, then V_t is a partial isometry with initial and final subspaces $\mathscr{H}_u$ and $V_t\mathscr{H}_u \subset \mathscr{H}_u$, respectively, in contradiction with the postulate (u2).

Proof of Theorem 4.1. Part (a) is simple: $V(e) = I_u$ holds obviously; further,

$$\sum_{i,j=1}^{n} (\varphi_i, V(g_i^{-1}g_j)\varphi_j) = \sum_{i,j=1}^{n} (\varphi_i, U(g_i^{-1}g_j)\varphi_j)$$

$$= \sum_{i,j=1}^{n} (\varphi_i, U(g_i)^* U(g_j)\varphi_j) = \left\| \sum_{j=1}^{n} U(g_j)\varphi_j \right\|^2 \geqslant 0$$

and the continuity follows from Proposition 2.1. On the contrary, part (b) is highly non-trivial, especially the existence proof based on the so-called *Naimark–Sz.-Nagy construction.* It starts from the set $\mathscr{N}$ of all pairs $\Phi = \{M_\varphi, \varphi\}$, where M_φ is a finite subset of G and φ is a mapping $G \to \mathscr{H}_u$ such that $\varphi(g) = 0$ unless $g \in M_\varphi$. The set $\mathscr{N}$ has the natural vector-space structure, $\alpha\Phi + \Psi \equiv \{M_\varphi \cup M_\psi, \alpha\varphi + \psi\}$, and one can define a sesquilinear form $\langle\,.\,,\,.\,\rangle$ on it by

$$\langle \Phi, \Psi \rangle = \sum_{g \in M_\varphi} \sum_{h \in M_\psi} (\varphi(g), V(g^{-1}h)\psi(h));$$

this is non-negative owing to the positive definiteness of V. The Schwarz inequality $|\langle \Phi, \Psi \rangle|^2 \leqslant \langle \Phi, \Phi \rangle \langle \Psi, \Psi \rangle$ shows that $\mathscr{K} = \{\Phi : \langle \Phi, \Phi \rangle = 0\}$ is a linear manifold in $\mathcal{N}$. Thus we can define an inner product in the factor space $\tilde{\mathcal{N}} = \mathcal{N}/\mathcal{K}$ by $\langle\tilde{\Phi}, \tilde{\Psi}\rangle = \langle\Phi, \Psi\rangle$, where Φ, Ψ are some representatives of the equivalence classes $\tilde{\Phi}$, $\tilde{\Psi}$. The Hilbert space $\mathcal{H}$ is obtained by standard completition of $\tilde{\mathcal{N}}$ with respect to the norm generated by $\langle\,.\,,\,.\,\rangle$.

Further, we have to embed $\mathcal{H}_{\mathrm{u}}$ into $\mathcal{H}$. For arbitrary $g \in G$, $\varphi \in \mathcal{H}_{\mathrm{u}}$ we denote by $\Phi(g, \varphi)$ the element of $\mathcal{N}$ with M_φ consisting only of the point g, which is mapped by φ to the vector φ. The mapping $i : \mathcal{H}_{\mathrm{u}} \to \tilde{\mathcal{N}}$ defined by $i(\varphi) = \tilde{\Phi}(e, \varphi)$ can easily be seen to preserve the Hilbert-space structure: it is linear, and

$$\langle i(\varphi), i(\psi)\rangle = \langle\tilde{\Phi}(e, \varphi), \tilde{\Psi}(e, \psi)\rangle = (\varphi(e), V(e)\psi(e)) = (\varphi, \psi)$$

for each $\varphi, \psi \in \mathcal{H}_{\mathrm{u}}$, so it is also injective. Thus i embeds $\mathcal{H}_{\mathrm{u}}$ isomorphically into $\tilde{\mathcal{N}} \subset \mathcal{H}$.

The following step consists of constructing the representation U. We define $\Phi_g \equiv \{gM_\varphi, \varphi_g\}$, where $\varphi_g(h) = \varphi(g^{-1}h)$ for each $g \in G$ and $\Phi \in \mathcal{N}$. The mapping $g \mapsto \Phi_g$ is clearly linear, $\Phi_e = \Phi$ and $(\Phi_g)_h = \Phi_{hg}$ for all $g, h \in G$. Thus the operators $\tilde{U}(g) : \tilde{U}(g)\tilde{\Phi} = \tilde{\Phi}_g$ form a representation of G, which is furthermore isometric

$$\langle\tilde{U}(g)\tilde{\Phi}, \tilde{U}(g)\tilde{\Psi}\rangle = \langle\Phi_g, \Psi_g\rangle = \sum_{h \in gM_\varphi} \sum_{f \in gM_\psi} (\varphi(g^{-1}h), V(h^{-1}f)\psi(g^{-1}f))$$

$$= \sum_{m \in M_\varphi} \sum_{n \in M_\psi} (\varphi(m), V(m^{-1}n)\varphi(n)) = \langle\tilde{\Phi}, \tilde{\Psi}\rangle.$$

Since $\tilde{\mathcal{N}}$ is dense in $\mathcal{H}$, each $\tilde{U}(g)$ extends by continuity to an operator $U(g)$ on $\mathcal{H}$ such that $U(e) = I$, $U(h)U(g) = U(hg)$ and $\langle U(g)\tilde{\Phi}, U(g)\tilde{\Psi}\rangle = \langle\tilde{\Phi}, \tilde{\Psi}\rangle$ for all $g, h \in G$, $\tilde{\Phi}, \tilde{\Psi} \in \mathcal{H}$. Assume, further, any $\varphi, \psi \in \mathcal{H}_{\mathrm{u}}$, then

$$\langle i(\varphi), U(g)i(\psi)\rangle = \langle\Phi(e, \varphi), \Psi_g(e, \psi)\rangle = \langle\Phi(e, \varphi), \Psi(g, \psi)\rangle = (\varphi, V(g)\psi),$$

i.e. $U(\,.\,)$ is the desired unitary representation, $V(g) = \mathrm{pr}\; U(g)$ for each $g \in G$.

Every element $\tilde{\Phi}$ of $\tilde{\mathcal{N}}$ can be expressed as a finite sum,

$$\tilde{\Phi} = \sum_{j=1}^{n} \tilde{\Phi}(g_j, \varphi_j) = \sum_{j=1}^{n} \tilde{U}(g_j)\tilde{\Phi}(e, \varphi_j) = \sum_{j=1}^{n} U(g_j)i(\varphi_j),$$

and therefore $U(\,.\,)$ obeys the condition (4.3). Let $\mathcal{H}'$, $U'(\,.\,)$ belong to some other minimal unitary dilation of $V(\,.\,)$ and consider the vectors

$$\varphi', \psi' \in \mathcal{N}' \equiv \Big[\bigcup_{g \in G} U'(g)\mathcal{H}_{\mathrm{u}}\Big]_{\mathrm{lin}},$$

$$\varphi' = \sum_{j=1}^{n} U'(g_j)\varphi_j, \qquad \psi' = \sum_{k=1}^{m} U'(g_k)\psi_k.$$

There is a natural mapping $\tilde{S} : \mathcal{N} \to \mathcal{N}'$, $\tilde{S}\tilde{\Phi} = \varphi'$ for $\tilde{\Phi} = \Sigma_{j=1}^{n} \tilde{\Phi}(g_j, \varphi_j)$, which obeys

$$\begin{aligned}(\tilde{S}\varphi', U'(g)\tilde{S}\psi')_{\mathcal{H}'} &= \sum_{j=1}^{n} \sum_{k=1}^{m} (U'(g_j)\varphi_j, U'(g)U'(g_k)\varphi_k)_{\mathcal{H}'} \\ &= \sum_{j=1}^{n} \sum_{k=1}^{m} (\varphi_j, U'(g_j^{-1}gg_k)\varphi_k)_{\mathcal{H}'} \\ &= \sum_{j=1}^{n} \sum_{k=1}^{m} (\varphi_j, V(g_j^{-1}gg_k)\varphi_k) \\ &= \sum_{j=1}^{n} \sum_{k=1}^{m} \langle i(\varphi_j), U(g_j^{-1}gg_k)i(\varphi_k)\rangle = \langle\tilde{\Phi}, U(g)\tilde{\Psi}\rangle\end{aligned}$$

for all $\varphi', \psi' \in \mathcal{N}'$ and $g \in G$. In particular, the last relation with $g = e$ shows that $\tilde{S}$ is an isometric isomorphism of the inner-product spaces $\tilde{\mathcal{N}}, \mathcal{N}'$, and since the latter are dense in $\mathcal{H}, \mathcal{H}'$, respectively, S extends continuously to a unitary operator $S : \mathcal{H} \to \mathcal{H}'$. Moreover, one has $U(g) = S^{-1} U'(g)S$ for each $g \in G$, i.e. the representations U, U' are unitarily equivalent.

Finally, let G be a topological group and $V(\,.\,)$ weakly continuous. One has to check weak continuity of $U(\,.\,)$, i.e. that $g \mapsto \langle\tilde{\Phi}, U(g)\tilde{\Psi}\rangle$ is continuous for all $\tilde{\Phi}, \tilde{\Psi} \in \mathcal{H}$. Since $U(g)$ has a bound independent of g, $\|U(g)\| = 1$, and linear combinations of $\tilde{\Phi}(h, \varphi)$ are dense in $\mathcal{H}$, it is sufficient to verify continuity of $g \mapsto \langle\tilde{\Phi}(h, \varphi), U(g)\tilde{\Psi}(f, \psi)\rangle$ for each $\varphi, \psi \in \mathcal{H}_u$, $h, f \in G$; but the latter equals

$$\langle U(h)i(\varphi), U(g)U(f)i(\psi)\rangle = (\varphi, V(h^{-1}gf)\psi),$$

which is continuous with respect to g due to the assumptions. ■

The minimality condition (4.3) has a natural physical meaning if G is a one-parameter group of time translations. Suppose that (4.3) is not fulfilled, then the isolated system possesses states which are disjoint with those of the unstable subsystem being considered; in particular, there are vectors in $\mathcal{H}$ which are orthogonal to each vector state which started its evolution inside $\mathcal{H}_u$. Consequently, in such a case the chosen isolated system is too large for description of the particular decay process. On the other hand, it is often useful to separate the parts of the reduced propagator which are independent of the rest. A relevant assertion is presented below; it is again formulated for a general topological group G.

PROPOSITION 1.4.3. *Let G be a topological group and suppose that $V_i : G \to \mathcal{B}(\mathcal{G}_i)$, $i = 1, 2$, are weakly continuous functions of the positive type with $V_i(e) = I_{\mathcal{G}_i}$, then the same is true for the function $V : G \to \mathcal{B}(\mathcal{G}_1 \oplus \mathcal{G}_2)$, $V(g) = V_1(g) \oplus V_2(g)$. If $\{\mathcal{H}_i, U_i, E_i\}$ are minimal unitary dilations (m.u.d.) of the functions V_i, $i = 1, 2$, then the m.u.d. of V equals $\{\mathcal{H}_1 \oplus \mathcal{H}_2, U, E_1 \oplus E_2\}$,* *where* $U(g) := U_1(g) \oplus U_2(g)$ *for all* $g \in G$. *Conversely, let* $V : G \to \mathscr{B}(\mathscr{G})$ *be weakly continuous and positive definite with* $V(e) = I$ *and let its m.u.d. be* $\{\mathscr{H}, U, E\}$. *If the function* V *is reduced by a projection* $P \in \mathscr{B}(\mathscr{G})$, *then the functions* $V_1 : V_1(g) = \mathrm{pr}_{P\mathscr{G}}\, V(g)$ *and* $V_2 : V_2(g) = \mathrm{pr}_{\bar{P}\mathscr{G}}\, V(g)$, *where* $\bar{P} = I_{\mathscr{G}} - P$, *also fulfil the assumptions of the theorem. Moreover, there is a projection* F *on* $\mathscr{H}$ *which reduces* $U(g)$ *for each* $g \in G$, *such that* $\mathrm{pr}_{\mathscr{G}}\, EF = P$ *and* $\{F\mathscr{H}, FU(\,.\,), P\}$, $\{\bar{F}\mathscr{H}, \bar{F}U(\,.\,), \bar{P}\}$, $\bar{F} = I_{\mathscr{H}} - F$, *are m.u.d.'s of the functions* V_1, V_2, *respectively.*

The proof is simple and we omit it. The assertion generalizes easily to the case of an n-fold orthogonal sum (see Problem 1). Consider the particular case when the reduced propagator decomposes into an orthogonal sum and one of its parts is unitary so that its minimal unitary dilation is trivial. Then the above assertion shows that one can extract this 'stable part' of the problem and study the other part separately when it is convenient. On the other hand, caution is needed. Often the propagator U, or equivalently the total Hamiltonian H, is reduced by a projection F; however, the proposition may be applied only if F also commutes with the corresponding projection E_{u}. Furthermore, the reduced propagator can have invariant subspaces – in particular, one-dimensional ones which correspond to eventual common eigenvectors of the operators V_t, $t \geqslant 0$ – but the corresponding projections do not in general reduce V.

We shall now derive a useful equivalent expression for the solution to the inverse-decay problem. First we shall generalize the classical theorem of Bochner and Khinchin (cf., e.g., Akhiezer and Glazman (1966, §VI.70) or Reed and Simon (1975, theorem IX.9)) to the case of operator-valued functions:

PROPOSITION 1.4.4. *A function* $V : \mathbb{R} \to \mathscr{B}(\mathscr{H}_{\mathrm{u}})$ *is weakly continuous and positive definite iff there is a bounded positive-operator-valued function* $F : \mathbb{R} \to \mathscr{B}_+(\mathscr{H}_{\mathrm{u}})$ *which is weakly continuous on the right, and such that* V_t *is expressed as*

$$(\varphi, V_t\psi) = \int_{\mathbb{R}} \mathrm{e}^{-i\lambda t}\, \mathrm{d}(\varphi, F_\lambda \psi), \qquad \varphi, \psi \in \mathscr{H}_{\mathrm{u}}. \tag{4.4a}$$

Proof. If V is given by (4.4a), then we have for any natural n and arbitrary $t_1, \dots, t_n \in \mathbb{R}$, $\varphi_1, \dots, \varphi_n \in \mathscr{H}_{\mathrm{u}}$ the following inequality

$$\sum_{i,j=1}^{n} (\varphi_i, V_{t_j - t_i}\varphi_j) = \sum_{i,j=1}^{n} \int_{\mathbb{R}} \mathrm{e}^{-i\lambda(t_j - t_i)}\, \mathrm{d}(\varphi_i, F_\lambda \varphi_j)$$
$$= \int_{\mathbb{R}} \mathrm{d}\left(\sum_{i=1}^{n} \mathrm{e}^{-i\lambda t_i}\varphi_i, F_\lambda \sum_{j=1}^{n} \mathrm{e}^{-i\lambda t_j}\varphi_j \right) \geqslant 0,$$

which proves positive definiteness. Further, $t \mapsto (\psi, V_t\psi)$ is continuous for each $\psi \in \mathscr{H}_{\mathrm{u}}$ (as the Fourier transform of a finite Lebesgue–Stieltjes measure on $\mathbb{R}$) so the polarization identity implies weak continuity. Conversely, for any $\psi \in \mathscr{H}_{\mathrm{u}}$

the function $t \mapsto (\psi, V_t\psi)$ is continuous and of the positive type, so according to the classical Bochner–Khinchin theorem it expresses as the Fourier transform of a Lebesgue–Stieltjes measure corresponding to some non-decreasing right-continuous function ω_ψ. Thus we can define

$$(\varphi, F_\lambda \psi) = \tfrac{1}{4} \sum_{k=0}^{3} (-i)^k \omega_{\psi + i^k \varphi}(\lambda)$$

for each $\lambda \in \mathbb{R}$, $\varphi, \psi \in \mathcal{H}_u$. Boundedness of V_0 ensures that the last relation defines a bounded operator F_λ to every $\lambda \in \mathbb{R}$. One can check easily that the function $\lambda \mapsto F_\lambda$ has the necessary properties. ∎

The equivalent expression obtained can be rewritten as $V_t = \int_{\mathbb{R}} e^{-i\lambda t} dF_\lambda$ if the integral is understood in the weak sense. It is desirable to have a stronger result; one can obtain it simply from the above theorem for the particular case in which we are interested:

PROPOSITION 1.4.5. *A function $V : \mathbb{R} \to \mathcal{B}(\mathcal{H}_u)$ is weakly continuous and positive definite with $V_0 = I_u$ iff there is a positive-operator-valued measure F on $\mathbb{R}$ such that the equality*

$$V_t \psi = \int_{\mathbb{R}} e^{-i\lambda t} dF_\lambda \psi \tag{4.4b}$$

holds for all $t \geq 0$ and $\psi \in \mathcal{H}_u$.

Proof. According to Theorem 4.1, the function V fulfils the stated assumptions iff there is a Hilbert space $\mathcal{H} \supset \mathcal{H}_u$ and a continuous unitary group $\{U_t : t \in \mathbb{R}\} \subset \mathcal{B}(\mathcal{H})$ so that $V_t = \text{pr}\, U_t$ for all $t \geq 0$. Further, there is a projection-valued measure E_H on $\mathbb{R}$ such that $U_t\psi = \int_{\mathbb{R}} e^{-i\lambda t} dE_\lambda \psi$ for all $t \in \mathbb{R}$ and $\psi \in \mathcal{H}$, where $E_\lambda \equiv E_H((-\infty, \lambda])$ – cf., e.g., Akhiezer and Glazman (1966, §VI.73). Let $\mathcal{A}_H$ be the the σ-algebra of E_H-measurable sets, then we define

$$F(M) = \text{pr}\, E_H(M), \qquad M \in \mathcal{A}_H. \tag{4.5}$$

The function $F : \mathcal{A}_H \to \mathcal{B}_+(\mathcal{H}_u)$ is clearly σ-additive and fulfils $F(\phi) = 0$, $F(\mathbb{R}) = I_u$, i.e. it is a positive operator-valued measure. Finally, $F(\,.\,)\psi$ is a vector-valued measure for each $\psi \in \mathcal{H}_u$ and it fulfils $V_t\psi = E_u U_t \psi = \int_{\mathbb{R}} e^{-i\lambda t} dF_\lambda$ (Dunford and Schwartz, 1958, theorem IV.10.8f), i.e. the relation (4.4b). ∎

Let us turn now to the case which is physically the most interesting. In the introduction, we mentioned the importance of the semigroup condition (1.1); further, we found in Section 2 that the reduced propagator must be strongly continuous and contraction-valued. Thus we are interested primarily in *continuous contractive semigroups* $\{V_t : t \geq 0\}$. In order to show that they obey the assumptions of Theorem 4.1, let us first prove the following assertion:

LEMMA 1.4.6. *Let $T \in \mathscr{B}(\mathscr{H})$ be a contraction and define $T(0) = I$, $T(n) = T^n$ and $T(-n) = T(n)^*$ for $n = 1, 2, \ldots$, then the function $T(\,.\,)$ is positive definite on the additive group $\mathbb{Z}$ of integers.*

Proof. One has to check that $\Sigma_{m,\, n \in \mathbb{Z}} (\varphi_m, T(n - m)\varphi_n) \geqslant 0$ for each sequence $\{\varphi_n\}_{n \in \mathbb{Z}} \subset \mathscr{H}$ with a finite number of non-zero elements. This is equivalent to the requirement

$$\sum_{n=0}^{\infty} \sum_{m=0}^{\infty} (\varphi_m, T(n - m)\varphi_n) \geqslant 0$$

for all finitely non-zero sequences $\{\varphi_n\}_{n=0}^{\infty} \subset \mathscr{H}$, because one can always choose $\{\varphi_n'\}_{n \in \mathbb{Z}}$ with $\varphi_n' = \varphi_{n+p}$ so that $\varphi_n' = 0$ for all n negative and $T((n - p) - (m - p)) = T(n - m)$. Further, we notice that if we define $\psi_n = \Sigma_{m=n}^{\infty} T^{m-n} \varphi_m$ for all $n \geqslant 0$, then $\{\varphi_n\}_{n=0}^{\infty} \mapsto \{\psi_n\}_{n=0}^{\infty}$ maps the set of all finitely non-zero sequences bijectively onto itself, the inverse being given by $\varphi_n = \psi_n - T\psi_{n+1}$, $n \geqslant 0$. Thus another equivalent reformulation of the above requirement reads

$$\sum_{n=0}^{\infty} \sum_{m=0}^{\infty} (\psi_m - T\psi_{m+1}, T(n - m)(\psi_n - T\psi_{n+1}))$$

$$\equiv \sum_{n=0}^{\infty} \sum_{m=0}^{\infty} (\psi_m, D_{mn}\psi_n) \geqslant 0.$$

It is an easy exercise to express D_{mn} using the definition of $T(\,.\,)$; one obtains the condition

$$\|\psi_0\|^2 + \sum_{n=1}^{\infty} (\psi_n, (I - T^*T)\psi_n) \geqslant 0,$$

which is fulfilled, because $\|T^*T\| = \|T\|^2 \leqslant 1$ due to the assumption. ■

THEOREM 1.4.7 (Sz.-Nagy). *Let $\{V_t : t \geqslant 0\}$ be a continuous contractive semi-group on $\mathscr{H}$, then its extension to $\mathbb{R}$ defined by $V_{-t} = V_t^*$ is a continuous function of the positive type.*

Proof. Since $B \mapsto B^*$ is weakly continuous, the extension of V is weakly continuous in $\mathbb{R}$. We have to check validity of (4.1b). Suppose first that all the $\{t_j\}$ are commensurable, i.e. $t_j = k_j\tau$ with k_j integer, $j = 1, \ldots, n$; then the preceding lemma applied to $T = V_\tau$ shows that (4.1b) holds. In the general case, one chooses sequences $\{t_j^{(m)}\}_{m=1}^{\infty}$, $j = 1, \ldots, n$, such that $t_j^{(m)} \to t_j$ as $m \to \infty$ and $t_1^{(m)}, \ldots, t_n^{(m)}$ are commensurable for each m; the assertion then follows from the continuity of V. ■

EXAMPLE 1.4.8. Consider the simplest possible *semigroup, which acts on a one-dimensional space* $\mathscr{H}_u$ (spanned by a vector ψ) as $V_t\psi = e^{-i\alpha t}\psi$. Contractivity requires $\mathrm{Im}\,\alpha \leqslant 0$; for α real, $\{V_t : t \in \mathbb{R}\}$ is a unitary group so its unitary dilation is trivial. Assume now $\mathrm{Im}\,\alpha < 0$, then we choose $\mathscr{H} = L^2(\mathbb{R})$ and H on $\mathscr{H}$ acting as a multiplication operator, $(H\varphi)(\lambda) = \lambda\varphi(\lambda)$, and further we identify the vector ψ which spans $\mathscr{H}_u$ with $\psi(\lambda) = (-\mathrm{Im}\,\alpha/\pi)^{1/2}(\lambda - \alpha)^{-1}$. It holds

$$(\psi, e^{-iHt}\psi) = \int_{\mathbb{R}} e^{-i\lambda t}|\lambda - \alpha|^{-2}\, d\lambda = e^{-i\alpha t} \qquad \text{for} \quad t \geqslant 0$$

so the described triple $\{\mathscr{H}, U, E_u\}$ represents a *unitary dilation* of V, which is moreover *minimal*: if, for some $\varphi \in \mathscr{H}$, the equality $(\varphi, U_t\psi) = \int_{\mathbb{R}} e^{-i\lambda t}\overline{\varphi(\lambda)}\,\psi(\lambda)\, d\lambda = 0$ holds for all real t, then injectivity of the Fourier transformation on $L(\mathbb{R})$ implies $\overline{\varphi(\lambda)}\,(\lambda - \alpha)^{-1} = 0$ almost everywhere (a.e.) in $\mathbb{R}$, and therefore $\varphi = 0$.

The minimal unitary dilation found in this example has two remarkable properties; namely, that spectrum of the Hamiltonian H is absolutely continuous and that it covers the whole real line. As we shall see in the next section, this is also true for each continuous contractive semigroup, with the possible exception of those which are partly (in the sense of Proposition 4.3) or completely unitary.

1.5. Semiboundedness and Other Properties of the Energy Spectrum

We have already mentioned that the exponential shape of the decay law, or, more generally, semigroup nature of the reduced propagator, cannot be exact under the natural physical assumption that the total Hamiltonian H is semibounded. The present section is devoted primarily to this problem, though we shall also derive some other properties of the energy spectrum. We start by stating the following general result:

THEOREM 1.5.1 (Sinha). *Let* $\{V_t : t \geqslant 0\}$ *be a reduced propagator on* $\mathscr{H}_u$ *and* $\{\mathscr{H}, U, E_u\}$ *its minimal unitary dilation with* $U_t = \exp(-iHt)$, $t \in \mathbb{R}$. *Assume that there is a non-negative* T_r *such that* $V_tV_s = V_{t+s}$ *holds for all* $t \geqslant 0$, $s \geqslant T_r$, *then* H *has the whole real axis in its spectrum,* $\sigma(H) = \mathbb{R}$.

The theorem particularly asserts (for $T_r = 0$) that a reduced propagator cannot be a continuous contractive semigroup unless the total energy of the system is below unbounded. The last named result was derived by various authors (cf. notes to this section); in view of the discussion presented in Section 3, it is equivalent to the fact that there is no regeneration. The above theorem therefore says that the *regeneration cannot cease* after some (finite, but arbitrary) relaxation time T_r.

Before proving the theorem we present some technical lemmas concerning Fourier transforms of finite (complex) Lebesgue–Stieltjes measures on $\mathbb{R}$. Let ω be such a measure (generated by a right-continuous function of bounded variation which, for the sake of simplicity, is denoted again by ω) and $f(t) = \int_{\mathbb{R}} e^{-i\lambda t}\, d\omega(\lambda)$. The function f is continuous (in view of the dominated-convergence theorem) and bounded (because ω has a bounded variation); further, we have:

LEMMA 1.5.2. *If* $\lim_{|t|\to\infty} f(t) = 0$, *then* ω *is continuous.*

Proof. For any $\epsilon > 0$, there is a positive T_0 such that $|f(t)e^{-i\lambda t}| = |f(t)| < \epsilon$ for each $\lambda \in \mathbb{R}$ and all $|t| > T_0$. Further, there is a constant C such that $|f(t)| \leqslant C$ for all $t \in [-T_0, T_0]$; thus $M_\lambda(T) \equiv (2T)^{-1} \int_{-T}^{T} f(t) e^{i\lambda t}\, dt$ can be estimated as follows

$$|M_\lambda(T)| \leqslant \frac{1}{2T}\left(\int_{-T}^{-T_0} + \int_{-T_0}^{T_0} + \int_{T_0}^{T}\right) |f(t)|\, dt \leqslant C\frac{T_0}{T} + \epsilon\left(1 - \frac{T_0}{T}\right).$$

We see that for $T \to \infty$ the limit is bounded by ϵ, and since ϵ is arbitrary positive, we get $\lim_{T\to\infty} M_\lambda(T) = 0$ for each $\lambda \in \mathbb{R}$. Continuity of ω then follows from the second formula in (A.3). ∎

LEMMA 1.5.3 (Riemann and Lebesgue). *Let* ω *be absolutely continuous,* $\omega(\lambda) = \int_{-\infty}^{\lambda} g(\xi)\, d\xi$ *with* $g \in L(\mathbb{R})$, *then* $\lim_{|t|\to\infty} f(t) = 0$.

This assertion is well known so we omit the proof (given, e.g., in Kolmogorov and Fomin (1976, §VIII.4) or Reed and Simon (1975, theorem IX.7)). It provides no information, however, on how fast the function f decays at infinity. A sort of converse to the Riemann–Lebesgue lemma can be proved if we accept an integral restriction on the decay rate:

LEMMA 1.5.4. *If* $f \in L(\mathbb{R})$, *then the corresponding Lebesgue–Stieltjes measure* ω *is absolutely continuous with respect to the Lebesgue measure on* $\mathbb{R}$.

Proof. Let us first check the continuity of ω. Using the notation from the proof of the preceding lemma, we get $|M_\lambda(T)| \leqslant (2T)^{-1} \int_{\mathbb{R}} |f(t)|\, dt$, i.e. $\lim_{T\to\infty} M_\lambda(T) = 0$ so ω is continuous. The first of the inversion formulae (A.3) then reads

$$\omega(\lambda) = \text{const} + \frac{1}{2\pi} \lim_{C\to\infty} \int_{-C}^{C} f(t) \frac{e^{i\lambda t} - 1}{it}\, dt.$$

Further, $f \in L(\mathbb{R})$ and the function $t \mapsto (e^{i\lambda t} - 1)/it$ is continuous and bounded, so the dominated-convergence theorem gives

$$\omega(\lambda) = \text{const} + \frac{1}{2\pi} \int_{\mathbb{R}} f(t) \frac{e^{i\lambda t} - 1}{it}\, dt. \tag{5.1}$$

This equality further implies

$$\left| \frac{\omega(\xi) - \omega(\lambda)}{\xi - \lambda} - \frac{1}{2\pi} \int_{\mathbb{R}} f(t) e^{i\lambda t}\, dt \right| \leqslant \frac{1}{2\pi} \int_{\mathbb{R}} |f(t)|\, |h_t(\xi - \lambda)|\, dt,$$

where $h_t(\zeta) = (e^{it\zeta} - 1)/it - 1$. We have $\lim_{\zeta\to 0} h_t(\zeta) = 0$ for each $t \in \mathbb{R}$ and also $|h_t(\zeta)| \leqslant 2$ for all $t, \zeta \in \mathbb{R}$; thus according to the dominated-convergence theorem, the function ω is differentiable everywhere in $\mathbb{R}$ and

$$\frac{d\omega(\lambda)}{d\lambda} = \frac{1}{2\pi}\int_{\mathbb{R}} f(t)\, e^{i\lambda t}\, dt.$$

The function $g : g(\lambda) = d\omega(\lambda)/d\lambda$ is clearly locally integrable and $\omega(\zeta) - \omega(\lambda) = \int_\lambda^\zeta g(\zeta)\, d\zeta$. There is a unique Lebesgue–Stieltjes measure (denoted again by ω) such that $\omega([\lambda, \xi])$ is given by the above expression, namely $\omega : \omega(M) = \int_M g(\zeta) d\zeta$. The function g is bounded since $f \in L(\mathbb{R})$ so $|\omega(M)| \leqslant (2\pi)^{-1}\, \|f\|_1\, m(M)$, where m is the Lebesgue measure, and therefore ω is absolutely continuous with respect to m. ■

LEMMA 1.5.5. *Consider the following requirements on the function f:*

(a) $|f(t)| \leqslant C|t|^{-1-\epsilon}$ *for some positive C, ϵ and all $|t|$ large enough,*
(b) $|f(t)| \leqslant C e^{-\beta |t|}$ *for some positive C, β and all $|t|$ large enough,*
(c) *there is a positive B such that $f(t) = 0$ for $|t| > B$.*

Any of these conditions implies that ω corresponding to f is absolutely continuous. Moreover, if (b) *or* (c) *is valid, then the support of ω equals $\mathbb{R}$.*

Proof. Since f is continuous and bounded, any of the conditions ensures its integrability, and therefore absolute continuity by virtue of the previous lemma. Further, ω is given by the inversion formula (5.1); we can extend it to the complex plane as follows

$$\omega(z) = \text{const} + \frac{1}{2\pi}\int_{\mathbb{R}} f(t)\, \frac{e^{itz} - 1}{it}\, dt.$$

If (b) or (c) is valid, then this integral is well-defined in the open strip $|\operatorname{Im} z| < \beta$ or in $\mathbb{C}$, respectively, and defines there an analytic function which coincides with the original one on the real axis. Consequently, ω cannot be constant in any open subset of $\mathbb{R}$. ■

Proof of Theorem 5.1. In view of the relation (3.6), the assumption can be rephrased as

$$E_u U_s E_d U_t E_u = 0, \quad s \geqslant 0, t \geqslant T_r. \tag{5.2}$$

Let us consider the following expression:

$$\begin{aligned}(E_d U_t E_u)^* U_s (E_d U_t E_u) &= E_u U_{-t} E_d U_s E_d U_t E_u \\ &= E_u U_{-t} U_s E_d U_t E_u - E_u U_{-t} E_u U_s E_d U_t E_u.\end{aligned}$$

Using (5.2), we can reduce it to

$$E_u U_{s-t} E_d U_t E_u = 0 \quad \text{for } s \geqslant t \geqslant T_r.$$

Similarly we obtain

$$\begin{aligned}(E_d U_t E_u)^* U_{-s} (E_d U_t E_u) &= E_u U_{-t} E_d U_{-s} E_d U_t E_u \\ &= E_u U_{-t} E_d U_{t-s} E_u - E_u U_{-t} E_d U_{-s} E_u U_t E_u \\ &= (E_u U_{s-t} E_d U_t E_u)^* - (E_u U_s E_d U_t E_u)^* U_t E_u \\ &= 0 \quad \text{for } s \geqslant t \geqslant T_r.\end{aligned}$$

Thus for any $\psi \in \mathscr{H}$, we have

$$(E_d U_t E_u \psi, U_s E_d U_t E_u \psi) = 0, \quad |s| \geqslant t \geqslant T_r;$$

this can be further rewritten as

$$\int_{\mathbb{R}} e^{-i\lambda s} \, d\|E_\lambda E_d U_t E_u \psi\|^2 = 0, \quad |s| \geqslant t \geqslant T_r. \tag{5.3}$$

However $M \mapsto \|E_H(M)\varphi\|^2$ is a finite (positive) Lebesgue–Stieltjes measure for each $\varphi \in \mathscr{H}$. Since V is supposed to be a reduced propagator, the postulates (u1) and (u2) imply that there are $\psi \in \mathscr{H}_u$ and $t \geqslant T_r$ such that $\varphi = E_d U_t \psi$ is non-zero. Thus the support of the corresponding Lebesgue–Stieltjes measure equals $\mathbb{R}$ by virtue of Lemma 5.5(c), and the same holds, of course, for the spectral measure $E_H(\,.\,)$ of H. ■

REMARK 1.5.6. The assumption of minimality of $\{\mathscr{H}, U, E_u\}$ is not necessary, because in view of Proposition 4.3 we can always restrict our attention to the minimal unitary dilation, and all the proof needs is to find a vector ψ such that the corresponding measure in (5.3) is non-zero. On the other hand, the fact that V is a reduced propagator, i.e. that the operators V_t are non-unitary, is essential; the argument clearly fails if V is a unitary group so that $E_d = 0$ holds for the minimal unitary dilation.

By the same method, one can show that the decay laws of real unstable systems cannot be exponential, even in the asymptotic sense only:

THEOREM 1.5.7. *Assume ρ to be a state of an unstable system to which some positive C, β correspond so that $|P_\rho(t)| \leqslant C e^{-\beta t}$ for all $t \geqslant 0$, then the spectrum of the corresponding Hamiltonian equals* $\mathbb{R}$.

Proof. We found in the proof of Theorem 3.1 that

$$0 \leqslant \sum_k w_k |(\varphi_k, V_t \varphi_k)|^2 \leqslant P_\rho(t).$$

Since the eigenvalues w_k are non-negative, $\Sigma_k w_k = 1$, there is at least one non-zero vector $\varphi \in \mathscr{H}_u$ such that $p(t) = |(\varphi, V_t\varphi)|^2 = p(-t) \leqslant C' e^{-\beta t}$ for all $t \geqslant 0$, i.e. $|(\varphi, U_t\varphi)| \leqslant C' \exp(-\frac{1}{2}\beta|t|)$ for all $t \in \mathbb{R}$. On the other hand, we have

$$(\varphi, U_t\varphi) = \int_{\mathbb{R}} e^{-i\lambda t}\, d(\varphi, E_\lambda\varphi) \tag{5.4a}$$

so the assertion follows from Lemma 5.5(b). ∎

The above lemmas may be also used to establish sufficient conditions for continuity and absolute continuity of the energy spectrum (i.e. of $\sigma(H)$, where H refers to the minimal unitary dilation of V):

THEOREM 1.5.8. (a) *Let* $\text{w-lim}_{t\to\infty} V_t = 0$, *then* $\sigma(H)$ *is continuous.*

(b) *If the function* $t \mapsto (\psi, V_t\psi)$ *belongs to* $L(\mathbb{R}_+)$ *for each* $\psi \in \mathscr{H}_u$, *then* $\text{w-lim}_{t\to\infty} V_t = 0$ *and* $\sigma(H)$ *is absolutely continuous.*

Proof. In view of the relation (2.2), we have $(\varphi, V_t\varphi) = (\varphi, V_{-t}\varphi)$ for all $\varphi \in \mathscr{H}_u$ and $t \geqslant 0$; further, $(\varphi, V_t\varphi) = (\varphi, U_t\varphi)$ so

$$(\varphi, V_t\varphi) = \int_{\mathbb{R}} e^{-i\lambda t}\, d(\varphi, E_\lambda\varphi) \tag{5.4b}$$

converges to zero as $|t| \to \infty$. Then Lemma 5.2 implies that $\lambda \mapsto \|E_\lambda\varphi\|^2$ is continuous for each $\varphi \in \mathscr{H}_u$, i.e. that $\mathscr{H}_u$ is contained in $\mathscr{H}_c$, the continuous subspace with respect to H – cf., e.g., Kato (1966a, §X.1). This subspace is, however, preserved by $U_t = \exp(-iHt)$, so the minimality condition (4.3) yields

$$\mathscr{H} = \overline{\left[\bigcup_{t\in\mathbb{R}} U_t\mathscr{H}_u\right]}_{\text{lin}} \subset \mathscr{H}_c \subset \mathscr{H},$$

i.e. the assertion (a). If $t \mapsto (\varphi, V_t\varphi)$ is integrable on $\mathbb{R}_+$, then in the same way as above we can establish its integrability on the whole real axis. Then for each $\varphi \in \mathscr{H}_u$, the function $\lambda \mapsto \|E_\lambda\varphi\|^2$ is absolutely continuous by Lemma 5.4, i.e. $\mathscr{H}_u \subset \mathscr{H}_{ac}$. Since the absolutely continuous subspace with respect to H reduces again the operator H, the same argument as above shows that $\mathscr{H}_{ac} = \mathscr{H}$. Finally $(\varphi, V_t\varphi)$ converges to zero as $t \to \infty$ for each $\varphi \in \mathscr{H}_u$ in view of Lemma 5.3 so the polarization identity implies $\text{w-lim}_{t\to\infty} V_t = 0$. ∎

COROLLARY 1.5.9. *Let* $\mathscr{G}$ *be a linear manifold in* $\mathscr{H}_u$ *such that the function* $t \mapsto (\varphi, V_t\psi)$ *belongs to* $L(\mathbb{R}_+)$ *for each* $\varphi, \psi \in \mathscr{G}$, *then the functions* $g_{\varphi,\psi}$:

$$g_{\varphi,\psi}(\lambda) = (2\pi)^{-1} \int_{\mathbb{R}} e^{i\lambda t} (\varphi, V_t\psi)\, dt \tag{5.5a}$$

belong to $L^2(\mathbb{R}) \cap L(\mathbb{R})$; *they are non-negative almost everywhere in* $\mathbb{R}$ *for* $\varphi = \psi$. *Moreover, we have*

$$(\varphi, V_t\psi) = \int_{\mathbb{R}} e^{-i\lambda t} g_{\varphi,\psi}(\lambda)\, d\lambda, \tag{5.5b}$$

for all $\varphi, \psi \in \mathscr{G}$ *and* $t \geqslant 0$.

Proof. The relation (2.2) implies $(\varphi, V_t\psi) = \overline{(\psi, V_{-t}\varphi)}$ for all $\varphi, \psi \in \mathscr{G}$ and $t \geqslant 0$ so each function $t \mapsto (\varphi, V_t\psi)$ belongs to $L(\mathbb{R})$, and therefore to $L^2(\mathbb{R})$ too because of its boundedness; then its Fourier transform (5.5a) belongs to $L^2(\mathbb{R})$. As in the above proof, Lemma 5.4 implies that the Lebesgue–Stieltjes measure generated by the function $\lambda \mapsto (\varphi, E_\lambda\psi)$ is absolutely continuous with respect to m, the Lebesgue measure on $\mathbb{R}$, and one can see easily (cf. proof of Lemma 5.4) that $(\varphi, E_H(M)\psi) = \int_M g_{\varphi,\psi}(\lambda)\, d\lambda$. This equality shows that $g_{\varphi,\varphi}$ must be non-negative except possibly an m-zero set. Further, it implies $\|\varphi\|^2 = \int_{\mathbb{R}} g_{\varphi,\varphi}(\lambda)\, d\lambda$ so $g_{\varphi,\varphi} \in L(\mathbb{R})$. Finally, for each $\varphi, \psi \in \mathscr{G}$ the function $g_{\varphi,\psi}$ is a linear combination of four integrable functions (due to the polarization identity), thus $g_{\varphi,\psi} \in L(\mathbb{R})$ and the relation (5.5b) holds. ■

Let us present one more condition under which the energy spectrum is unbounded below:

THEOREM 1.5.10. *Assume that there is a state* ρ *of the unstable system under consideration such that*

(a) *for each* $\psi \in \operatorname{Ran}\rho \subset \mathscr{H}_u$, *the function* $t \mapsto (\psi, V_t\psi)$ *belongs to* $L(\mathbb{R}_+)$;
(b) *for the decay law* $P_\rho(\,.\,)$, *the following integral diverges*

$$\int_{\mathbb{R}} \frac{\ln P_\rho(t)}{1+t^2}\, dt = -\infty; \tag{5.6}$$

then the spectrum of the corresponding Hamiltonian is not semibounded.

Proof. As in the proof of Theorem 5.7, there is a non-zero vector $\varphi \in \operatorname{Ran}\rho \subset \mathscr{H}_u$ and a positive b such that $|(\varphi, V_t\varphi)|^2 \leqslant bP_\rho(t)$. This inequality together with (5.6) imply that the integral

$$\int_{\mathbb{R}} \frac{\ln|(\varphi, V_t\varphi)|}{1+t^2}\, dt \leqslant \tfrac{1}{2}\int_{\mathbb{R}} \frac{\ln P_\rho(t)}{1+t^2}\, dt + \frac{\pi}{2}\ln b$$

also diverges. In view of the assumption (a), we can apply Corollary 5.9 to the one-dimensional subspace spanned by φ, and express therefore $(\varphi, V_t\varphi) = \int_{\mathbb{R}} e^{-i\lambda t} g(\lambda)\, d\lambda$ with $g \in L^2(\mathbb{R}) \cap L(\mathbb{R})$. Further, $|(\varphi, V_t\varphi)|$ is bounded so the theorem of Paley and Wiener (A.17) asserts that if g would have a semibounded support, then the above integral must converge. Since $(\varphi, E_H(M)\varphi) = \int_M g(\lambda)\, d\lambda$, the spectral measure of H cannot have a semibounded support. ∎

REMARK 1.5.11. Assumptions of the above theorem simplify particularly if ρ is a pure state represented by some unit vector. Let us further stress limitations of the theorem, which are not always respected in various applications of the Paley–Wiener theorem appearing in the physical literature. Firstly, this theorem concerns L^2-functions so in order to use it, one has to ensure the square integrability. Furthermore, if $f(t) = \int_{\mathbb{R}} e^{-i\lambda t} g(\lambda)\, d\lambda$ with $g \in L^2(\mathbb{R}) \cap L(\mathbb{R})$ and $\int_{\mathbb{R}} |\ln|f(t)||(1+t^2)^{-1}\, dt < \infty$, then the theorem does *not* imply g to have a semibounded support. Consider the following *example*: the support of $g : g(\lambda) = e^{-|\lambda|}$ is the whole real axis, but $f(t) = (1+t^2)^{-1}$ so the above integral converges. The Paley–Wiener theorem asserts only that there is a function in $L^2(\mathbb{R})$ with a semibounded support such that modulus of its Fourier transform coincides a.e. with that of f. This is true, of course, because one can choose for instance $g_+ : g_+(\lambda) = \lambda e^{-\lambda}\Theta(\lambda)$, where $\Theta = \chi_{[0,\infty)}$ is the Heaviside function, so that $f_+(t) = (1+t^2)^{-1} \exp(-2i \operatorname{arctg} t)$ and $|f_+(t)| = f(t)$ for all $t \in \mathbb{R}$.

Among the above proved assertions, Theorems 5.1, 5.7 and 5.10 yield various necessary conditions for semiboundedness of the total Hamiltonian in terms of the reduced propagator or the decay laws. We are also naturally interested in some sufficient condition. In order to formulate it, let us first show how one can find for a given reduced propagator V the corresponding positive-operator-valued measure F, whose existence is established by Proposition 4.5:

PROPOSITION 1.5.12. *Let $V : \mathbb{R} \to \mathscr{B}(\mathscr{H}_u)$ be weakly continuous and positive definite with $V_0 = I_u$, then the positive-operator-valued measure F appearing in the equality* (4.4b) *is given by*

$$\tfrac{1}{2}[F_{\lambda+0} + F_{\lambda-0}]\psi = \tilde{F}_0\psi + \frac{1}{\pi}\lim_{\eta\to 0+}\int_0^{\lambda} \operatorname{Re}\Phi(\xi + i\eta)\psi\, d\xi, \tag{5.7a}$$

$$\Phi(\xi + i\eta)\psi = \int_0^{\infty} e^{it(\xi+i\eta)} V_t\psi\, dt, \qquad \eta > 0, \tag{5.7b}$$

for each $\psi \in \mathscr{H}_u$, where $\operatorname{Re}\Phi \equiv \frac{1}{2}(\Phi + \Phi^*)$ *and $\tilde{F}_0$ is a (λ-independent) operator from $\mathscr{B}(\mathscr{H}_u)$, which can be fixed, e.g., by the normalization condition* $\text{s-lim}_{\lambda\to-\infty} F_\lambda = 0$. *Furthermore, we have*

$$(F_{\lambda+0} - F_{\lambda-0})\psi = \lim_{T\to\infty}\frac{1}{T}\int_0^T \operatorname{Re}\{V_t e^{i\lambda t}\}\psi\, dt \tag{5.8}$$

for each $\psi \in \mathscr{H}_u$ and $\lambda \in \mathbb{R}$.

Proof. Since a vector-valued function $f(\,.\,)$ is integrable iff the same is true for $\|f(\,.\,)\|$ (Dunford and Schwartz, 1958, theorem III.2.22), existence of the integrals in (5.7), (5.8) can be easily established. Further, F exists due to Proposition 4.5, so applying the Stieltjes inversion formulae (A.3) to (4.4a) we get the 'weak form' of the above relations, which follow then from the Riesz lemma. ■

Let us mention yet another form of the relation (5.7a). Consider arbitrary real $\lambda < \mu$, then we have $F([\lambda, \mu]) = F_{\mu+0} - F_{\lambda-0}$ and $F((\lambda, \mu)) = F_{\mu-0} - F_{\lambda+0}$ so (5.7a) can be rewritten as

$$\tfrac{1}{2}\{F([\lambda, \mu]) + F((\lambda, \mu))\}\psi = \frac{1}{\pi} \lim_{\eta \to 0+} \int_{\lambda}^{\mu} \mathrm{Re}\, \Phi(\xi + i\eta)\psi \, d\xi \tag{5.7c}$$

for any $\psi \in \mathscr{H}_u$. Since there is one-to-one correspondence between the functions V which obey the assumptions of Proposition 5.12 (in particular, the reduced propagators) and the positive-operator-valued measures F, we can introduce the following notion: the **energy support** $\sigma[V]$ of V is the support of the corresponding measure F, i.e. the smallest set $M \subset \mathbb{R}$ for which $F(\mathbb{R} \backslash M) = 0$. Equivalently, $\sigma[V] = \{\lambda \in \mathbb{R} : F_{\lambda+\epsilon} - F_{\lambda-\epsilon} \neq 0 \text{ for each } \epsilon > 0\}$. With these prerequisites, we can formulate an assertion which gives particularly the above-mentioned sufficient condition:

THEOREM 1.5.13. *Let* $V : \mathbb{R} \to \mathscr{B}(\mathscr{H}_u)$ *be weakly continuous and positive definite with* $V_0 = I_u$, *then* $\sigma[V] \subset \sigma(H)$ *holds. If* H *refers to the minimal unitary dilation of* V, *then both the sets coincide,*

$$\sigma(H) = \sigma[V]. \tag{5.9}$$

In particular, if the energy support of V *is bounded below, so is the spectrum of Hamiltonian.*

Proof. Let us consider an arbitrary fixed $\lambda \in \mathbb{R}$ and denote $\Delta_\epsilon = (\lambda - \epsilon, \lambda + \epsilon)$. According to (4.5), $E_H(\Delta_\epsilon) = 0$ implies $F(\Delta_\epsilon) = 0$ so $\sigma[V] \subset \sigma(H)$. Conversely, let $\lambda \notin \sigma[V]$ so that $F(\Delta_\eta) = 0$ for some $\eta > 0$. Let $\varphi = U_t\psi$ with $t \in \mathbb{R}$ and $\psi \in \mathscr{H}_u$, then $(\varphi, E_H(\Delta_\eta)\varphi) = (\psi, U_t^* E_H(\Delta_\eta)U_t\psi)$. However, the projection $E_H(\Delta_\eta)$ belongs to the spectral family of H so it commutes with $U_t = \exp(-iHt)$; thus the unitarity of U_t implies $(\varphi, E_H(\Delta_\eta)\varphi) = (\psi, E_H(\Delta_\eta)\psi) = (\psi, F(\Delta_\eta)\psi) = 0$. Since $E_H(\Delta_\eta)$ is a projection, we obtain

$$E_H(\Delta_\eta)\varphi = 0 \tag{*}$$

for all $\varphi \in \Phi \equiv \bigcup_{t \in \mathbb{R}} U_t \mathscr{H}_u$; further, linearity and boundedness of $E_H(\Delta_\eta)$ implies that (*) holds for any φ from the closure of Φ_{lin}, the linear span of Φ. However, $\overline{\Phi_{\text{lin}}} = \mathscr{H}$ in view of minimality condition (4.3), and therefore $E_H(\Delta_\eta) = 0$, i.e. $\lambda \notin \sigma(H)$ so that $\sigma(H) \subset \sigma[V]$. ■

In the case of a finite-dimensional $\mathscr{H}_u$, all the information needed for a decision about semiboundedness of the corresponding Hamiltonian can be concentrated in one complex function only:

COROLLARY 1.5.14. *Let V be as in Theorem* 5.13 *and* dim $\mathcal{H}_u < \infty$, *then* $\sigma(H) = \sigma[v]$, *where* $v : v(t) = \mathrm{Tr}\, V_t$, $t \in \mathbb{R}$, *and H refers to the minimal unitary dilation of V.*

Proof. The function v is obviously continuous, positive definite and obeys $v(t) = \int_{\mathbb{R}} e^{-i\lambda t}\, d\,\mathrm{Tr}\, F_\lambda$. The inclusion $\sigma[v] \subset \sigma[V]$ is trivial; on the other hand, each $F(M)$ is positive so $\mathrm{Tr}\, F(M) = 0$ implies $F(M) = 0$, and therefore $\sigma[v] \supset \sigma[V]$. ■

Thus we have obtained a way in which spectrum of the Hamiltonian can be obtained from the knowledge of the reduced propagator. On the other hand, the situation is less satisfactory when the starting information consists of the decay laws, which are often the only directly measured quantities. A knowledge of the decay laws for sufficiently many states of the considered unstable system gives us the decay function $P(\,.\,)$ (cf. Section 2); thus the problem is essentially 'To what extent does this function determine the reduced propagator?' In other words, we have a *strongly continuous positive-operator-valued function* $P : \mathbb{R} \to \mathcal{B}_+(\mathcal{H}_u)$ with $P(0) = I_u$, and we ask for solutions $V : \mathbb{R} \to \mathcal{B}(\mathcal{H}_u)$ of the equation

$$V_t^* V_t = P(t), \tag{5.10}$$

which would be weakly continuous and positive definite with $V_0 = I_u$. Notice that $P(\,.\,)$ itself *need not be of the positive type* for dim $\mathcal{H}_u > 1$: in such a case the condition analogous to (4.2) would have to hold,

$$V_t^* V_t = P(t) = P(-t)^* = (V_{-t}^* V_{-t})^* = V_t V_t^*,$$

but the operators V_t are not necessarily normal. For instance, consider the following matrix semigroup (cf. Example 5.16 below):

$$V_t = e^{-t} \begin{pmatrix} 1 & 2a(1 - e^{-t}) \\ 0 & e^{-t} \end{pmatrix}, \quad a \in \mathbb{R},\ t \geqslant 0; \tag{5.11}$$

in this case the operators V_t and V_t^* do not commute for any non-zero t unless $a = 0$.

It is clear that if a solution to equation (5.10) exists, it need not be unique:

(i) A function $V' : t \mapsto W_t V_t$, where $\{W_t : t \in \mathbb{R}\}$ is an arbitrary continuous one-parameter unitary group $\subset \mathcal{B}(\mathcal{H}_u)$, solves (5.10) together with V; moreover, if the operators V_t, W_s commute for all $t, s \in \mathbb{R}$, then positive definiteness of V implies positive definiteness of V'. In particular, consider two reduced propagators V and V' to which some Hamiltonians H, H' correspond. If the operator $G = H - H'$ is also self-adjoint, reduced by $\mathcal{H}_u$, and commutes with one of the operators H, H' (hence, with both of them), then the reduced propagators V, V' give the same decay function.

(ii) If V is positive definite so is $V' : V_t' = V_t^*$; further, V' solves (5.10) together with V in the case that all the operators V_t are normal.

(iii) The above-named possibilities do not exhaust fully the non-uniqueness, even in the simplest case of one-dimensional $\mathcal{H}_u$, when equation (5.10) reduces to $|v(t)|^2 = p(t)$ with p continuous and such that $0 \leqslant p(t) \leqslant p(0) = 1$. Remember the example mentioned in Remark 5.11: there are two different (continuous, positive definite) functions – namely $v_1(t) = (1 + t^2)^{-1}$ and $v_2(t) = (1 + t^2)^{-1} \exp(-2i \operatorname{arctg} t)$ – which give the same $p(t)$.

Thus we have obtained a good illustration of the problems connected with any attempt to formulate a sufficient condition on the decay laws, under which they would correspond to a semibounded energy spectrum. Let us now conclude the section by some examples:

EXAMPLE 1.5.15. Due to Theorem 5.7, exponentiality of any decay law must be violated, particularly *for large times*. Historically, this effect of energy semiboundedness was recognized first (see Khalfin (1957) and notes to this chapter). Consider the simplest case of one-dimensional $\mathcal{H}_u$ (Example 2.4) and ψ, H such that the energy distribution $d(\psi, E_\lambda \psi) = g(\lambda)\, d\lambda$ has the Breit–Wigner shape cut sharply below some threshold energy λ_t:

$$g(\lambda) = \frac{\Gamma N}{2\pi} \frac{\Theta(\lambda - \lambda_t)}{(\lambda - \lambda_0)^2 + \frac{1}{4}\Gamma^2} . \tag{5.12a}$$

One can set $\lambda_t = 0$ without loss of generality, then the normalization factor $N = [\frac{1}{2} + (1/\pi) \operatorname{arctg}(2\lambda_0/\Gamma)]^{-1}$. The function $v(t) = \int_{\mathbb{R}} e^{-i\lambda t} g(\lambda)\, d\lambda$ is evaluated by contour integration (cf. Figure 1); after the limit $R \to \infty$ we get for $t \geqslant 0$

$$v(t) = N \exp(-i\lambda_0 t - \tfrac{1}{2}\Gamma t) - \frac{iN\Gamma}{2\pi} \int_0^\infty \frac{e^{-zt}\, dz}{(iz + \lambda_0)^2 + \frac{1}{4}\Gamma^2} ,$$

and this further gives

$$v(t) = N \exp(-i\lambda_0 t - \tfrac{1}{2}\Gamma t) - \\ - \frac{iN}{2\pi} \sum_{\alpha = \pm} \alpha \exp(-i\lambda_0 t - \tfrac{1}{2}\alpha\Gamma t)\, \mathrm{Ei}(i\lambda_0 t + \tfrac{1}{2}\alpha\Gamma t). \tag{5.12b}$$

Using now the first-order asymptotic expansion for Ei(.) (Gradshtein and Ryzhik, 1971, 8.215) together with a simple estimate of the remainder terms (based on $(\cos \varphi/2)^{-1} + (\cos(\pi/2 - \varphi/2))^{-1} < 4/\sin \varphi$, where $\varphi = \arg(\frac{1}{2}\Gamma + i\lambda_0)$), we obtain

$$v(t) = N \exp(-i\lambda_0 t - \tfrac{1}{2}\Gamma t) - \frac{i}{2\pi} \frac{N\Gamma}{\lambda_0^2 + \frac{1}{4}\Gamma^2} t^{-1} + R(t), \tag{5.12c}$$

where the remainder term obeys $|R(t)| < (2/\pi) N (\lambda_0 t)^{-2}$. It is clear from (5.12b) that the decay law $P_\psi(t) = |v(t)|^2$ violates exponentiality for all times. We are particularly interested in the asymptotic region characterized by the time t_c after

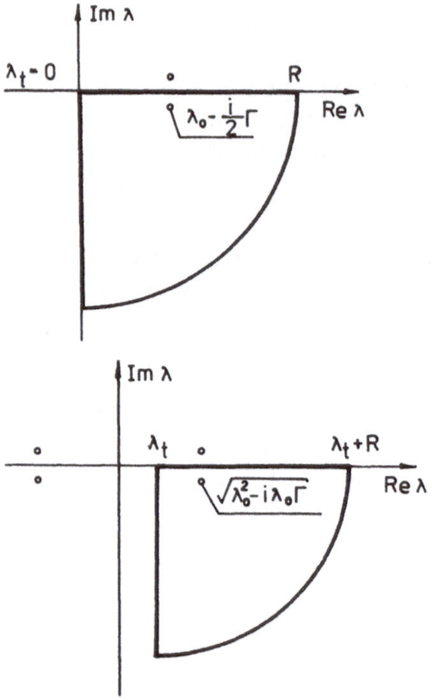

1.1. Integration contours used in Example 1.5.15.

which the power-like term in (5.12c) dominates the exponential one. If $\Gamma \ll \lambda_0$, we have the following equation for t_c:

$$\tfrac{1}{2}\Gamma t_c \exp(-\tfrac{1}{2}\Gamma t_c) = \pi^{-1}\left(\frac{\Gamma}{2(\lambda_0 - \lambda_t)}\right)^2. \tag{5.13}$$

For correctness of interpretation, the remainder term should not exceed the power-like term; it is true if $\Gamma t_c \gtrsim 4$.

Before solving (5.13), we shall comment on another decay-law formula. In their attempt to construct a quantum field theory of unstable particles, Matthews and Salam (1959) used the following energy distribution function

$$g(\lambda) = \frac{2\Gamma N}{\pi} \frac{\lambda_0 \lambda \Theta(\lambda - \lambda_t)}{(\lambda^2 - \lambda_0^2)^2 + \lambda_0^2 \Gamma^2}, \tag{5.14a}$$

where the normalization factor $N = [\frac{1}{2} + (1/\pi)\,\mathrm{arctg}((\lambda_0^2 - \lambda_t^2)/\lambda_0 \Gamma)]^{-1}$. The corresponding function v can be obtained again with the help of contour integration (cf. Figure 1): for $t \geqslant 0$ we get

$$v(t) = N \exp(-iS_- t) + \frac{iN}{2\pi} \sum_{\alpha,\beta=\pm} \alpha \exp(i\beta S_\alpha t)\mathrm{Ei}(-i\lambda_t t - i\beta S_\alpha t), \tag{5.14b}$$

where $S_{\pm} = (\lambda_0^2 \pm i\lambda_0\Gamma)^{1/2}$. In the same way as above, we obtain the following asymptotic expression

$$v(t) = N\exp[-i(\lambda_0^2 - i\lambda_0\Gamma)^{1/2}t] + \frac{2iN}{\pi t}\frac{\lambda_0\lambda_t\Gamma\exp(-i\lambda_t t)}{(\lambda_0^2 - \lambda_t^2)^2 + \lambda_0^2\Gamma^2} + R(t). \tag{5.14c}$$

where $|R(t)| < (4/\pi)N(\lambda_0 - \lambda_t)^{-2}t^{-2}$. Physically the most interesting situation is the one with $\Gamma \ll \lambda_0 - \lambda_t \ll \lambda_0$; in such a case $S_- \approx \lambda_0 - (i/2)\Gamma$ and the time t_c is again given by equation (5.13) (its proper interpretation now needs $\Gamma t_c \gtrsim 8$).

For $r \equiv \frac{1}{2}\Gamma/(\lambda_0 - \lambda_t)$ sufficiently small, equation (5.13) has two positive solutions; we are interested in the greater one. Numerically obtained t_c in dependence on r are plotted on Figure 2. In view of the sharp cut-off, the decay law oscillates rapidly (with a frequency $(\lambda_0 - \lambda_t)/2\pi$) around t_c; this is sketched on Figure 3 for $P_\psi(\,.\,)$ referring to (5.12c) and the particular value $r = 10^{-9}$.

Concluding this example, we see that the large-time deviations from the exponential decay law are manifested after some critical time t_c, and that the values of $P_\psi(t_c)$ *do not allow us to check this effect experimentally* (unless by some lucky coincidence the parameter r can be made sufficiently large). The energy semiboundedness need not be expressed, of course, by a 'sharp' truncation as above, but this circumstance can hardly change the order of magnitude of the discussed effect.

EXAMPLE 1.5.16. Let us now return to the semigroup given by (5.11), i.e.

$$V_t = \begin{pmatrix} e^{-t} & 2a\,e^{-t}(1-e^{-t}) \\ 0 & e^{-2t} \end{pmatrix}, \qquad t \geq 0,$$

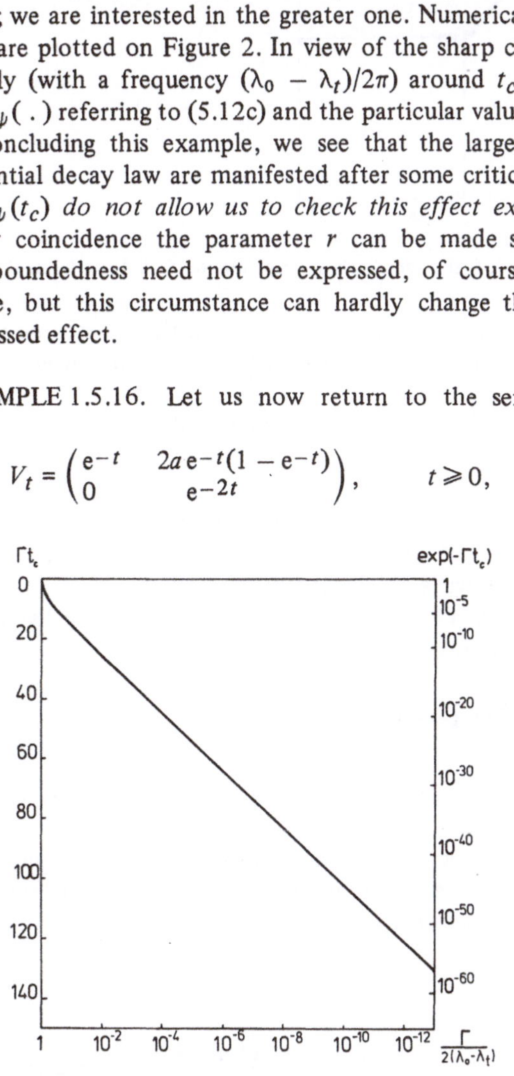

1.2. Large-time non-exponentiality of the decay law: solution to equation (1.5.13).

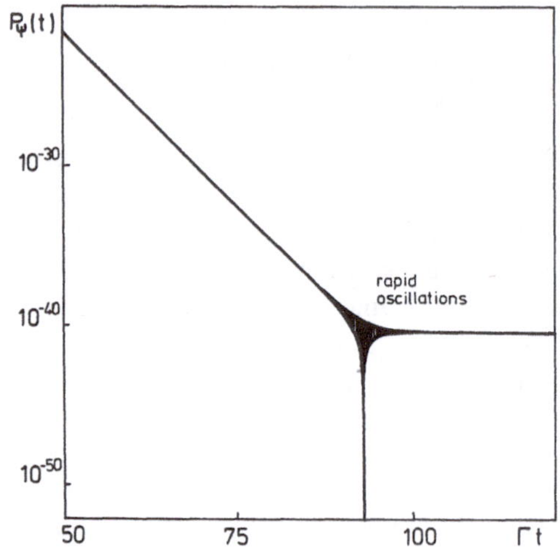

1.3. Large-time behaviour of the decay law referring to (1.5.12c) with $\Gamma = 2 \times 10^{-9}(\lambda_0 - \lambda_f)$.

and discuss its minimal unitary dilation. We shall start with the restrictions on the parameter a set by the requirement of contractivity, which can be rewritten as

$$e^{-2t}|\psi_1 + 2a(1-e^{-t})\psi_2|^2 + e^{-4t}|\psi_2|^2 \leqslant |\psi_1|^2 + |\psi_2|^2$$

for all complex ψ_1, ψ_2. Since this inequality is valid for $\psi_2 = 0$ automatically, we may assume ψ_2 non-zero and $\psi_1 = \beta\psi_2$. Abbreviating $x = e^{-t}$, we get $x^4 + x^2[\beta + 2a(1-x)]^2 \leqslant 1 + \beta^2$; without loss of generality we can consider positive a, β only. Further, we rewrite the inequality into the form $\beta^2(1-x^2) - 4a\beta x^2(1-x) + 1 - x^4 - 4a^2x^2(1-x)^2 \geqslant 0$. Its l.h.s. is non-negative iff the same is true for its minimum with respect to β:

$$\frac{1-x}{1+x}\,[-4a^2x^2 + (1+x)^2(1+x^2)] \geqslant 0,$$

and since $x \in (0, 1]$, we obtain $a^2 \leqslant \frac{1}{4}x^{-2}(1+x)^2(1+x^2)$; this inequality holds for all $x \in (0, 1]$ iff $a^2 \leqslant 2$.

In order to find the minimal unitary dilation, let us first write down the corresponding measure F (cf. (4.4)). Since all the functions involved are square-integrable, we can use the inverse Fourier transformation to obtain

$$G(\lambda) \equiv \frac{dF_\lambda}{d\lambda} = \frac{1}{\pi}\begin{pmatrix} (\lambda^2+1)^{-1} & -a(\lambda+i)^{-1}(\lambda+2i)^{-1} \\ -a(\lambda-i)^{-1}(\lambda-2i)^{-1} & 2(\lambda^2+4)^{-1} \end{pmatrix}, \quad \lambda \in \mathbb{R}. \qquad (5.15)$$

Now we can distinguish the following three cases:

(i) $a = 0$. The considered semigroup can be expressed as an orthogonal sum, so in view of Proposition 4.3 and Example 4.8, the minimal unitary dilation can be constructed in $\mathscr{H} = L^2(\mathbb{R}; \mathbb{C}^2)$ as a direct sum of m.u.d.'s of the respective one-dimensional semigroups.

(ii) $|a| < \sqrt{2}$. We take again $\mathscr{H} = L^2(\mathbb{R}; \mathbb{C}^2)$; further, $H : (H\Psi)(\lambda) = \{\lambda\psi_1(\lambda), \lambda\psi_2(\lambda)\}$ and $\mathscr{H}_u$ is the two-dimensional subspace in $\mathscr{H}$ spanned by the orthonormal vectors

$$\begin{aligned} &\Psi^{(1)} : \Psi^{(1)}(\lambda) = \{-i\pi^{-1/2}(\lambda - i)^{-1}, 0\}, \\ &\Psi^{(2)} : \Psi^{(2)}(\lambda) = \{ia\pi^{-1/2}(\lambda + 2i)^{-1}, (2 - a^2)^{1/2}\pi^{-1/2}(\lambda + 2i)^{-1}\}. \end{aligned} \tag{5.16}$$

One verifies readily that

$$\begin{aligned} (\Psi^{(j)}, U_t\Psi^{(k)}) &= \sum_{n=1}^{2} \int_{\mathbb{R}} e^{-i\lambda t}\, \overline{\psi}_n^{(j)}(\lambda)\psi_n^{(k)}(\lambda)\, d\lambda \\ &= \int_{\mathbb{R}} e^{-i\lambda t}\, G_{jk}(\lambda)\, d\lambda, \end{aligned} \tag{5.17}$$

so the described triple $\{\mathscr{H}, U, E_u\}$ represents a unitary dilation of the given semigroup; its minimality can be checked in the same way as in Example 4.8. In the particular case $a = 0$, we return with (5.16) to (i),

(iii) $|a| = \sqrt{2}$. In this case the minimal unitary dilation 'loses one half of its dimension': we have $\mathscr{H} = L^2(\mathbb{R})$ with $(H\psi)(\lambda) = \lambda\psi(\lambda)$ and $\mathscr{H}_u$ being spanned by the vectors $\psi^{(1)} : \psi^{(1)}(\lambda) = -i\pi^{-1/2}(\lambda - i)^{-1}$ and $\psi^{(2)} : \psi^{(2)}(\lambda) = i(2/\pi)^{1/2}(\lambda + 2i)^{-1}$.

Let us show *how the above unitary dilations can be constructed*. This, of course, is not necessary once we have found them; however, the same method applies also to other finite-dimensional semigroups and positive definite functions. Consider the Hilbert space $\mathscr{H} = \Sigma_{n=1}^{N} \oplus L^2(\mathbb{R}, \mu_n)$, where $N = 1, 2, \ldots$ or infinity, and μ_n are finite positive measures on $\mathbb{R}$, each of them being generated by the Lebesgue measure and a function ω_n. In other words, the scalar product of the vectors $\Psi = \{\psi_n\}_{n=1}^{N}$ and $\Phi = \{\varphi_n\}_{n=1}^{N} \in \mathscr{H}$ is given by

$$(\Psi, \Phi) = \sum_{n=1}^{N} \int_{\mathbb{R}} \overline{\psi_n(\lambda)}\varphi_n(\lambda)\omega_n(\lambda)\, d\lambda.$$

Consider further the operator H on $\mathscr{H}$, $(H\Psi)(\lambda) = \{\lambda\psi_n(\lambda)\}_{n=1}^{N}$. Comparison with the multiplication-operator form of the spectral theorem (Reed and Simon, 1972, theorems VII.3 and VIII.4) shows, of course, that this is not the most general form of a self-adjoint operator on $\mathscr{H}$; however, the described class is reasonably large, especially when spectrum of H is often expected to be continuous. Now a relation analogous to (5.17) gives the equations

$$\sum_{n=1}^{N} \bar{\psi}_n^{(j)}(\lambda)\psi_n^{(k)}(\lambda)\omega_n(\lambda) = G_{jk}(\lambda), \quad \lambda \in \mathbb{R} \tag{5.18}$$

for components of the vectors $\Psi^{(j)}$ which span $\mathscr{H}_u$.

In our example, equations (5.18) can be solved with $N = 1$ for $|a| = \sqrt{2}$, and with $N = 2$ for $|a| < \sqrt{2}$. It is also illustrative to see how the method fails in the case $|a| > \sqrt{2}$ when the considered semigroup is not contractive so that no unitary dilation exists. Positivity of the functions ω_n together with the Hölder inequality and the relations (5.15), (5.18) imply

$$\begin{aligned}\left|\frac{a}{\pi}((\lambda+i)(\lambda+2i))^{-1}\right| &\leqslant \sum_{n=1}^{N} |\psi_n^{(1)}(\lambda)|\,|\psi_n^{(2)}(\lambda)|\,\omega_n(\lambda)\\ &\leqslant \left(\sum_{n=1}^{N} |\psi_n^{(1)}(\lambda)|^2\omega_n(\lambda)\right)^{1/2}\left(\sum_{n=1}^{N} |\psi_n^{(2)}(\lambda)|^2\omega_n(\lambda)\right)^{1/2}\\ &= \left(\frac{1}{\pi}(\lambda^2+1)^{-1}\right)^{1/2}\left(\frac{2}{\pi}(\lambda^2+4)\right)^{1/2},\end{aligned}$$

and therefore a solution to equations (5.18) can exist only if $|a| \leqslant \sqrt{2}$.

The most important conclusion drawn from the present example is that a *solution to the inverse decay problem need not depend 'continuously' on initial data*: though the semigroup under consideration depends continuously on the parameter a, the spectrum of H has multiplicity 2 for $|a| < \sqrt{2}$, while its value at the points $|a| = \sqrt{2}$ equals 1.

There is yet another point to mention. If $|a| = \sqrt{2}$, then there is a vector in $\mathscr{H}_u$, namely $\psi = 3^{-1/2}(2^{1/2}\psi^{(1)} + \operatorname{sg} a\,\psi^{(2)})$, $\psi(\lambda) = (6/\pi)^{1/2}((\lambda+2i)(\lambda-i))^{-1}$, which represents a finite-energy state, and even belongs to $D(H)$. In view of Theorem 3.1, the initial decay rate $\dot{P}_\psi(0) = (d/dt)\|V_t\psi\|^2\,|_{t=0}$ is zero, despite the naive expectation that each state must decay with a non-zero decay rate from the beginning when $\{V_t : t \geqslant 0\}$ is a semigroup and $\lim_{t\to\infty}\|V_t\| = 0$. We have $\dot{P}_\psi(0) = 2\,\mathrm{Im}(\psi, H\psi) = -2\,\mathrm{Re}(\psi, B\psi)$, where $B = \begin{pmatrix}1 & -2a\\ 0 & 2\end{pmatrix}$ is the generator of V, $V_t = e^{-Bt}$. We have already mentioned that V_t are not normal (unless $a = 0$ or $t = 0$) so B too is not normal; this is obvious from its matrix form. The above-mentioned fact then represents a manifestation of this non-normality, which enables the numerical range of B to exceed the convex hull of the spectrum $\sigma(B)$: we can verify readily that $(\psi, B\psi) = 2(2/3)^{1/2}(2^{1/2} - |a|) = 0$ for $|a| = \sqrt{2}$, while $\sigma(B) = \{1, 2\}$.

EXAMPLE 1.5.17. Let us now mention some relations between the *lifetime* defined by the relation (2.6) and the properties of the energy spectrum. Suppose first that the assumptions of Corollary 5.9 are fulfilled for $\mathscr{G} = \mathscr{H}_u$ and $\dim \mathscr{H}_u = n < \infty$.

In such a case, the spectrum of H is absolutely continuous due to Theorem 5.8(b); further, the decay law P_ρ can be rewritten with the aid of (3.5) into the form

$$P_\rho(t) = \sum_{j,k=1}^{n} w_k \, |(\psi_j, V_t\varphi_k)|^2, \tag{5.19}$$

where $\{\psi_j\}$ is some orthonormal basis in $\mathcal{H}_u$. All the functions $t \mapsto (\psi_j, V_t\varphi_k)$ here are bounded and integrable due to the assumption, hence they are also square-integrable in $\mathbb{R}_+$ and the lifetime T_ρ is finite. In this deduction, finiteness of the sum (5.19) is essential; one can easily construct an example showing that the assertion may fail for $\dim \mathcal{H}_u = \infty$.

On the other hand, the fact that $\sigma(H)$ is absolutely continuous does not ensure finiteness of the lifetime even in the simplest case $\dim \mathcal{H}_u = 1$. In order to see this, consider $\mathcal{H} = L^2([-1, 1])$ and $H : (H\varphi)(\lambda) = \lambda\varphi(\lambda)$, so $\sigma(H)$ is absolutely continuous. Further, let $\mathcal{H}_u$ be spanned by the vector $\psi : \psi(\lambda) = \pi^{-1/2}(1 - \lambda^2)^{-1/4}$; then using the notation from Example 2.4 we get

$$v(t) = \frac{2}{\pi}\int_0^1 (1 - \lambda^2)^{-1/2} \cos \lambda t \, d\lambda = J_0(t).$$

Consequently,

$$|v(t)|^2 = \frac{2}{\pi t} \cos^2\left(t - \frac{\pi}{4}\right) + O(t^{-2})$$

for large values of t so $T_\psi = \int_0^\infty |v(t)|^2 \, dt = \infty$. The reason is obviously that we have chosen the function $g_{\psi,\psi}$ in (5.5b) which belongs to $L(\mathbb{R})$ but not to $L^2(\mathbb{R})$. In fact, we can express T_ψ from $g_{\psi,\psi}(\,.\,)$ directly if the latter is square-integrable and $\dim \mathcal{H}_u = 1$: in this case $P_\psi(-t) = P_\psi(t)$ so the Parseval identity implies

$$T_\psi = \pi \int_{\mathbb{R}} |g_{\psi,\psi}(\lambda)|^2 \, d\lambda. \tag{5.20}$$

Notice that one cannot generalize (5.20) to the case $\dim \mathcal{H}_u > 1$ unless the decay law is an even function of time, but this is not true in general: in Example 2.5, for instance, we have $P_\rho(t) = 1$ for each ρ, $\operatorname{Ran} \rho \subset \mathcal{H}_u$, and $t \leqslant 0$.

1.6. Bounded-Energy Approximation

We have found up to this point that the inevitable violations of the semigroup condition (1.1) are likely to be negligible from the experimental point of view, at least with respect to the large-time behaviour of the decay laws. We shall return to this 'practical' task in the next chapter; now we are going to discuss other side of the problem, namely we shall ask the way in which the methods involving the (below) unbounded energies are related to the exact description of an unstable system.

Consider, for instance, the inverse-decay problem. One can choose as a reduced propagator either a continuous contractive semigroup or some continuous operator-valued function of the positive type which is in some sense near to this semigroup; however, relations between the corresponding minimal unitary dilations are not *a priori* clear. Let us recall Example 5.16 which shows that a small change in initial data can eventually affect solution to the inverse-decay problem substantially.

Our treatment is based on the notion of **bounded-energy state**: a density matrix ρ is said to describe such a state if the measure μ_ρ (cf. (3.2)) has a bounded support. In other words, there is a positive b such that the function (3.1) obeys

$$\mu_\rho(\lambda) = \begin{cases} 0 & \text{for } \lambda \leqslant -b \\ 1 & \text{for } \lambda \geqslant b. \end{cases}$$

The set of all bounded-energy states (to a given Hamiltonian) will be denoted as $B(H)$. It is important that this set is large enough in the following sense:

PROPOSITION 1.6.1. *For any state ρ, there is a one-parameter family $\{\rho_b\} \subset B(H)$ such that* $\lim_{b\to\infty} \operatorname{Tr} |\rho - \rho_b| = 0$.

Proof. Consider the projections $E^b \equiv E_H(\Delta_b)$, where $\Delta_b = (-b, b)$ for each $b > 0$, and define

$$\rho_b = N_b E^b \rho E^b, \qquad N_b^{-1} = \operatorname{Tr}(\rho E^b) \tag{6.1}$$

for those b for which $N_b^{-1} > 0$. Obviously, $\rho_b \in B(H)$. According to the definition, $\text{s-lim}_{b\to\infty} E^b = I$, so $\lim_{b\to\infty} N_b = 1$ follows from Proposition 2.2(b). Further, the assertion (a) of the same proposition yields the following estimate:

$$\begin{aligned}\operatorname{Tr} |\rho - \rho_b| &\leqslant \operatorname{Tr} |\rho - E^b \rho| + \operatorname{Tr} |E^b(\rho - \rho E^b)| + |1 - N_b| \operatorname{Tr} |E^b \rho E^b| \\ &\leqslant 2 \operatorname{Tr} |\rho - E^b \rho| + N_b - 1.\end{aligned}$$

According to the polar-decomposition theorem, there is a partial isometry W such that $|(I - E^b)\rho| = W^*(I - E^b)\rho$. Finally, $\text{s-lim}_{b\to\infty} W^*(I - E^b) = 0$, so the result follows. ∎

It holds $B(H) \subset M(H)$; in particular, any vector ψ which represents a pure bounded-energy state belongs to $D(H) \subset Q(H)$ as well as to the domain of H^n for every n natural. This makes it possible to prove an assertion which strengthens Theorem 3.1 for $\rho \in B(H)$. Let us recall that the main point in the proof of this theorem was to establish the existence of the first derivative $\dot{P}_\rho(0)$. A much stronger result holds for the bounded-energy states:

THEOREM 1.6.2. *If $\rho \in B(H)$, then the decay law P_ρ is a restriction to $\mathbb{R}_+$ of an entire function.*

Proof. If $\rho \in B(H)$, there is a positive b such that $\rho = E^b \rho E^b$. The decay law (2.3a) then reads as $P_\rho(t) = \mathrm{Tr}\{U_b(t)^* E_u U_b(t)\rho\}$, where

$$U_b : U_b(t) = E^b U_t = \exp(-iH_b t), \quad H_b = E^b H. \tag{6.2}$$

The last operator is bounded, $\|H_b\| \leqslant b$, and therefore all derivatives of the function $U_b(\,.\,)$ are also bounded. By a simple induction, one finds the derivatives

$$P_\rho^{(n)}(t) = \sum_{k=0}^{n} \binom{n}{k} i^{2k-n}\, \mathrm{Tr}\{H_b^k E_u H_b^{n-k} U_b(t)\rho U_b(t)^*\}$$

so Proposition 2.2(a) implies

$$|P_\rho^{(n)}(t)| \leqslant \sum_{k=0}^{n} \binom{n}{k} \|H_b^k\|\, \|H_b^{n-k}\| \leqslant \sum_{k=0}^{n} \binom{n}{k} b^n = (2b)^n.$$

Let us further define the function $f : \mathbb{C} \to \mathbb{C}$ by

$$f(z) = \sum_{n=0}^{\infty} P_\rho^{(n)}(0) \frac{z^n}{n!}, \tag{*}$$

then the last inequality gives $|f(z)| \leqslant e^{2b|z|}$ so the series has an infinite radius of convergence and f is an entire function. It remains to prove that (*) is the Taylor series of P_ρ for real t. A short calculation shows that

$$P_\rho(t) - \sum_{n=0}^{N} P_\rho^{(n)}(0) \frac{t^n}{n!} = \mathrm{Tr}\{E_u [U_b(t) - U_b^N(t)]\rho U_b(t)^*\} +$$

$$+ \mathrm{Tr}\left\{E_u \sum_{r=0}^{N} \frac{1}{r!}(-iH_b t)^r \rho[U_b(t)^* - U_b^{N-r}(t)^*]\right\},$$

where $U_b^N(t) = \Sigma_{r=0}^N (r!)^{-1}(-iH_b t)^r$. Thus we obtain the following estimate

$$\left|P_\rho(t) - \sum_{n=0}^{N} P_\rho^{(n)}(0)\frac{t^n}{n!}\right| \leqslant \|U_b(t) - U_b^N(t)\| +$$

$$+ \left\| \sum_{r=0}^{N} [U_b(t)^* - U_b^{N-r}(t)^*]\, E_u \frac{1}{r!}(-iH_b t)^r \right\|$$

$$\leqslant \|U_b(t) - U_b^N(t)\| +$$

$$+ \sum_{r=0}^{N} \|U_b(t) - U_b^{N-r}(t)\| \frac{(b|t|)^r}{r!}.$$

Furthermore,

$$\|U_b(t) - U_b^{N-r}(t)\| \leqslant 1 + \sum_{j=0}^{N-r} (j!)^{-1} \|H_b t\|^j \leqslant 1 + e^{b|t|},$$

so the second term in the above estimate can be majorized by the series $\Sigma_{r=0}^{\infty} (r!)^{-1} (1 + e^{b|t|})(b|t|)^r$, and therefore converges uniformly with respect to N. Boundedness of H_b implies $U_b(t) = \text{u-lim}_{N\to\infty} U_b^N(t)$ so that finally we get the relation

$$\lim_{N\to\infty} \left| P_\rho(t) - \sum_{n=0}^{N} P_\rho^{(n)}(0) \frac{t^n}{n!} \right|$$

$$\leqslant \lim_{N\to\infty} \|U_b(t) - U_b^N(t)\| + \sum_{r=0}^{\infty} \frac{(b|t|)^r}{r!} \lim_{N\to\infty} \|U_b(t) - U_b^{N-r}(t)\| = 0,$$

which means $P_\rho(t) = f(t)$ for all real t. ■

Proposition 6.1 shows that each state ρ can be approximated by bounded-energy states, in particular, by the family $\{\rho_b\}$ defined by (6.1). Now we shall try to use this fact to construct an approximative description of an unstable system; the physical meaning of the approximation will be discussed later.

Let ρ be a state of the given unstable system, $\text{Ran}\,\rho \subset \mathscr{H}_u$. Notice first that ρ_b in general need not fulfil $\text{Ran}\,\rho_b \subset \mathscr{H}_u$. This means that in constructing the approximation we cannot keep $\mathscr{H}_u$ as the state space, otherwise interpretative difficulties would arise, e.g. $P_{\rho_b}(0) \neq 1$. Thus we choose (for a given b) $\mathscr{H}_u^b := \overline{E^b \mathscr{H}_u} = \text{Ran}\,\overline{E^b E_u}$ as the **approximative state Hilbert space**; the corresponding projection is denoted by E_u^b. Clearly, $\text{Ran}\,\rho_b \subset \text{Ran}\,E^b E_u \subset \mathscr{H}_u^b$. Similarly, we introduce the **approximative reduced propagator** V_b by $V_b(t) = E_u^b U_t E_u^b$ for all $t \in \mathbb{R}$. The subspaces $\mathscr{H}_u, \mathscr{H}_u^b \subset \mathscr{H}$ are, in general, different, because the projections E_u, E^b need not fulfil $E_u \leqslant E^b$. The main problem is in which sense the one-parameter family of subspaces $\{\mathscr{H}_u^b\}$ approximates $\mathscr{H}_u$. Depending on the answer, the following assertions hold:

PROPOSITION 1.6.3. *If* $\text{s-lim}_{b\to\infty} E_u^b = E_u$, *then* $\text{s-lim}_{b\to\infty} V_b(t) = V_t$ *and* $\lim_{b\to\infty} P_{\rho_b}(t) = P_\rho(t)$ *for all* $t \in \mathbb{R}$. *Moreover, if* $\{E_u^b\}$ *converges to* E_u *in the operator norm as* $b \to \infty$, *then the relations* $\text{u-lim}_{b\to\infty} V_b(t) = V_t$ *and* $\lim_{b\to\infty} P_{\rho_b}(t) = P_\rho(t)$ *hold uniformly in* $\mathbb{R}$.

Proof. The relation $\text{s-lim}_{b\to\infty} V_b(t) = V_t$ verifies easily. It further implies $\text{s-lim}_{b\to\infty} P_b(t) = P(t)$, where $P_b(t) := V_b(t)^* V_b(t)$; this relation together with the following simple estimate

$$|P_{\rho_b}(t) - P_\rho(t)| \leqslant \text{Tr}\,|\rho_b - \rho| + |\text{Tr}((P_b(t) - P(t))\rho)|,$$

Proposition 6.1 and Proposition 2.2(b) prove the remaining part of the first assertion. The second one is due to the estimates

$$\begin{aligned}\|V_b(t) - V_t\| &\leqslant 2\,\|E_u^b - E_u\|,\\ |P_{\rho_b}(t) - P_\rho(t)| &\leqslant \operatorname{Tr}|\rho_b - \rho| + \|P_b(t) - P(t)\|\\ &\leqslant \operatorname{Tr}|\rho_b - \rho| + 4\,\|E_u^b - E_u\|.\end{aligned}$$ ■

Both the assumptions about the convergence of $\{E_u^b\}$ are valid trivially for $\dim \mathscr{H}_u = 1$, and this result can be extended to an arbitrary finite dimension:

THEOREM 6.4. *If* $\dim \mathscr{H}_u < \infty$, *then* $\text{u-lim}_{b\to\infty} E_u^b = E_u$.

We shall first prove the following auxiliary assertion:

LEMMA 6.5. *The inequality* $\dim \mathscr{H}_u^b \leqslant \dim \mathscr{H}_u$ *holds for all* $b > 0$. *Moreover, if* $\dim \mathscr{H}_u < \infty$, *then for all b large enough*, $\dim \mathscr{H}_u^b = \dim \mathscr{H}_u$.

Proof. The linear dependence of vectors $\chi_1, \ldots, \chi_n \in \mathscr{H}_u$ implies the linear dependence of $E^b\chi_1, \ldots, E^b\chi_n$, so by negation we obtain $\dim \mathscr{H}_u^b = \dim \overline{E^b\mathscr{H}_u} \leqslant \dim \mathscr{H}_u$. Further, let $\{\varphi_1, \ldots, \varphi_n\}$ denote an orthonormal basis in $\mathscr{H}_u$. The Gram determinant $\Gamma(E^b\varphi_1, \ldots, E^b\varphi_n)$ is, due to the definition of E^b, a continuous function of b. Since $\lim_{b\to\infty} \Gamma(E^b\varphi_1, \ldots, E^b\varphi_n) = 1$, the vectors $E^b\varphi_1, \ldots, E^b\varphi_n$ must be linearly independent for all sufficiently large b. ■

Proof of the theorem. Let $\{\varphi_1, \ldots, \varphi_N\}$ be an orthonormal basis in $\mathscr{H}_u$. According to Lemma 6.5, we can choose b such that the vectors $E^b\varphi_1, \ldots, E^b\varphi_N$ span the N-dimensional subspace $\operatorname{Ran} E^bE_u = \mathscr{H}_u^b$ in $\mathscr{H}$. We denote by $\psi_1^b, \ldots, \psi_N^b$ the basis obtained from $E^b\varphi_1, \ldots, E^b\varphi_N$ by Gram–Schmidt orthogonalization:

$$\psi_n^b = \frac{\tilde{\psi}_n}{\|\tilde{\psi}_n\|}, \qquad \tilde{\psi}_n = E^b\varphi_n - \sum_{k=1}^{n-1} (\psi_k^b, E^b\varphi_n)\psi_k^b,$$

$n = 1, 2, \ldots, N$. Using the fact that $E^b\psi_k^b = \psi_k^b$ together with orthogonality of $\varphi_1, \ldots, \varphi_N$, we get

$$\tilde{\psi}_n = E^b\varphi_n - \sum_{k=1}^{n-1} (\psi_k^b - \varphi_k, \varphi_n)\psi_k^b \tag{*}$$

so the following estimate is possible

$$\|E^b\varphi_n\| - \sum_{k=1}^{n-1} \|\psi_k^b - \varphi_k\| \leqslant \|\tilde{\psi}_n\| \leqslant \|E^b\varphi_n\| + \sum_{k=1}^{n-1} \|\psi_k^b - \varphi_k\|. \tag{**}$$

Since $\text{s-lim}_{b\to\infty} E^b = I$, one can choose a positive b_0 to each $\delta \in (0, 2^{2-3N})$ such that for all $b > b_0$, the inequalities

$$\|\varphi_n - E^b\varphi_n\| < \delta, \qquad n = 1, \ldots, N \tag{6.3a}$$

hold; we shall show that in such a case

$$\|\psi_n^b - \varphi_n\| < 2^{3n-1}\,\delta, \qquad n = 1, \ldots, N. \tag{6.3b}$$

The inequalities (6.3a) give $\big|1 - \|E^b\varphi_n\|\big| \leqslant \|\varphi_n - E^b\varphi_n\| < \delta$; further, (*) and (**) yield

$$\|\psi_n^b - \varphi_n\| \leqslant \frac{1}{\|\tilde{\psi}_n\|}\left\{\big|1 - \|\tilde{\psi}_n\|\big| + \|\varphi_n - E^b\varphi_n\| + \sum_{k=1}^{n-1}\|\psi_k^b - \varphi_k\|\right\}$$

$$< \frac{\big|1 - \|E^b\varphi_n\|\big| + \delta + 2\sum_{k=1}^{n-1}\|\psi_k^b - \varphi_k\|}{\|E^b\varphi_n\| - \sum_{k=1}^{n-1}\|\psi_k^b - \varphi_k\|}$$

$$< 2\left\{\delta + \sum_{k=1}^{n-1}\|\psi_k^b - \varphi_k\|\right\}\left\{1 - \delta - \sum_{k=1}^{n-1}\|\psi_k^b - \varphi_k\|\right\}^{-1}.$$

Using this estimate, one obtains (6.3b) from (6.3a) by induction. Thus, for an arbitrary vector $\psi \in \mathscr{H}_{\mathrm{u}}$, the inequality

$$\|E_{\mathrm{u}}^b\psi - E_{\mathrm{u}}\psi\| = \left\|\sum_{k=1}^{N}(\psi_n^b, \psi)\psi_n^b - \sum_{n=1}^{N}(\varphi_n, \psi)\varphi_n\right\|$$

$$\leqslant \sum_{n=1}^{N}|(\psi_n^b, \psi)|\,\|\psi_n^b - \varphi_n\| + \sum_{n=1}^{N}|(\psi_n^b - \varphi_n, \psi)|$$

$$\leqslant 2\|\psi\|\sum_{n=1}^{N}\|\psi_n^b - \varphi_n\|$$

holds; combining it with (6.3b), we obtain $\|E_{\mathrm{u}}^b - E_{\mathrm{u}}\| < 2^{3N+1}\,\delta$ for all b large enough, so the assertion follows. ■

The case $\dim \mathscr{H}_{\mathrm{u}} = \infty$ is more complicated, because in general, $\{E_{\mathrm{u}}^b\}$ *need not converge in the operator norm.* In order to illustrate this, let us first prove some simple relations between the considered projections. If $\varphi \in \operatorname{Ran} E^b E_{\mathrm{u}}$, then $E^b\varphi = \varphi$ so $\operatorname{Ran} E^b E_{\mathrm{u}} \subset \mathscr{H}^b \equiv E^b\mathscr{H}$; further, closedness of $\mathscr{H}^b$ implies $\mathscr{H}_{\mathrm{u}}^b \subset \mathscr{H}^b$. This inclusion is equivalent to $E_{\mathrm{u}}^b \leqslant E^b$ or

$$E_{\mathrm{u}}^b E^b = E^b E_{\mathrm{u}}^b = E_{\mathrm{u}}^b. \tag{6.4}$$

Consider now a vector $\psi \in \mathscr{H}_{\mathrm{u}}$. We have $E^b\psi = E^b E_{\mathrm{u}}\psi \in \mathscr{H}_{\mathrm{u}}^b$ so $E_{\mathrm{u}}^b E^b\psi = E^b\psi$, and therefore the relation (6.4) gives $E_{\mathrm{u}}^b\psi = E^b\psi$. Consequently, we obtain

$$(E_{\mathrm{u}}^b - E_{\mathrm{u}})\psi = E^b\psi - \psi, \qquad \psi \in \mathscr{H}_{\mathrm{u}}. \tag{6.5}$$

EXAMPLE 1.6.6. Let $\mathscr{H}, \mathscr{H}_{\mathrm{u}}$, H be as in Example 2.5, and choose the family of unit vectors $\psi_a \in \mathscr{H}_{\mathrm{u}}$, $\psi_a(x) = a^{-1}\chi_{[-a^2, 0]}(x), a > 0$. It is easy to see that

$$\|E^b\psi_a\|^2 = \frac{1}{\pi}\int_{-y_a}^{y_a} \frac{\sin^2 y}{y^2}\,\mathrm{d}y,$$

where $y_a = \frac{1}{2}ba^2$, and a rough estimate then gives $\|E^b\psi_a\|^2 \leqslant \pi^{-1}ba^2$. Using this, together with the relation (6.5), we get

$$\|E_{\mathrm{u}}^b\psi_a - E_{\mathrm{u}}\psi_a\| = \|E^b\psi_a - \psi_a\| \geqslant 1 - \left(\frac{b}{\pi}\right)^{1/2} a.$$

Since a is arbitrary positive, we obtain $\|E_{\mathrm{u}}^b - E_{\mathrm{u}}\| = 1$ in this case; thus $\{E_{\mathrm{u}}^b\}$ does not converge to E_{u} in the operator norm.

The relation (6.5) also implies $\lim_{b\to\infty} E_{\mathrm{u}}^b\psi = E_{\mathrm{u}}\psi$ for each $\psi \in \mathscr{H}_{\mathrm{u}}$, i.e. $\text{s-lim}_{b\to\infty} E_{\mathrm{u}}^b E_{\mathrm{u}} = E_{\mathrm{u}}$; on the other hand, it is not clear whether $\text{s-lim}_{b\to\infty} E_{\mathrm{u}}^b E_{\mathrm{d}} = 0$ holds in general. A lesson drawn from the above considerations is that for an infinite-dimensional $\mathscr{H}_{\mathrm{u}}$, one has to check convergence of $\{E_{\mathrm{u}}^b\}$, which is needed for application of Proposition 6.3, separately in each particular case of interest.

We can also answer (for the reduced evolution operators V_t and $V_b(t)$) the question formulated in the introduction to this section:

THEOREM 1.6.7. *Let* $\{\mathscr{H}, U, E_{\mathrm{u}}\}$ *be the minimal unitary dilation of* V, *then for each* $b > 0$, *the m.u.d. of the approximative reduced propagator* $V_b(\,.\,)$ *equals* $\{\mathscr{H}^b, U_b, E_{\mathrm{u}}^b\}$, *where* $U_b(\,.\,)$ *is given by* (6.2).

Proof. According to the definition, V_b has a unitary dilation, namely $\{\mathscr{H}, U, E_{\mathrm{u}}^b\}$. In general, it is not minimal: the relation (6.4) shows that $\{\mathscr{H}^b, U_b, E_{\mathrm{u}}^b\}$ is also a unitary dilation of V_b ($U_b(\,.\,)$ has the necessary properties, because E^b belongs to the spectral family of H). We shall prove that the latter is minimal. Suppose that there is a non-zero $\chi \in \mathscr{H}^b$ such that $(\chi, U_b(t)\psi) = 0$ holds for all $\psi \in \mathscr{H}_{\mathrm{u}}^b$ and $t \in \mathbb{R}$. This must be particularly true for $\psi \in \operatorname{Ran} E^b E_{\mathrm{u}}$, so $0 = (\chi, E^b U_t E^b \varphi) = (E^b\chi, U_t\varphi)$ holds for any $\varphi \in \mathscr{H}_{\mathrm{u}}$. However, $E^b\chi = \chi$ since $\chi \in \mathscr{H}^b$, and therefore we finally get $(\chi, U_t\varphi) = 0$ for all $\varphi \in \mathscr{H}_{\mathrm{u}}$ and $t \in \mathbb{R}$, in contradiction with the assumption that the vectors $U_t\varphi$ with $\varphi \in \mathscr{H}_{\mathrm{u}}$ and $t \in \mathbb{R}$ span the Hilbert space $\mathscr{H}$. ∎

Let us now turn to the physical interpretation of the described approximation, which can be understood in the following simple way: the system prepared in a state ρ, $\operatorname{Ran}\rho \subset \mathscr{H}_{\mathrm{u}}$, passes through an *energy filter* which is open for the values from Δ_b. The output of this operation is the state ρ_b given by (6.1), $\operatorname{Ran}\rho_b \subset \mathscr{H}_{\mathrm{u}}^b$. The filtering may be applied at any time instant between the preparation of the state ρ and the next measurement performed on the system, because the corresponding projection E^b commutes with the propagator. Without loss of generality, we can assume this operation to follow immediately after the preparation of ρ, so that together they prepare the state ρ_b (at a given instant).

Now the assertions proved above yield the correspondence between the description of an unstable system prepared in this way and the one related to the case when the energy filter is absent. The reason why the presented considerations are not only academical is that the 'unphysical' semigroup condition is in a very good agreement with experimental experience. However, any reduced propagator $V : \mathbb{R}_+ \to \mathscr{B}(\mathscr{H}_u)$ with the semigroup property possesses a minimal unitary dilation due to Corollary 4.2 and Theorem 4.7 so one can construct the approximating family $\{V_b(\,.\,) : b > 0\}$ whose members have not the unpleasant property, because $\sigma[V_b] = \sigma(H_b) \subset \Delta_b$. Moreover, if $\{E_u^b\}$ approximates (strongly, or even in the operator norm) the projection E_u – in particular, if $\mathscr{H}_u$ is finite-dimensional – then the reduced propagators $V_b(\,.\,)$ are near to the original semigroup V in the sense of Proposition 6.3, and the corresponding relations hold for the decay laws. Finally, Theorem 6.7 shows that the solution to the inverse-decay problem exhibits no pathology as $b \to \infty$.

Conversely, the semigroup condition (1.1) can be viewed as an approximation to the true physical description of the time evolution of unstable systems. One cannot, of course, exclude the possibility that some decay processes will be discovered for which (1.1) would not be fulfilled even approximately. However, this should not create problems for the theory: there is a lot of place for such cases in the kinematical framework developed in the preceding sections. The approximation elaborated here helps us to understand why the use of the semigroup condition is admissible despite its unphysical property.

REMARK 1.6.8. In the above discussion, we again neglect the fact that the energy distribution of the actual unstable states presumably has no sharp cut-off that is peculiar for the states ρ_b (for more details, see the notes to Section 2.3). Let us further stress that the choice $\Delta_b = (-b, b)$ is only a matter of convenience. In fact, Theorem 6.2 was the only assertion in which the boundedness of Δ_b was used substantially. In the remainder of this section, we can employ any other family Δ_b of Borel sets $\subset \mathbb{R}$ such that $\text{s-lim}_{b\to\infty} E_H(\Delta_b) = I$. In particular, choosing $\Delta_b = (-b, \infty)$ we can introduce naturally the notion of the *semibounded-energy state* (notice that the set of these states contains $B(H)$, but it is neither contained in $M(H)$ nor is the opposite inclusion valid) and construct the *semibounded-energy approximation* with the physical meaning analogous to that discussed above.

Let us finally mention that the problem treated in this section can also be viewed in a more general context. Arguing for another problem – namely, why one does not really need all the inequivalent irreducible representations of the algebra of observables that arise for systems with infinitely many degrees of freedom – Haag and Kastler (1964) pointed out the following fact: in any real experiment, we do not determine a state as a *point* in the state space, but as some **-weak neighbourhood*. By the 'state space' we mean a Banach space of normal linear functionals on the algebra of observables (equivalently, the trace ideal $\mathscr{J}_1 \subset \mathscr{B}(\mathscr{H})$ or its subspace determined by eventual superselection rules) in which states are the normalized elements of the positive cone. Thus an experiment does not tell us that the system is in a certain state ρ, but rather that it is in a state ρ' such that for some Hermitean operators $A_1, \ldots, A_n$ and positive numbers $\epsilon_1, \ldots, \epsilon_n$ the inequalities

$$|\mathrm{Tr}((\rho' - \rho)A_k)| < \epsilon_k, \quad k = 1, 2, \ldots, n, \tag{6.6a}$$

hold. The family $\{\rho_b\}$ approximates ρ in the sense of the trace norm topology, which is, of course, finer than the $*$-weak topology on the state space: Proposition 2.2(a) gives

$$|\mathrm{Tr}((\rho_b - \rho)A)| \leqslant \|A\| \,\mathrm{Tr}\, |\rho_b - \rho| \tag{6.6b}$$

for any Hermitean A, so $\lim_{b\to\infty} \mathrm{Tr}((\rho_b - \rho)A) = 0$. This means that for a sufficiently large b, one cannot distinguish ρ_b from ρ. In particular, it is impossible to decide experimentally whether a given state is a bounded-energy state or not; this strengthens the conclusion drawn from the discussion of physical realizability of infinite-energy states presented in Section 1.3. Moreover, we shall see in Section 2.3 that from the experimentalist's point of view it is difficult to distinguish ρ_b from ρ at all.

Notes to Chapter 1

SECTION 1.1. Radioactive decay was historically the first effect in which the indeterministic behaviour of individual physical objects, so peculiar for the systems described by quantum theory, was found (von Laue, 1958). As for the radiation damping, it is manifested macroscopically by a non-zero width of spectral lines and finite coherence length of light. The classical explanation says that the electron oscillations are damped by a weak force, which is due to the interaction of the electron with its own electromagnetic field. This (essentially correct) idea then leads to a formula for the intensity distribution (Heitler,. 1954, §I.4) which is similar for $|\omega - \omega_0| \ll \omega_0$ to the one assumed by the Weisskopf–Wigner theory. However, the line width here is given by $\gamma = (2e^2/3mc^3)\omega_0^2$; it is at this point that the classical theory fails.

• Beside the mentioned paper of Martinson, there is another extensive review of beam-foil spectroscopy by Andrä (1974). When speaking about non-exponential violations of a decay law, one usually has in mind a time plot 'cleaned' from the influence of various kinematical factors. For instance, if the non-decay probability is measured in a beam of unstable particles (in dependence on position of the detector), then the experimentally determined decay curve is affected, e.g. by energy spread of the beam or by energy loss when the particles are detected in a bubble chamber (cf. Newton (1966, §19.1) and Section 2.3 below).

• The kinematical concept of unstable systems in the form treated here was first formulated in the papers of Horwitz, LaVita, Marchand and Williams in 1971. However, there are much older papers in which properties of decaying quantum-mechanical systems are discussed independently of the concrete dynamics. This approach can be traced back at least to Krylov and Fock (1947); we can also refer to the papers of Khalfin (1957a, b), Newton (1961), Beskow and Nilsson (1967), Jersák (1969, 1970), and Horwitz and Marchand (1969a, b, 1971). Further contributions to this problem are, for example, due to Sinha (1972), Fonda and his group (1972, 1977, 1978), Fleming (1973), Alda *et al.* (1974), Misra and Sinha (1977) and others. More detailed references to some other problems mentioned in the introduction will be given later.

SECTION 1.2. Unless stated otherwise, by subspace of a Hilbert space we always mean a closed subspace. The operator 'pr' is sometimes called *compression* – cf. Sz.-Nagy and Foias (1970, notes to chap. I), however, caution is needed because this notion can also have another meaning (Friedman, 1972). For basic properties of the trace-class operators see, e.g. Reed and Simon (1972, §VI.5), Kuo (1975, §I.1), or Blank and Exner (1978, §3.5). Each trace-class operator may be written as a sum of four positive trace-class operators. Then part (b) of Proposition 2.2 can be rephrased as follows: each normal state on $\mathscr{B}(\mathscr{H})$ is weakly *sequentially* continuous. Notice that the one-parameter family $\{B_\alpha\}$ cannot by replaced by a general net of operators, because a state is normal iff it is σ-weakly continuous, and the σ-weak (or ultraweak) topology is stronger than the weak one – cf., e.g., Dixmier (1969, §§I.3–4, III.6) or Bratelli and Robinson (1979, prop. 2.4.2 and theorem 2.4.21).

• The concept of 'integrity' is due to Lurçat (1968). As mentioned above, the postulates (u1, u2) characterizing unstable quantum systems appeared first about ten years ago (see Williams (1971), and also Horwitz *et al.* (1971) or Havlíček and Exner (1973)), but they represented nothing more than formalization of an approach used long before. The term 'reduced propagator' is probably not flawless, because the subspace $\mathscr{H}_u$ does not reduce the full propagator, but we have not found a better one. Sometimes 'contracted' is used instead of 'reduced', e.g., in Horwitz *et al.* (1971), but this, too, is not ideal.

• The notion of sojourn time is due to Ekstein and Siegert (1971). It is of some interest because of the peculiar role played by time in quantum theory: remember the well-known Pauli argument showing that an operator representing the time observable cannot exist as far as a below-bounded energy is required. There are many attempts to bypass this difficulty, of course, but we shall not comment on them here. The concept of sojourn time, elaborated more rigorously by Sinha (1979), illustrates that in some specific situations the 'time observable' can be given with reasonable meaning. Notice also that there is a close relation to the concept of time delay in scattering, which shall be mentioned in notes to Section 3.2.

• The theory of open systems coupled to a reservoir in such a way that $\mathscr{H} = \mathscr{H}_{\text{open}} \otimes \mathscr{H}_{\text{res}}$ refers to an isolated system is exposed in a monograph by E. B. Davies (1976). The lines of investigation of the open-system evolution in Davies are in many aspects similar to the program carried out here for the unstable systems. In particular, the problem of repeated measurements is treated there in a much greater generality than in our Chapter 2. The dynamics of such open systems is usually studied by means of the so-called *master equations* which are obtained by reducing the unitary time evolution on $\mathscr{H}$ to $\mathscr{H}_{\text{open}}$. One is interested primarily in the cases when the memory effects are absent and the equations are *Markovian*, i.e. such that their solution is an operator semigroup on $\mathscr{H}_{\text{open}}$. Under some additional conditions (strong continuity, positivity and trace preservation, etc.), these semigroups are usually referred to as *dynamical semigroups*. There are various situations in which the general master equations turn out to be Markovian, such as the weak-coupling limit; for a review of this subject see Gorini *et al.* (1978). Particular models of decay processes in this formalism were worked out, e.g., by Davies and Eckmann (1975), Alicki (1978) and Arai (1981). The scattering theory for dynamical semigroups was discussed by Alicki (1981); for non-linear dynamical semigroups see Alicki and Messer (1983).

• The problem of quantization of dissipative systems has been considered by many authors, mostly with emphasis on simple systems such as free particle in a viscous medium, damped harmonic oscillator, etc. Sufficiently complete information can be derived from the reviews by Haase (1975), Messer (1979) or Dekker (1981); see also Burzlaff (1979), Lemos (1981) and Caldirola (1983). The quantization is often performed with account of a random force. For application of stochastic methods to dissipative systems see, e.g., Yasue (1978) and Jona-Lasinio *et al.* (1982). A review of the phenomenological description of nuclear reactions by means of frictional forces is due to Haase (1978).

• Recently Gisin (1981, 1982) advocated the use of another non-linear equation for a description of dissipative systems, namely

$$i\dot{\varphi}_t = H\varphi_t + i(\langle B\rangle_{\varphi_t} - B)\varphi_t,$$

where H, B are some self-adjoint operators on the appropriate state Hilbert space. There is a simple connection between such an evolution and the one governed by the semigroup $\{e^{-i(H-iB)t}\}$ (cf. Problem 56). The equation may be used in the treatment of relaxation phenomena, provided one is able to give a reasonable physical meaning to the norm preservation of φ_t.

SECTION 1.3. Non-exponential behaviour of the decay laws for small times was observed in model calculations concerning escape of a particle confined initially inside a potential barrier (e.g. Winter (1961), Frey and Thiele (1968)). However, much earlier Mandel'shtam and Tamm (1945) noticed that in the case of one-dimensional $\mathscr{H}_u$ the decay law must obey $P_\psi(t) \geqslant |\cos((\Delta H)_\psi t)|^2$. This fact was later rediscovered by Petzold (1959b), Beskow and Nilsson (1967) and Fleming (1973) – cf. also Fonda (1977), Fonda *et al.* (1978) or Peres (1980). The formal deduction presented in the opening to this section is due to Horwitz and Marchand (1969b, 1971); in the paper of Horwitz *et al.* (1971) the equality $\dot{P}_\psi(0) = 0$ is proved under the assumptions that dim $\mathscr{H}_u < \infty$, $\{V_t\}$ is a (continuous) semigroup and ψ has a finite energy. Theorem 3.1 shows that only the last one is essential; the proof was given by Havlíček and Exner (1973) for pure states and by Exner (1976a) in the general case. In these papers, also, the problem of physical realizability of the infinite-energy states is discussed.

• Remark 3.2 shows that the condition yielded by Theorem 3.1 is not necessary for the initial decay rate to be zero. On the other hand, it is not clear whether a semibounded-energy state (cf. Remark 6.8) with infinite mean value of energy and zero initial decay rate can exist. Lemma 3.3 remains valid when $\mathscr{H}$ is replaced by a Banach space X. The proof is adapted from Chernoff (1974, §2.5), where a much more sophisticated result is proved – namely, that one may assume continuity of $F(\,.\,)$ instead of $C(\,.\,)$. Under some additional requirements, the assertion holds generally for Lipschitz mappings on a complete metric space; we limit ourselves with the Banach-space case: Let $\{F_t : t > 0\}$ be a strongly continuous family $\subset \mathscr{B}(X)$, and assume that $C_t := \text{s-lim}_{n\to\infty} (F_{t/n})^n$ exists. Then $\{C_t : t > 0\}$ is strongly continuous and fulfils $C_t C_s = C_{t+s}$ for all $t, s > 0$.

• Theorem 3.4 is due to Misra and Sinha (1977). The connection between regeneration of the decaying system and non-semigroup character of the reduced propagator was recognized first by Jersák (1969a; he is usually referred to incorrectly as 'Ersak') and its present formulation (without dimensional restrictions) is due to Sinha (1972); see also the numerically solved model of Winter (1961) and the paper of Fonda and Ghirardi (1972). There are some confused statements in the literature, mostly in connection with the so-called *Jersák's equation*. The latter is nothing more than a matrix form of the equality $U_t\psi = V_t\psi + E_d U_t \psi$ for $\psi \in \mathscr{H}_u$, when $\mathscr{H}_u$ is assumed to be finite-dimensional. Jersák argues for neglect of the regeneration (expressed by $E_u U_s E_d U_t \psi = 0$ for all $t, s \geqslant 0$) and asserts that in such a case $\{V_t : t \geqslant 0\}$ obeys the semigroup condition. It should be stressed, however, that the *approximative* character of the semigroup reduced evolution is explicitly stated in the considered paper (with reference to the Khalfin's (1957) argument based on the Paley–Wiener theorem). The only essential error there is contained in the conclusion that the difficulty might be overcome by choosing an infinite-dimensional $\mathscr{H}_u$.

SECTION 1.4. The theory of unitary dilations was developed mainly for the analysis of non-self-adjoint operators; a leading role in this was played by B.Sz.-Nagy. Our Theorem 4.1, which is just one among many nice results of this theory, together with its constructive proof, is originally due to Naimark (1943); an earlier proof was given by the same author in 1940. It is worth mentioning that the same construction for the particular case of one-dimensional $\mathscr{H}_u$ was published independently by Gel'fand and Raikov (1943). In Naimark's paper, the described construction is used to prove an assertion analogous to our Proposition 4.4 with a general commutative topological group G instead of $\mathbb{R}$ (in such a case, the exponential integrand in (4.4a) has to be replaced by $\chi(g)$, where χ is an element of $\hat{G}$, the dual space of the group G), then Theorem 4.1 appears as a corollary to this result. Further generalizations to the cases when G is non-commutative or a topological $*$-semigroup, and the Hilbert space $\mathscr{H}$ may be real, belong to Sz.-Nagy (1953b, 1955), though Naimark was clearly aware of the first of them. For a more detailed history of the problem and related questions see Sz.-Nagy and Foias (1970; notes to the first chapter).

• Applications of the unitary-dilations theory to the treatment of unstable quantum systems was suggested first by Horwitz and Marchand (1971) who also invented the name 'inverse-decay problem'. The idea was pursued further in the paper written by these authors together with LaVita (1971), and formulated independently by Williams (1971). Let us remark that beside the one discussed here, there are various other physical applications of the unitary dilations ranging from the classical scattering theory of acoustical waves (Lax and Phillips, 1967, chap. 3) to the theory of lossy electrical networks (Helton, 1972).

• Notice that if H has an eigenvector $\varphi \in \mathcal{H}_u$, then the corresponding one-dimensional projection reduces V_t too and Proposition 4.3 may be applied. The Hamiltonian obviously cannot have eigenvectors in $\mathcal{H}_d$, at least if it corresponds to the minimal unitary dilation of V. On the other hand, the minimality condition does not exclude existence of eigenvectors with non-zero orthogonal projections to both $\mathcal{H}_u$ and $\mathcal{H}_d$ as the following simplest example shows: $\mathcal{H} = \mathbb{C}^2$ and

$$U_t = \begin{pmatrix} \cos t & \sin t \\ -\sin t & \cos t \end{pmatrix}, \qquad E_u = \begin{pmatrix} 1 & 0 \\ 0 & 0 \end{pmatrix},$$

where eigenvectors of H are $\varphi_\pm = 2^{-1/2}\{1, \pm i\}$. However, in applications to unstable systems such situations do not occur as shown by Theorem 5.8(a).

• As mentioned above, Proposition 4.4 was first obtained by Naimark (1943). The presented proof is due to Exner (1976b), and Proposition 4.5 is taken from Exner (1977d). The fact that the continuous contractive semigroups are of the positive type (Theorem 4.7), which proves existence of their minimal unitary dilations, was established first by Sz.-Nagy (1953a, b); our proof follows the monograph of Sz.-Nagy and Foias. Notice further that in the case of a contractive semigroup $\{V_t : t \geqslant 0\}$, strong continuity follows from w-$\lim_{t\to 0+} V_t = I$. It is an easy exercise; however, the same assertion holds in general for any operator semigroup on a Banach space (cf. (A.2)).

SECTION 1.5. In a relativistic theory, Hamiltonian must not only be semibounded, but also positive. Moreover, since each non-relativistic theory should be a reasonable limit of some relativistic one, we can regard the 'natural physical assumption' of energy semiboundedness as a consequence of the spectrality postulate.

• Lemma 5.2 is due to Sinha (1972). His paper also contains an assertion adapted from Lax and Phillips (1967, lemma V.2.3), which represents a partial converse to this lemma: if ω is continuous, then there is at least one sequence $\{t_j\}_{j=1}$ tending to infinity such that $\lim_{j\to\infty} f(t_j) = 0$. One should notice, however, that the first of the two proofs presented by Lax and Phillips (via prop. V.2.1) has to be improved.

• Among various conditions presented in this section which imply $\sigma(H)$ to be either not semibounded, or even equal to $\mathbb{R}$, the most general one (expressed by Theorem 5.1) is adopted from the paper of Sinha (1972); the proof is modified. Various particular cases of this assertion can be found in the literature: see, e.g., Williams (1971) or Alda *et al.* (1974) for the semigroup case ($T_r = 0$), or Horwitz *et al.* (1971) for the case when V is a finite-dimensional semigroup obeying s-$\lim_{t\to\infty} V_t = 0$. Theorem 5.7 covers and generalizes many of earlier results concerning asymptotic behaviour of the decay laws – cf. references given below in the remark to Example 5.15. Another closely related result is the theorem proved in the appendix to Williams' (1971) paper (see also Hack (1982)), which asserts that $(\varphi, V_t\psi)$ cannot decay exponentially as $t \to \infty$ for any non-zero $\varphi, \psi \in \mathcal{H}_u$ unless $\sigma(H) = \mathbb{R}$ (Problem 4). Theorem 5.8(a) belongs to Horwitz *et al.* (1971); however, essentially the same assertion can already be found in the paper of Krylov and Fock (1947) with reference to a theorem due to Bernstein. A sort of converse, namely that $\lim_{t\to\infty} (\varphi, V_t\psi) = 0$ for $\varphi, \psi \in \mathcal{H}_{ac}$, the absolutely continuous subspace with respect to H, follows from the Riemann–Lebesgue lemma.

Part (b) of Theorem 5.8, as well as the presented proof, are due to Sinha (1972); Corollary 5.9 is taken from Exner (1977d). The equality (5.20) was noticed by Fleming (1973). A necessary condition for energy semiboundedness based on the Paley–Wiener theorem was first derived by Khalfin (1957), which was then followed by many other authors – cf. again the notes to Example 5.15. Theorem 5.10 represents a generalization to these results.

• The sufficient condition for energy semiboundedness given by Proposition 5.12 and Theorem 5.13 is due to Exner (1976b). As for the difficulties tied with the reconstruction of V_t from $P(t)$, some information could be obtained from the behaviour of $P(\,.\,)$ in the lower complex halfplane of t, at least for dim $\mathscr{H}_u = 1$. Such considerations were presented by Khalfin (1957); the problem reduces essentially to another reformulation of the Paley–Wiener theorem, which relates the L^2-functions with semibounded support $\subset \mathbb{R}$ to the elements of Hardy–Lebesgue class $H^2(0)$ (Yosida, 1966, theorem VI.4.2). However, behaviour of the decay laws for complex times is not measurable and, moreover, the non-uniqueness discussed under (i) and (ii) remains untouched.

• Notice also that in practice one meets various other reconstruction problems. For instance, one can measure the time plot of the transition probability $|(\psi,\, U_t\varphi)|^2$ for $\varphi \in \mathscr{H}_u$ and a fixed vector $\psi \in \mathscr{H}_d$, and use it to get some information about φ and H. Such a situation occurs in beam-foil spectroscopy, where φ is typically a superposition of two or more excited atomic levels. If these levels are mutually close, the transition probability exhibits oscillations in time (sometimes called quantum beats) from which the energy splitting of the levels can be extracted – see, e.g.,Macek (1970), Alguard and Drake (1973) or Andrä (1974).

• Conclusions from Example 5.15 depend essentially on the value of the parameter $r = \frac{1}{2}\Gamma/(\lambda_0 - \lambda_t)$. There are, of course, cases in which Γ is such that r can be made large, especially when the unstable objects under consideration are resonances in some scattering process, but then the time region involved is too short to be reached experimentally. This fact is noticed in most of the standard quantum-mechanical textbooks (e.g. Messiah (1959, chaps X, XXI)) for non-relativistic potential scattering. A more general heuristic analysis can be found in the book of Newton (1966, §19.2) or in Terentiev (1972). The large-time non-exponentialities were observed by Hellund (1953) and Namiki and Mugibayashi (1953) in the Weisskopf–Wigner description of radiation damping, and by Okabayashi and Sato (1957) in the Lee model. Khalfin (1957) was the first who related this effect to the energy semiboundedness, and the problem was further studied, e.g. by Höhler (1958), Lévy (1959b), Matthews and Salam (1959), Petzold (1959a), Newton (1961), Nussenzweig (1961), Winter (1961), Kerler and Petzold (1965), Terentiev (1965, 1972), Alzetta and d'Ambrogio (1966), Demuth (1976) and Peres (1980). Let us further mention that in the discussed example, the non-exponential corrections to the decay law are present during the whole process of decay, though they are manifested in some time regions only, in particular, for times larger than t_c. It is also clear from the example that a similar behaviour has also to be expected for other modifications of the Breit–Wigner energy distribution. This fact is properly reflected in Newton's book, which stresses the *approximative* validity of the exponential decay law within a certain intermediate interval of times, as well as in the papers quoted above; this is worth mentioning, for there is a folklore statement that 'the exponential decay law holds between some positive t_1, t_2' contained in a good many papers. There is yet another point which should be mentioned: the region of approximative validity of the exponential decay law (characterized particularly by the time t_c in our example) depends essentially on the behaviour of $g(\,.\,)$ near the point λ_0. The truncated Breit–Wigner distribution (5.12a) as well as the function (5.14a) can be continued (with the exception of the halfline $\lambda \leqslant \lambda_t$) into meromorphic functions of complex λ. In the first case, this continued function has a simple pole at $\lambda = \lambda_0 - (i/2)\Gamma$ (in the case of resonance scattering, the latter is commonly related to a pole of the scattering amplitude on the 'unphysical' Riemannian sheet; for a more detailed discussion see Sections 3.1–3.3), and a similar situation arises for (5.14a). Now if we take instead a higher-order pole, then the peak of g changes form and the decay law is substantially altered: neglecting the correction terms due to energy semiboundedness, it assumes the shape of decreasing exponential function multiplied by a polynomial. Such decay laws were studied intensively in the sixties (cf., e.g., Goldberger and Watson (1964), Bell and Goebel (1965), Newton (1966, § 19.3) or Jersák (1969a, 1970)); however, until now we were unable to relate them to any known decay process existing in Nature.

• The semigroup discussed in Example 5.16 was treated by Horwitz *et al.* (1971, example 3) for the particular value $a = \sqrt{2}$. Their aim was to show the existence of a finite-energy state in the considered two-dimensional space $\mathscr{H}_u$, which would have zero initial decay rate. However, they failed to notice the most interesting feature of this example; namely, the non-stability of the solution to the inverse-decay problem in dependence on the parameter a.

SECTION 1.6. The material in this section comes from Exner (1980). The idea of energy (or mass) filtering in a somewhat different form can be found in the paper of Schwinger (1960). The argument by Haag and Kastler is also discussed in Haag (1972) or in Bratelli and Robinson (1979, p. 14). Notice that the definition of an $*$-weak neighbourhood does not require the operators $A_1, \ldots, A_n$ to be Hermitean. However, each of them can be written as a linear combination of two Hermitean operators, so we may actually use the formulation of (6.6a) with the Hermitean A_k which is physically transparent.

Chapter 2

Repeated Measurements on Unstable System

"The 'path' emerges only if we observe it."

W. HEISENBERG

The observation of an unstable system often includes a premeditated interaction with its environment. In many cases, such as the detection of particles by trace-monitoring devices, the interaction with the surrounding medium can be described as a sequence of measurements performed on the system under consideration. The study of such situations is the main subject of this chapter. First we formulate the basic assumptions and derive the integral equations which yield the decay law of a system which suffers repeated measurements, in general, randomly and non-homogeneously distributed. The existence of the solution to these equations, together with its simple properties, are established. Section 2 is devoted to a discussion of some important examples. The results show that the measured decay laws behave asymptotically exponentially in a certain sense and, moreover, that the measured decay rate may be influenced by structure of the measuring device, if the decay law of the same system undisturbed by the measurements is not purely exponential. This conclusion might be used for search of the non-exponentialities; in order to get some quantitative estimates of the effect, we treat in Section 3 a simple model describing the decay of charged kaons in a bubble chamber. It appears that even in an optimal experimental arrangement, the effect is so very small that it can hardly be observed. Section 4 is devoted to the treatment of the limiting case when the measurement frequency tends to infinity, which is sometimes used to describe the behaviour of a system exposed to continual observation. This interpretation leads to some counterintuitive conclusions, which will be discussed and clarified.

2.1. Decay Law in the Presence of Repeated Measurements

The basic scheme used for description of measurements was formulated by the founding fathers of quantum mechanics: a microsystem evolves freely until a time t when an appropriate measuring apparatus is switched on; then, as a result of their interaction, the state of the considered microsystem changes instantaneously. Perhaps no other postulate of the quantum theory has been discussed so intensively and permanently; however, until now a satisfactory theory of measurement is missing. Beside this 'big' problem, there are smaller ones worthy of attention. One of them concerns the fact that in experimental nuclear and high-energy physics some measuring devices are used fairly often, for which the above-mentioned scheme seems to be too narrow, e.g., bubble chambers, photoemulsion detectors and spark or streamer chambers. Interacting with such an apparatus, the particle or nucleus under consideration is obviously subjected to *a sequence of measurements which follow quickly one after another*; at least in principle we are able to register the output of each of them (from appearance and localization of the bubbles, sparks, etc.).

Processes with frequently repeated measurements are particularly interesting in connection with the time evolution of unstable systems, because the decay law in such circumstances must be reinterpreted, and intuitively it can be expected to decrease approximately exponentially if the measurements are homogeneously distributed. This is seen from the following qualitative argument based on the assumption that a measurement, in which the system is found undecayed, restores its original state. The classical derivation of the decay-law exponentiality due to von Schweidler relies on the premise that the radioactive nucleus does not remember its history. On the other hand, any non-exponentiality in the time plot of non-decay probability means existence of such a memory, which could be naturally weakened or even destroyed by frequently repeated measurements. It would certainly be an exaggeration to declare the decay-law exponentiality to be an apparatus effect because, in the first place, there are other experiments which yield exponential decay curves without involving repeated measurements and, secondly, the deviations from the exponential decay law (due to energy semi-boundedness and similar effects) are all the same expected to be small in the cases of interest.

Nevertheless, the influence of repeated measurements on the decay law is worthy of study, and we shall treat this problem on different levels in the following three sections. In order to avoid frequent repeating of the words 'unstable micro-object' and 'measuring device', we replace them temporarily by 'particle' and 'chamber', respectively; we keep in mind, however, that the presented considerations are not limited to unstable elementary particles and bubble or spark chambers only. Since each measurement means a reduction of the state after which the free evolution (and in particular, the decay) starts anew, we must first know the result of this reduction. Consider first the situation when $\dim \mathscr{H}_u = 1$, i.e. the particle is described up to a phase factor by a single vector ψ in the state Hilbert space of the 'large' system. The *'yes–no'* experiment connected with the decay law (cf. Section 1.2) then corresponds to the one-dimensional projection E_ψ containing ψ in its range. Suppose now that the state of the particle evolved into φ and then the *'yes–no'* experiment was performed. If the result is positive – i.e. the particle is found undecayed – then its state immediately after the measurement is described again by the vector ψ.

In general, however, the space $\mathscr{H}_u$ need not be one-dimensional, even if we factor out the degrees of freedom, which are irrelevant for description of the decay process under consideration. We have seen in Section 1.6 that there are limits beyond which the state cannot be fixed experimentally. Furthermore, at least for relativistic particles, one cannot hope to bypass the space–time-born part of $\mathscr{H}_u$ fully (cf. Section 3.5), to say nothing about the internal degrees of freedom. Nevertheless, we shall accept the following assumption as a starting point:

(r) *If the particle is found undecayed in the considered particular process of measurement, then the reduction results into its original state.*

We shall return to the above mentioned problems later, here we limit ourselves with the following remark. Validity of the assumption (r) is closely tied to the problem of identity of the considered particle. In fact, an experimentalist certainly has to believe that a pion which just made a bubble in his 'Gargamelle' was the same one (localized roughly within the bubble diameter) as before, neglecting the slightly reduced velocity.

Let us turn now to the chamber structure. We start with the following observation: for most of the 'chambers' named in the introductory paragraph, the measurements not only follow quickly one after another, but they are also performed *at random time instants*. A possible exception is represented by the spark chambers where the measuring points are fixed by positions of the wires, but even in these cases there is some randomness because in a real spark chamber a track is rarely marked by discharges on all the 'engaged' wires. Thus, it is natural to describe the chamber structure by a family of positive numbers $W(s, t)$ which denote the probabilities that a particle prepared (by a measurement in the chamber) at an instant s is not measured during the time interval $(s, t]$. The function $W(\,.\,,\,.\,)$ is supposed to be defined on an appropriate square $J \times J \subset \mathbb{R}^2$, and the interpretation requires it to be non-increasing with respect to the second argument.

$$0 \leqslant W(s, u) \leqslant W(s, t) \leqslant W(s, s) = 1 \qquad \text{for} \quad s \leqslant t \leqslant u. \tag{1.1a}$$

For each $s \in J$, the function $W(s,\,.\,)$ is obviously right-continuous with a bounded variation, and therefore defines a positive Lebesgue–Stieltjes measure ν_s on (s, ∞), such that

$$W(s, t) = 1 - \int_s^t \mathrm{d}\nu_s(\xi) = \int_t^\infty \mathrm{d}\nu_s(\xi). \tag{1.1b}$$

The situation is a bit more complicated when the dependence on the first argument is considered, because $W(s, t)$ need not make sense for some values of s. For instance, a sharply localized particle travelling through a spark chamber cannot be measured unless it moves near some of the wires. In the remaining part of the definition interval J, the function $W(\,.\,, t)$ is naturally expected to be right-continuous and non-decreasing,

$$0 \leqslant W(r, t) \leqslant W(s, t) \leqslant W(t, t) = 1 \qquad \text{if} \quad r \leqslant s \leqslant t \tag{1.2a}$$

and the involved probabilities make sense. However, $W(\,.\,,t)$ can be defined as assuming a constant value in the open sets $\subset J$ during which the measurement cannot occur, in such a way that the monotonicity is preserved. We shall employ in the following either the values $W(s, t)$ for the points for which the measurement is possible, or essentially the derivative of $W(\,.\,,t)$; thus the use of $W(\,.\,,t)$ as a function on $(-\infty, t] \cap J$ would do no harm. By the same argument as above, to each $t \in J$ a positive Lebesgue–Stieltjes measure μ_t on $(-\infty, t)$ corresponds such that

$$W(s, t) = 1 - \int_s^t \mathrm{d}\mu_t(\xi) = \int_{-\infty}^s \mathrm{d}\mu_t(\xi). \tag{1.2b}$$

Summing up the above discussion, we can say:

(c) *The chamber structure is described by a function* $W(\,.\,,.\,)$ *assuming values from* $[0, 1]$, *which is right-continuous in both arguments and non-decreasing (non-increasing) with respect to the first (second) one. Equivalently, one can use the one-parameter families of positive measures* $\{\mu_t\}$ *and* $\{\nu_s\}$ *related to W through* (1.1b), (1.2b).

Below we shall treat some more special classes of the functions W.

Concerning the decay laws, we are interested mainly in the probability $E(s, t)$ that the particle prepared at an instant s (and suffering then repeated measurements characterized by the function $W(\,.\,,.\,)$) will survive up to a time t. It admits two interpretations (cf. Figure 1):

(i) There is a probability that a particle which is found undecayed in all measurements realized before the time t will also be found undecayed when measured at the instant t. However, this measurement need not occur automatically so we must ensure it eventually by a suitable detector. Such a probability will be denoted by $E_0(s, t)$,

(ii) There is a probability that the particle will be found undecayed in all measurements realized in the chamber until the instant t. For this quantity we reserve the symbol $E(s, t)$.

In fact, we are interested mostly in the probability $E(s, t)$. On the other hand, $E_0(s, t)$ is a useful auxiliary quantity, which is moreover expected to be near to $E(s, t)$ in the cases of practical interest when the density of measurements is large enough.

In order to formulate equations for the above mentioned survival probabilities, we have to know in the first place how the particles would decay if they were undisturbed by the repeated measurements. In such a case, the non-decay probability is given by the decay law (1.2.3); within the present context we shall refer to it as to the **primary decay law** dropping usually the subscripts ρ, ψ. Consider now the probability $E(s, t)$ to which the following processes contribute: either the particle is not measured within $(s, t]$ or it is measured (finitely many times)

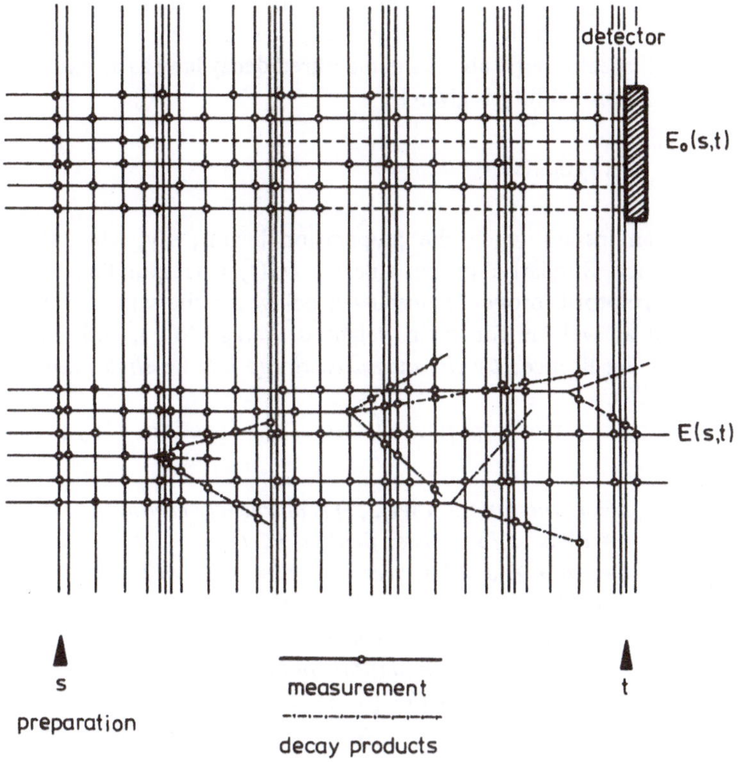

2.1. Two possible interpretations of the survival probability.

with the last measurement performed at some $\xi \in (s, t]$. Consequently, we have the relation

$$E(s, t) = W(s, t) + \int_s^t E_0(s, \xi - 0)\, d\mu_t(\xi). \tag{1.3a}$$

In an analogous way, we obtain for $E_0(s, t)$ the following integral equation

$$E_0(s, t) = W(s, t)P(t - s) + \int_s^t E_0(s, \xi - 0)P(t - \xi)\, d\mu_t(\xi). \tag{1.3b}$$

The left limit appears in the integrals due to the following reason. In order to avoid ambiguities, we shall always *understand* E, E_0 *as right-continuous functions* in both arguments. Such a convention is reasonable, because the non-integral terms on the r.h.s. of equations (1.3) are right-continuous. In this case, however, we must insert $E_0(s, \xi - 0)$ into the integrals in order to prevent double-counting.

Further, we have to mention that in general the survival probabilities are not directly measured, because the preparation of the particle is usually realized by the last measurement performed in the chamber not later than some chosen time, conventionally $s = 0$. In such a case, the decay law measured in the considered chamber (we shall speak briefly about the **measured decay law**) as an experimentally determined survival probability is given by

$$N(t) = \int_{\tau}^{0} E(s, t)\, d\mu_0(s), \tag{1.4}$$

where τ marks an instant when the preparation begins; say, when the particle enters the chamber. Similarly, one can define $N_0(t)$ corresponding to $E_0(\,.\,, t)$. However, the asymptotic properties discussed below are the same for each of the functions $E(s, .)$ as well as for their weighted average $N(\,.\,)$, and therefore we accept the following licence: *by the measured decay law we shall understand the function* $E(\,.\,) = E(0, .)$.

We shall now specify two important classes of the functions W by means of the assumptions:

(c1) *For each* $t \in J$, *the function* $W(\,.\,, t)$ *is absolutely continuous so a positive* $m_t : m_t(s) = \partial W(s, t)/\partial s$ *exists; further,* $(s, t) \mapsto m_t(s)$ *is measurable and bounded a.e. in* $J \times J$. *Moreover, the functions* $W(0, .)$ *and* $t \mapsto m_t(s)$ *for almost all* $s \in J$ *are assumed to be continuous.*

(c2) $W(\,.\,, .)$ *is a step function in each of the arguments: there is a sequence* $\{t_i\} \subset \mathbb{R}$ *with no accumulation point (except at infinity) and a family of numbers* $W[m, n] \in [0, 1]$ *such that* $W(s, t) = W[m, n]$ *for* $t_m \leqslant s < t_{m+1} \leqslant t_n \leqslant t < t_{n+1}$ *and* $W(s, t) = W[m, m] = 1$ *for* $t_m \leqslant s \leqslant t < t_{m+1}$.

PROPOSITION 2.1.1. *Assume* (r) *and* (c1), *then there is a unique measured decay law which is locally integrable and such that the inequalities*

$$W(0, t)P(t) \leqslant E_0(t) \leqslant E(t), \tag{1.5a}$$

$$W(0, t) \leqslant E(t) \leqslant 1 \tag{1.5b}$$

hold, where $E_0(t) \equiv E_0(0, t)$. *Furthermore, both the functions* $E(\,.\,)$ *and* $E_0(\,.\,)$ *are continuous and equal to* 1 *at* $t = 0$.

Proof. Under the assumption (c1), equation (1.3b) has a unique solution in the class of continuous functions. In order to check it, one has only to apply the well-known existence result for Volterra integral equations (e.g. Kolmogorov and Fomin (1976, §IX.2) or Vladimirov (1971, §IV.17): owing to (c1) and Proposition 2.3, the kernel $K : K(t, \xi) = P(t - \xi)m_t(\xi)$ is measurable and bounded a.e.; further, it is continuous and therefore measurable so that the inhomogeneous term in (1.3b) is integrable in any finite interval $\subset \mathbb{R}_+$. Consequently, the equation

$$E_0(t) = W(0, t)P(t) + \int_0^t P(t - \xi)m_t(\xi)E_0(\xi)\, d\xi$$

has a unique locally integrable solution. This further implies the existence of $E(\,.\,)$; notice that the integral term in (1.3a) is already absolutely continuous with respect to t. Moreover, one has the following uniformly convergent expansions

$$E_0(t) = W(0,t)P(t) + \sum_{n=1}^{\infty} \int_0^t ds_1 \int_0^{s_1} ds_2 \ldots \int_0^{s_{n-1}} ds_n\, W(0,s_n) \times$$
$$\times P(s_n) m_{s_{n-1}}(s_n) P(s_{n-1} - s_n) \ldots m_{s_1}(s_2) P(s_1 - s_2) m_t(s_1) P(t - s_1), \quad (1.6a)$$

$$E(t) = W(0,t) + \sum_{n=1}^{\infty} \int_0^t ds_1 \int_0^{s_1} ds_2 \ldots \int_0^{s_{n-1}} ds_n\, W(0,s_n) P(s_n) \times$$
$$\times m_{s_{n-1}}(s_n) P(s_{n-1} - s_n) \ldots m_{s_1}(s_2) P(s_1 - s_2) m_t(s_1). \quad (1.6b)$$

which have a transparent physical meaning: the nth terms correspond to the processes when the particle is measured n times during $(0, t)$ and always found to be undecayed. The integrated functions are non-negative so the first inequalities in (1.5a, b) hold; the second in (1.5a) follows from the term-by-term comparison of (1.5a), (1.5b) and the fact that $P(\,.\,)$ is bounded by 1 from above. Using it again, we see that $E(t)$ can be estimated from above by the expression (1.6b) with $\tilde{P}(s_n) = \cdots = \tilde{P}(s_1 - s_2) = 1$, which in turn is obtained from the equations (1.3) with $\tilde{P}(t) \equiv 1$, but (1.3b) is solved in such a case by $\tilde{E}_0(t) \equiv 1$, so (1.2b) yields $E(t) \leqslant 1$. Further, $W(0,\,.\,)$ is continuous, so in view of the above remark the same is true for $E(\,.\,)$. The integrand in each term of (1.6a) is continuous a.e. and has a t-independent bound, so the integrals are continuous with respect to t and uniform convergence of the series implies continuity of $E_0(\,.\,)$. Finally, the uniqueness was established in the class of continuous functions; however, Proposition 1.2.3 together with (c1) imply that the r.h.s. of (1.3b) is continuous with respect to t, so equations (1.3) have no discontinuous solution. ■

PROPOSITION 2.1.2. *Assume* (r) *and* (c2), *then there is a unique measured decay law. Both* $E_0(\,.\,)$ *and* $E(\,.\,)$ *are continuous in each open interval* (t_n, t_{n+1}) *(according to the above-mentioned agreement, they are right-continuous at the points* t_n*). Moreover,* $E(\,.\,)$ *is a step function and the inequalities* (1.5) *hold again.*

Proof. Without loss of generality, we can set $s = t_0 = 0$; further, it is useful to introduce

$$M_n[m] := W(t_m, t_n) - W(t_m - 0, t_n) = W[m,n] - W[m-1,n]. \quad (1.7a)$$

Then equations (1.3) can be rewritten for $t \in [t_n, t_{n+1})$ as follows

$$E(t) = W(0,t) + \sum_{i=1}^{n} E_0(t_i - 0) M_n[i], \quad (1.7b)$$

$$E_0(t) = W(0,t)P(t) + \sum_{i=1}^{n} E_0(t_i - 0) P(t - t_i) M_n[i]. \quad (1.7c)$$

It is an easy exercise to see that the solution is

$$E_0(t) = W(0, t)P(t) + \sum_{k=1}^{n} \sum_{n \geqslant i_1 > \cdots > i_k \geqslant 1} W[0, i_k - 1]P(t_{i_k}) \times$$

$$\times M_{i_{k-1}}[i_k]P(t_{i_{k-1}} - t_{i_k}) \ldots M_{i_1}[i_2]P(t_{i_1} - t_{i_2})M_n[i_1]P(t - t_{i_1}), \quad (1.8a)$$

$$E(t) = W(0, t) + \sum_{k=1}^{n} \sum_{n \geqslant i_1 > \cdots > i_k \geqslant 1} W[0, i_k - 1]P(t_{i_k})M_{i_{k-1}}[i_k] \times$$

$$\times P(t_{i_{k-1}} - t_{i_k}) \ldots M_{i_1}[i_2]P(t_{i_1} - t_{i_2})M_n[i_1] \quad (1.8b)$$

and to verify the stated properties of $E_0(\,.\,)$ and $E(\,.\,)$. ■

We are now going to discuss a more special requirement on the function $W(\,.\,,\,.\,)$; namely, that the probability of non-measuring the particle during $(s, t]$ *does not depend on the instant u when it was prepared.* In such a case, the condition

$$W(u, s)W(s, t) = W(u, t), \qquad u \leqslant s \leqslant t \quad (1.9)$$

must hold in J. Though it seems reasonable at a glance, a deeper inspection shows that (1.9) is valid, at most, approximately in real 'chambers'. As an example, consider a nearly monochromatic beam of particles passing through a bubble chamber. Each particle loses energy by ionization so its velocity slowly decreases. Since the chamber is homogeneous, a natural choice of the function W obeying (1.9) is $W(s, t) = \exp(-\lambda(t - s))$. In such a case, however, the mean distance between the bubbles (measurements) would equal $\lambda^{-1}v$ – i.e. it would decrease linearly with velocity along the particle track – while in the real bubble chambers it is proportional to the square of the velocity (cf. Section 3 below). In order to avoid contradiction, one must assume that λ varies with energy, and therefore also with time, $\lambda = \lambda(s)$, so the condition (1.9) is necessarily violated. This effect is even more conspicuous when the detection in photoemulsions is considered.

On the other hand, the functions W of this type are worthy of study, because they represent a good approximation to real situation in some cases, and besides, the condition (1.9) simplifies manipulation with equations (1.3).

We shall restrict our attention to the 'continuous' chamber structures and adopt the following assumption:

(c3) *There is a function* $\varphi : J \to \mathbb{R}_+$, *which is measurable and bounded a.e., such that* $W(s, t) = \exp\{-\int_s^t \varphi(\xi)\,d\xi\}$ *for all relevant s, t.*

PROPOSITION 2.1.3. *Assume* (r) *and* (c3), *then the conclusions of Proposition* 1.1 *hold and, moreover, the measured decay law is absolutely continuous and non-increasing.*

Proof. We have $m_t(s) = \varphi(s)W(s, t)$ so all the assumptions of Proposition 1.1 are fulfilled. We have noticed proving this assertion that the integral term on the r.h.s. of (1.3a) is absolutely continuous in t. Introducing now the function $g : [s, \infty) \to \mathbb{R}_+$ by $g(\xi) = \int_s^\xi \varphi(\xi)\,d\xi$, one verifies readily with the help of (A.4), (A.5) that

$$1 - \int_s^t e^{-g(\xi)}\,\varphi(\xi)\,d\xi = 1 - \int_0^{g(t)} e^{-u}\,du = e^{-g(t)}.$$

Consequently, $W(s, t) = 1 - \int_s^t \varphi(\xi)W(s, \xi)\,d\xi$ and the function $W(s, .)$ is absolutely continuous too; hence, the absolute continuity of $E(\,.\,)$ follows. Further, it is useful to introduce the function $f : f(t) = W(0, t)^{-1}E_0(t)$ (notice that $W(0, t)$ is always non-zero under the assumption (c3)) which allows us to cast equations (1.3) into the form

$$E(t) = W(0, t)\left\{1 + \int_0^t f(s)\varphi(s)\,ds\right\}, \tag{1.10a}$$

$$f(t) = P(t) + \int_0^t P(t - s)\varphi(s)\,ds. \tag{1.10b}$$

Finally, the derivative of $E(\,.\,)$ (which exists a.e. in view of the above considerations) can be calculated easily

$$\dot{E}(t) = -\varphi(t)W(0, t)\left\{1 + \int_0^t f(s)\varphi(s)\,ds\right\} + W(0, t)f(t)\varphi(t)$$

so the relations (1.5a) and (1.10) give $\dot{E}(t) = -\varphi(t)[E(t) - E_0(t)] \leqslant 0$. ∎

Discussion of equations (1.3) will be continued below; this section is concluded by the following important comment:

REMARK 2.1.4. There is one more assumption contained implicitly in the above-presented considerations. In accord with the general quantum-theoretical scheme mentioned in the opening of this section, we assumed that each measurement would occur at some time instant. However, each individual act of registering the 'particle' in those 'chambers' we have in mind leans on a specific physical process, most frequently ionization of some atom (which subsequently causes, e.g., evaporation of the liquid hydrogen or the excretion of some silver in its vicinity). Such processes can be described microphysically; in particular, their duration can be estimated. It is clear that the arguments which led us to equations (1.3) may be applied only if the particle decays slowly enough in the time scale given by characteristic time of the measuring interaction, and if the measurements are not too frequent.

2.2. Periodically Structured Measuring Devices

We are now going to discuss in more detail the behaviour of decay laws measured in chambers whose structure obeys the condition (1.9). We have presented above a heuristic argument showing that repeated measurements may suppress the non-exponentialities; the density of measurements must, of course, be approximately constant in the time scale given by the lifetime of the particle. This is granted for most of the considered 'chambers' in which the 'measuring environment' is reasonably homogeneous. On the other hand, the spark chambers represent an example of the measuring device structured approximatively homogeneously in a long-time scale, but in no case locally. This is why we shall concentrate here on the case of *periodically structured chambers*, specified by the following assumption:

(c4) *There is a positive t_p such that $W(s, t) = W(s + kt_p, t + kt_p)$, $k = 1, 2, \ldots,$ holds for all relevant values of the arguments.*

For the 'continuous' chamber structures, the assumptions (c3) and (c4) are together equivalent to the relation

$$\varphi(t) = \varphi(t + kt_p), \qquad k = 1, 2, \ldots \tag{2.1}$$

valid almost everywhere in the considered time region.

We shall need a quantity characterizing how dense the measurements are. For this purpose, we introduce the **measurement frequency** by

$$\nu_\varphi = t_p^{-1} \int_0^{t_p} \varphi(\xi)\, d\xi. \tag{2.2}$$

Some comments on this definition should be made, because ν_φ need not coincide with the inverse mean time between the measurements. The latter cannot be defined in a straighforward manner, since occurrence of any pair of neighbouring measurements is influenced by the preceding history of the particle in the chamber. Consider the following situation: the particle is prepared at s, then after some fixed time $u \geqslant s$, $k + 1$ measurements occur. We denote by $T_k(s, u)$ the mean time interval between the last pair of measurements. It is easy to see that the assumption (c3) implies

$$T_k(s, u) = \int_u^\infty ds_1 \int_{s_1}^\infty ds_2 \ldots \int_{s_k}^\infty ds_{k+1}\, \varphi(s_1) \ldots \varphi(s_k) \times \\ \times \varphi(s_{k+1})\, (s_{k+1} - s_k) W(s, s_{k+1}). \tag{2.3}$$

PROPOSITION 2.2.1. *If* (c3) *and* (c4) *hold, then the mean time* (2.3) *is equal to*

$$T_k(s, u) = [1 - \exp(-\nu_\varphi t_p)]^{-1} \int_u^{u+t_p} du_1 \int_{u_1}^{u_1+t_p} du_2 \ldots \\ \ldots \int_{u_k}^{u_k+t_p} du_{k+1}\, \varphi(u_1) \ldots \varphi(u_k) W(s, u_{k+1}) \tag{2.4a}$$

and the following inequalities hold,

$$f(\nu_\varphi t_p)^{k+1} \exp(-(k+1)\nu_\varphi t_p) \leqslant \nu_\varphi T_k(s, s) \leqslant f(\nu_\varphi t_p)^{k+1}, \tag{2.4b}$$

where $f(x) = x(1 - e^{-x})^{-1}$.

Proof. We observe first that the inequalities

$$\nu_\varphi(t - s - t_p) \leqslant \int_s^t \varphi(\xi + u)\,\mathrm{d}\xi \leqslant \nu_\varphi(t - s + t_p) \tag{2.5}$$

hold in view of (2.1) and (2.2) for all u and $s \leqslant t$. Thus $W(s, .)$ may be estimated by decreasing exponentials from both sides. In particular, it implies $\lim_{t\to\infty} tW(s, t) = 0$, so one can write $W(s, s_{k+1}) = W(s, s_k)W(s_k, s_{k+1})$ in (2.3) and integrate *per partes*, obtaining in this way

$$T_k(s, u) = \int_u^\infty \mathrm{d}s_1 \int_{s_1}^\infty \mathrm{d}s_2 \ldots \int_{s_k}^\infty \mathrm{d}s_{k+1}\, \varphi(s_1) \ldots \varphi(s_k) W(s, s_{k+1}).$$

Further, the relations $T_k(s + nt_p, u + nt_p) = T_k(s, u)$ and $T_k(s, u) = W(s, r)T_k(r, u)$ are valid for all natural n and $s \leqslant r \leqslant u$, as can easily be seen from (2.1) and (1.9). Thus we have

$$\begin{aligned} T_k(s, u) &= \int_u^\infty \mathrm{d}s_1\, \varphi(s_1) T_{k-1}(s, s_1) \\ &= \sum_{n=0}^\infty \int_{u+nt_p}^{u+(n+1)t_p} \mathrm{d}s_1\, \varphi(s_1) W(s, s + nt_p) T_{k-1}(s + nt_p, s_1) \\ &= \sum_{n=0}^\infty \exp(-n\nu_\varphi t_p) \int_u^{u+t_p} \mathrm{d}u_1\, \varphi(u_1) T_{k-1}(s, u_1); \end{aligned}$$

that is,

$$T_k(s, u) = [1 - \exp(-\nu_\varphi t_p)]^{-1} \int_u^{u+t_p} \mathrm{d}u_1\, \varphi(u_1) T_{k-1}(s, u_1). \tag{2.6a}$$

In the same way, one obtains

$$T_0(s, u) = [1 - \exp(-\nu_\varphi t_p)]^{-1} \int_u^{u+t_p} \mathrm{d}y\, W(s, y), \tag{2.6b}$$

then (2.4a) follows from (2.6a, b) by induction. The integral in (2.4a) can be estimated with the help of the inequalities $W(s, u)\exp(-(k + 1)\nu_\varphi t_p) \leqslant W(s, u_{k+1}) \leqslant W(s, u)$, so the relations (2.1), (2.2) readily yield (2.4b). ■

We see that one can speak about the mean time between measurements without any reservation in the case of a homogeneous chamber, $W(s, t) = \exp(-\lambda(t - s))$. Then an arbitrary positive number may play role of t_p so the inequalities (2.4b) give $T_k(s, s) = \nu_\varphi^{-1} = \lambda^{-1}$ for all k, s. If the chamber structure is inhomogeneous but *remains to be fine enough* in the sense that $\nu_\varphi t_p \ll 1$, then the inequalities (2.4b) can still be used: it holds $f(x) = 1 + \frac{1}{2}x + O(x^2)$, and therefore

$$|\nu_\varphi T_k(s, s) - 1| \leqslant \tfrac{1}{2}(k + 1)\nu_\varphi t_p + O(\nu_\varphi^2 t_p^2).$$

Consequently, for a fixed k which is not very large we have $\nu_\varphi T_k(s, s) \approx 1$ so ν_φ may be understood again as the measurement frequency.

On the other hand, for chambers with a rough structure, $\nu_\varphi t_p \gtrsim 1$, the above argument clearly fails, because (2.4b) does not represent an actual restriction on $\nu_\varphi T_k(s, s)$ even for $k = 0$. Moreover, it is easy to see that there are functions φ in this case such that the corresponding $\nu_\varphi T_0(s, s)$ would differ substantially from one:

EXAMPLE 2.2.2. Assume that the chamber structure is described by the function φ:

$$\varphi(t) = \begin{cases} 0 & \ldots \quad 0 < t < t_p - \tau \pmod{t_p} \\ \alpha & \ldots \quad t_p - \tau \leqslant t \leqslant t_p \pmod{t_p} \end{cases} \tag{2.7}$$

where α, τ are some positive numbers, $\tau < t_p$ (cf. Figure 2). Obviously $\nu_\varphi = \alpha\tau/t_p$; further, $T_0(s, s)$ can be calculated from (2.6b). It is sufficient to consider $t_p - \tau \leqslant s \leqslant t_p$, because the particle can be measured only there (mod t_p). We easily find

$$T_0(s, s) = \frac{\alpha\tau}{1 - e^{-\alpha\tau}} \left(1 - \frac{\tau}{t_p}\right) \exp(-\alpha(t_p - s)) + \frac{\tau}{t_p}.$$

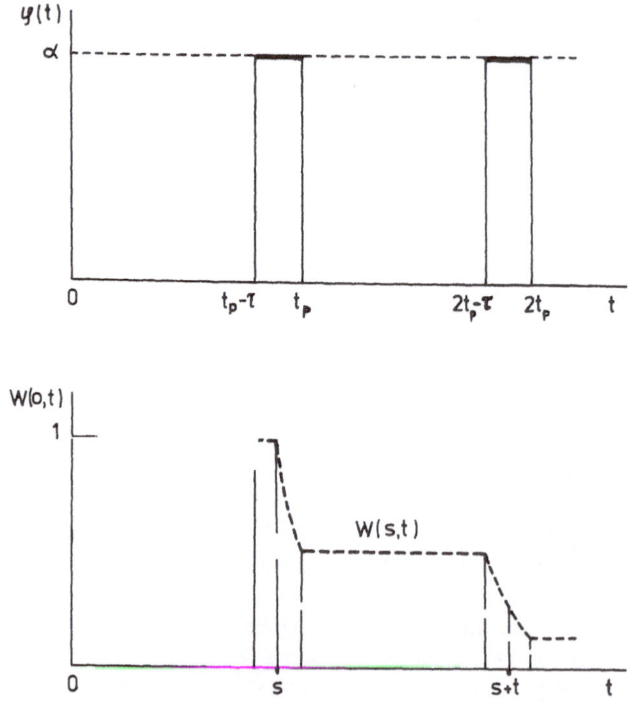

2.2. The functions φ and $W(s, .)$ considered in Example 2.2.2.

In particular, if the peaks of φ are tall and narrow in the sense that $\alpha\tau \gg 1$ and $\tau \ll t_p$, then this quantity oscillates between the values

$$\nu_\varphi T_0(t_p - \tau, t_p - \tau) \approx \alpha\tau\, e^{-\alpha\tau} + \frac{\tau}{t_p} \ll 1,$$

$$\nu_\varphi T_0(t_p, t_p) \approx \alpha\tau \gg 1.$$

Nevertheless, ν_φ still preserves the meaning of the measurement frequency in the following sense. It is not hard to find the probability that the particle (prepared at some $s \in (0, t_p - \tau)$) is measured just k times within $[t_p - \tau, t_p]$; it equals $w_k = e^{-\alpha\tau}(\alpha\tau)^k/k!$ so $\langle k \rangle = \Sigma_{k=1}^{\infty} k w_k = \alpha\tau$, i.e. in the mean $\alpha\tau$ measurements occur during the period t_p. Notice, however, that this argument fails if W is assumed to obey the assumption (c2), because then at most one measurement is possible at each point t_m.

After these preliminaries, let us return to equations (1.10). Since we are interested in the influence of the primary-decay-law non-exponentialities on the measured decay law, it is natural to start with the solution corresponding to the purely exponential $P(\,.\,)$ (see also Problem 5):

THEOREM 2.2.3. *Let* (r), (c3) *and* (c4) *be valid and* $P(t) = \exp(-\Gamma t)$, $\Gamma > 0$, *for all* $t \geqslant 0$, *then the measured decay law* $E(\,.\,)$ *is a non-decreasing absolutely continuous function, which can be expressed as*

$$E(t) = F(t)\exp(-\Gamma t) \tag{2.8}$$

with $F(t) \geqslant 1$ *for all* $t \geqslant 0$. *If* $\nu_\varphi > \Gamma$, *then the function* F *is asymptotically periodic with the period* t_p, *the asymptotic region being given by* $t \gg (\nu_\varphi - \Gamma)^{-1}$.

Proof. Absolute continuity and monotonicity are due to Proposition 1.3. One can check easily that (1.10b) is now solved by $f(t) = W(0, t)^{-1} e^{-\Gamma t}$, so that (1.10a) together with (1.9) give

$$E(t) = W(0, t) + \int_0^t \varphi(s) W(s, t)\, e^{-\Gamma s}\, ds.$$

Integration *per partes* yields

$$E(t) = e^{-\Gamma t} + \Gamma \int_0^t W(s, t)\, e^{-\Gamma s}\, ds$$

so we can write (2.8) with $F(t) = 1 + \Gamma \int_0^t W(t - u, t)\, e^{\Gamma u}\, du \geqslant 1$. The inequalities (2.5) imply $\exp(-\nu_\varphi(u + t_p)) \leqslant W(t - u, t) \leqslant \exp(-\nu_\varphi(u - t_p))$, and therefore the following estimate is possible

$$\frac{\Gamma \exp(-\nu_\varphi t_p)}{\nu_\varphi - \Gamma}\,[1 - \exp((\Gamma - \nu_\varphi)t)] \leqslant F(t) - 1$$

$$\leqslant \frac{\Gamma \exp(\nu_\varphi t_p)}{\nu_\varphi - \Gamma}\,[1 - \exp((\Gamma - \nu_\varphi)t]\,. \tag{2.9a}$$

In particular, we have

$$1 + \frac{\Gamma \exp(-\nu_\varphi t_p)}{\nu_\varphi - \Gamma} \leqslant \liminf_{t\to\infty} F(t) \leqslant \limsup_{t\to\infty} F(t)$$
$$\leqslant 1 + \frac{\Gamma \exp(\nu_\varphi t_p)}{\nu_\varphi - \Gamma}. \tag{2.9b}$$

Further, the above expression for $F(t)$, together with (c4), yield

$$F(t + t_p) - F(t) = \int_t^{t+t_p} W(t - u, t)\, e^{\Gamma u}\, du,$$

and using again an estimate based on the inequalities (2.5), we obtain

$$\lim_{t\to\infty} [F(t + t_p) - F(t)]\, e^{\alpha t} = 0 \tag{2.9c}$$

for all $t \geqslant 0$ and an arbitrary $\alpha < \nu_\varphi - \Gamma$. ■

Notice that in the cases of practical interest, we have usually $\nu_\varphi \gg \Gamma$, so the asymptotic region starts after few ν_φ^{-1} from the beginning and covers the whole interval where the decay law is actually measured.

Now let us look at the changes that will occur when the primary decay law is allowed to be non-exponential. Instead of attempting to solve equations (1.3) for a general function W which obey (1.9), we shall discuss in detail two typical examples. The first concerns the *spark-chamber-type measuring devices*, which have a discrete structure (see the assumption (c2) above) such that its period is equal to the distance of neighbouring points, i.e. there are positive t_p, a such that

$$t_k = kt_p, \tag{2.10a}$$
$$W[m, n] = e^{-a(n-m)} \tag{2.10b}$$

for all $k, m, n = 0, 1, 2, \ldots$. Of course, the assumption (2.10) carries a certain amount of idealization. In the first place, it leans on the condition (1.9), which was shown to hold only approximately. Another idealization concerns the strict localization of measurements to the points t_k, while in fact each of these points have some vicinity of a definite size where the measurement can be performed. Thus instead of (2.10), one might use a function W of the type discussed in Example 2.2. On the other hand, the discrete ansatz corresponds well to the real spark-chamber structure in the respect that at each 'measuring point' at most one measurement can occur.

In order to find the measured decay law, we must first solve equation (1.7c) with $M_n[m] = e^{-a}\, bW[m, n]$, where we have denoted $b = e^a - 1$. We are especially interested in the values $E_0(t_n - 0)$; it is useful to introduce $f_n = E_0(t_n - 0)/W[0, n - 1]$ which allow us to rewrite (1.7c) into the form

$$f_n = P(nt_p) + b \sum_{j=1}^{n} P(jt_p) f_{n-j}. \tag{2.11a}$$

The measured decay law is then determined by (1.7b) to be

$$E(t_n) = W[0,n]\left\{1 + b\sum_{j=1}^{n} f_j\right\} \tag{2.11b}$$

(remember that according to Proposition 1.2, $E(\,.\,)$ is a right-continuous step function). We are now going to prove the following assertion:

THEOREM 2.2.4. *Assume that the chamber structure is given by* (2.10), *and let the conditions* $0 < P(t_p) < 1$ *and*

$$\lim_{s\to 0} b \sum_{k=1}^{\infty} P(kt_p)\mathrm{e}^{-ks} > 1 \tag{2.12}$$

hold. Then the measured decay law is equal to

$$E(nt_p) = \left\{\frac{Ab}{1-\exp(-\nu_0 t_p)} + R_n\right\}\exp(-n\gamma t_p), \tag{2.13a}$$

where

$$A = \left[b^2 \sum_{k=1}^{\infty} kP(kt_p)\exp(-k\nu_0 t_p)\right]^{-1}, \tag{2.13b}$$

$$\gamma = at_p^{-1} - \nu_0 \tag{2.13c}$$

and ν_0 *is determined uniquely by the equation*

$$b\sum_{k=1}^{\infty} P(kt_p)\exp(-k\nu_0 t_p) = 1 \tag{2.13d}$$

and obeys $0 < \nu_0 < at_p^{-1}$. *Finally, there is a non-negative* $\nu_1 < \nu_0$ *such that for each* $\alpha < \nu_0 - \nu_1$, *then remainder terms satisfy*

$$\lim_{n\to\infty} R_n \exp(n\alpha t_p) = 0. \tag{2.13e}$$

The core of the proof is contained in solving equation (2.11a):

LEMMA 2.2.5. *Let the conditions* $P(t_p) > 0$ *and* (2.12) *be satisfied, then* (2.11a) *is solved by*

$$f_n = (A + B_n)\exp(n\nu_0 t_p), \tag{2.14a}$$

where A, ν_0 *are as introduced above, and for each* $\alpha < \nu_0 - \nu_1$, *the remainder terms obey*

$$\lim_{n\to\infty} B_n \exp(n\alpha t_p) = 0. \tag{2.14b}$$

Proof. Let us denote for brevity $P_k \equiv P(kt_p)$ and introduce the discrete Laplace transform

$$\tilde{P}(s) = \sum_{k=1}^{\infty} P_k \, \mathrm{e}^{-ks}, \qquad \tilde{B}(s) = \sum_{k=1}^{\infty} B_k \, \mathrm{e}^{-ks} \tag{2.15}$$

for those $s \in \mathbb{C}$ for which the series (absolutely) converge. Inserting $f_n = (A + B_n)\mathrm{e}^{nz}$ into (2.11a), we obtain

$$B_n = P_n \, \mathrm{e}^{-nz} - A + b \sum_{j=1}^{n-1} P_j(A + B_{n-j})\mathrm{e}^{-jz}.$$

We now multiply both sides by e^{-ns} and sum over n; a simple transformation of the summation indices gives

$$\tilde{B}(s) = \tilde{P}(s+z) - \frac{A\,\mathrm{e}^{-s}}{1-\mathrm{e}^{-s}} + Ab\tilde{P}(s+z)\frac{\mathrm{e}^{-s}}{1-\mathrm{e}^{-s}} + b\tilde{P}(s+z)\tilde{B}(s)$$

so

$$\tilde{B}(s) = \frac{\tilde{P}(s+z)}{1 - b\tilde{P}(s+z)} - \frac{A}{\mathrm{e}^s - 1}. \tag{2.16}$$

Let us now list some simple properties of the function $\tilde{P}(\,.\,)$. This is defined and analytic (at least) in the halfplane $M_+ = \{s : \mathrm{Re}\, s > 0\}$. The restriction of $\tilde{P}(\,.\,)$ to the positive real axis is a decreasing function with $\lim_{s\to\infty} \tilde{P}(s) = 0$. Moreover, $|\tilde{P}(s)| \leqslant \tilde{P}(\mathrm{Re}\, s)$ and the relations $\overline{\tilde{P}(s)} = \tilde{P}(\bar{s})$ and $\tilde{P}(s) = \tilde{P}(s + 2\pi i m)$ hold for all $s \in M_+$ and each m integer. This implies:

(i) The equation $b\tilde{P}(z) = 1$ has just one real solution z_0. If we set $z = z_0 \equiv \nu_0 t_p$ in (2.16), then both the terms have a simple pole at $s = 0$, and these poles compensate mutually when $A = (-b^2\tilde{P}'(z_0))^{-1}$.

(ii) There are no solutions of the equation $b\tilde{P}(z) = 1$ with $\mathrm{Re}\, z > z_0$. If z is a solution, then $\bar{z}$ and $z + 2\pi i m$ are also solutions; thus, it is enough to find the solutions contained in $N = \{z \in M_+ : 0 \leqslant \mathrm{Im}\, z < 2\pi\}$. The function $b\tilde{P}(\,.\,) - 1$ is analytic and non-constant in M_+ so it can have at most a finite number of zeros in any bounded region $\subset M_+$; in particular, in N. Let us denote these solutions $z^{(1)}, \ldots, z^{(k)}$, and set $z_1 \equiv \nu_1 t_p = \max_{1 \leqslant j \leqslant k} \mathrm{Re}\, z^{(j)}$, eventually $z_1 = 0$ if $b\tilde{P}(z) = 1$ has no non-real solutions.

According to (2.16) and implication (i), the function $\tilde{B}(\,.\,)$ is meromorphic in $\{s : \mathrm{Re}\, s > -z_0\}$. Further if A, ν_0 are given by (2.13b, d), then (ii) shows that $\tilde{B}(\,.\,)$ is analytic in $\{s : \mathrm{Re}\, s > z_1 - z_0\}$, so choosing $s = -\alpha t_p$ we arrive at (2.14b).

It remains to prove that $\nu_1 < \nu_0$. Assume that they are equal, i.e. that the equation $b\tilde{P}(z) = 1$ has a solution $z_\eta = z_0 + i\eta$ with $0 \leqslant \eta < 2\pi$. In such a case, we get

$$\mathrm{Re}\, b\tilde{P}(z_\eta) = b \sum_{k=1}^{\infty} P_k \exp(-kz_0) \cos k\eta = b\tilde{P}(z_0)$$

so $\cos k\eta = 1$ must hold for those k for which P_k is non-zero, but it is impossible since P_1 is supposed to be positive. ∎

Proof of the theorem. One has only to insert (2.14a) into (2.11b) and to express the remainder term

$$E(nt_p) = e^{-na} \left\{ 1 + b \sum_{k=1}^{n} (A + B_k) \exp(kz_0) \right\}$$

$$= \exp(-n\gamma t_p) \left\{ \exp(-nz_0) + Ab \sum_{j=0}^{n-1} \exp(-jz_0) + b \sum_{j=0}^{n-1} B_{n-j} \exp(-jz_0) \right\}$$

so

$$R_n = \left(1 + \frac{Ab}{1 - \exp(-z_0)} \right) \exp(-nz_0) + b \sum_{j=0}^{n-1} B_{n-j} \exp(-jz_0).$$

Now the above lemma asserts particularly that for an arbitrary $\alpha < \nu_0 - \nu_1$, the sequence $\{B_n \exp(n\alpha t_p)\}$ is bounded, i.e. there is a positive $B(\alpha)$ such that $|B_n| \leqslant B(\alpha) \exp(-n\alpha t_p)$. This makes it possible to estimate the second term in the above expression

$$|R_n^{(2)}| \leqslant bB(\alpha) \exp(-n\alpha t_p) \sum_{j=0}^{n-1} \exp(-j(\nu_0 - \alpha)) \leqslant \frac{bB(\alpha) \exp(-n\alpha t_p)}{1 - \exp(\alpha - \nu_0)}$$

so (2.13e) follows easily. It remains to prove that $\nu_0 < at_p^{-1}$, i.e. $\gamma > 0$. Equation (2.13d) can be rewritten with the help of (2.13c) as

$$(e^a - 1)^{-1} = \sum_{k=1}^{\infty} P(kt_p) \exp(k(\gamma t_p - a)); \tag{2.17}$$

Obviously γ may be zero only if all $P_k = 1$, which is excluded by the assumption. ∎

The obtained measured decay law (2.13) is at a glance closely similar to the one corresponding to purely exponential $P(\,.\,)$. It can be again expressed as a decreasing exponential times a function which is asymptotically t_p-periodic. However, there is an important difference: in the previous case the decay rates of $E(\,.\,)$ and $P(\,.\,)$

coincide, while now γ depends in general (as a solution to equation (2.17)) on the parameters a, t_p. As we shall see later, in some cases even γ must depend on the chamber structure if only the primary decay law violates exponentiality.

Consider now two extreme situations:

(i) *Highly efficient chamber.* The particle is measured at nearly all points t_k, or equivalently, $a \gg 1$. Notice that this is the mentioned case when $\nu_\varphi = at_p^{-1}$ loses the meaning of measuring frequency, which is characterized better by t_p^{-1} here. The basic equation can now be rewritten as

$$(1 - e^{-a})^{-1} = \left\{ P(t_p) + \sum_{k=2}^{\infty} P(kt_p) \exp((k-1)(\gamma t_p - a)) \right\} \exp(\gamma t_p). \quad (2.17a)$$

Suppose first that the measurement is performed with certainty at each point t_k, then we let $a \to \infty$ in (2.17a) and obtain

$$\gamma = -\frac{1}{t_p} \ln P(t_p). \quad (2.18a)$$

This is not surprising because the measured decay law assumes the form

$$E(t) = P(t_p)^n, \quad t_n \leqslant t < t_{n+1}. \quad (2.18b)$$

The relation (2.18a) shows how the non-exponentialities of $P(\,.\,)$ are manifested in dependence of γ on t_p. Suppose, further, a to be finite but very large so that we can neglect all terms on the r.h.s. of (2.17a) except the first two. In this way we obtain a quadratic equation for $\exp(\gamma t_p)$; its positive solution can be expressed as

$$\exp(\gamma t_p) = \frac{1 + e^{-a}}{P(t_p)} \left\{ 1 - \frac{P(2t_p)}{P(t_p)^2} e^{-a} + O(e^{-2a}) \right\}$$

so

$$\gamma = -\frac{1}{t_p} \ln P(t_p) + \frac{e^{-a}}{t_p} \left[1 - \frac{P(2t_p)}{P(t_p)^2} \right] + O(e^{-2a}).$$

We see that γ depends now not only on t_p, but also on a in the case that $P(2t_p) \neq P(t_p)^2$.

(ii) *The homogeneous-chamber limit.* We have argued above that if the chamber structure is 'continuous' and fine enough, then the quantity ν_φ defined by (2.2) may be interpreted as a measurement frequency. The central trick there was based on the fact that $W(\,.\,,.\,)$ obeyed the condition (1.9). Thus Proposition 2.1 can be adapted easily for the case when W is given by (2.10) and the conclusion about the measurement frequency $\nu_\varphi = a t_p^{-1}$ is preserved. Equation (2.17) now reads as follows:

$$\frac{t_p}{\exp(\nu_\varphi t_p) - 1} = \sum_{k=1}^{\infty} P(kt_p) \exp((\gamma - \nu_\varphi)t_p) \cdot t_p,$$

and if we keep ν_φ fixed and let $t_p \to 0$, it becomes

$$\nu_\varphi^{-1} = \int_0^\infty P(t) \exp((\gamma - \nu_\varphi)t)\, dt. \tag{2.19a}$$

Before proceeding further, let us notice that the above derived behaviour of the measured decay law – namely, that it is expressed as a product of decaying exponential and an asymptotically t_p-periodic function – seems to be peculiar for all periodically structured chambers in the case that the density of measurements is large enough. In order to illustrate this, we shall present the following heuristic argument limiting ourselves again to the 'continuous' chamber structures. We express solution of equation (1.10b) in the form $f(t) = g(t) \exp(\nu_0 t)$, where $\nu_0 = \inf\{\nu : \limsup_{t\to\infty} f(t) e^{-\nu t} = 0\}$, so the function g is of the zero order of growth and obeys the integral equation

$$g(t) = P(t) \exp(-\nu_0 t) + \int_0^t P(s) \exp(-\nu_0 s) \varphi(t-s) g(t-s)\, ds.$$

The inequality (1.5b) shows that $0 \leqslant \nu_0 \leqslant \nu_\varphi$. Suppose that P is a decreasing function (at least at large in a suitable time scale) and, on the other hand, that the measurement frequency ν_φ is sufficiently large, then both these inequalities may be expected to be sharp. In such a case, the last equation acquires for large t the following form

$$g(t) \approx \int_0^t P(s) \exp(-\nu_0 s) \varphi(t-s) g(t-s)\, ds$$

and the functions $t \mapsto g(t + nt_p)$ must also obey this asymptotic equation due to the assumed periodicity of φ; this indicates that $E(\,.\,)$ might behave asymptotically as described above.

However, we are not going to pursue this path further and to convert the presented consideration into a rigorous proof, because it would probably yield no new physical information. Instead we shall treat in detail the case of a *homogeneous chamber* in which the function W is determined by a single positive parameter λ (which coincides with the measurement frequency ν_φ)

$$W(s, t) = \exp(-\lambda(t - s)). \tag{2.20}$$

Since any positive number can now play the role of the period t_p, the measured decay law is expected to decrease asymptotically exponentially.

THEOREM 2.2.6. *Let the primary lifetime T be finite and $\lambda T > 1$. Then the measured decay law corresponding to* (2.20) *is of the form*

$$E(t) = \left\{ \frac{\lambda A}{\lambda - \gamma} + R(t) \right\} \exp(-\gamma t), \tag{2.21a}$$

where

$$A = \left[\lambda^2 \int_0^\infty tP(t) \exp(-\nu_0 t) \, dt \right]^{-1} \tag{2.21b}$$

and γ is determined uniquely by the equation

$$\lambda \int_0^\infty P(t) \exp((\gamma - \lambda)t) \, dt = 1 \tag{2.19b}$$

and obeys $0 < \gamma < \lambda$. Finally, there is a non-negative $\nu_1 < \nu_0 \equiv \lambda - \gamma$ such that

$$\lim_{t \to \infty} R(t) \, e^{\alpha t} = 0 \tag{2.21c}$$

for each $\alpha < \nu_0 - \nu_1$.

Equations (1.10) in the considered case read

$$E(t) = \left\{ 1 + \lambda \int_0^t f(s) \, ds \right\} e^{-\lambda t}, \tag{2.22a}$$

$$f(t) = P(t) + \lambda \int_0^t P(t - s) f(s) \, ds. \tag{2.22b}$$

We start again by solving the second of them:

LEMMA 2.2.7. *In the assumptions of Theorem* 2.6, *equation* (2.22b) *is solved by*

$$f(t) = (A + B(t)) \exp(\nu_0 t) \tag{2.23}$$

where A, ν_0 are as introduced above and $B(\,.\,)$ is a continuous function such that for all $\alpha < \nu_0 - \nu_1$, the function $t \mapsto B(t) e^{\alpha t}$ belongs to $L^2(\mathbb{R}_+)$.

Proof. The proof is divided into three steps.

STEP 1: By tildes we denote the Laplace transforms, $\widetilde{P}(s) = \int_0^\infty e^{-st} P(t) \, dt$, etc. The central trick is based on the observation that equation (2.22b) is of the convolution type, similarly as in Lemma 2.5, where we have applied the discrete Laplace transformation (2.15) to the 'discrete convolution' in (2.11a). Substituting (2.23) into (2.22b), and taking the Laplace transform, we obtain

$$As^{-1} + \widetilde{B}(s) = \widetilde{P}(\nu_0 + s) + \lambda A s^{-1} \widetilde{P}(\nu_0 + s) + \lambda \widetilde{P}(\nu_0 + s) \widetilde{B}(s)$$

so

$$\widetilde{B}(s) = \frac{\widetilde{P}(\nu_0 + s)}{1 - \lambda \widetilde{P}(\nu_0 + s)} - \frac{A}{s} \equiv \widetilde{B}_0(s) - \frac{A}{s}. \tag{2.24}$$

Let us collect the needed properties of the function $\widetilde{P}$: since $\widetilde{P}(0) = T$ is assumed to be finite (see (1.2.6)), $\widetilde{P}(\,.\,)$ is defined in the halfplane $\bar{M}_+ = \{s : \mathrm{Re}\, s \geqslant 0\}$ and analytic in its interior M_+. Its restriction to the positive real axis is decreasing and $\lim_{s\to\infty} \widetilde{P}(s) = 0$. Moreover, the relations $\overline{\widetilde{P}(s)} = \widetilde{P}(\bar{s})$ and $|\widetilde{P}(s)| \leqslant \widetilde{P}(\mathrm{Re}\, s)$ hold for all $s \in \bar{M}_+$.

We assume $\lambda T > 1$, so the equation $\lambda\widetilde{P}(\nu) = 1$ has a unique positive real solution ν_0. Further, we introduce

$$\nu_1 = \sup\{\{0\} \cup \{\mathrm{Re}\, \eta : \lambda P(\eta) = 1\}\};$$

the properties of $\widetilde{P}$ imply $0 \leqslant \nu_1 \leqslant \nu_0$. The function $\widetilde{P}$ is analytic and non-constant in M_+, so it can have there isolated zeros only. Consequently, $\widetilde{B}$ is meromorphic in the halfplane $\{s : \mathrm{Re}\, s > -\nu_0\}$. The simple pole which it has at $s = 0$ can be removed by a proper choice of the constant A; in this way one obtains (2.21b). The function $\widetilde{B}$ is then analytic in $\{s : \mathrm{Re}\, s > -(\nu_0 - \nu_1)\}$. The original function B is continuous due to Proposition 1.1 and one can express it using the Mellin transformation (Sveshnikov and Tikhonov, 1971, §8.2)

$$B(t) = \frac{1}{2\pi i} \lim_{n\to\infty} \int_{-\alpha - in}^{-\alpha + in} \mathrm{e}^{st}\, \widetilde{B}(s)\, \mathrm{d}s, \qquad t \geqslant 0,$$

where $\alpha < \nu_0 - \nu_1$. If at the same time $\alpha > 0$, we have $\int_{\mathbb{R}} \mathrm{e}^{i\xi t} (\xi + i\alpha)^{-1}\, \mathrm{d}\xi = 0$ for $t > 0$, and therefore

$$B(t) = \frac{\mathrm{e}^{-\alpha t}}{2\pi} \lim_{n\to\infty} \int_{-n}^{n} \mathrm{e}^{i\xi t}\, \widetilde{B}_0(-\alpha + i\xi)\, \mathrm{d}\xi \equiv \frac{\mathrm{e}^{-\alpha t}}{2\pi}\, b(\alpha, t) \tag{2.25}$$

holds for all $t > 0$.

STEP 2: We have to check that ν_1 is strictly less than ν_0. Writing $\nu = u + iv$, we can cast the considered equation into the form

$$R(u, v) := \lambda \int_0^\infty \mathrm{e}^{-ut}\, P(t) \cos vt\, \mathrm{d}t = 1,$$

$$I(u, v) := \lambda \int_0^\infty \mathrm{e}^{-ut}\, P(t) \sin vt\, \mathrm{d}t = 0.$$

In view of the properties of $\widetilde{P}(\,.\,)$, we can restrict our attention to $(u, v) \in [0, \nu_0] \times \mathbb{R}_+$. Since $\widetilde{P}(\,.\,)$ is assumed to belong to $L(\mathbb{R}_+)$, the Riemann–Lebesgue lemma implies $\lim_{v\to\infty} R(u, v) = 0$ for each $u \in [0, \nu_0]$. We shall check that the convergence is *uniform* with respect to u.

Consider first the expressions

$$R(u, v; \tau) := \lambda \int_0^\tau \mathrm{e}^{-ut}\, P(t) \cos vt\, \mathrm{d}t$$

with some fixed positive τ. An easy application of the dominated-convergence theorem yields

$$R(u, v; \tau) = \sum_{k=0}^{\infty} \frac{(-u)^k}{k!} R_k(v; \tau), \tag{$*$}$$

where $R_k(v; \tau) = \lambda \int_0^\tau t^k P(t) \cos vt \, dt$. For $(u, v) \in [0, \nu_0] \times \mathbb{R}_+$ and an arbitrary n, the inequalities

$$\left| \sum_{k=0}^{n} \frac{(-u)^k}{k!} R_k(v; \tau) \right| \leqslant \sum_{k=0}^{n} \frac{\nu_0^k}{k!} |R_k(v; \tau)|$$

$$< \lambda \sum_{k=0}^{n} \frac{\nu_0^k}{k!} \frac{\tau^{k+1}}{k+1} < \lambda\tau \exp(\nu_0 \tau) \tag{$**$}$$

hold, which show that the series ($*$) converges uniformly with respect to v. Further, $\lim_{v \to \infty} |R_k(v; \tau)| = 0$ and $R(u, v; \tau)$ is in view of ($**$) majorized by the u-independent v-uniformly convergent series $\Sigma_{k=0}^{\infty} (k!)^{-1} \nu_0^k R_k(v; \tau)$; it means that $\lim_{v \to \infty} R(u, v; \tau) = 0$ uniformly in u.

Now the assumed integrability of P implies that to any $\epsilon > 0$, there is $\tau(\epsilon)$ such that $\int_{\tau(\epsilon)}^{\infty} P(t) \, dt < \frac{1}{2}\epsilon$. The assertion we have just proved implies the existence of $v_0 = v_0(\epsilon, \tau(\epsilon))$ such that $|R(u, v; \tau(\epsilon))| < \frac{1}{2}\epsilon$ for all $v > v_0$ and $u \in [0, \nu_0]$. Consequently, we have

$$|R(u, v)| \leqslant |R(u, v; \tau(\epsilon))| + \left| \int_{\tau(\epsilon)}^{\infty} e^{-ut} P(t) \cos vt \, dt \right|$$

$$< \tfrac{1}{2}\epsilon + \int_{\tau(\epsilon)}^{\infty} P(t) \, dt < \epsilon$$

for all $v > v_0$ and v_0 is independent of u; thus the uniform convergence is proved.

In particular, if $\epsilon = 1$ then the last result implies the existence of some v_0 such that the equation $\lambda \widetilde{P}(\nu) = 1$ has no solutions outside $[0, \nu_0] \times [-v_0, v_0]$. Consequently, there is at most a finite number of solutions. Suppose that some of them have a real part which is equal to ν_0, and differs from ν_0, then the equality $R(\nu_0, v) = R(\nu_0, 0) = 1$ would have to hold for some $v > 0$, i.e.

$$2\lambda \int_0^\infty P(t) \exp(-\nu_0 t) \sin^2(\tfrac{1}{2} vt) \, dt = 0.$$

However, this is impossible in view of Proposition 1.2.3, thus $\operatorname{Re} \nu < \nu_0$ holds for each non-real solution, so by definition $\nu_1 < \nu_0$.

STEP 3: Let us now return to the relation (2.25). For a fixed $\alpha \in (0, \nu_0 - \nu_1)$, the function $C_\alpha : C_\alpha(\xi) = |1 - \lambda \widetilde{P}(\nu_0 - \alpha + i\xi)|$ *is bounded below by a positive number*. This can be seen as follows: the Riemann–Lebesgue lemma gives

$\lim_{|\xi|\to\infty} \tilde{P}(\nu_0 - \alpha + i\xi) = 0$ so $|C_\alpha(\xi)| \geqslant \frac{1}{2}$ holds outside some finite closed interval $J \subset \mathbb{R}$. Continuity of $C_\alpha(\,.\,)$ implies the existence $\xi_0 \in J$ where it assumes a minimal value; on the other hand, $C_\alpha(\xi)$ is always positive because $\nu_0 - \alpha > \nu_1$. Consequently, $C_\alpha(\xi_0) > 0$ and

$$C_\alpha(\xi) \geqslant K_\alpha \equiv \min\{\tfrac{1}{2}, C_\alpha(\xi_0)\} > 0$$

for each $\xi \in \mathbb{R}$. This yields the following estimate

$$|\tilde{B}_0(-\alpha + i\xi)| \leqslant K_\alpha^{-1}\, |\tilde{P}(\nu_0 - \alpha + i\xi)|.$$

Function $\xi \mapsto P(\nu_0 - \alpha + i\xi)$ as the Fourier transform of $t \mapsto \Theta(t)P(t)\exp(-(\nu_0 - \alpha)t)$ belongs to $L^2(\mathbb{R})$, and the same is true for $\xi \mapsto \tilde{B}_0(-\alpha + i\xi)$ due to the last inequality. The Fourier–Plancherel theorem shows that the function g: $g(t) = \text{l.i.m.}_{n\to\infty}\, g_n(t)$, where

$$g_n(t) = \int_{-n}^{n} e^{i\xi t}\, \tilde{B}_0(-\alpha + i\xi)\, d\xi,$$

belongs to $L^2(\mathbb{R})$. Since $\{g_n\}$ converges in the mean, there is a subsequence $\{g_{n_k}\}$ which converges to g pointwise a.e. in $\mathbb{R}$. However, the limit of $\{g_n(t)\}$ exists for each $t > 0$ according to (2.25), so $g(\,.\,) = b(\alpha,\,.\,)$ a.e. in $(0, \infty)$. Thus the function $b(\alpha,\,.\,)$ is square-integrable and the proof is completed. ■

Proof of the theorem. Positivity of ν_0 gives $\gamma < \lambda$. On the other hand, equation (2.19b) shows that γ is positive unless $\lambda\tilde{P}(\lambda) = 1$, that is, $\lambda \int_0^\infty (1 - P(t))\, e^{-\lambda t}\, dt = 0$, but such a possibility is excluded by Proposition 1.2.3 and the assumed integrability of $P(\,.\,)$. Now one has to insert (2.23) into (2.22a), and a simple rearrangement yields (2.21a) with

$$R(t) = \left[1 - \frac{\lambda A}{\lambda - \gamma}\right] \exp(-\nu_0 t) + \lambda \int_0^t B(s) \exp(-\nu_0(t - s))\, ds.$$

Since $\nu_0 - \alpha > \nu_1 \geqslant 0$, we need in order to prove (2.21c) only to estimate

$$h(\alpha, t) := \int_0^t B(s)\, e^{\alpha s} \exp(-(\nu_0 - \alpha)(t - s))\, ds.$$

This is easily achieved by virtue of Lemma 2.7 and the Hölder inequality:

$$\begin{aligned} |h(\alpha, t)| &\leqslant \left(\int_0^t \exp(-2(\nu_0 - \alpha)u)\, du\right)^{1/2} \left(\int_0^t |B(s)\, e^{\alpha s}|^2\, ds\right)^{1/2} \\ &< \frac{1}{2\pi}\, [2(\nu_0 - \alpha)]^{-1/2}\, \|b(\alpha,\,.\,)\|_2, \end{aligned}$$

where $b(\alpha,\,.\,)$ is the function introduced by the relation (2.25). Thus for any fixed $\alpha \in (0, \nu_0 - \nu_\Gamma)$, the function $t \mapsto R(t)\, e^{\alpha t}$ is bounded by some constant $C(\alpha)$ so $|R(t)\, e^{\alpha t}| \leqslant C(\frac{1}{2}\alpha) \exp(-\frac{1}{2}\alpha t)$ and (2.21c) follows. ■

The exponential decrement γ of the measured decay law obeys equation (2.19b) and we see that it coincides with (2.19a). The above theorem establishes the existence of a unique solution to this equation. We can write it in a more explicit way if we denote by R the inverse function to the restriction of $\tilde{P}(\,.\,)$ to the positive real axis, $R = (\tilde{P} \upharpoonright \mathbb{R}_+)^{-1}$. The above-mentioned properties of $\tilde{P}(\,.\,)$ show that R is a decreasing function which maps $(0, T]$ onto $\mathbb{R}_+$. Now the solution γ of the equation $\lambda\tilde{P}(\lambda - \gamma) = 1$ can be expressed as

$$\gamma(\lambda) = \lambda - R\left(\frac{1}{\lambda}\right). \tag{2.26}$$

We have seen above that γ may depend on the chamber structure, in particular, on the density of measurements, when $P(\,.\,)$ is not purely exponential. In the present case, even a stronger assertion holds:

PROPOSITION 2.2.8. *Let the assumptions of Theorem* 2.6 *be valid for a homogeneous chamber. If there is an open interval* (λ_1, λ_2) *of measurement frequencies such that* γ *is independent of* λ *there,* $\gamma(\lambda) = \Gamma$ *for* $\lambda \in (\lambda_1, \lambda_2)$, *then the primary decay law is purely exponential,* $P(t) = \exp(-\Gamma t)$ *for* $t \geqslant 0$.

Proof. Equation (2.26) gives $R(x) = x^{-1} - \Gamma$ for $\lambda_2^{-1} < x < \lambda_1^{-1}$, and therefore $\tilde{P}(\lambda) = (\lambda + \Gamma)^{-1}$ for $\lambda \in (\lambda_1, \lambda_2)$. However, $\tilde{P}(\,.\,)$ is analytic in M_+ so the above expression can be continued there and the assertion follows. ∎

2.3. A Model: Charged Kaons in a Bubble Chamber

"There is also no objection to a mathematician's doing physics, provided he is qualified. The prime example was von Neumann – when he did physics, he talked, thought and calculated like a physicist (but faster)."

D. R. RICHTMYER

Considerations of the preceding section show that the measured decay law can be affected by the structure of the measuring device, in particular, that it is likely to vary with the measurement frequency. However, the results we have obtained are not suitable for quantitative estimates which are needed for decision whether the effect is actually observable or not. To this purpose, we shall treat here a simple model describing unstable particles which decay within a bubble chamber; numerical results will be found for the charged kaons in a hydrogen chamber.

We have argued in Section 1 that an appropriate choice of the function W for a real homogeneous chamber is

$$W(s, t) = \exp[-\lambda(s)\,(t - s)]\,. \tag{3.1}$$

The function $\lambda(\,.\,)$ is, of course, slowly varying in the time scale given by λ^{-1}, i.e. we assume $|d\lambda(t)^{-1}/dt| \ll 1$. We have

$$m_t(s) = \lambda(s)\left[1 - \frac{\lambda'(s)}{\lambda(s)}\,(t - s)\right] W(s, t),$$

but $W(s, t)$ decreases exponentially with t so we can neglect it unless $t - s \lesssim \lambda(s)^{-1}$. The condition imposed on λ then implies $m_t(s) \approx \lambda(s)W(s, t)$. If the exact equality would hold in this relation, then equations (1.3) read as

$$E(t) = W(0, t) + \int_0^t \lambda(s)E_0(s)W(s, t)\, ds,$$

$$E_0(t) = W(0, t)P(t) + \int_0^t \lambda(s)E_0(s)P(t - s)W(s, t)\, ds.$$

Now we notice that for t large enough (after few $\lambda(0)^{-1}$) the l.h.s. values are determined essentially by the integrals only. Further, if $P(\,.\,)$, $E_0(\,.\,)$ are also slowly varying in the above described sense, then the values of the integrals will not change substantially when the function (3.1) is replaced by another with the following properties:

(i) $W(s, t)$ differs essentially from zero only if $t - s \lesssim \lambda(s)^{-1}$,
(ii) $\int_s^\infty W(s, t)\, dt = \lambda(s)^{-1}$, i.e. the new function W gives the same mean distances between the bubbles.

These considerations show that we can choose for model purposes the function W in the form

$$W(s, t) = \begin{cases} 1 & \ldots \quad s \leqslant t < s + \lambda(s)^{-1} \\ 0 & \ldots \quad s + \lambda(s)^{-1} \leqslant t. \end{cases} \tag{3.2}$$

The existence of the measured decay law in such a case is established by Proposition 1.2. In order to see how $E(\,.\,)$ looks when the primary decay law slightly violates the exponentiality, let us write the latter as

$$P(t) = N^2(e^{-\Gamma t} + M(t)). \tag{3.3}$$

We shall specify $M(\,.\,)$ later; for the present we assume only $|M(t)| \ll 1$ for all $t \geqslant 0$. Suppose that a nearly monochromatic beam of particles goes through the chamber, then it is natural to express all quantities in dependence not on the time but rather on the particle position directly. Of course, one must take the relativistic effects into account. While the chamber characteristics are expressed in the laboratory time, the values of the primary decay law refer to the proper time of the particle (see (3.5.17) and the following remarks). Let it move between two measurements located at x_1, x_2 with a velocity βc, then the corresponding difference of the proper times is

$$t_2' - t_1' = \frac{x_2 - x_1}{c\beta\gamma},$$

where $\gamma = (1 - \beta^2)^{-1/2}$ is the contraction factor (primed quantities refer to the rest system of the particle).

Let us return to the measured decay law. Our model assumption can be reformulated equivalently as follows: the measurements are performed at the points

$x_1, x_2, x_3, \ldots$, and $x_{j+1} - x_j = L(x_j)$, where L is a slowly varying function, $|dL(x)^{-1}/dx| \ll 1$. Suppose now that the particle is initially prepared at $x_0 = 0$, then the measured decay law assumes the form

$$E(x) = \prod_{j=1}^{n} P(\Delta t'_{j-1}), \quad x \in [x_n, x_{n+1}), \tag{3.4}$$

where $\Delta t'_j = L(x_j)/c\beta(x_j)\gamma(x_j)$. Substitution from (3.3) to (3.4), together with a simple rearrangement, yield

$$\ln E(x) = \sum_{j=1}^{n} \ln(\exp(-\Gamma\,\Delta t'_{j-1}) +$$

$$+ \sum_{j=1}^{n} \{2 \ln N + \ln[1 + M(\Delta t'_{j-1}) \exp(\Gamma\,\Delta t'_{j-1})]\}.$$

Now, since $P(\,.\,)$ is assumed to be slowly varying (in the time scale given by $L(x)^{-1}$), we can replace the summation by integration in the last expression. The first term gives

$$\sum_{j=1}^{n} \frac{-\Gamma(x_j - x_{j-1})}{c\beta(x_{j-1})\gamma(x_{j-1})} \approx -\Gamma \int_0^x \frac{dy}{c\beta(y)\gamma(y)};$$

it corresponds to the exponential part of $E(\,.\,)$ modified by the relations between the proper and laboratory times,

$$F(x) = \exp\left\{-\Gamma \int_0^x \frac{dy}{c\beta(y)\gamma(y)}\right\}. \tag{3.5a}$$

The second term can be handled in an analogous way, so we obtain

$$E(x) = F(x)\, e^{K(x)}, \tag{3.5b}$$

where

$$K(x) = \int_0^x \left\{2 \ln N + M\left(\frac{L(y)}{c\beta(y)\gamma(y)}\right) \exp\left(\frac{\Gamma L(y)}{c\beta(y)\gamma(y)}\right)\right\} \frac{dy}{L(y)}. \tag{3.5c}$$

Now one has to specify the primary decay law. We choose a simple modification of the exponential law obtained by a symmetric cut-off in the Breit–Wigner energy distribution, $P(t) = |v(t)|^2$, with

$$v(t) = \frac{N\delta}{\pi} \int_{-\epsilon}^{\epsilon} \frac{e^{-i\lambda t}}{\lambda^2 + \delta^2}\, d\lambda. \tag{3.6}$$

Here we have denoted $\delta = \frac{1}{2}\Gamma$; without loss of generality we can assume that the peak is centred at $\lambda = 0$. The normalization factor N is given by

$N^{-1} = (2/\pi)\,\mathrm{arctg}(\epsilon/\delta)$. The primary decay law corresponding to (3.6) admits a natural interpretation (cf. Section 1.6); we shall discuss this later. The function v can be written as $v(t) = N(e^{-\delta t} + r(t))$ with

$$r(t) = -\frac{2\delta}{\pi}\int_{\epsilon}^{\infty}\frac{\cos\lambda t}{\lambda^2+\delta^2}\,d\lambda,$$

and $M(t) = r(t)^2 + 2e^{-\lambda t}r(t)$. We are interested, in fact, only in those values of the parameters which give $\delta/\epsilon \ll 1$ and $\delta t \ll 1$ (in the particular example discussed below, $\delta/\epsilon \lesssim 10^{-11}$ and $\delta t \lesssim 10^{-4}$), and therefore the last integral can be evaluated easily to be

$$r(t) = -\frac{2\delta}{\pi\epsilon}\,[\cos(\epsilon t) + \epsilon t\,\mathrm{si}(\epsilon t)] + O(\delta^2/\epsilon^2).$$

Consequently, the function M (up to the higher-order terms in δ/ϵ and δt) is given by

$$M(t) = -\frac{4\delta}{\pi\epsilon}\,[\cos(\epsilon t) + \epsilon t\,\mathrm{si}(\epsilon t)]\,; \tag{3.7}$$

properties of the integral sinus show that $M(\,.\,)$ is oscillating with decreasing amplitude, $\lim_{t\to\infty} M(t) = 0$ (cf. Figure 3), and obeys

$$|M(t)| \leqslant \frac{4\delta}{\pi\epsilon}\,. \tag{3.7a}$$

Besides the primary decay law, we have to specify the function L^{-1} which describes the density of bubbles on the track. This quantity is known to be proportional to the inverse squared velocity of the particle,

$$L = C\beta^2. \tag{3.8}$$

The value of the constant C depends on the concrete expertimental arrangement. For hydrogen chambers, one usually has $C^{-1} = (14^{+4}_{-6})\,\mathrm{cm}^{-1}$ so we choose for numerical analysis $C = 0.071$ cm. Dependence of β on x is determined by the energy loss of the particle going through the liquid hydrogen. Since our aim is nothing more than to estimate the order of magnitude of the effect, we may replace the exact formula (see Bricman *et al.* (1978)) by the simple approximation

$$-\frac{d\mathscr{E}}{dx} = D\beta^{-2}, \tag{3.9}$$

which allows us to find an analytic expression for $L(\,.\,)$. As mentioned above, we are going to present numerical results for the charged kaons which have

$$m = 493\ \mathrm{MeV}/c^2, \qquad \Gamma^{-1} = 1.24\times 10^{-8}\ \mathrm{sec}.$$

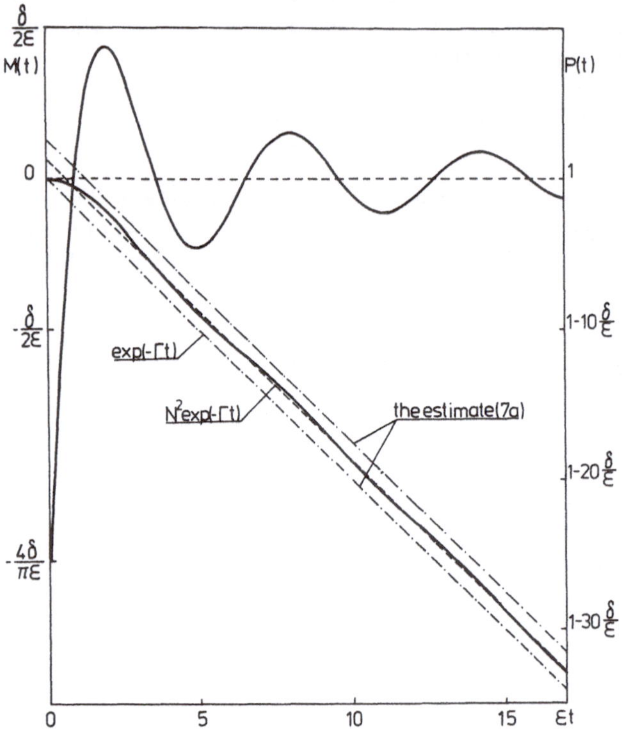

2.3. The function $M(\,.\,)$ according to (2.3.7) and the corresponding primary decay law for small times (provided $\delta/\epsilon \ll 1$).

In Table 1 we compare the tabulated values of the energy loss and range of the kaons with those obtained from formula (3.9) with D = 0.155 MeV/cm. Thus the approximation (3.9) means errors $\lesssim 20\%$ in the interval $0.2 \lesssim \beta \lesssim 0.9$; this is sufficient for the purposes of the estimate.

TABLE 2.1.

Momentum [MeV/c]	$-d\mathscr{E}/dx$ [MeV/cm]		Range [cm]	
	tab.	from (9)	tab.	from (9)
100	3.40	3.93	1.50	1.29
200	1.10	1.10	19.0	18.4
400	0.45	0.39	190	204
600	0.32	0.26	570	666
1000	0.25	0.19	1800	2230

Let us now find the functions $\beta(\,.\,)$ and $L(\,.\,)$. We express both sides of equation (3.9) in terms of the contraction factor γ, thus obtaining a first-order differential equation which can easily be integrated:

$$\gamma + \frac{1}{\gamma} = \gamma_0 + \frac{1}{\gamma_0} - \frac{Dx}{mc^2} \equiv g(p_0, x), \tag{3.10a}$$

where γ_0 is the contraction factor corresponding to the initial momentum p_0. This formula gives the expression for the range

$$x_r = \frac{mc^2}{D}(\gamma_0 + \gamma_0^{-1} - 2),$$

which we have used in Table 1. Further, it is not difficult to extract β from here: we obtain

$$L(x)^{-1} = \frac{1}{C\beta(x)^2} = \frac{1}{2C}\left\{1 + \frac{g(p_0, x)}{(g(p_0, x)^2 - 4)^{1/2}}\right\}. \tag{3.10b}$$

We shall now evaluate the correction term (3.5c). The function L which has to be inserted into this formula is given by (3.10b), with the exception of the points near to the track end (where $\beta \lesssim 0.2$; Table 1 shows that this concerns the last 1–2 cm of track). Further, $\Gamma\Delta t' = \Gamma L/c\beta\gamma = \Gamma C\beta/c\gamma$, so substitution of the numerical value gives

$$\Gamma\Delta t'(x) \approx 2\beta(x)\,[1 - \beta(x)^2]^{1/2} \times 10^{-4}.$$

Thus, $\Gamma\Delta t' \approx 10^{-4}$ in the considered region of β, and since $\epsilon/\delta \gtrsim 10^{11}$, we have $\epsilon\Delta t' \gtrsim 10^7$. In such a case, the second term in the curly bracket in (3.5c) can be neglected comparing to $2 \ln N \approx 4\delta/\pi\epsilon$, and the relation (3.5c) simplifies to the form

$$K(x) = \frac{4\delta}{\pi\epsilon}\int_0^x \frac{dy}{L(y)}. \tag{3.11a}$$

Now substituting from (3.10b) to (3.11a), and performing the integration, we obtain

$$K(x) = \frac{2}{\pi C}\frac{\delta}{\epsilon}\left\{x + \frac{mc^2}{D}\left[(g(p_0, 0)^2 - 4)^{1/2} - (g(p_0, x)^2 - 4)^{1/2}\right]\right\}. \tag{3.11b}$$

This correction terms for several values of the initial momentum is plotted on Figure 4. Figure 5 shows the full measured decay law; the curves for $F(\,.\,)$ are obtained by a numerical integration.

One is interested mainly in the kind of experiment in which the corrections might be measured, at least in principle. It is certainly hopeless to try to fulfil this task by a direct measurement, because we do not know $F(\,.\,)$ with sufficient accuracy. One must therefore compare the measured decay laws obtained under different measuring conditions. The conclusions of the preceding section do not help here. Of course, the corrections depend on the measurement frequency, in

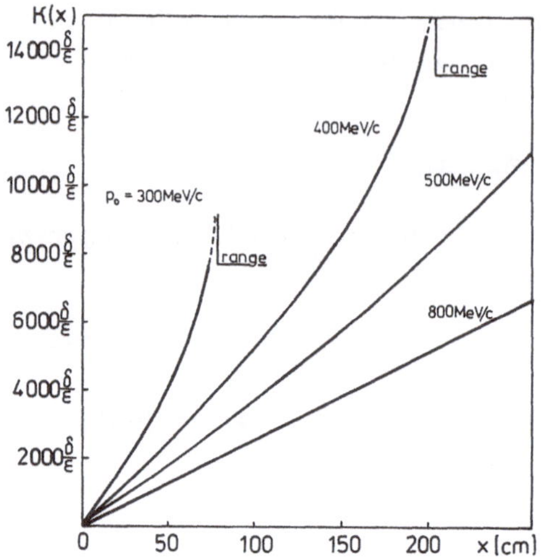

2.4. Corrections to the measured decay law due to repeated measurements as determined by (2.3.11b).

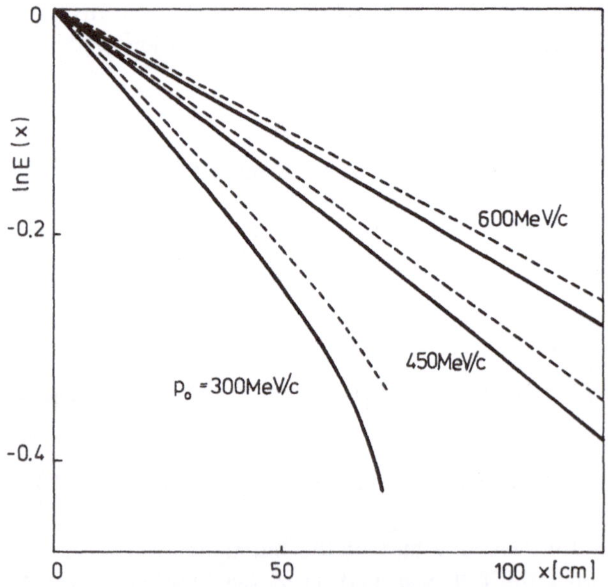

2.5. The measured decay law given by (2.3.5) and (2.5.11). Corrections corresponding to a physically reasonable choice cannot be distinguished in this scale; for comparison the dashed curves show the function $E(\,.\,)$ with $\delta/\epsilon = 7 \times 10^{-6}$.

particular, through the overall parameter C in (3.11b), but variation of this quantity is strictly limited.

On the other hand, the corrections depend on primary-decay-law non-exponentialities through the cut-off parameter ϵ, which can be fitted experimentally in the process of preparation of the kaon beam. The same idea as in Section 1.6 can be used: before entering the chamber, the beam passes through an energy filter which has a 'slit' 2ϵ wide. Since we are trying to estimate the order of magnitude of the effect only, we may ignore again the fact that a real energy filter would not cause a sharp truncation of the energy distribution.

The scheme of an appropriate experiment is sketched on Figure 6. The simplest way to realize the energy filter consists of using some mass-spectrometer-type device. We choose two planes in the chamber, perpendicular to the beam direction, and from the number of kaons that have crossed the first plane, we shall count the kaons (undecayed and unscattered) that will also cross the second. Since the cross-section of kaon scattering on the hydrogen nuclei is a slowly and smoothly varying function of energy, the ratio of scattered kaons is likely to be independent of ϵ and we can neglect them. Then the number $N(\epsilon, x)$ of tracks is proportional to $E(x)$ related to the particular value of ϵ. Thus, comparing the number of kaons that will survive up to x for two different values of ϵ, we can suppress the dependence on $F(\,.\,)$. In particular, for $\epsilon_1 \ll \epsilon_2$ we get

$$\frac{N(\epsilon_1, x) - N(\epsilon_2, x)}{N(\epsilon_2, x)} \approx K(\epsilon_1, x).$$

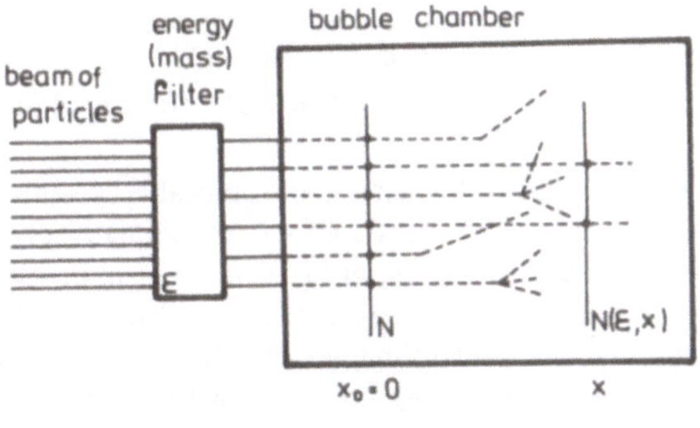

2.6. Scheme of an experiment in which the corrections can be measured.

Let us now ask whether the corrections could be actually observed in the described experiment. Figure 4 shows that $K(x)$ can reach values about $10^4 \delta/\epsilon$ in the existing hydrogen chambers. Thus the lowest possible value of energy resolution (characterized by ϵ) is decisive. The kaon mass is known up to 10^{-2} MeV, the best mass spectrometers have resolution of the order of keV. Now $\Gamma = 5.32 \times 10^{-14}$ MeV so that $\delta/\epsilon \lesssim 10^{-11}$. Consequently, one has to expect $K(x) \lesssim 10^{-7}$ in the particular case under consideration, and such an accuracy can hardly be reached experimentally.

Of course, kaons are not the only possibility. Any attempt to maximize $K(x)$, however, must take two contradictory requirements into account. First, the considered particle has to live long enough in order to undergo sufficiently many measurements during its lifetime. Besides, in view of Remark 1.4, we are limited to the cases when the measurements (based on the hydrogen-atom ionization) can be interpreted as instantaneous processes. On the other hand, the particle must decay fast enough, otherwise δ/ϵ would be too small. Somewhat larger corrections may be expected when we do not restrict ourselves to the unstable elementary particles, and replace the kaons by suitable unstable nuclei, because in this case the ionization (identified with the density of measurements) is increased by the factor Z^2 as can be seen from the Bethe–Bloch formula (see Bricman *et al.* (1978)). However, even in this case an experimental check seems to be difficult.

Concluding the above considerations, we can say that the observation of an unstable particle in the bubble chamber may change the measured lifetime (so that it increases with the measurement frequency), but *the effect is very small under realistic physical conditions*. These results are particularly interesting in connection with the considerations of Section 1.6, because they illustrate well the influence of energy filtering on the decay.

2.4. Limit of Continual Observation and the 'Zeno's Paradox'

"We put the kettle on to boil, up in the nose of the boat, and went down to the stern and pretended to take no notice of it . . . That is the only way to get a kettle to boil up the river. If it sees that you are waiting for it and are anxious, it will never even sing."

J. K. JEROME, *Three Men in a Boat*

In the preceding sections, we have studied the behaviour of unstable systems which are exposed to some sequences of measurements, either intentionally organized or randomly distributed. In every case, these measuring sequences can be varied by an experimentalist within certain limits. We have seen that the observed behaviour of an unstable system depends (at least, in principle) on the way in which it is measured; in particular, that the measured decay law varies with the measurement frequency, though possibly insignificantly.

The present section is devoted to the important limiting case, when the density of measurements tends to infinity. Such a situation intuitively corresponds to continual observation of the unstable system under consideration which is certainly worthy of study. It appears, however, that this limit of continual observation leads to counterintuitive conclusions, namely that a perpetually observed unstable system would not decay at all. This effect is sometimes called the *quantum Zeno's paradox*, for it recalls the well-known Zeno's aporia which denies the motion of a flying arrow. But all the same, the arrows fly (and pierce) and the unstable particles decay even if they are surrounded by an array of switched-on detectors. Thus the new paradox must be explained, as was the old one. This problem will be discussed below, but first let us deduce the paradoxical results.

We start with the following simple observation: when the particle is measured at t/n, $2t/n$, . . . , t, then the measured decay law according to (2.18b) equals

$P(t/n)^n$. Now if $\dot{P}(0+) = 0$, then $\lim_{n\to\infty} P(t/n)^n = 1$ for all t positive. This assertion remains in essence valid for randomly distributed measurements in a homogeneous chamber (see also Problem 6).

PROPOSITION 2.4.1. *Let there exist $b > 0$ such that the primary decay law $P_\rho(\,.\,)$ is twice continuously differentiable in $[0, b]$ with the initial decay rate $\dot{P}_\rho(+0) = \Gamma_0 \geqslant 0$. Then the measured decay rate $\gamma(\lambda)$ defined by* (2.19) *and* (2.26) *obeys* $\lim_{\lambda\to\infty} \gamma(\lambda) \leqslant \Gamma_0$; *in particular, this limit is zero if* $\Gamma_0 = 0$.

Proof. We shall use the notation introduced in Section 2. Suppose that two primary decay laws P_1, P_2 obey $P_1(t) \geqslant P_2(t)$ for all $t \geqslant 0$, then $\tilde{P}_1(s) \geqslant \tilde{P}_2(s)$ holds for all $s \geqslant 0$. The functions $\tilde{P}_i(\,.\,)$ are decreasing in $\mathbb{R}_+$ so their inverses obey $R_1(x) \geqslant R_2(x)$ for each $x \in (0, T_2]$. The relation (2.26) then gives

$$\gamma_1(\lambda) \leqslant \gamma_2(\lambda), \qquad \lambda > 0. \tag{4.1a}$$

Notice that we can also obtain this inequality in another way: the relation between the primary decay laws implies, in view of (1.6b), that $E_2(\,.\,)$ is majorized by $E_1(\,.\,)$, and the asymptotic behaviour of these functions established by Theorem 2.6 again yields (4.1a).

Similarly, we obtain the following estimate: let there exist $s_0 \geqslant 0$ and a decreasing function $g : [s_0, \infty) \to (0, g(s_0)]$ such that $\tilde{P}(s) \geqslant g(s)$ for all $s \geqslant s_0$; then, repeating the above considerations, we get

$$\gamma(\lambda) \leqslant \lambda - g^{-1}\left(\frac{1}{\lambda}\right), \qquad \lambda > \frac{1}{g(s_0)}, \tag{4.1b}$$

where $g^{-1} : (0, g(s_0)] \to [s_0, \infty)$ is the inverse function to g.

Now the smoothness assumptions for small times together with Proposition 1.2.3 imply the existence of a positive c such that $P(t) \geqslant P_2(t)$, where

$$P_2(t) = \begin{cases} 1 - \Gamma_0 t - ct^2 & \ldots \quad 0 \leqslant t \leqslant t_0 = \dfrac{1}{2c}[(\Gamma_0^2 + 4c)^{1/2} - \Gamma_0] \\ 0 & \ldots \quad t \geqslant t_0 \end{cases}$$

The function $\tilde{P}_2$ is easily calculated to be $\tilde{P}_2(s) = s^{-1} - \Gamma_0 s^{-2} - 2cs^{-3} + (\Gamma_0 + 2ct_0 + 2cs^{-1})s^{-2}\exp(-st_0)$, so one has

$$\tilde{P}_2(s) > g(s) := \frac{1}{s} - \frac{\Gamma_0}{s^2} - \frac{2c}{s^3}.$$

For all s large enough, $g(\,.\,)$ is decreasing, so the inverse function exists. If we set $g^{-1}(x) = z(x)/x$, then $z(\,.\,)$ is obtained by solving the cubic equation $z^3 - z^2 + \Gamma_0 xz + 2cx^2 = 0$. This has three real roots for small x and we have to choose the one which tends to 1 as $x \to 0$. It equals

$$z(x) = \tfrac{1}{3} + \tfrac{2}{3}(1 - 3\Gamma_0 x)^{1/2} \cos\left\{\tfrac{1}{3}\arccos\left(\frac{1 - \frac{9}{2}\Gamma_0 x - 27cx^2}{(1 - 3\Gamma_0 x)^{3/2}}\right)\right\}$$

so the relation (4.1b) gives $\gamma(\lambda) \leqslant G(\lambda^{-1})$ with

$$G(x) \doteq \frac{2}{3x}\left[1-(1-3\Gamma_0 x)^{1/2}\cos\left\{\tfrac{1}{3}\arccos\left(\frac{1-\frac{9}{2}\Gamma_0 x-27cx^2}{(1-3\Gamma_0 x)^{3/2}}\right)\right\}\right].$$

One finds easily the relation $\lim_{x\to 0+} G(x) = \Gamma_0$ which yields $0 \leqslant \lim_{\lambda\to\infty}\gamma(\lambda) \leqslant \Gamma_0$, i.e. the desired result. ∎

Combining this result with Theorem 1.3.1 and Proposition 1.6.1, we see that for a very large set of states of the unstable system, the *decay disappears in the limit of continual observation*. In particular, for pure states this means that the corresponding vectors stay confined within $\mathscr{H}_u$. It is even natural to expect their evolution to be governed by a unitary group on $\mathscr{H}_u$. It can be illustrated as follows: if the non-decay probability is measured repeatedly with positive results at 0, t/n, $2t/n, \dots, t$, then the transition from the initial to the final state vector is mediated by the operator $(E_u U_{t/n} E_u)^n$. Now the following assertion is valid:

PROPOSITION 2.4.2. *Let* $\mathscr{H}_u = E_u\mathscr{H} \subset D(H)$, *then*

$$\operatorname*{s-lim}_{n\to\infty} (E_u U_{t/n} E_u)^n = \exp(-iE_u H E_u t)E_u \tag{4.2}$$

holds for each $t \geqslant 0$ (*in fact, for* $t < 0$ *as well*).

Proof. The function $F : \mathbb{R} \to \mathscr{B}(\mathscr{H}_u)$, $F(t) = V_t$, is strongly continuous and $F(0) = I_u$. Further $F'(0)\varphi = -iE_u H E_u \varphi$ for each $\varphi \in \mathscr{H}_u$ in view of $\mathscr{H}_u \subset D(H)$, so $iF'(0)$ is symmetric and defined everywhere in $\mathscr{H}_u$. Consequently, $iF'(0)$ is Hermitean due to the Hellinger–Toeplitz theorem. Then it generates a unitary group, and the result follows from the second of the Chernoff theorems quoted in (A.8). ∎

Now we are going to formulate the main result of this section. First we notice that, in general, $\mathscr{H}_u$ is not contained in the domain of H and the above limit need not exist. Moreover, it is possible that the limit exists but it does not give a unitary group:

EXAMPLE 2.4.3. Consider the situation described in Example 1.2.5. The reduced propagator $\{V_t\}$ in this case clearly has the semigroup property, $E_u U_t E_u U_s E_u = E_u U_{t+s} E_u$ for $t, s \geqslant 0$, so

$$\operatorname*{s-lim}_{n\to\infty} (E_u U_{t/n} E_u)^n = E_u U_t E_u .$$

Hence the limit exists, but its restriction to $\mathscr{H}_u$, which coincides with V_t, is in no case unitary.

Thus we have to look for the conditions under which the assertion of Proposition 4.2 would be essentially preserved. Before doing that, we reformulate the problem in a somewhat more general way. Considerations of the preceding sections suggest that it is too restrictive to work with the equidistantly distributed sequences of measurements only. Besides, a more general formulation will be useful for our future purposes. Thus we consider all *partitions* $\sigma = \{\tau_j : 0 = \tau_0 < \tau_1 < \cdots < \tau_n = t\}$ of the interval $[0, t]$; we abbreviate $\delta_j = \tau_{j+1} - \tau_j$ and $\delta(\sigma) = \max_{0 \leqslant j \leqslant n-1} \delta_j$. Then we can define

$$U(t, 0; E_u, \sigma) = E_u \exp(-iH\delta_{n-1}) E_u \exp(-iH\delta_{n-2}) \ldots \exp(-iH\delta_0) E_u \tag{4.3a}$$

as the evolution operator of the system which has suffered at time instants $0, \tau_1, \tau_2, \ldots, t$ the non-decay measurements, each of them with a positive result. In the limit of continual observation, the evolution is governed by

$$U(t, 0; E_u) := \operatorname*{s\text{-}lim}_{\delta(\sigma) \to 0} U(t, 0; E_u, \sigma). \tag{4.3b}$$

The limit has the obvious meaning: to every $\epsilon > 0$ and $\varphi \in \mathcal{H}$, there is some $\eta > 0$ such that $\|U(t, 0; E_u)\varphi - U(t, 0; E_u, \sigma)\varphi\| < \epsilon$ holds for each partition σ with $\delta(\sigma) < \eta$. The definition (4.3) can also, if necessary, be extended to negative t, i.e. to give the operator which mediates the transition from the final to the initial state. In such a case, of course, the factors $\exp(-iH\delta_j)$ in (4.3a) have to be replaced by $\exp(iH\delta_j)$, where the δ_j's refer to partitions of the interval $[0, -t]$.

THEOREM 2.4.4. *Assume that $U(t, 0; E_u)$ exists for all $t > 0$ and the Hamiltonian H is semibounded. Further, let an antiunitary operator Θ on $\mathcal{H}$ exist such that*

$$\Theta E_u \Theta^{-1} = E_u \text{ and } \Theta U_t \Theta^{-1} = U_{-t} \qquad \text{for all} \quad t \in \mathbb{R}. \tag{4.4}$$

Then there is a projection $P \leqslant E_u$ and a semibounded self-adjoint operator A such that $A = PAP$ and

$$U(t, 0; E_u) = e^{-iAt} P \tag{4.5}$$

for all non-zero t. Furthermore, Ran $P = \bar{D}$, *where $D = \mathcal{H}_u \cap Q(H)$, and A is uniquely determined by its restriction to $P\mathcal{H}$, which is associated with the closed densely defined form $q : D \to \mathbb{R}$, $q(\varphi) = \|(H + \gamma)^{1/2}\varphi\|^2 - \gamma\|\varphi\|^2$ for some $\gamma \geqslant -\inf \delta(H)$.*

REMARK 2.4.5. (a) When realistic physical situations are considered, then the assumption about Θ can be reduced to the question of invariance of the theory with respect to the time inversion (or CPT-operation). Moreover, this assumption is not always necessary, which is clear from the following remark.

(b) Example 4.3 shows that the semiboundedness of H is substantial. In this example, $\sigma(H) = \mathbb{R}$ and the conclusions of the theorem are violated. In fact, the assumption about the existence of Θ is also violated in this case, but it is employed in the proof only to ensure the existence of $U(t, 0; E_u)$ for t negative, which can be checked trivially here. On the other hand, the semiboundedness of H will play a central role.

(c) It is natural to set $U(0, 0; E_u) = E_u$, then $U(\,.\,, 0; E_u)$ might be discontinuous at $t = 0$ (see Remark 4.9 below). Such a possibility is avoided if one accepts the additional assumption $\text{s-lim}_{t \to 0+} U(t, 0; E_u) = E_u$ (Misra and Sudarshan, 1977), or, equivalently, $D(H)$ dense in $\mathscr{H}_u$ (Exner, 1982a).

A brief inspection reveals a close similarity between the problem at hand and the Kato product formula (cf. (A.9)). However, the latter applies to the self-adjoint semigroups only, so one has to combine it with some sort of analytical continuation to get the necessary result. In order to realize this idea, we shall first prove the following three lemmas:

LEMMA 2.4.6. *Let H and E be a positive self-adjoint operator and a projection on $\mathscr{H}$, respectively, and $F_n(z) := (E \exp(-iHz)E)^n$. For each n, the function $F_n(\,.\,)$ is defined and strongly continuous in $\{z : \operatorname{Im} z \leqslant 0\}$; further, it is strongly analytic in the open lower halfplane of z. The following integral relations hold:*

$$\frac{(z-i)^2}{2\pi i} \int_{\mathbb{R}} \frac{F_n(t)}{(t-i)^2(z-t)}\, dt = F_n(z), \quad \operatorname{Im} z < 0, \tag{4.6a}$$

$$\frac{1}{2\pi i} \int_{\mathbb{R}} \frac{F_n(t)}{(t-i)^2(z-t)}\, dt = 0, \quad \operatorname{Im} z > 0. \tag{4.6b}$$

Proof. Consider first $F_1(z) = E e^{-iHz} E$. Its existence and continuity with respect to z follows from the functional-calculus rules for self-adjoint operators, which further imply that the functions $z \mapsto (\varphi, F_1(z)\psi)$ are differentiable at each point z with $\operatorname{Im} z < 0$. It means that $F_1(\,.\,)$ is weakly analytic, hence also strongly analytic (Reed and Simon, 1972, theorem VI.4) in the open lower halfplane. We have $F_n(z) = F_1(z/n)^n$ and as these operators are uniformly bounded, one can easily check the strong continuity of $F_n(\,.\,)$ and the existence of a strong derivative

$$F_n'(z) = \frac{1}{n} \sum_{k=0}^{n-1} F_1(z/n)^k F_1'(z/n) F_1(z/n)^{n-k-1},$$

i.e. the strong analycity of $F_n(\,.\,)$ in $\{z : \operatorname{Im} z < 0\}$.

Now, one has to apply the Cauchy formula (for vector-valued functions, cf. Rudin (1973, theorem 3.31)) to the function $z \mapsto F_n(z)(z-i)^{-2}$, which is also analytic in $\{z : \operatorname{Im} z < 0\}$. For any z with $\operatorname{Im} z < 0$, we have

$$\frac{F_n(z)}{(z-i)^2} = -\frac{1}{2\pi i} \oint_C \frac{F_n(\zeta)}{(\zeta - i)^2(\zeta - z)}\, d\zeta,$$

where C is an arbitrary simple closed curve, which lies fully in the open lower halfplane and encircles the point z anticlockwise. In particular, we choose C which consists of the line segment connecting the points $\pm R - i\epsilon$ and of the semicircular arc in the lower halfplane. Since the norm of the integrated function falls quickly enough with $|\zeta|$, one can perform the limit $R \to \infty$, thus obtaining

$$F_n(z) = -\frac{(z-i)^2}{2\pi i} \int_{\mathbb{R}} \frac{F_n(t - i\epsilon)}{(t - i - i\epsilon)^2 (t - i\epsilon - z)}\, dt, \quad \operatorname{Im} z < -\epsilon < 0.$$

As shown above, $\text{s-lim}_{\epsilon \to 0+} F_n(t - i\epsilon) = F_n(t)$. Further, we choose a fixed positive $\delta < -\text{Im } z$, then for all $\epsilon \in (0, \delta)$ we have the estimate

$$\left\| \frac{F_n(t - i\epsilon)}{(t - i - i\epsilon)^2 (t - i\epsilon - z)} \right\| < -(1 + t^2)^{-1} (\delta + \text{Im } z)^{-1}. \tag{4.7}$$

This makes it possible to use the dominated-convergence theorem (for vector-valued functions, cf. Dunford and Schwartz (1958, cor. III.5.16)), which yields the integral representation (4.6a). The formula (4.6b) is proved in the same way. ∎

LEMMA 2.4.7. *In the same notation as above, suppose that* $G(t) = \text{s-lim}_{n \to \infty} F_n(t)$ *exists for all real* t. *Then*

(a) $G(z) := \text{s-lim}_{n \to \infty} F_n(z)$ *exists if* $\text{Im } z < 0$;
(b) *the function* $G(\,.\,)$ *is strongly analytic in the open lower halfplane and fulfils there* $G(z_1) G(z_2) = G(z_1 + z_2)$;
(c) $G(z) = e^{-iAz} P$ *for* $\text{Im } z < 0$, *where* A, P *are the operators described in Theorem* 4.4 *with* $\gamma = 0$.

Notice that the statement of this lemma can be expressed concisely as follows: $\{G(z) \upharpoonright \bar{D} : \text{Im } z < 0\}$ is a holomorphic semigroup with the angle $\pi/2$ – cf. Reed and Simon (1975, §X.8).

Proof. We start from the integral representation (4.6a). The limit $\text{s-lim}_{n \to \infty} F_n(t)$ exists for all t due to the assumption; further, the estimate (4.7) with $\epsilon = \delta = 0$ makes it possible to use the dominated-convergence theorem, which shows that for $\text{Im } z < 0$, the strong limit of $\{F_n(z)\}$ exists and equals

$$G(z) = \frac{(z - i)^2}{2\pi i} \int_{\mathbb{R}} \frac{G(t)}{(t - i)^2 (z - t)} \, dt, \quad \text{Im } z < 0. \tag{4.8}$$

This proves (a). All the functions $F_n(\,.\,)$ are strongly analytic in the open lower halfplane so $G(\,.\,)$ is also analytic there according to the Vitali theorem (Hille and Phillips, 1957, theorem 3.18.1). Now $G(-it)$ can be expressed by means of the Kato product formula (A.9) to be

$$G(-it) \equiv \text{s-lim}_{n \to \infty} (E\, e^{-Ht/n} E)^n = e^{-tA} P$$

for each t positive. The same argument as in the preceding proof shows that $z \mapsto e^{-iAz} P$ is strongly analytic in $\{z : \text{Im } z < 0\}$. It means that we have two analytic vector-valued functions, which coincide on a halfline, hence $G(z) = e^{-iAz} P$ for all z with $\text{Im } z < 0$ (Hille and Phillips, 1957, theorem 3.11.5). Finally, the functional-calculus rules give $\exp(-iAz_1) P \exp(-iAz_2) P = \exp(-iA(z_1 + z_2)) P$, i.e. the remaining part of assertion (b). ∎

LEMMA 2.4.8. *In the same notation as above,* $\text{s-lim}_{\eta \to 0+} G(s - i\eta) = G(s)$ *holds for all non-zero* $s \in \mathbb{R}$.

Proof. For $z = s - i\eta$, the relation (4.8) gives

$$G(s - i\eta) = \frac{(s - i - i\eta)^2}{2\pi i} \int_{\mathbb{R}} \frac{G(t)}{(t - i)^2 (s - i\eta - t)} \, dt, \quad \eta > 0.$$

On the other hand, (4.6b), together with the dominated-convergence theorem, yield

$$0 = \frac{(s - i - i\eta)^2}{2\pi i} \int_{\mathbb{R}} \frac{G(t)}{(t - i)^2 (s + i\eta - t)} \, dt, \quad \eta > 0.$$

Subtracting these two relations, we get the following integral representation for $G(s - i\eta)$:

$$G(s - i\eta) = \frac{(s - i - i\eta)^2}{\pi} \int_{\mathbb{R}} \frac{G(t)}{(t - i)^2} \frac{\eta}{(t - s)^2 + \eta^2} \, dt, \quad \eta > 0.$$

Now one has to employ the result of Chernoff (1974) quoted in the notes to Section 1.3: since the families $\{F_1(\pm t) : t > 0\}$ are strongly continuous and $G(t)$ is assumed to exist for all real t, it must be strongly continuous too, with the possible exception of the point $t = 0$. Using the dominated-convergence theorem once again, we get

$$\lim_{\eta \to 0+} G(s - i\eta)\varphi = (s - i)^2 \lim_{\eta \to 0+} \frac{1}{\pi} \int_{\mathbb{R}} \frac{G(s + x\eta)\varphi}{(s + x\eta - i)^2} \frac{dx}{1 + x^2} = G(s)\varphi$$

for all $\varphi \in \mathscr{H}$ and $s \neq 0$, i.e. the desired result. ∎

Proof of Theorem 4.4. Notice first that we may limit ourselves to positive H only. In the opposite case, we choose $\gamma > 0$ so that $\tilde{H} = H + \gamma \geqslant 0$, and obviously $\tilde{U}(t, 0; E_u) = e^{-i\gamma t} U(t, 0; E_u)$. The existence of $U(t, 0; E_u)$ for t negative follows from the assumption (4.4):

$$U(-t, 0; E_u) \equiv \operatorname*{s-lim}_{\delta(\sigma) \to 0} E_u \exp(iH\delta_{n-1})E_u \ldots E_u \exp(iH\delta_0)E_u$$

$$= \operatorname*{s-lim}_{\delta(\sigma) \to 0} \Theta U(t, 0; E_u, \sigma)\Theta^{-1} = \Theta U(t, 0; E_u)\Theta^{-1}.$$

Since $U(t, 0; E_u)$ exists, $G(t) = \text{s-lim}_{n \to \infty} (E\, e^{-iHt/n} E)^n$ exists *a fortiori* for all t (as the limit with respect to special partitions only). Combining now the assertions of Lemmas 4.7(c) and 4.8, we obtain

$$G(t) = \operatorname*{s-lim}_{\eta \to 0+} G(t - i\eta) = \operatorname*{s-lim}_{\eta \to 0+} e^{-iA(t - i\eta)} P = e^{-iAt} P$$

for all non-zero t. ∎

REMARK 2.4.9. The fact that the projection P may be strictly less than E_u (if only $D(H)$ is not dense in $\mathcal{H}_u$) seems a little curious. Remember, however, that the existence of $U(\,.\,,0;E_u)$ represents one of the *assumptions* of the theorem. Consider the following *example*: dim $\mathcal{H}_u = 1$, and $\mathcal{H}_u$ is spanned by a vector ψ such that the function $g = g_{\psi,\psi}$ (cf. Corollary 1.5.9) is given by $g(\lambda) = (2/\pi)(1+\lambda^2)^{-1}\Theta(\lambda)$. Obviously, $\psi \notin Q(H)$ so $P = 0$. Further, $(\psi, (E_u U_{t/n} E_u)^n \psi) = v(t/n)^n$, where

$$v(t) = (\psi, U_t\psi) = \int_0^\infty e^{-i\lambda t} g(\lambda)\, d\lambda.$$

This function can be calculated to be $v(t) = e^{-t} - (i/\pi)(e^{-t}\,\mathrm{Ei}(t) - e^t\,\mathrm{Ei}(-t))$; using the formula

$$\mathrm{Ei}(\pm x) = \gamma + \ln x + \sum_{k=1}^{\infty} \frac{(\pm x)^k}{k \cdot k!}, \qquad x > 0$$

(Gradshtein and Ryzhik, 1971, 8.214), one finds readily that it behaves like

$$v(t) = e^{-t}\left[1 - \frac{2i}{\pi}(t \ln t + O(t))\right]$$

for small t. Now it is an easy exercise to show that $\lim_{n\to\infty} \arg(v(t/n)^n)$ is dominated by term $\lim_{n\to\infty}(2t/\pi)\ln(n/t)$, so it does not exist. Consequently, the operators $U(t, 0; E_u)$, $t > 0$, do not exist in the present case. The question, 'under the assumptions of the theorem, can $U(\,.\,,0;E_u)$ exist such that the corresponding P is strictly less than E_u?' is left open.

It is interesting to notice that Theorem 4.4 can be used for a short proof of an assertion which represents a special case of Theorem 1.5.1.

COROLLARY 2.4.10. *Let H be a positive self-adjoint operator on $\mathcal{H}$, and assume that $\{V_t = \mathrm{pr}\,\exp(-iHt) : t \geq 0\}$ corresponding to a subspace $\mathcal{H}_u \subset \mathcal{H}$ is a semigroup, then $U_t = \exp(-iHt)$ is reduced by $\mathcal{H}_u$ for all $t \in \mathbb{R}$. In particular, if $\{V_t : t \geq 0\}$ is a reduced propagator with the semigroup property, then H is not semibounded.*

Proof. The semigroup property gives $(V_{t/n})^n = V_t$ so $G(t) = \text{s-lim}_{n\to\infty}(E_u U_{t/n} E_u)^n = E_u U_t E_u$. Similarly, $U(t, 0; E_u)$ exists for all $t \geq 0$, and in view of (1.2.2) for t negative as well. Thus the assumptions of Theorem 4.4 are fulfilled with the exception of (4.4), but the latter was used only to ensure the existence of $U(t, 0; E_u)$ for $t < 0$. According to the theorem, $E_u U_t E_u = e^{-iAt} P$ for all non-zero t; further, the l.h.s. of the last equality is strongly continuous at $t = 0$, so $P = E_u$. Consequently, $\{V_t : t \in \mathbb{R}\}$ is a unitary group on $\mathcal{H}_u$, and the result follows from the equivalence of the assertions (i), (iii) from Section 1.2. ∎

Let us now turn to the real meaning of the results described above. Should we conclude that some of the basic principles of the quantum theory are wrong, or that this theory cannot pretend for some reason to yield a description of systems under continual observation? Or have we simply neglected some natural physical restriction which may not be overlooked?

We have already mentioned (in Remark 1.4) that the considerations of the present chapter carry a certain amount of idealization contained in the assumption that the successively performed measurements are instantaneous events. We know that the actual duration of the measuring interaction is finite; moreover, that it is bounded to a definite time region peculiar for electromagnetic interactions on which nearly all measuring instruments rely. Let us stress that here we have in mind the duration of the interaction between the unstable system and a microscopic part of the apparatus (e.g. ionization of an atom), and not the registration time, which is usually even much longer. Now it is clear that the formalism elaborated in this chapter may be used legitimately only if the mean interval between the neighbouring measurements is much longer than the measuring time. In this sense, the evolution operator $U(t, 0; E_u)$ defined by (4.3b) *lacks an operational meaning*, and consequently, the 'Zeno's paradox' in its pure form is avoided.

One can, however, formulate a more general question. The above argument does not exclude the possibility that an unstable system and an apparatus could exist such that the result of repeated measurements would be near to the 'Zeno's limit', i.e. that the escape from $\mathcal{H}_u$ characterized by the norm decrease would be small for a very long time. In particular, one may ask whether some cases exist in which the mean life of an unstable system is lengthened substantially by virtue of repeated measurements. The model example treated in the preceding section shows that in the bubble-chamber experiments the lifetime can be actually affected, but that its increase is minimal and can hardly be expected to be distinguished experimentally.

In this connection, it is useful to comment on some limits for application of the considered formalism. It is well suited to the measuring processes in the track-monitoring devices (bubble, cloud, spark or streamer chambers, photoemulsion detectors, etc.) if the function describing the density of measurements is chosen properly. On the other hand, consider the situation when a sample of a radioactive material is prepared at $t = 0$ and then left until some later time, when we determine by radiochemical methods the fraction of the original material preserved and/or the amount of the decay products (the time interval involved here is much longer than in the 'tracking' methods, but otherwise it varies from seconds in some transuranium experiments to thousands of years in the C^{14}-archaeology). Some authors claim that this case fits into the above scheme too, because the restructuration of the electronic shells, which follows after the (α- or β-)decay of the nucleus, constitutes itself an act of measurement (cf., e.g., Fonda *et al.* (1978)). Such a point of view is not acceptable. The scheme discussed above relies on the assumption that we have a *sequence of well-defined individual acts of measurement* and that we register (at least, in principle) *the result of each of them*. If this is not true, then it makes no sense to speak about the successively repeated reduction of the state vector. The radiochemical experiments clearly do not fit into this conception. If we consider a particular atom from the sample, then it is natural to describe it by a superposition of the 'decayed' and undecayed' states until the beginning of the chemical analysis when it must 'decide' whether it has decayed or not. On the other hand, if the system under consideration is the nucleus alone 'watched' by its electronic shells, then we should interpret it rather as the 'true' continual observation mentioned below.

The following is another often discussed situation: a detector ascertains perpetually whether a particle is present within some volume V (or, on the contrary, the particle is localized initially outside V and the detector measures its time of arrival – cf. Allcock (1969)). Here again the considered scheme must not be applied: the observation is actually performed continually during the evolution of the particle, but the switched-on detector is not equivalent to a well-defined sequence of instantaneous measurements. If we try to interpret its action through the limit of such sequences, we immediately obtain the 'Zeno's paradox' as we shall show later.

Does this mean that the continual inspection of a microsystem is impossible? Certainly not. Consider, for instance, the following simple example: a single, free neutron is placed into a switched-on detector of charged particles (scintillation counter, proportional chamber, etc.), and we ascertain the instant when the signal occurs. As in the case of the perpetual position measurement, one cannot view the situation as the limiting case of successively performed instantaneous measurements. A consistent description should start from analysis of the *quantum* system consisting of the observed object (the neutron + the $(pe\bar{\nu})$ states resulting from the decay) coupled to the many-body system of the detecting-medium atoms. The essential dynamical properties of such a system are known, so one can in principle determine its time evolution – in particular, to find a probability that a medium will contain ionized atoms at a given time. In this way, one might hope to find (a probability density of) the decay time for the continually observed neutron (we take no account of the great practical difficulties connected with such a programme, or the fact that from the practitioner's point of view the result would be – similar to the primary decay law – indistinguishable from the exponential function with the well-known decay rate). At the same time, a more important problem about the mechanism of reduction (what does happen with the wave function at the instant when the counter has clicked?) remains unanswered. However, this goes beyond the scope of the present discussion.

Concluding these considerations, we can say that the 'Zeno's paradox' is likely to cause no interpretation troubles to the quantum theory. If the scheme elaborated here applies – i.e. if we have a sequence of well-defined measuring events which are interpretable as instantaneous ones and their outcomes can be registered – then the effect of repeated measurements for all we know is even too small to be observed. On the other hand, the cases where a true continual observation occurs have to be treated in a different way, so the paradox is avoided again.

At the same time, Theorem 4.4 and the related assertions proved in this section represent interesting mathematical results. Moreover, we shall see later that some of them may be used for other purposes. Before concluding the section, we shall return therefore to the consequences of the main result for the 'continual position measurement'.

EXAMPLE 2.4.11. Assume that the dynamics of the considered system is governed by a Schrödinger operator $H = H_0 + V$, where $H_0 = -\frac{1}{2}\Delta$ on $\mathcal{H} = L^2(\mathbb{R}^d)$ and V is a bounded (real-valued) potential, $V : (V\psi)(x) = v(x)\psi(x)$ with $v \in L^\infty(\mathbb{R}^d)$. Further, let M be an open region in $\mathbb{R}^d$ (consisting of a finite or infinite number of connected components). We denote by E_M the projection on $L^2(M) \subset \mathcal{H}$; further, V_M is the operator or multiplication by $v_M \equiv v \upharpoonright M$ on $L^2(M)$, and $H_0^M = -\frac{1}{2}\Delta_D^M$ corresponds to the Dirichlet Laplacian of the region M (Reed and Simon, 1978, §XIII.15). We shall prove the following assertion:

If $U(t, 0; E_M)$ exists for all $t > 0$, then it also exists for each real t and

$$U(t, 0; E_M) = \exp(-iH_M t)E_M \tag{4.9}$$

holds, where $H_M = H_0^M + V_M$.

Proof. The proof goes as follows: H is clearly semibounded; further, the complex conjugation on $L^2(\mathbb{R}^d)$ has the property (4.4) (which follows, e.g., from (1.2.7) and Trotter's product formula), so Theorem 4.4 may be applied. We have $Q(H) \supset D(H) \supset C_0^\infty(\mathbb{R}^d) \supset C_0^\infty(M)$, hence $Q(H)$ is dense in $L^2(M)$ and $P = E_M$. The operator A is associated with $q : D \to \mathbb{R}$, where $D = Q(H) \cap L^2(M)$ and $q(\varphi) = \|(H + \gamma)^{1/2}\varphi\|^2 - \gamma\|\varphi\|^2$ for some γ such that $H + \gamma \geqslant 0$. If $\varphi \in C_0^\infty(M)$, then

$$q(\varphi) = (\varphi, H\varphi) = \frac{1}{2}\int_M |(\nabla\varphi)(x)|^2\, dx + (\varphi, V_M\varphi). \tag{4.10}$$

The Dirichlet Laplacian is associated with a quadratic form which is the closure of the form defined on $C_0^\infty(M)$ by $\varphi \mapsto \int_M |(\nabla\varphi)(x)|^2\, dx$. Its form domain consists of all $\varphi \in L^2(M)$ such that $\nabla\varphi \in L^2(\mathbb{R}^d)$, where $\nabla\varphi$ means gradient in the weak (distributional) sense; however, this set is simply D. Further, we notice that the form sum of H_0^M and V_M coincides with their operator sum because V_M is bounded (which follows, e.g., from theorem X.17 in Reed and Simon (1975)). Then H_M is associated with the quadratic form q_1 defined on $D = Q(H_0^M)$, which is the closure of the form defined on $C_0^\infty(M)$ by the r.h.s. of (4.10). This means that $q = q_1$, and therefore $A = H_M$. ■

Notes to Chapter 2

SECTION 2.1. The quantum theory of measurement is far from being complete and well understood. An excellent exposition to the basic notions and problems can be found in Jauch (1969, chap. 11); for further formal development of the theory we can refer, e.g., to Davies (1976, chaps. 2–4) or Kholevo (1980), and for a discussion of the various physical aspects of quantum measurements, to d'Espagnat (1971). In these books we can also find an extensive bibliography. Fortunately, conceptual questions do not arise in the situation studied in this chapter, because we assume that the system undergoes a sequence of well-defined instantaneous measurements (*yes–no* experiments), though possibly a randomly distributed one. Of course, one must be aware that the scheme elaborated here has a limited applicability; when these physical limits are not respected, we come to paradoxical results, as discussed in Section 4.

• The study of microsystems interacting with their environment through successively repeated measurements goes back to Heisenberg (1927), who explained in this way how the notion of a 'path' could be understood within quantum mechanics. The influence of the environment on specific irreversible processes (sequential γ-decays of excited nuclei) was discussed by Coester (1954). Another model is that of Rau (1963), who considered repeated measurements performed on systems of coupled spins or harmonic oscillators which relax by means of interaction with a thermal bath. The first model-independent study of a decaying particle exposed to (equidistantly distributed) repeated measurements is due to Beskow and Nilsson (1967).

• Ekstein and Siegert (1971) were first to take into account the fact that in most of the real chambers the measurements were randomly distributed. They proved that the decay law measured in such circumstances had an exponential upper bound. The idea was pursued further and developed considerably by the Trieste group (see, especially, Degasperis *et al.* (1974), Ali *et al.* (1975), Fonda *et al.* (1973, 1978); the last paper contains a more complete bibliography). All these papers rest on the assumption of constant measurement frequency which is, however, too restrictive from the point of view of the real track-monitoring devices. The results of this section apply to rather general chamber structures; they come essentially from Exner (1976a), and from Dolejší and Exner (1977).

• Instead of equations (1.3) for the decay laws, one can write operator integral equations for the density matrix of the unstable system which suffer repeated measurements (Problem 7). For a constant measurement frequency, this was done by Ali *et al.* (1975), who simultaneously discussed the case when the unstable system suffered repreated measurements, but their results were not registered (cf. also Fonda *et al.* (1974), Ali and Ghirardi (1974)). Reformulation of these results for more general chamber structures is straightforward. Furthermore, expression of the problem in terms of the density matrices makes more transparent its relations to the theory of quantum stochastic processes (Davies, 1969; 1976, chap. 5).

SECTION 2.2. As for Theorem 2.4, the assumption $0 < P(t_p) < 1$ was used because it enabled us to formulate the proof conveniently, and at the same time it represented a very weak restriction from the viewpoint of the proposed physical application. However, it is not actually needed. As we have seen, positivity of γ is ensured if only one natural k exists such that $P(kt_p) < 1$. On the other hand, the assumption that $P(t_p)$ is non-zero was used in Lemma 2.5 to check the positivity of $\nu_0 - \nu_1$. In fact, the argument fails only if there is a natural $r > 1$ such that $P(kt_p) = 0$ unless $k = jr$, $j = 1, 2, \ldots$. Even in this highly pathological case, the conclusions of the lemma remain valid when applied to the sequence $\{f_{jr}\}_{j=1}$. The proof of Theorem 2.4 is inspired by the homogeneous-chamber case; it is taken from Exner (1977b).

• Theorem 2.6 was obtained essentially by Fonda *et al.* (1973), who started from the equation

$$E(t) = e^{-\lambda t} + \lambda \int_0^\infty e^{-\lambda s} P(s) E(t-s)\, ds.$$

It differs, of course, from equations (2.22) but their solutions coincide; it can be checked by term-by-term comparison of the Volterra iteration series. The disadvantage of the above equation is that it cannot be generalized to more complicated chamber structures, because its deduction employs implicitly the assumption $E(s, t) = E(0, t-s)$ – cf. Fonda *et al.* (1978, §8.3) which is valid in the homogeneous chambers only. The proof of Theorem 2.6 is based on the observation that equation (2.22b) is of the convolution type which can be solved by means of Laplace transformation; this idea goes back to Ekstein and Siegert (1971).

• The dependence of the measured decay law on the measurement frequency was first mentioned explicitly by Beskow and Nilsson (1967). Equation (2.19) for the measured decay rate $\gamma(\lambda)$ is due to Fonda *et al.* (1973); its discrete analogy (2.17) comes from Exner (1977b). Proposition 2.8 and the heuristic argument preceding Theorem 2.6 can be found in Exner (1977c).

SECTION 2.3. The model discussed here is due to Dolejší and Exner (1977). The treatment of Degasperis *et al.* (1974) yields qualitatively similar results provided the applicability conditions of the formalism are respected (see the comments on this point in Section 4). Let us also mention a simple argument due to Beskow and Nilsson (1967) which leads to the conclusion that the measured decay law can differ significantly from the primary one if the number of measurements can be matched with E_1/Γ, where the parameter E_1 characterizes modification of the Breit–Wigner distribution in the same manner as our ϵ. The considerations of the present section suggest that such a situation can hardly be achieved.

• The energetic filtering as used here (and particularly in Section 1.6) is, of course, idealized. In fact, one can never realize the corresponding *yes–no* experiment (which relies on sharp truncation of the energetic distribution) exactly, or even approximately when the 'energy slit' should be made maximally narrow. A possible way out of this difficulty consists of employing the generalized observables which are described by positive-operator-valued (rather than projection-valued) measures on the set of measurable values (for the present status of the theory of generalized observables see, e.g., the monographs by Davies (1976) and Kholevo (1980)). In particular, our energy filter can be described by $\varphi(H)$, where H as usual is the Hamiltonian and $\varphi : \mathbb{R} \to [0, 1]$ is a function with the following properties:

(i) $\varphi(\lambda) = 1$ for $|\lambda| \leqslant \epsilon$,
(ii) $\varphi(\lambda) = 0$ for $|\lambda| \geqslant \epsilon + \eta$, where $\delta \ll \eta \lesssim \epsilon$,
(iii) $\varphi(\lambda)$ changes sufficiently smoothly from 0 to 1 for $\epsilon \leqslant |\lambda| \leqslant \epsilon + \eta$.

The parameter η obviously characterizes smearing of the energy slit. Notice that in the paper by Schwinger mentioned in the notes to Section 1.6, the energy filtering was described in a closely similar way. The function v given by (3.6) has now to be replaced by

$$v(t) = (N'\delta/\pi) \int_{\mathbb{R}} e^{-i\lambda t} (\lambda^2 + \delta^2)^{-1} \varphi(\lambda)\, d\lambda,$$

but it is easy to see that this modification cannot change qualitatively the conclusions of the present section.

SECTION 2.4. The fact that the decay of an unstable system can stop in the limit of continual observation was first discussed, to our knowledge, by Beskow and Nilsson (1967) who used the simple argument presented in the opening to this section. However, the effect was essentially known earlier as the 'Turing paradox' – cf. Yourgrau (1968). The term 'quantum Zeno's paradox' was invented by Misra and Sudarshan (1977).

• Proposition 4.2 was formulated particularly for dim $\mathscr{H}_u = 1$ by Friedman (1976) and for a finite-dimensional $\mathscr{H}_u$ by Exner (1982a). The main result of this section, Theorem 4.4, is essentially due to Misra and Sudarshan (1977). However, the use of the results of Chernoff and Kato makes it possible to simplify the original proof, and at the same time to get a stronger result, namely the explicit form of the operators P, A, which is not limited by the assumption that $G(\,.\,)$ is continuous at $t = 0$. Various particular cases of Theorem 4.4 are known, mainly for the case of the 'continual observation' localizing a particle into some volume of the configuration space – cf. Friedman (1972, 1976), and also the heuristic analysis contained in the papers by Allcock (1969), and by Bloch and Burba (1974). The assertion proved in Example 4.11 generalizes partly the result of Friedman (1972).

• As for the description of the 'true' continual observations, two model examples were discussed recently by Kraus (1981). In both of them, the considered (quantum) system consists of a simple (non-realistic) unstable system coupled to a 'counter'. It appears that here, too, the decay can be prevented if the counter 'discharges' quickly enough, but that it can probably never happen under realistic physical conditions. The nature of this 'watchdog effect' has only to be clarified. A more formal approach to the problem of continual observation can be based on the generalized observables. Such an attempt was undertaken by Barchielli *et al.* (1982); for a more rigorous formulation see Lupieri (1983). These papers point out the connection between the continual observation and Feynman path integrals, which is interesting to compare with the considerations of Section 6.3 below. Let us also mention that the study of the 'Zeno's paradox' has inspired a search for some non-standard models in the quantum theory of measurement – cf. Sherry and Sudarshan (1978, 1979), Sherry *et al.* (1979).

• The small-time behaviour of decay laws and the 'Zeno's limit' have been revisited recently in connection with the widely discussed possibility of proton non-stability. Khalfin (1982) suggested that this decay might be suppressed by the small-time non-exponentiality. However, considerations of the preceding section show that the latter can play a substantial role only for the times $\lesssim \lambda^{-1}$, where λ characterizes the region in which the energy distribution may be approximated by the Breit–Wigner function. Thus the suppression is not only practically excluded, but also it cannot be achieved by repeated measurements, as pointed out by Chiu *et al.* (1982) in a reply to Khalfin. There are also speculations about suppression of the proton decay in heavy nuclei (Horwitz and Katznelson, 1983a), but they rely on the dubious assumption that interactions with 'fellow nucleons' can constitute a non-decay measurement.

Chapter 3

Dynamics and Symmetries

"It has been clear for many years that the Fermi golden rule is the right answer; what has been unclear is the right question."

B. SIMON

Considerations of the previous two chapters yield a framework in which concrete decay processes may be studied. Here we are going to discuss the dynamical mechanism which applies to a major part of them – namely, the perturbation theory of embedded eigenvalues. First, we introduce the reduced resolvent and show the principal role played by its analytical structure, especially by poles of its analytic continuation. Section 2 is devoted to a detailed analysis of the simplest possible case (the so-called Friedrichs model) in which the perturbation is a finite-rank operator. We use this model for illustrating various aspects of the problem, in particular, its relation to the resonance-scattering effect. We then discuss, in Section 3, a more general situation where the perturbation is a bounded operator and the embedded eigenvalue is not necessarily simple. We find the analytical structure of the reduced resolvent and derive a rigorous version of the Fermi golden rule. In Section 4 we discuss briefly how one may use symmetries to simplify the treatment of decay processes. The last section is devoted to space-time transformations. We show there the main features of the relativistically invariant description of decays. Particular attention is paid to the representations of the Poincaré group which may be associated with unstable elementary particles. Comments are also made on the problem of identity of unstable systems.

3.1. Poles of the Reduced Resolvent

In order to describe specific unstable systems within the general kinematical framework developed in the first chapter, one must be able to express the propagator U_t, or at least its projection V_t to the subspace $\mathscr{H}_{\mathrm{u}}$. This task can be rarely accomplished exactly; we are usually forced to work with some approximation to the solution of the dynamical problem.

Fortunately, a vast majority of the decay processes which we meet in quantum microphysics can be dealt with by means of perturbation techniques. The total Hamiltonian H typically decomposes into the sum of an 'unperturbed' part H_0 which is reduced by $\mathscr{H}_u$, and of the interaction term V which is therefore responsible for the decay. Of course, such a separation of the 'non-interacting' dynamics may cause some technical troubles, and sometimes even conceptual ones as in the axiomatic quantum field theory; but it represents a reasonable starting point for the discussion and is appropriate for most practical purposes.

We know from Section 1.5 that the spectrum of the total Hamiltonian should be expected to be continuous provided we consider only its part relevant to the decay. On the other hand, the spectrum of H_0 typically consists of both eigenvalues and a continuous part. As an example, consider the electromagnetic deexcitation of atoms, i.e. the classical problem of radiation damping. Let the atom be described by a Hamiltonian H_a on $\mathscr{H}_a$. If we assume for simplicity the center-of-mass motion removed, then H_a has a set of eigenvalues λ_n, $n = 0, 1, 2, \ldots$, corresponding to eigenspaces $\mathscr{H}_a^{(n)} \subset \mathscr{H}_a$. Let $\mathscr{F}$ and H_{em} on $\mathscr{F}$ be Fock space of the electromagnetic field and Hamiltonian of this field, respectively. We choose a fixed excited level λ_n, $n > 0$, and construct the subspace $\mathscr{H}_u^{(n)} = \mathscr{H}_a^{(n)} \otimes \{\Psi_0\}_{lin}$ in $\mathscr{H} = \mathscr{H}_a \otimes \mathscr{F}$, where Ψ_0 is the vacuum vector in $\mathscr{F}$. If the interaction between the atom and the electromagnetic field were absent, the composed system would be described by

$$H_0 = H_a \otimes I_{\mathscr{F}} + I_a \otimes H_{em}.$$

This operator clearly has the same eigenvalues as H_a; in particular, $\mathscr{H}_u^{(n)}$ is its eigenspace referring to λ_n. On the other hand, the part of H_0 corresponding to $(\Sigma_n^{\oplus} \mathscr{H}_a^{(n)} \otimes \{\Psi_0\}_{lin})^{\perp}$ is expected to have a continuous spectrum and, moreover, λ_n is embedded in it. The last assertion is explained loosely by the fact that the energy of the system (the atom in some lower energetic state λ_m + one photon) may vary continuously in (λ_m, ∞), in particular, around λ_n. With a little effort, this argument can be reformulated more rigorously so that the sharp-energy photons are avoided.

The interaction V between the atom and the electromagnetic field now represents a perturbation to H_0. Needless to say, actually it can never be switched off so H_0 describes no real system. It is most important to note that *the perturbation substantially changes the character of the spectrum.* The probability of finding an atom in the nth excited state decreases to zero with time, even quickly in the macroscopic scale. Thus in view of Theorem 1.5.8, the spectrum of H is continuous, of course, with the exception of the ground state eigenvalue λ_0. On the other hand, it is clear that the eigenvalues of H_0 cannot 'dissolve' fully, at least if V is small in some well-defined sense. It appears that in such cases H remembers the eigenvalues that have disappeared, because its spectrum is concentrated near them: loosely speaking, the states which have their energy support disjoint with some small interval around λ_n are nearly orthogonal to $\mathscr{H}_u^{(n)}$. These intervals of spectral concentration (which are closely connected to the line widths, and of the same order of magnitude) are usually much less than the spacing between the energy levels, with the exclusion of the highly excited states. This is the property that justifies the use of the approximation which describes the atom by the Hamiltonian H_a on $\mathscr{H}_a$, neglecting the presence of the electromagnetic field.

Thus the main object of our interest will be the *perturbation theory of eigenvalues embedded into the continuous spectrum.* It is discussed in the next two sections. We do not pretend to fully expose the methods of the theory; rather we are going to discuss some simple cases which allow us to illustrate the peculiar features of the solutions.

Let us start with some preliminary considerations. As usual, we have a pair of Hilbert spaces $\mathscr{H}_u \subset \mathscr{H}$ and a self-adjoint operator H on $\mathscr{H}$. We define the **reduced resolvent** of H by

$$R_u(z, H) = \mathrm{pr}_u(H - z)^{-1}.$$

As a projection of the full resolvent, $R_u(z, H)$ is defined for $\mathrm{Im}\, z \neq 0$ with eventual singularities on the real axis. We know that the reduced propagator is a Fourier transform of the positive-operator-valued measure F given by (1.5.7). The reduced resolvent yields an equivalent expression of this measure. To this end, we have to calculate

$$\Phi(\xi + i\eta)\psi = \int_0^\infty e^{it(\xi + i\eta)} E_u U_t \psi \, dt$$

for $\psi \in \mathscr{H}_u$. Since the projection E_u may be interchanged with the integral (Dunford and Schwartz, 1958, theorem III.2.19), the functional-calculus rules give $\Phi(\xi + i\eta) = iE_u(H - \xi - i\eta)^{-1}$. Substituting this into (1.5.7c), we get

$$\tfrac{1}{2}\{F([\lambda, \mu]) + F((\lambda, \mu))\}\psi$$

$$= \frac{1}{2\pi i} \lim_{\eta \to 0+} \int_\lambda^\mu [R_u(\xi + i\eta, H) - R_u(\xi - i\eta, H)]\, \psi \, d\xi \qquad (1.1)$$

for all $\lambda < \mu$ and $\psi \in \mathscr{H}_u$. An alternate way of obtaining (1.1) consists of applying the operator $\mathrm{pr}_u \in \mathscr{B}(\mathscr{H}, \mathscr{H}_u)$ to the Stone formula – cf. Reed and Simon (1972, theorem VII.13).

The correspondence between the spectral properties of a self-adjoint operator H and the analytic structure of its resolvent is well known: the latter has poles at the eigenvalues of H and a cut along its continuous spectrum. The consideration presented below employ substantially analytic properties of the reduced resolvent. The function $R_u(\,.\,, H)$ is, of course, strongly analytic in both the upper and lower complex halfplanes, because the same is true for the full resolvent, $R(z, H) = (H - z)^{-1}$. However, there is an important difference. The operator-valued function $R(\,.\,, H)$ cannot be continued analytically across the continuous spectrum of H as the following generic example shows: let $Q : (Q\psi)(x) = x\psi(x)$ on $L^2(\mathbb{R})$ and choose a vector ψ such that $\psi(x) \neq 0$ everywhere in $\mathbb{R}$, then $\lim_{\eta \to 0} R(\xi + i\eta, Q)\psi$ makes sense for no $\xi \in \mathbb{R}$.

On the other hand, it may well happen that at the same time analytic continuation of $R_u(\,.\,, H)$ across the continuous spectrum is possible. More explicitly, there can exist a region Ω_- in the lower halfplane containing an open interval $J \subset \sigma_c(H)$ in its boundary, and a $\mathscr{B}(\mathscr{H}_u)$-valued analytic function F on $\Omega = \Omega_- \cup J \cup \{z : \operatorname{Im} z > 0\}$ such that $F(z) = R_u(z, H)$ if $\operatorname{Im} z > 0$. Similarly, one can consider the continuation of $R_u(\,.\,, H)$ up from the lower halfplane, but only the former case is actually needed, as we shall see later. It is clear that $F(z)$ cannot coincide with $R_u(z, H)$ for $z \in \Omega_-$; otherwise the formula (1.1) would imply $F(J) = 0$, but then Theorem 1.5.13 gives $E_H(J) = 0$, which is impossible since $J \subset \sigma_c(H)$. Thus the analytic continuation leads us to another Riemannian sheet of the multivalued function $R_u(\,.\,, H)$; for simplicity we shal speak about the second sheet and write $R_u^{II}(\,.\,, H) = F(\,.\,)$.

The function $R_u^{II}(\,.\,, H)$ is especially interesting when it is meromorphic, i.e. if it has isolated poles in $\Omega_- \cup J$. The reason is twofold. Firstly, the meromorphic character of the second-sheet reduced resolvent is typical for perturbations of embedded eigenvalues, where the poles are obtained by shifting the eigenvalue poles of $R_u(\,.\,, H_0)$. Secondly, if the reduced resolvent is used to express the reduced propagator (through the relations (1.1) and (1.4.4)), then the second-sheet poles from Ω_- determine a semigroup part of $\{V_t\}$ which we know is usually the dominating one.

In order to illustrate the last assertion, consider the simplest case when $R_u^{II}(\,.\,, H)$ has just one simple pole at $z_p \in \Omega_-$, $z_p = \lambda_p - i\delta_p$ (cf. Figure 1). Then there are $B \in \mathscr{B}(\mathscr{H}_u)$ and a $\mathscr{B}(\mathscr{H}_u)$-valued analytic function $C(\,.\,)$ on Ω such that

$$R_u^{II}(z, H) = \frac{B}{z_p - z} + C(z), \quad z \in \Omega \backslash \{z_p\}. \tag{1.2}$$

We have $R(z, H)^* = R(\bar{z}, H)$ if only $z \notin \sigma_p(H)$ (Akhiezer and Glazman, 1966, §IV.49) so

$$R_u(z, H)^* = R_u(\bar{z}, H), \quad z \notin \sigma_p(H). \tag{1.3}$$

The relations (1.1)–(1.3) together imply

$$\tfrac{1}{2}\{F([\lambda, \mu]) + F((\lambda, \mu))\}\psi$$

$$= \frac{1}{2\pi i} \lim_{\eta \to 0+} \int_\lambda^\mu \left[\frac{B^*}{\xi - i\eta - \bar{z}_p} - \frac{B}{\xi + i\eta - z_p} + 2i \operatorname{Im} C(\xi + i\eta)\right] \psi \, d\xi$$

for each $\psi \in$ $_u$. Norm of the integrated function is majorized by

$$2\{\|B\|((\xi - \lambda_p)^2 + (\delta_p + \eta)^2)^{-1/2} + \|C(\xi + i\eta)\|\}\|\psi\|$$
$$\leqslant 2\{\|B\|\delta_p^{-1} + \|C(\xi + i\eta)\|\}\|\psi\|;$$

further, the analyticity of $C(\,.\,)$ implies that $\|C(\,.\,)\|$ is continuous in Ω, and therefore bounded in $[\lambda, \mu] \times [-\eta_0, \eta_0]$ for all finite $[\lambda, \mu] \subset J$ and some $\eta_0 > 0$. Hence we may perform the limit $\eta \to 0+$ by means of the dominated-convergence

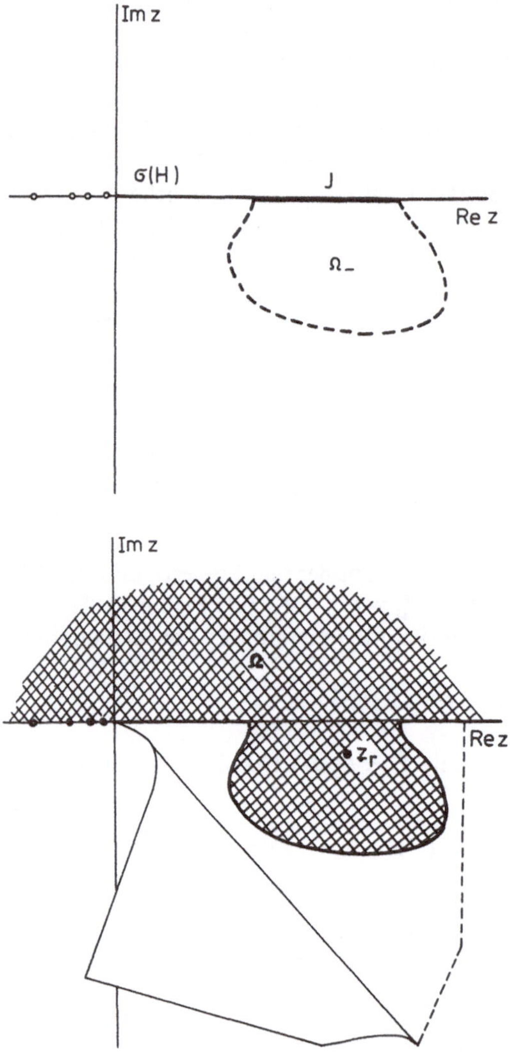

3.1. Analytic continuation of the reduced resolvent. On the lower picture, the first sheet is partly removed to reveal the pole of $R_u^{II}(\,.\,, H)$.

theorem. The resulting $F(\,.\,)$ is easily seen to be absolutely continuous with respect to the Lebesgue measure, and the Radon–Nikodým derivative equals

$$\left(\frac{\mathrm{d}F_\lambda}{\mathrm{d}\lambda}\right)\psi = \left(\frac{\mathrm{d}F_\lambda^p}{\mathrm{d}\lambda}\right)\psi + \frac{1}{\pi}\,\mathrm{Im}\,C(\lambda)\psi, \tag{1.4a}$$

$$\frac{\mathrm{d}F_\lambda^p}{\mathrm{d}\lambda} = \frac{1}{2\pi i}\left[\frac{B^*}{\lambda-\lambda_p-i\delta_p} - \frac{B}{\lambda-\lambda_p+i\delta_p}\right]. \tag{1.4b}$$

Now the reduced propagator is given by (1.4.4), where the integration goes in fact over $\sigma(H)$ only, in view of Theorem 1.5.13. Suppose that $R_u(\,.\,, H)$ can be continued analytically through the whole continuous spectrum, i.e. that $\bar{J} = \sigma_c(H)$, then we obtain

$$V_t\psi = \int_{\mathbb{R}} e^{-i\lambda t}\, dF_\lambda^p\,\psi - \int_{\mathbb{R}\setminus\sigma(H)} e^{-i\lambda t}\, dF_\lambda^p\,\psi + \frac{1}{\pi}\int_{\sigma(H)} e^{-i\lambda t}\, \mathrm{Im}\, C(\lambda)\psi\, d\lambda.$$

The first term can be evaluated by contour integration. Standard estimates show that the integral over the arc closing the integration contour in the lower halfplane is zero in the limit of infinite radius. Then the second part of (1.4b) contributes by the residue term while the first gives zero: we have

$$V_t\psi = B\exp(-i\lambda_p t - \delta_p t) + \frac{1}{\pi}\int_{\mathbb{R}\setminus\sigma(H)} e^{-i\lambda t}\, \mathrm{Im}\, \frac{B}{\lambda - \lambda_p + i\delta_p}\,\psi\, d\lambda +$$
$$+ \frac{1}{\pi}\int_{\sigma(H)} e^{-i\lambda t}\, \mathrm{Im}\, C(\lambda)\psi\, d\lambda \tag{1.5}$$

for $t > 0$ and $\psi \in \mathscr{H}_u$.

Usually the first term in (1.5) dominates, as far as the 'true' unstable system are considered which decay slowly enough so that their time evolution is observable. In such cases, typically:

(i) the pole is close to the real axis, in the sense that there is an interval $J_\epsilon = (\lambda_p - \epsilon, \lambda_p + \epsilon)$ contained in $\sigma_c(H)$ and $\epsilon \gg \delta_p$; then the second term in (1.5) may be estimated similarly as in Example 1.5.15 and Section 2.3;

(ii) the 'background' term $(1/\pi)\,\mathrm{Im}\, C(\,.\,)$ is much smaller than the pole term within J_ϵ (and it varies slowly in the scale given by δ_p) so the third term in (1.5) may also be neglected.

Thus for practical purposes one may replace (1.2) by $R_u^{II}(z, H) = (z_p - z)^{-1}E_u$ and assume $\sigma_c(H) = J = \mathbb{R}$. This simplified description, which can be called the **pole-and-real-axis approximation**, yields an exponential decay law which is determined fully by the position of the pole. Similarly, one can proceed if $R_u^{II}(\,.\,, H)$ has a finite number of simple poles in Ω_-. However, notice that the reduced propagator thus obtained then may not fulfil the semigroup condition (1.1.1) unless products of the residuum operators belonging to different poles are zero.

3.2. Friedrichs Model

"Please try to illustrate your assertion on an example which would involve 2 × 2 matrices only."

M. Havlíček

In this section we are going to discuss in detail the simplest non-trivial case of an embedded-eigenvalue perturbation. We shall assume that

(i) H_0 is a self-adjoint operator with simple continuous spectrum supported by a halfline and an eigenvalue of multiplicity 1, embedded in $\sigma_c(H_0)$;
(ii) the perturbation V connects the eigenvector of H_0 with the continuum states, but not the latter between themselves.

Without loss of generality, we can consider the following model: the state Hilbert space $\mathscr{H}$ is the direct sum $\mathscr{H}_u \oplus \mathscr{H}_d$, where

$$\mathscr{H}_u = \mathbb{C}, \qquad \mathscr{H}_d = L^2(\mathbb{R}_+), \tag{2.1}$$

further the unperturbed Hamiltonian and the perturbation are defined by

$$H_0 : \quad H_0\{\alpha, f\} = \{\lambda_0\alpha, Qf\}, \quad \lambda_0 > 0, \tag{2.2a}$$

$$V : \quad V\{\alpha, f\} = \{\lambda_u\alpha + (v, f), \alpha v\}, \tag{2.2b}$$

where λ_u is a real number, v is some function from $L^2(\mathbb{R}_+)$, and as usual, $(Qf)(\lambda) = \lambda f(\lambda)$. The operator H_0 has a simple eigenvalue λ_0 corresponding to $\psi_u = \{1, 0\}$ which is embedded in $\sigma_c(H_0) = \mathbb{R}_+$. The perturbation V is easily seen to be Hermitean and of rank 2; it fulfils the requirement (ii) or

$$E_d V E_d = 0. \tag{2.3}$$

This relation is sometimes called **Friedrichs condition**; it is the essential ingredient of the model which makes it algebraically soluble.

In order to make the dependence on the coupling strength more transparent, we shall treat the one-parameter family of self-adjoint operators H_g:

$$H_g = H_0 + gV. \tag{2.4}$$

PROPOSITION 3.2.1. *Assume that H_0 is reduced by E_u and the condition* (2.3) *is fulfilled, then the reduced resolvent of the operator* (2.4) *is given by*

$$R_u(z, H_g) = [\mathrm{pr}_u(-z + H_0 + gV - g^2 V E_d R(z, H_0) E_d V)]^{-1}. \tag{2.5a}$$

In particular, if the operators H_0 and V are defined by the relations (2.2), *then $R_u(z, H_g)$ acts on $\mathscr{H}_u$ as multiplication by*

$$r_u(z, H_g) = \left(-z + \lambda_0 + g\lambda_u + g^2 \int_0^\infty \frac{|v(\lambda)|^2}{z - \lambda}\, d\lambda\right)^{-1}. \tag{2.5b}$$

Proof. We start from the second resolvent identity (Weidmann, 1980, theorem 5.13)

$$R(z, H) = R(z, H_0) - gR(z, H_0)VR(z, H),$$

which implies

$$E_u R(z, H) E_u = E_u R(z, H_0) E_u - gE_u R(z, H_0) E_u V E_u R(z, H) E_u - \\ - gE_u R(z, H_0) E_u V E_d R(z, H) E_u,$$

where we have used the commutativity of E_u and $R(z, H_0)$, and

$$E_d R(z, H) E_u = -g E_d R(z, H_0) E_d V E_u R(z, H) E_u,$$

where (2.3) is also taken into account. Now one has to substitute from the second relation to the first, and to multiply the resulting identity by $(H_0 - z)E_u$ from the left. Since $(H_0 - z)E_u R(z, H_0) = E_u$, we get after a simple rearrangement the relation

$$[(H_0 - z)E_u + gE_u V E_u - g^2 E_u V E_d R(z, H_0) E_d V E_u] E_u R(z, H) E_u = E_u,$$

which yields (2.5a). Substituting there the operators (2.2), we obtain (2.5b). ∎

In order to find the analytic structure of $r_u(\,.\,, H_g)$, more information about the function v is needed. We adopt the following assumption:

(a) $|v(\,.\,)|^2$ can be continued analytically across $J = (0, \infty)$, i.e. there is an (open) region $\Omega \subset \mathbb{C}$ containing J and a meromorphic function $f : \Omega \to \mathbb{C}$ such that $|v(\lambda)|^2 = f(\lambda)$ for $\lambda \in J$. For notational convenience, we shall write $|v(z)|^2 = f(z)$ also for non-real z.

PROPOSITION 3.2.2. *Suppose that* H_0, V, H_g *are given by* (2.2), (2.4), *and the assumption* (a) *is fulfilled, then*

(a) $\sigma_c(H_g) = [0, \infty)$ *for any* g,

(b) *the function* $r_u(\,.\,, H_g)$ *can be continued analytically across* $(0, \infty)$ *and*

$$r_u(z, H_g) = (-z + w(z, g))^{-1}, \tag{2.6}$$

where

$$w(\lambda, g) = \lambda_0 + g\lambda_u + g^2 \mathscr{P}\int_0^\infty \frac{|v(\xi)|^2}{\lambda - \xi}\, d\xi - \pi i g^2 |v(\lambda)|^2, \lambda > 0 \tag{2.6a}$$

$$w(z, g) = \lambda_0 + g\lambda_u + g^2 \int_0^\infty \frac{|v(\xi)|^2}{z - \xi}\, d\xi - $$

$$- 2\pi i g^2 |v(z)|^2, z \in \Omega, \operatorname{Im} z < 0. \tag{2.6b}$$

Proof. (a) The essential spectrum of a self-adjoint operator is invariant under (relatively) compact perturbations (Reed and Simon, 1978, §XIII.4), thus *a fortiori* $\sigma_{ess}(H_0) = \mathbb{R}_+ = \sigma_{ess}(H_g)$. It can easily be checked that a positive λ_1 can be a (simple) eigenvalue of H_g only if $|v(\lambda_1)|^2 = 0$. The analycity assumption then implies that H_g has, at most, isolated simple eigenvalues in $\mathbb{R}_+$ with no accumulation point except at infinity, and therefore $\sigma_c(H_g) = \sigma_{ess}(H_g) = \mathbb{R}_+$.

(b) The question clearly reduces to analycity of the function $w(\,.\,, g)$ which is defined by (2.5b), (2.6) in the upper halfplane, and by (2.6b) in $\Omega_- = \{z \in \Omega :$

$\operatorname{Im} z < 0\}$. In both these regions, $w(\,.\,, g)$ is obviously analytic. Further, one has to find $\lim_{\epsilon\to 0+} w(\lambda \pm i\epsilon, g)$. The only non-trivial term is the integral one, which we divide as follows:

$$\lim_{\epsilon\to 0+}\left(\int_0^{\lambda-R} + \int_{\lambda-R}^{\lambda+R} + \int_{\lambda+R}^{\infty}\right) \frac{|v(\xi)|^2}{\lambda \pm i\epsilon - \xi}\, d\xi$$

for some positive R. The limit may be performed under the integral sign in the first and the third term, because $|v(\,.\,)|^2$ is integrable. Smoothness of $|v(\,.\,)|^2$ makes it possible to handle the second integral by means of the well-known distribution-theory trick (see, e.g. Vladimirov (1971, §II.5.8)), so we obtain

$$\lim_{\epsilon\to 0+}\int_0^{\infty} \frac{|v(\xi)|^2}{\lambda \pm i\epsilon - \xi}\, d\xi = \mp\pi i\, |v(\lambda)|^2 + \mathscr{P}\int_0^{\infty} \frac{|v(\xi)|^2}{\lambda - \xi}\, d\xi, \tag{2.7a}$$

where $\mathscr{P}$ denotes conventionally the principal value of the integral,

$$I(\lambda, v) \equiv \mathscr{P}\int_0^{\infty} \frac{|v(\xi)|^2}{\lambda - \xi}\, d\xi := \lim_{\eta\to 0+}\left(\int_0^{\lambda-\eta} + \int_{\lambda+\eta}^{\infty}\right) \frac{|v(\xi)|^2}{\lambda - \xi}\, d\xi. \tag{2.7b}$$

Let us now check the continuity of $I(\,.\,, v)$. To a given $\lambda > 0$, we choose some positive $R < \lambda$ and a function $\varphi \in C_0^{\infty}(J_R)$, $J_R = [-\frac{1}{2}R, \frac{1}{2}R]$, which is even with respect to the origin and such that $\varphi(0) = 1$. Let $|\mu - \lambda| \leqslant \frac{1}{2}R$, then one easily finds $I(\mu, v) = \int_0^{\infty} h(\mu, \xi)\, d\xi$, where

$$h(\mu, \xi) := \frac{|v(\xi)|^2 - |v(\mu)|^2 \varphi(\xi - \mu)}{\mu - \xi}.$$

The function $h(\mu, \,.\,)$ is not only integrable but it can be majorized independently of μ. In order to see this, one has to realize that outside $[\lambda - R, \lambda + R] \supset [\mu - \frac{1}{2}R, \mu + \frac{1}{2}R]$ we have $h(\mu, \xi) = |v(\xi)|^2(\mu - \xi)^{-1}$. On the other hand, $h(\,.\,, .\,)$ is continuous on $[\lambda - \frac{1}{2}R, \lambda + \frac{1}{2}R] \times [-R, +R]$ so its modulus is bounded there by some constant C_R. Consequently,

$$|h(\mu, \xi)| \leqslant h_0(\xi) := \begin{cases} \dfrac{2}{R}\, |v(\xi)|^2 & \dots \quad |\xi - \lambda| > R \\ C_R & \dots \quad |\xi - \lambda| \leqslant R \end{cases}$$

for each $\mu \in [\lambda - \frac{1}{2}R, \lambda + \frac{1}{2}R]$ and the dominated-convergence theorem implies that $I(\,.\,, v)$ is continuous at an arbitrary point $\lambda \in (0, \infty)$. The first term on the r.h.s. of (2.7a) is also continuous. This means that the limits $\lim_{\epsilon\to 0+} w(\lambda \pm i\epsilon, g)$ coincide and their common value is a continuous function of λ. It allows us to use the edge-of-wedge theorem (Streater and Wightman, 1964, theorem 2–13), according to which the function $w(\,.\,, g)$ defined by (2.5b) and (2.6) is continued analytically from the upper halfplane across $(0, \infty)$ to Ω_-. ■

Now we come to the main result which establishes the meromorphic structure of the reduced resolvent for the Friedrichs model, providing the coupling is sufficiently weak:

THEOREM 3.2.3. *Let the assumptions of Proposition* 2.2 *be valid and* $v(\lambda_0) \neq 0$. *Then for all non-zero real g which are small enough, the second-sheet continuation of the reduced resolvent* (2.5b) *has a simple pole at a point* $z_p(g)$ *from* $\Omega_- = \{z \in \Omega : \operatorname{Im} z < 0\}$. *Moreover,* $z_p(\,.\,)$ *is the restriction to* $\mathbb{R}$ *of a function analytic in some complex neighbourhood of* 0, *and for its real and imaginary parts,* $z_p(g) = \lambda_p(g) - i\delta_p(g)$, *the following relations hold*

$$\lambda_p(g) = \lambda_0 + g\lambda_u + g^2 I(\lambda_0, v) + g^3\lambda_u I'(\lambda_0, v) + O(g^4), \tag{2.8a}$$

$$\delta_p(g) = g^2\pi\,|v(\lambda_0)|^2 + 2g^3\lambda_u \operatorname{Re}(\overline{v(\lambda_0)}v'(\lambda_0)) + O(g^4), \tag{2.8b}$$

where $I(\,.\,, v)$ *is defined by* (2.7b) *and the primes denote derivatives with respect to* λ.

REMARK 3.2.4. The considerations preceding the theorem do not employ the fact that g is real. Of course, H_g with a complex g need not be self-adjoint, but otherwise Propositions 2.1 and 2.2 remain to hold true. The proof presented below also works for complex values of the coupling constant, $r_u^{II}(\,.\,, H_g)$ still has a simple pole for small enough g, and its position depends analytically on the coupling constant. However, for non-real g the relations (2.8) are no longer valid. Notice also that the existence of the pole and analycity of $z_p(\,.\,)$ will be established without reference to the assumption $v(\lambda_0) \neq 0$.

Proof. We define the function $f : \mathbb{C} \times M_w \to \mathbb{C}$ by $f(g, z) = z - w(z, g)$, where $M_w = \Omega \cup \{z : \operatorname{Im} z > 0\}$ is the analycity domain of $w(\,.\,, g)$. The function f has all partial derivatives in both g, z; further, we have $f(0, \lambda_0) = 0$ and $(\partial f/\partial z)(0, \lambda_0) = 1$. This makes it possible to apply the implicit-function theorem (Schwartz, 1967, theorems III.28, III.31), according to which there is a neighbourhood $U' \subset \mathbb{C}$ of zero and a unique analytic function $z_p : U' \to \mathbb{C}$ such that $f(g, z_p(g)) = 0$ for all $g \in U'$, i.e. $z_p(g) = w(z_p(g), g)$. Continuity of partial derivatives of f implies particularly that $(\partial f/\partial z)(\,.\,, z_p(\,.\,))$ is continuous is U', and therefore $(\partial f/\partial z)(g, z_p(g))$ is non-zero for g from some neighbourhood $U \subset U'$ of zero. Consequently, $r_u(\,.\,, H_g)$ has a simple pole at $z_p(g)$ and its position depends analytically on g within U. The remaining part of the proof consists of evaluating the implicit-function derivatives, which is left to the reader (Problem 10). ■

Once the existence of the pole is established, one may start discussing the legitimacy of the pole-and-real-axis approximation, i.e. to compare for given $v(\,.\,)$ and g the remainder terms in (1.5) to the pole one (by a suitable estimation procedure). However, we are not going to pursue this line further (cf. Problem 12). Instead, we shall use the model under consideration to illustrate some peculiar features connected with the embedded-eigenvalue perturbations.

Notice first that Theorem 2.3 establishes the so-called **Fermi golden rule** for our particular case. Actually, in the pole-and-real-axis approximation the eigenvector ψ_u corresponding to λ_0 decays exponentially 'into the continuum' under influence of the perturbation gV, and the decay rate equals

$$\Gamma(g) = 2\delta_p(g) = 2\pi g^2\,|v(\lambda_0)|^2 + O(g^3). \tag{2.9a}$$

Now the Fermi rule claims that in the lowest-order approximation, the decay rate is given by

$$\Gamma(g) = 2\pi g^2 \frac{\mathrm{d}}{\mathrm{d}\lambda} (V\psi_u, E_\lambda^0 P_c(H_0) V\psi_u) \big|_{\lambda=\lambda_0}, \tag{2.9b}$$

where $P_c(H_0)$ is the projection referring to the continuous subspace of H_0 and $\{E_\lambda^0\}$ is the corresponding decomposition of unity. In order to see that (2.9b) is actually the popular rule, one must realize that formally $(\mathrm{d}/\mathrm{d}\lambda)E_\lambda^0 P_c(H_0) = |\lambda\rangle\langle\lambda|$. Remember also that (2.9b) is in fact divided by $\hbar$ which was set equal to 1 in our considerations. Now one has to substitute for ψ_u and V in (2.9b) to obtain the relation

$$\Gamma(g) = 2\pi g^2 \frac{\mathrm{d}}{\mathrm{d}\lambda} \langle \{\lambda_u, v\}, E_\lambda^0 P_c(H_0)\{\lambda_u, v\}\rangle \big|_{\lambda=\lambda_0}$$
$$= 2\pi g^2 \frac{\mathrm{d}}{\mathrm{d}\lambda} \int_0^\lambda |v(\xi)|^2 \,\mathrm{d}\xi \,\big|_{\lambda=\lambda_0},$$

which yields the first term of (2.9a).

However, the Fermi rule should not be applied mechanically, otherwise is gold might appear to be mixed with brass (by the expression of J. Howland). It concerns particularly the *threshold effects*:

EXAMPLE 3.2.5. Suppose that the eigenvalue λ_0 is at the threshold of the continuous spectrum, $\lambda_0 = 0$. We choose further $\lambda_u > 0$ and the following function v:

$$v_\alpha(\lambda) = \left[\frac{2}{\pi}\cos(\pi\alpha/2)\,\frac{\lambda^\alpha}{1+\lambda^2}\right]^{1/2}, \tag{2.10}$$

where α is a non-zero number from $(-1, 1)$. The corresponding function $w_\alpha(\,.\,, g)$ is easily calculated (Gradshtein and Ryzhik, 1971, 3.263) to be

$$w_\alpha(z, g) = g\lambda_u + g^2\,\frac{z - \mathrm{cotg}(\pi\alpha/2) + z^\alpha\,\mathrm{e}^{-i\pi\alpha}\mathrm{cosec}(\pi\alpha/2)}{1+z^2}\,. \tag{2.11}$$

This can be continued analytically across $(0, \infty)$, but not across some larger interval, because $z = 0$ is the branching point. The continuation is, of course, given again by (2.11). Now one has to solve the equation (Problem 11)

$$z_\alpha - w_\alpha(z_\alpha, g) = 0. \tag{2.12a}$$

In distinction to Theorem 2.3, the solution need not be in general an analytic function of g. Nevertheless, one can easily check that it behaves for small g like

$$z_\alpha(g) = g\lambda_u - g^2\,\mathrm{cotg}(\pi\alpha/2) + g^{2+\alpha}\lambda_u\,\mathrm{e}^{-i\pi\alpha}\,\mathrm{cosec}(\pi\alpha/2) + $$
$$+\,O(g^3), \quad \alpha > 0, \tag{2.12b}$$

$$z_\alpha(g) = g\lambda_u + g^{2+\alpha}\lambda_u\,\mathrm{e}^{-i\pi\alpha}\,\mathrm{cosec}(\pi\alpha/2) - g^2\,\mathrm{cotg}(\pi\alpha/2) + $$
$$+\,O(g^{3+2\alpha}), \quad \alpha < 0, \tag{2.12c}$$

where, as usual, $h(g) = O(g^\beta)$ means $\beta = \sup\{\gamma : \lim_{g\to 0} h(g)g^{-\gamma} = 0\}$. The third term in (2.12c) is, of course, drowned in the remainder term, when $\alpha \leqslant -\frac{1}{2}$. Existence and uniqueness of the solution *for small enough real* g may be obtained by the implicit-function theorem again; however, it must now be used in a slightly more sophisticated way; we leave this to the reader. The following two cases appear:

(i) If $g < 0$, then the solutions $z_\alpha(g)$ are real and negative. Of course, it does not follow from (2.12b, c), but it is sufficient to find the shape of $w_\alpha(\,.\,, g)$ to see that (2.12a) has a solution within $(-\infty, 0)$. Consequently, $r_u(\,.\,, H_g)$ has a *first-sheet* simple pole at $z_\alpha(g)$: the threshold eigenvalues *becomes isolated* by influence of the perturbation (cf. Figure 2),

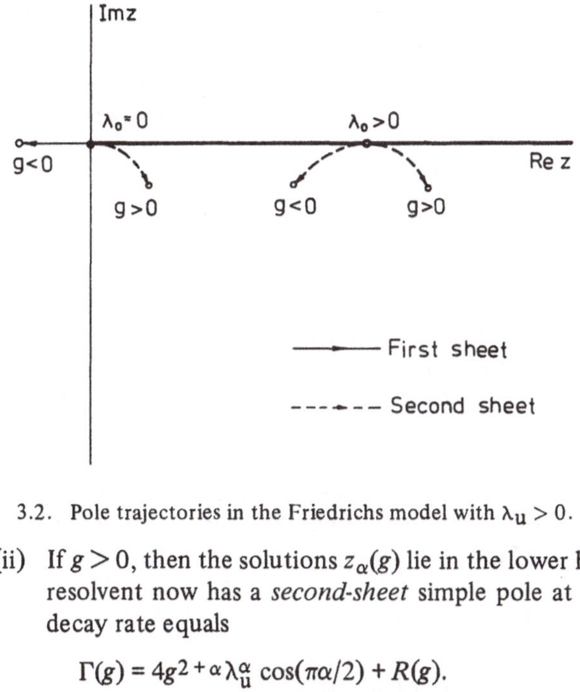

3.2. Pole trajectories in the Friedrichs model with $\lambda_u > 0$.

(ii) If $g > 0$, then the solutions $z_\alpha(g)$ lie in the lower halfplane, i.e. the reduced resolvent now has a *second-sheet* simple pole at $z_\alpha(g)$. The corresponding decay rate equals

$$\Gamma(g) = 4g^{2+\alpha}\lambda_u^\alpha \cos(\pi\alpha/2) + R(g). \tag{2.13}$$

For $\alpha > 0$, we have $R(g) = O(g^3)$ so (2.13) agrees with the prescription (2.9a) up to g^2, but this yields a trivial information only. On the other hand, $R(g) = O(g^{3+2\alpha})$ for $\alpha < 0$ and the contradiction between (2.9a) and (2.13) is evident.

Let us further comment briefly on the relations between the pole description of unstable systems and the resonances in scattering theory. The latter also uses a pair of operators as an input information; namely, the total Hamiltonian H and the free one H_0. The central question concerns the existence of the **wave operators**

$$\Omega_\pm(H, H_0) = \underset{t\to\pm\infty}{\text{s-lim}}\; U_t^* U_t^0 P_{ac}(H_0), \tag{2.14}$$

where $U_t = \exp(-iHt)$, $U_t^0 = \exp(-iH_0t)$, and their **completeness**

$$\text{Ran}\ \Omega_+(H, H_0) = \text{Ran}\ \Omega_-(H, H_0) = P_{ac}(H)\mathcal{H} \tag{2.15}$$

(which is equivalent to the **asymptotic completeness** when the singularly continuous spectrum of H is void). If the wave operators exist and are complete, we may define the **scattering operator** (or ***S*-matrix**) by

$$S = \Omega_+^*\Omega_-. \tag{2.16}$$

The wave operators are intertwining between the group $\{U_t\}$ and $\{U_t^0\}$, as well as between their generators. It follows easily that S commutes with $H_0P_{ac}(H_0)$, the absolutely continuous part of H_0 (Amrein *et al.*, 1977, prop. 4.7). This fact together with existence of the spectral representation for H_0 imply that S is *decomposable* (Amrein *et al.*, 1977, prop. 5.27), i.e. that there is a Hilbert space $L^2(\Lambda; \mathcal{H}_0)$ of $\mathcal{H}_0$-valued functions on $\Lambda \equiv \sigma_{ac}(H_0)$, a unitary operator $\mathcal{U}$ from $\mathcal{H}_{ac}^0 \equiv \text{Ran}\, P_{ac}(H_0)$ to $L^2(\Lambda; \mathcal{H}_0)$, and a function $S : \Lambda \to \mathcal{B}(\mathcal{H}_0)$ so that

$$(\mathcal{U} H_0P_{ac}(H_0)\psi)(\lambda) = \lambda(\mathcal{U}\psi)(\lambda), \tag{2.17a}$$

$$(\mathcal{U} S\psi)(\lambda) = S(\lambda)(\mathcal{U}\psi)(\lambda). \tag{2.17b}$$

If the wave operators are complete, then $S \upharpoonright \mathcal{H}_{ac}^0$ is unitary. The same is true for almost all operators $S(\lambda)$ (Amrein *et al.*, 1977, §5.7), which are called the **on-shell scattering operators** or ***S*-matrix at energy** λ. Similarly the ***R*-matrix**, $R \equiv S - P_{ac}(H_0)$, is decomposable and $R(\lambda) = S(\lambda) - I_0$; the operators $R(\lambda)$ are closely tied to the scattering amplitude. Another related operator is the **phase shift** $\delta(\lambda)$, $S(\lambda) = e^{2i\delta(\lambda)}$.

Most frequently H may be expressed as the sum of H_0 and some interaction term V. Then there are various ways to express the operators $\Omega_\pm$, S, R by means of V and the resolvents of the involved Hamiltonians; e.g. the Lippmann–Schwinger equations (Amrein *et al.*, 1977, prop. 6.5). Another equation of this type reads (Amrein *et al.*, 1977, prop. 6.11):

$$S = \Big\{ I + \underset{\eta\to 0+}{\text{s-lim}}\ \underset{\delta\to 0+}{\text{s-lim}} \int_{\mathbb{R}} (R(\lambda - i\eta, H_0) - R(\lambda + i\eta, H_0)) \times$$
$$\times [V - VR(\lambda + i\delta, H)V]\, dE_\lambda^0 \Big\} P_{ac}(H_0) \tag{2.18}$$

provided that the wave operators exist, are complete, the operator $H = H_0 + V$ is self-adjoint and $D(V) \supset D(H_0)$. Thus the scattering amplitude can be often calculated by means of $T_z \equiv V - VR(z, H)V$: performing the limit in (2.18) formally, we see that $R(\lambda)$ *is proprotional to the on-shell part of* $T_{\lambda+i0}$. Suppose that the scattering amplitude (or equivalently the function $R(\,.\,)$) possesses an analytic continuation across a part of the real axis, with a pole at a point $z_p = \lambda_p - i\delta_p$ of the 'second' Riemannian sheet; then the scattering system (H, H_0) is said to have **resonance** of the width $\Gamma = 2\delta_p$ at an energy λ_p. The above considerations reveal an intimate connection between the scattering resonances and the analytic properties of certain matrix elements of $R(\,.\,, H)$.

Let us now return to the Friedrichs model and discuss how it behaves as the scattering system with the free Hamiltonian H_0 and the total one given by (2.4). Since V is of rank 2, existence and (asymptotic) completeness of the wave operators follows from the Kato–Rosenblum theorem (Reed and Simon, 1979, theorem XI.8). Further the spectrum of $H_0 \upharpoonright \mathscr{H}^0_{ac}$ is simple so the auxiliary Hilbert space $\mathscr{H}_0$ is one-dimensional and the S-matrix acts as the operator of multiplication by some function $S(\,.\,)$ on $\mathscr{H}^0_{ac} = \mathscr{H}_d = L^2(\mathbb{R}_+)$. Of course, $|S(\lambda)| = 1$ for almost all non-negative λ. We shall express the function $R(\,.\,) = S(\,.\,) - 1$ explicitly:

PROPOSITION 3.2.6. *Let the assumptions of Theorem 2.3 be valid. Moreover, let the function $v(\,.\,)$ be piecewise continuous and bounded in $\mathbb{R}_+$, then the 'scattering amplitude' equals*

$$R(\lambda) = 2\pi i g^2 \lim_{\delta \to 0+} |v(\lambda)|^2\, r_u(\lambda + i\delta, H_g). \tag{2.19}$$

The scattering system (H_g, H_0) has for small enough g just one resonance; its position λ_p and width are given by the relations (2.8).

Proof. We choose $h, f \in C_0^\infty(\mathbb{R}_+)$ and express (h, Rf) from (2.18). The first term in the square bracket disappears, because $P_{ac}(H_0) = E_d$ and the Friedrichs condition (2.3) holds, so we get

$$(h, Rf) = \lim_{\eta \to 0+} \lim_{\delta \to 0+} \int_0^\infty \overline{(VR(\lambda + i\delta, H)^* V[R(\lambda - i\eta, H_0)^* -}$$

$$\overline{- R(\lambda + i\eta, H_0)^*]h)\,(\lambda)} f(\lambda)\, d\lambda.$$

Now $(VR(z)^*Vf)(\lambda) = \langle \psi_0, R(z)^*Vf\rangle v(\lambda) = (v, f)\overline{r_u(z)}\, v(\lambda)$, and the action of $R(z, H_0)^*$ may also be written explicitly, thus a simple calculation yields

$$(h, Rf) = \lim_{\eta \to 0+} \lim_{\delta \to 0+} \int_0^\infty \int_0^\infty \overline{v(\lambda)}\, v(\xi) r_u(\lambda + i\delta, H_g) \times$$

$$\times \frac{2i\eta}{(\lambda - \xi)^2 + \eta^2}\, \overline{h(\xi)}\, f(\lambda)\, d\lambda\, d\xi.$$

Since $r_u(\,.\,)$ is analytic in some neighbourhood of $(0, \infty)$ (and therefore locally bounded there), bounded at the origin, and h, f have compact supports due to the assumption, we may use the Lebesgue dominated-convergence theorem and the Fubini theorem, obtaining in this way

$$(h, Rf) = \lim_{\eta \to 0+} \int_0^\infty d\lambda\, \overline{v(\lambda)}\, f(\lambda) r_u(\lambda + i0, H_g) \times$$

$$\times \int_0^\infty d\xi \frac{2i\eta}{(\lambda - \xi)^2 + \eta^2}\, v(\xi)\overline{h(\xi)}.$$

The limit is performed by double application of the dominated-convergence theorem. Since $v(\,.\,)$ is bounded, so is $v(\,.\,)\overline{h(\,.\,)}$, and therefore

$$\lim_{\eta\to 0+}\int_0^\infty \frac{\eta v(\xi)\overline{h(\xi)}}{(\lambda-\xi)^2+\eta^2}\,d\xi = \lim_{\eta\to 0+}\int_{-\lambda/\eta}^\infty \frac{v(\lambda+x\eta)\overline{h(\lambda+x\eta)}}{1+x^2}\,dx$$
$$= \pi v(\lambda)\overline{h(\lambda)}$$

except for eventual points where v is discontinuous. Further, $|v(\xi)\overline{h(\xi)}|\leqslant C$ for $\xi\in\mathbb{R}_+$ implies $|\int_0^\infty \eta v(\xi)\overline{h(\xi)}\,[(\lambda-\xi)^2+\eta^2]^{-1}\,d\xi|\leqslant \pi C$, so the dominated-convergence theorem may also be applied to the first integral. Thus, we obtain

$$(h, Rf) = 2\pi i\int_0^\infty \overline{h(\lambda)}\,|v(\lambda)|^2\, r_u(\lambda+i0, H_g)\, f(\lambda)\,d\lambda.$$

The operator R is bounded and $C_0^\infty(\mathbb{R}_+)$ is dense in $\mathscr{H}_d = L^2(\mathbb{R}_+)$, so the last relation extends by continuity, giving (2.19). The existence and position of the resonance then follow from Theorem 2.3. ∎

In order to determine the scattering amplitude and the cross-section, knowledge of H, H_0 alone is not sufficient. One needs some information about the underlying kinematics from which the multiplicative factors stem, distinguishing $R(\lambda)$ from the scattering amplitude at energy λ (Amrein *et al.*, 1977, §§7.3, 7.4). However, these kinematical factors cannot alter the resonance character of $R(\,.\,)$, so for our present purpose we may neglect them and write the cross-section at energy λ as

$$\sigma(\lambda) = \tfrac{1}{4}|R(\lambda)|^2 = -\tfrac{1}{2}\operatorname{Re} R(\lambda). \qquad (2.20a)$$

The second equality here follows from unitarity of the S-matrix. The function obtained by substituting from (2.19) to (2.20a) has a peculiar behaviour: it exhibits a bump around $\lambda_p(g)$ which consists of the Breit–Wigner term resulting from the pole part of $r_u(\,.\,, H_g)$, and the background part, both multiplied by $|v(\lambda)|^2$. At the same time, if the pole part dominates around λ_p, i.e. $R(\lambda)\approx 2i\delta_p(\lambda_p - i\delta_p - \lambda)^{-1}$ for $|\lambda-\lambda_p|\lesssim\delta_p$, then the phase shift

$$\delta(\lambda)\approx \operatorname{arctg}\frac{\delta_p}{\lambda_p-\lambda} \qquad (2.20b)$$

grows rapidly there and increases on about π. This is an other characteristic feature, which is useful particularly for distinguishing the true resonances from accidental peaks in the cross-section connected, for example, with maxima of the function v.

Concluding this section, let us mention the **spectral concentration**, i.e. the way in which the family of operators H_g, $g\neq 0$, remembers the 'dissolved' eigenvalue. For the Friedrichs model, it can be formulated exactly as follows:

PROPOSITION 3.2.7. *Let the assumptions of Theorem* 2.3 *be valid, and denote* $\Delta_g = (\lambda_0 - \alpha g, \lambda_0 + \alpha g)$ *with* $\alpha > |\lambda_u|$, *then*

$$\operatorname*{s-lim}_{g\to 0} E_{H_g}(\Delta_g) = E_u. \tag{2.21}$$

The proof is left to the reader (Problem 13). ∎

3.3. Bounded Perturbations of Embedded Eigenvalues

After the detailed discussion of the model example, we are going now to treat more general perturbations. Let us start with the observation that some of the assumptions made in the previous section may be easily relaxed. This concerns the so-called **multidimensional Friedrichs model** which is defined as follows: the state Hilbert space is $\mathscr{H} = \mathscr{H}_u \oplus \mathscr{H}_d$ with

$$\mathscr{H}_u = \mathbb{C}^n, \ \mathscr{H}_d = L^2(\mathbb{R}; \mathscr{G}), \tag{3.1}$$

where $\mathscr{G}$ is some auxiliary Hilbert space (referring to degeneracy of the energy spectrum). We denote by $\alpha = (\alpha_1, \ldots, \alpha_n)$ the elements of $\mathbb{C}^n$ and $Q_{\mathscr{G}}$: $(Q_{\mathscr{G}}\psi)(\lambda) = \lambda\psi(\lambda)$ on $\mathscr{H}_d$, then the free Hamiltonian H_0 and the perturbation V are defined by

$$H_0\{\alpha, f\} = \{(\lambda_1\alpha_1, \ldots, \lambda_n\alpha_n), Q_{\mathscr{G}} f\}, \tag{3.2a}$$

$$V\{\alpha, f\} = \left\{M\alpha + ((v_1, f)_{\mathscr{G}}, \ldots, (v_n, f)_{\mathscr{G}}), \sum_{j=1}^{n} \alpha_j v_j\right\}, \tag{3.2b}$$

where λ_j, $j = 1, \ldots, n$, are mutually distinct real numbers, $M \in \mathscr{B}(\mathbb{C}^n)$ is a Hermitean operator, $M^* = M$, and $v_j \in L^2(\mathbb{R}; \mathscr{G})$. Thus H_0 has n simple eigenvalues embedded in $\sigma_{ac}(H_0) = \mathbb{R}$, which is of multiplicity dim $\mathscr{G}$. In fact, the choice of the continuous spectrum is merely a matter of convenience; we would obtain similar conclusions with the $\mathscr{G}$-valued functions of $\mathscr{H}_d$ supported by $\mathbb{R}_+$ or some other subinterval of the real axis. The operator (3.2b) obviously fulfils the Friedrichs condition (2.3) again. Now the following assertion is valid:

THEOREM 3.3.1. *Let the operators* H_0, V *be given by* (3.2) *and* $H_g = H_0 + gV$. *Assume that there is an open region* $\Omega \supset \mathbb{R}$ *in* $\mathbb{C}$ *on which the functions* $(v_j(\,.\,), v_k(\,.\,))_{\mathscr{G}}$ *are analytically continuable across* $\mathbb{R}$, *and that* $v_j(\lambda_j) \neq 0$ *for* $j, k = 1, \ldots, n$. *Then* $\sigma_c(H_g) = \mathbb{R}$ *for any* g, *and the reduced resolvent* $R_u(\,.\,, H_g)$ *may be continued analytically across this spectrum from the upper halfplane to* $\Omega_- = \{z \in \Omega : \operatorname{Im} z < 0\}$. *Moreover, for all sufficiently small non-zero* g, *the continued function* $R_u^{II}(\,.\,, H_g)$ *has just* n *simple poles* $z_j(g) \in \Omega_-$, $j = 1, \ldots, n$.

The functions $z_j(\,.\,)$ are analytic around $g = 0$, and for g real, $z_j(g) = \lambda_j(g) - i\delta_j(g)$ are given by the relations

$$\lambda_j(g) = \lambda_j + gM_{jj} + g^2 \left(\sum_{k \neq j} \frac{|M_{kj}|^2}{\lambda_k - \lambda_j} + I(\lambda_j, v_j) \right) + O(g^3), \tag{3.3a}$$

$$\delta_j(g) = \pi g^2 \, \|v_j(\lambda_j)\|^2 + O(g^3), \tag{3.3b}$$

where M_{kj} are the matrix elements of M in the basis formed by the eigenvectors of H_0 and $I(\lambda, v_j)$ is defined by (2.7b) *with replacement of $|v_j(\xi)|^2$ by $\|v_j(\xi)\|^2$*

The proof follows the same line as in the case of the original Friedrichs model, and we leave it to the reader (Problem 14). ■

This theorem could be used as the starting point for further investigations of the model, similar to those performed in the preceding section; however, we are not going to pursue this task. Instead, we shall try to draw a lesson for the treatment of yet more general perturbations. In order not to overload our considerations with technicalities, we restrict ourselves to the case of bounded perturbations; on the other hand, we would like to relax the remaining requirements of the above-discussed model, namely the Friedrichs condition and simplicity of embedded eigenvalues.

Consider a fixed eigenvalue λ_j of the multidimensional model. The relations (3.3) show that, up to second-order terms in g, the distinction from the one-dimensional case (which refers to the requirement $M_{jk} = 0$ unless $j = k$) is confined in

$$g^2 \sum_{k \neq j} \frac{|M_{jk}|^2}{\lambda_k - \lambda_j}$$

which is the familiar second-order term from the 'stationary' perturbation theory. It expresses a shift of the considered eigenvalue under influence of $M - M_{jj}$. This suggests that one may try *to solve the problem in two steps*, starting with a part of the perturbation which is reduced by E_u and therefore shifts the eigenvalues of H_0.

Let us formulate the problem in a slightly more general way so that the perturbation considered in the first step is not necessarily linear in the coupling constant:

(p1) *Let $\mathscr{H} = \mathscr{H}_u \oplus \mathscr{H}_d$ with* dim $\mathscr{H}_u = m < \infty$, *and let the operator H_0 be self-adjoint with $\mathscr{H}_u$ as an eigenspace, $H_0 E_u = \lambda_0 E_u$ for some $\lambda_0 \in \mathbb{R}$. Suppose further that an operator family $\mathscr{T} = \{T_g : |g| < g_0\}$, $g_0 > 0$, is given with the following properties: T_g is self-adjoint for real g, $T_0 = H_0$; further, $\mathscr{T}$ is reduced by $\mathscr{H}_u$, $E_u T_g \subset T_g E_u$ for $|g| < g_0$ and, moreover, $\mathscr{T}$ is a holomorhic family of the type* (A), *i.e. the domain $D(T_g)$ is independent of g and $g \mapsto T_g$ is strongly analytic* (cf. Kato (1966a, §VII.2)). *Finally, let $\lambda_0 \in \sigma(\mathrm{pr_d}\, T_g)$ for all $|g| < g_0$.*

Thus we shall treat a single eigenvalue of multiplicity m embedded into $\sigma(\mathrm{pr_d}\, T_g)$ (as shown below, the condition (a) will ensure via Proposition 3.2 that these spectra are continuous around λ_0). Of course, we might consider n distinct eigenvalues of multiplicities $m_1, \ldots, m_n$; however, such a generalization would mean rather a clumsy notation than some essentially new information, and we rest therefore on the hypothesis formulated above. The assumptions about $\mathscr{T}$ are easily seen to be fulfilled when the latter consists of the operators $T_g = H_0 + gV_1$, where V_1 is a bounded self-adjoint operator reduced by $\mathscr{H}_u$. We introduce the following abbreviations: $U_0 = \{g : |g| < g_0\}$, and $E_g(\,.\,)$ for $g \in U_0 \cap \mathbb{R}$ means the spectral measure $E_{T_g}(\,.\,)$.

In the second step, we shall consider a perturbation gV to T_g, where the following assumptions concerning V are made:

(p2) *$V \in \mathscr{B}(\mathscr{H})$ is Hermitean and factorizes, i.e. there are $A, B \in \mathscr{B}(\mathscr{H})$ such that $V = B^*A$. Consequently, we have $H_g = T_g + gB^*A$.*

Of course, there are many such factorizations. Below we shall restrict this arbitrariness by additional requirements; then the factorization will prove useful for finding suitable candidates for the poles of $R_u^{II}(\,.\,, H_g)$. The first of these requirements may be deemed as a sort of *minimality condition*:

(m) *$\mathscr{H}$ is spanned by the set $M_g(V) = \{E_g(\Delta)\psi : \psi \in \mathrm{Ran}\, V, \Delta \in \mathrm{Bor}(\mathbb{R})\}$ for any $g \in U_0 \cap \mathbb{R}$, where $\mathrm{Bor}(\mathbb{R})$ is the σ-algebra of Borel sets in $\mathbb{R}$.*

In order to realize the meaning of condition (m), suppose that it is not valid and denote by P_g, $\bar{P}_g$ the projections to the closed linear span of $M_g(V)$ and to its orthogonal complement in $\mathscr{H}$, respectively. If $\psi \in \bar{P}_g\mathscr{H}$, then

$$(\psi, E_g(\Delta)V\varphi) = 0, \qquad \Delta \in \mathrm{Bor}(\mathbb{R}), \qquad \varphi \in \mathscr{H}, \tag{3.4}$$

so $\psi \in \mathrm{Ker}\, VE_g(\Delta)$ for all $\Delta \in \mathrm{Bor}(\mathbb{R})$. Choosing particularly $\Delta = \mathbb{R}$ we see that V is reduced by P_g and its part in $\bar{P}_g\mathscr{H}$ is zero. The relation (3.4) further implies $E_g(\Delta)\bar{P}_g\mathscr{H} \subset \bar{P}_g\mathscr{H}$ for each $\Delta \in \mathrm{Bor}(\mathbb{R})$, so $E_g(\Delta)$ is also reduced by P_g, and consequently, T_g is reduced by P_g. Now if P_g is *assumed* independent of g, $P_g = P$, then we are able to fulfil (m) by considering projections of the operators to $P\mathscr{H}$ only; however, we have seen that their complements are irrelevant to our perturbation problem. It is illustrative to inspect how the condition (m) looks like in the Friedrichs model.

Let us now introduce some further notation. We shall use $\phi(z, g) = \phi_u(z, g) + \phi_d(z, g)$ with $\phi_j(z, g) = AE_j(T_g - z)^{-1}E_jB^* = AR_j(z, T_g)B^*$, $j = \mathrm{u, d}$. Particularly for $g = 0$, we have

$$\phi_u(z, 0) = \frac{AE_uB^*}{\lambda_0 - z}. \tag{3.5a}$$

The finite-dimensional perturbation theory gives the behaviour of $\phi_u(z,\,.\,)$ around $g = 0$: the fact that $\mathrm{pr}_u\, T_g$ are Hermitean for real g together with the analycity requirement on $\mathscr{T}$ imply (cf. Kato (1966a, theorem II.6.1)) that there is a neighbourhood $U_1 \subset U_0$ of 0 and analytic functions $\lambda_j : U_1 \to \mathbb{C}$ and $P_j : U_1 \to$ (projections on $\mathscr{H}_u$), $j = 1, \ldots, r$, such that $\mathrm{pr}_u\, T_g = \Sigma_{j=1}^{r} \lambda_j(g)P_j(g)$. Moreover, the functions $\lambda_j(\,.\,)$ are mutually distinct except at the point $g = 0$. Thus we obtain

$$\phi_u(z, g) = \sum_{j=1}^{r} \frac{AP_j(g)B^*}{\lambda_j(g) - z}, \tag{3.5b}$$

and the function ϕ_u is analytic in $(\mathbb{C} \times U_1)\backslash\Sigma_0$, where $\Sigma_0 = \{(z, g) : z = \lambda_j(g), g \in U_1, j = 1, \ldots, r\}$.

As for the other part of $\phi(z, g)$, we adopt the following assumption:

(a) *There is a neighbourhood Z_0 of λ_0 such that the function $\phi_d(\,.\,, g)$ has an analytic continuation from the upper (lower) halfplane to $Z_0^{\mp} = \{z \in Z_0 : \operatorname{Im} z \lessgtr 0\}$ for each $g \in U_0$, and that these continuations as functions of both z, g are analytic in $Z_0 \times U_0$.*

The first named continuation is again the more important; we denote this by ϕ_d^{II}. Notice that in the particular case when $A = B$, the relation $\phi_d(z, g)^* = \phi_d(\bar{z}, g)$, $z \notin \sigma_p(T_g)$, follows from (1.3), and the assumption (a) may concern the existence and analycity of $\phi_d^{II}(\,.\,,\,.\,)$ alone. Now the conditions (p1), (m) and (a) together imply that λ_0 is actually embedded in the continuous spectrum of the operators T_g (especially, of H_0).

PROPOSITION 3.3.2. *Assume* (p1), (m) *and* (a), *then the spectrum of $E_g(Z_0 \cap \mathbb{R})\mathrm{pr}_d\, T_g$ is absolutely continuous for each real $g \in U_0$, and contains λ_0 as its interior point.*

Proof. The operators $F_g(\Delta) = \mathrm{pr}_d\, E_g(\Delta)$ can be expressed by means of (1.1), so

$$\tfrac{1}{2}A\,\{F_g([\lambda, \mu]) + F_g((\lambda, \mu))\}B^*\psi$$

$$= \frac{1}{2\pi i} \lim_{\eta \to 0+} \int_{\lambda}^{\mu} [\phi_d(\xi + i\eta, g) - \phi_d(\xi - i\eta, g)]\,\psi\, d\xi$$

for all $\lambda < \mu$ and $\psi \in \mathscr{H}$. If $[\lambda, \mu] \subset Z_0$, then the limit of the integrand exists due to (a) and is absolutely continuous as a function of ξ. Thus

$$AE_dE_g((-\infty, \lambda))E_dB^*\psi = \frac{1}{2\pi i}\,[\phi_d(\lambda + i0, g) - \phi_d(\lambda - i0, g)]\,\psi$$

and the function $(\varphi, E_dE_g(-\infty, .\,))E_d\varphi)$ is absolutely continuous in $Z_0 \cap \mathbb{R}$ for each $\varphi \in \mathrm{Ran}\, V$, because $V = B^*A = A^*B$ so $\mathrm{Ran}\, V$ is contained in both the ranges of B^*, A^*. For an arbitrary interval $\Delta \subset \mathbb{R}$, $E_g(\Delta)$ commutes with E_d, so by (m) the functions $(\varphi, E_dE_g((-\infty, .\,))E_d\varphi)$ are absolutely continuous in $Z_0 \cap \mathbb{R}$ for all $\varphi \in \mathscr{H}$ and $g \in U_0 \cap \mathbb{R}$. Finally, these functions are restrictions to $\mathbb{R}$ of functions analytic in Z_0, and at least some of them are non-constant in $Z_0 \cap \mathbb{R}$ because $\lambda_0 \in \sigma(\mathrm{pr}_d\, T_g)$ due to (p1). ∎

Our aim is, of course, to find singularities of the analytically continued reduced resolvent $R_u^{II}(\,.\,, H_g)$. We shall approach this problem indirectly, treating first the operator-valued function

$$X : X(z, g) = AR(z, H_g)B^* \tag{3.6}$$

and the analytic continuation $X^{II}(\,.\,, g)$ of $X(\,.\,, g)$ from the upper halfplane to Z_0^-, because the singularities of $X^{II}(\,.\,, g)$ can be found algebraically.

THEOREM 3.3.3 (Baumgärtel). *Let* (p1), (p2), (m) *and* (a) *be valid. Then there is a neighbourhood* $Z \times U \subset Z_0 \times U_1$ *of* $(\lambda_0, 0)$ *such that* $X^{II}(\,.\,,\,.\,)$ *is meromorphic in* $Z \times U$ *with the manifold of singularities* $\Sigma = \{(z, g) : \gamma(z, g) = 0\}$, *where*

$$\gamma(z, g) = \det\{\mathrm{pr_u}(-z + T_g + gB^*G(z, g)A\}, \tag{3.7}$$

$$G(z, g) = (I + g\phi_d^{II}(z, g))^{-1}. \tag{3.8}$$

Proof. First, we express $X(z, g)$ by mean of $\phi(z, g)$. To this end, the identity

$$R(z, H_g) = R(z, T_g) - gR(z, T_g)B^*(I + gAR(z, T_g)B^*)^{-1}AR(z, T_g) \tag{3.9}$$

is used, which yields the relation

$$X(z, g) = \phi(z, g)\,(I + g\phi(z, g))^{-1}.$$

Now according to (3.5b) and (a), the function $\phi(\,.\,, g) = \phi_u(\,.\,, g) + \phi_d(\,.\,, g)$ has a meromorphic continuation $\phi^{II}(\,.\,, g)$ from the upper halfplane to Z_0^- for each $g \in U_1$. Replacing $\phi(z, g)$ by $\phi^{II}(z, g)$ in the last relation, we obtain a suitable candidate for $X^{II}(z, g)$:

$$X^{II}(z, g) = \begin{cases} g^{-1}\,[I - (I + g\phi^{II}(z, g))^{-1}] & \ldots \quad g \neq 0 \\ \phi^{II}(z, 0) & \ldots \quad g = 0 \end{cases} \tag{3.10}$$

In order to check that the function defined by (3.10) is actually the desired one, we must first determine its manifold of singularities Σ in $(Z_0 \times U_1)\backslash\Sigma_0$. Let us choose some neighbourhood $Z_1 \times U_2$ of $(\lambda_0, 0)$ such that its closure lies in $Z_0 \times U_1$, then $\sup\{\|\phi_d^{II}(z, g)\| : (z, g) \in Z_1 \times U_2\} = K < \infty$ in view of (a). It further implies that the Neumann series

$$(I + g\phi_d^{II}(z, g))^{-1} = \sum_{k=0}^{\infty} (-g\phi_d^{II}(z, g))^k \tag{3.8a}$$

converges in the operator norm uniformly in $Z_1 \times U_3$, where $U_3 = \{g \in U_2 : |g| < K^{-1}\}$, i.e. that the function G defined by (3.8) is analytic in $Z_1 \times U_3$. This function may be used for expression of $X^{II}(z, g)$; with the help of a simple operator identity one finds

$$X^{II}(z, g) = g^{-1}I - g^{-1}G(z, g)\,[I + g\phi_u(z, g)G(z, g)]^{-1}. \tag{3.11}$$

Thus for $g \neq 0$, *the function* X^{II} *has the same singularities as* $(I + g\phi_{\mathrm{u}}G)^{-1}$ within $Z_1 \times U_3$; for $g = 0$ we have $X^{\mathrm{II}}(z, 0) = \phi^{\mathrm{II}}(z, 0)$ with the only singularity at $z = \lambda_0$.

Obviously, $I + g\phi_{\mathrm{u}}(z, g)G(z, g)$ makes no sense for $z = \lambda_j(g)$ and g small enough, so we have to find the points of $(Z_1 \times U_3)\backslash\Sigma_0$ for which the operator

$$I + g\sum_{j=1}^{r} (\lambda_j(g) - z)^{-1} AP_j(g)B^*G(z, g) \tag{3.12}$$

is not invertible. An operator $I + C$ with C of the trace class is invertible iff $\det(I + C)$ is non-zero, where the latter is defined as the product of eigenvalues repeated according to their multiplicity. In our case, the operator C is even of a finite rank. Moreover, the determinant in question may be calculated from the $m \times m$ matrix only: we shall show that *the operator* (3.12) *is non-invertible iff*

$$\det\left\{\mathrm{pr}_{\mathrm{u}}\left(I + g\sum_{j=1}^{r} \frac{P_j(g)}{\lambda_j(g) - z} B^*G(z, g)A\right)\right\} = 0. \tag{3.13}$$

To this end, we check first that $A_{\mathrm{u}} = A \restriction \mathscr{H}_{\mathrm{u}}$ is regular. Suppose that $(\psi, V\varphi) = 0$ for some $\psi \in \mathscr{H}_{\mathrm{u}}$ and all $\varphi \in \mathscr{H}$. For a fixed $g \in U_0 \cap \mathbb{R}$, we have $\psi = E_g(\{\lambda_0\})\psi$ so $(\psi, E_g(\{\lambda_0\})V\varphi) = (E_g(\{\lambda_0\} \cap \Delta)\psi, V\varphi) = 0$ for each $\Delta \in \mathrm{Bor}(\mathbb{R})$ and $\psi = 0$ in view of (m). Further, V is Hermitean, and therefore no non-zero vector $\psi \in \mathscr{H}_{\mathrm{u}}$ may belong to $\mathrm{Ker}\, V \supset \mathrm{Ker}\, A$. This means that A maps $\mathscr{H}_{\mathrm{u}}$ onto some m-dimensional subspace, say $\mathscr{H}_A$, of $\mathscr{H}$. Now let us take a unitary operator $S \in \mathscr{B}(\mathscr{H}_{\mathrm{d}}, \mathscr{H}_A^{\perp})$ and construct $R_\eta = AE_{\mathrm{u}} + \eta SE_{\mathrm{d}}$ for $\eta > 0$. This operator is clearly invertible with $R_\eta^{-1} = A_{\mathrm{u}}^{-1}E_A + \eta^{-1}S^{-1}\bar{E}_A$. The operator $I + C$ defined by (3.12) is then non-invertible iff the determinant of

$$R_\eta^{-1}(I + C)R_\eta = I + R_\eta^{-1}(AE_{\mathrm{u}})\left(g\sum_{j=1}^{r} \frac{P_j(g)}{\lambda_j(g) - z} B^*G(z, g)\right)R_\eta$$

$$= I + \left(g\sum_{j=1}^{r} \frac{P_j(g)}{\lambda_j(g) - z} B^*G(z, g)\right)(AE_{\mathrm{u}} + \eta SE_{\mathrm{d}})$$

equals zero. However, the value of the determinant is not changed by similarly transformations, so the last assertion also holds for the limit $\eta \to 0+$. Finally, $\det(I + C_1 + \eta C_2)$ with C_1, C_2 of the trace class is continuous (even analytic) with respect to η (Dunford and Schwartz, 1963, lemma XI.9.17), and therefore non-invertibility of the operator under consideration is equivalent to

$$\det\left\{I + g\sum_{j=1}^{r} \frac{P_j(g)}{\lambda_j(g) - z} B^*G(z, g)AE_{\mathrm{u}}\right\} = 0. \tag{3.13a}$$

However, the operator in the curly bracket is reduced by E_{u} and its part in $\mathscr{H}_{\mathrm{d}}$ is I_{d}, so values of the determinants in (3.13) and (3.13a) coincide.

Now we may write $\mathrm{pr_u}\, I = E_\mathrm{u}$ in (3.13) as $\Sigma_{j=1}^r P_j(g)(\lambda_j(g) - z)^{-1} T_g$ and use $\det MN = \det M \,.\, \det N$; then the determinant (3.13) acquires the form of the product $\gamma_0(z, g)^{-1}\gamma(z, g)$, where

$$\gamma_0(z, g) = \prod_{j=1}^{r} (\lambda_j(g) - z)^{m_j} \tag{3.14}$$

with $m_j = \dim P_j(g)$ and $\gamma(z, g)$ is given by (3.7). Hence *the manifold of singularities of* $X^{\mathrm{II}}(\,.\,,.\,)$ *in* $(Z_1 \times U_3)\backslash\Sigma_0$ *is given by the condition* $\gamma(z, g) = 0$.

Further, we shall show that one need not exclude the points of Σ_0, because *each point* $(z, g) \in \Sigma_0$ *with* $\gamma(z, g) = 0$ *is a regularity point of* $X^{\mathrm{II}}(\,.\,,.\,)$ *within* $Z_1 \times U_3$. To this purpose, we choose some (arbitrary, but fixed) $g_0 \in U_3$ and j_0 such that $\gamma(\lambda_{j_0}(g_0), g_0) \neq 0$. The function $\gamma(\,.\,,.\,)$ is easily seen to be analytic in $Z_0 \times U_1$, so there is a neighbourhood $Z' \times U' \subset Z_1 \times U_3$ of $\{\lambda_{j_0}(g_0), g_0\}$ in which $|\gamma(z, g)| \geqslant K_1 > 0$. In view of (3.11), it is sufficient to check that $(I + g\phi_\mathrm{u} G)^{-1}$ is regular at the given point. The norm of this operator can be estimated as follows (cf. (A.10)):

$$\|(I + g\phi_\mathrm{u}(z, g)G(z, g))^{-1}\| \leqslant \frac{\Pi_{k=1}^m (1 + |g|\mu_k(\phi_\mathrm{u}(z, g)G(z, g)))}{|\det(I + g\phi_\mathrm{u}(z, g)G(z, g))|}, \tag{3.15}$$

where $\mu_k(\phi_\mathrm{u} G)$ are singular values of $\phi_\mathrm{u} G$ repeated according to multiplicity, and

$$\mu_k(\phi_\mathrm{u}(z, g)G(z, g)) \leqslant \|A\|\,\mu_k\left(\sum_{j=1}^{r} \frac{P_j(g)}{\lambda_j(g) - z}\right) \|B^*\|\,\|G(z, g)\|.$$

Further, $P_j(g) = U_j(g)P_j(0)U_j(g)^{-1}$, where the functions $U_j(\,.\,)$, $U_j(\,.\,)^{-1}$ are bounded (unitary-valued for real g) and analytic in U_1 (cf. Kato (1966, §II.6.2)), and $\|G(\,.\,,.\,)\|$ is bounded in $Z_1 \times U_3$ so there is a constant K_2 such that

$$\mu_k(\phi_\mathrm{u}(z, g)G(z, g)) \leqslant K_2\mu_k\left(\sum_{j=1}^{r} \frac{P_j(g)}{\lambda_j(g) - z}\right). \tag{3.16}$$

The singular values of the last operator are $|\lambda_j(g) - z|^{-1}$ (counted with multiplicities), and the determinant in the denominator of (3.15) equals $\gamma(z, g)/\gamma_0(z, g)$, so (3.15) and (3.16) give

$$\|(I + g\phi_\mathrm{u}(z, g)G(z, g))^{-1}\| \leqslant \left|\frac{\gamma_0(z, g)}{\gamma(z, g)}\right| \prod_{j=1}^{r} (1 + |g|\, K_2\, |\lambda_j(g) - z|^{-1})^{m_j}$$

$$\leqslant K_1^{-1} \prod_{j=1}^{r} (|\lambda_j(g) - z| + K_2\, |g|)^{m_j}.$$

Thus the function under consideration is analytic and bounded in $(Z' \times U')\backslash\Sigma_0$, so by the Riemann theorem (on extension of analytic functions, cf. Fuks (1962, theorem 6.7)), it is also analytic in $Z' \times U'$. Consequently, *$X^{\mathrm{II}}(\,.\,,.\,)$ is analytic in $(Z_1 \times U_3)\backslash\Sigma$*, where Σ is determined by the condition $\gamma(z, g) = 0$.

It remains to prove that X^{II} is *meromorphic*. In the same way as above, the norm $\|A(z, g)\|$, $A(z, g) = \gamma(z, g)(I + g\phi_{\mathrm{u}}(z, g)G(z, g))^{-1}$, can be estimated by some constant within $Z_1 \times U_3$ (now we need not exclude the points of Σ) so the function $A(\,.\,,.\,)$ is analytic there by the Riemann theorem. We have to show that $(z, g) \mapsto \gamma(z, g)X^{\mathrm{II}}(z, g)$ is analytic. Since $G(z, g) = I + g\widetilde{G}(z, g)$ with $\widetilde{G}(\,.\,,.\,)$ analytic in $Z_1 \times U_3$, it is sufficient in view of (3.11) to check that

$$g^{-1}\gamma(z,g)\;[I - (I + g\phi_{\mathrm{u}}(z,g)G(z,g))^{-1}]$$

is analytic with respect to (z, g). For a non-zero g, the analycity is obvious; as for $g = 0$, we may rewrite the last expression as

$$-\frac{A(z,g) - A(z,0)}{g} + \frac{\gamma(z,g) - \gamma(z,0)}{g},$$

because $A(z, 0) = \gamma(z, 0) = (\lambda_0 - z)^m$. However, $A(z,\,.\,)$ is analytic in U_3 and $\gamma(z,\,.\,)$ is a polynomial composed of analytical functions, so $\gamma(\,.\,,.\,)X^{\mathrm{II}}(\,.\,,.\,)$ is analytic in $Z_1 \times U_3$. Hence the theorem is proved for $Z \times U = Z_1 \times U_3$. ■

The next question concerns the structure of the manifold of singularities Σ. We know that in the Friedrichs model, the positions of the poles born by simple eigenvalues are analytical functions of the coupling constant. It turns out that this is not necessarily true when the embedded eigenvalue is not simple.

THEOREM 3.3.4 (Howland and Baumgärtel). *Under the assumptions of Theorem 3.3, singularities of $X^{\mathrm{II}}(\,.\,,.\,)$ are grouped into 'Puiseux cycles' $\{z_{j1}(g), \dots, z_{jp_j}(g)\}$, $j = 1, \dots, s$, with the multiplicities μ_j, $\sum_{j=1}^{s} \mu_j p_j = m$. For the members of the* jth *cycle, we have the 'Puiseux series'*

$$z_{jk}(g) = \lambda_0 + \sum_{n=1}^{\infty} \beta_{jn} \exp(2\pi ikn/p_j)g^{n/p_j}, \quad k = 1, \dots, p_j. \tag{3.17}$$

The inequality

$$\operatorname{Im} z_{jk}(g) \leqslant 0, \quad g \in U \cap \mathbb{R} \tag{3.18}$$

holds for all $j = 1, \dots, s$, $k = 1, \dots, p_j$. Furthermore, for each cycle we have the following alternative: either there is a positive integer q such that

$$z_{jk}(g) = \lambda_0 + \sum_{l=1}^{2q-1} \beta_{j,lp_j} g^l + \beta_{j,2qp_j} g^{2q} + O_k(g^{2q+1/p_j}), \tag{3.19}$$

where β_{j,lp_j}, $l = 1, \ldots, 2q-1$, *are real and* $\operatorname{Im} \beta_{j,2qp_j} < 0$, *or all the coefficients* β_{jn} *in* (3.17) *are real and* $p = 1$.

Proof. Since the singularities are determined by zeros of the function γ which is analytic in $Z \times U$, the grouping into cycles and the expansion (3.17) follow from the Weierstrass preparation theorem (cf. (A.11)). According to (3.6), $X^{\mathrm{II}}(\,.\,,.\,)$ has no singularities in the region, where the resolvent is analytic, so (3.18) is valid. This condition further implies

$$\operatorname{Im} \beta_{jn} \exp(2\pi ikn/p_j)\,(\operatorname{sgn} g)^{n/p_j} = \operatorname{Im} \beta_{jn} \exp\left[\frac{\pi in}{p_j}\left(2k + \frac{1-\operatorname{sgn} g}{2}\right)\right] \leqslant 0$$

for a fixed j, all $k = 1, \ldots, p_j$ and $g \in \mathbb{R}$, up to the first term in (3.17) which gives a non-real contribution. It follows that

$$\sin\left(\frac{\pi n}{p_j} r + \arg \beta_{jn}\right) \leqslant 0$$

must hold for $r = 0, 1, 2, \ldots$, but this condition requires either $n = qp_j$ with q integer and β_{jn} real or $n = 2qp_j$ and $\operatorname{Im} \beta_{jn} \leqslant 0$. If all the coefficients β_{jn} are real, then p_j may be 1 or 2, because of the primitive roots of unity appearing in (3.17). Suppose that $p = 2$ and consider the lowest non-zero term $\beta_{jn}(-1)^k g^{n/2}$ with n even. A change of sign of g multiplies it by i, so the imaginary part is positive for one of the values $k = 1, 2$; but this contradicts (3.18). ■

Now we shall return to the problem in which we are actually interested. Let us inspect how it is connected to the above considerations in the case when the Friedrichs condition (2.3) holds.

THEOREM 3.3.5 (Baumgärtel). *In addition to the hypotheses of Theorem* 3.3, *assume that* $E_{\mathrm{d}} V E_{\mathrm{d}} = 0$. *Then the (analytically continued) reduced resolvent of* H_g *is given by*

$$R_{\mathrm{u}}^{\mathrm{II}}(z, H_g) = (-z + W(z,g))^{-1}, \tag{3.20a}$$

$$W(z,g) = \operatorname{pr}_{\mathrm{u}}(T_g + gV - g^2 V R_{\mathrm{d}}^{\mathrm{II}}(z, T_g) V). \tag{3.20b}$$

This is a meromorphic function in $Z_0 \times U_0$ *and its singularities coincide with those of* $X^{\mathrm{II}}(\,.\,,.\,)$ *wherever the latter is meromorphic, independently of the factorization* $V = B^*A$.

Proof. The formulae (3.20) are obtained by an easy modification of the proof, which provide us with formula (2.5a). The function $(z,g) \mapsto -z + W(z,g)$ is analytic within $Z_0 \times U_0$ because of (p1), (a) and the relation

$$V R_{\mathrm{d}}^{\mathrm{II}}(z, T_g) V = B^* \phi_{\mathrm{d}}^{\mathrm{II}}(z,g) A. \tag{3.21}$$

Thus $(z,g) \mapsto R_{\mathrm{u}}^{\mathrm{II}}(z, H_g)$ is also analytic, with the exception of eventual singular points given by the condition

$$\det R_{\mathrm{u}}^{\mathrm{II}}(z, H_g) = 0. \tag{3.22}$$

Consider now a factorization $V = B^*A$. The condition $E_d B^* A E_d = 0$ implies $\phi_d(z, g)^2 = 0$, so the expansion (3.8a) is reduced to $G(z, g) = I - g\phi_d^{II}(z, g)$. Then the relations (3.7) and (3.21) show that the condition $\gamma(z, g) = 0$ coincides with (3.22) when both of them are applicable. ■

Thus the Friedrichs condition makes the solution easier. Moreover, at a glance it is a modest restriction only within the present context: having $H_g = H_0 + gU$ with a Hermitean $U \in \mathscr{B}(\mathscr{H})$, one may always set

$$T_g = H_0 + g(E_u U E_u + E_d U E_d), \tag{3.23a}$$

$$V = E_u U E_d + E_d U E_u \tag{3.23b}$$

so that $\mathrm{pr}_j\, V = 0$ for both $j = \mathrm{u}, \mathrm{d}$. Hence, one might ask why to bother with the factorizations at all. However, it may occur that verification of the conditions (m), (a) and partly (p1), too, is difficult with the decomposition (3.23). In such a case, there is an alternative way, namely to move the 'diagonal' terms or a part of them from T_g to V, and to find the singularities using a suitable factorization of the new V. Then, of course, one must show that the singularities do not depend on a special choice of the factorization and refer actually to the reduced resolvent, which, in general, is not less difficult. There is also a third possibility: to look for singularities of the reduced resolvent directly using specific properties of the operators involved. In particular, some methods were elaborated recently for Schrödinger operators; references are given in the notes to this section.

According to Theorem 3.4, degeneracy of the embedded eigenvalue means that the positions of the poles which it generates need not depend analytically on the coupling constant, and this non-analycity is expressed through the fractional-power series. This assertion is not void: one can construct a simple model in which the singularities are described by a non-analytic Puiseux series (Problem 15). Needless to say, the particular character of non-analycity discussed here is conditioned by the used assumptions. It is not necessarily true, for instance, when λ_0 is not actually embedded in the continuous spectrum: let us recall Example 2.5, where obviously the expansions (2.12b, c) are not of the Puiseux type if $|\alpha| \neq \frac{1}{2}$.

Another characteristic feature is that the degeneracy of the unperturbed eigenvalue may be removed by the perturbation. According to Theorem 3.4, the original pole in the reduced resolvent splits in general when the perturbation is 'switched on', and its daughter poles either stay on the real axis or travel to the second sheet. We shall consider this effect in the next section; here we limit ourselves to the simplest case where the splitting is maximal possible, and we derive the *Fermi golden rule* again.

Suppose that $g \in U_0 \cap \mathbb{R}$. The operators $\mathrm{pr}_u H_g$ may be diagonalized; further, (p1) implies that $g \mapsto H_g$ is analytic. Hence there is a neighbourhood $U_4 \subset U_0$ of 0 and analytic functions $\lambda^{(k)} : U_4 \to \mathbb{R}$, $\varphi_k : U_4 \to \mathscr{H}_u$, such that $H_g \varphi_k(g) = \lambda^{(k)}(g)\varphi_k(g)$ for $k = 1, \ldots, m$. Of course, the functions $\lambda^{(k)}$ coincide fully with the above used λ_j referring to T_g only if $\mathrm{pr_u}\, V = 0$. Let

$$\lambda^{(k)}(g) = \lambda_0 + \lambda_{k1} g + \sum_{n=2}^{\infty} \lambda_{kn} g^n, \tag{3.24a}$$

then we say that the perturbation *removes degeneracy in the first order* (respectively to g) if the numbers $\lambda_{11}, \ldots, \lambda_{m1}$ are mutually different.

THEOREM 3.3.6. *Let the hypotheses of Theorem* 3.3 *be valid, and assume further that the perturbation removes degeneracy of* λ_0 *in the first order. Then for all sufficiently small real* g, *the function* $X^{\mathrm{II}}(\,.\,, g)$ *defined by analytic continuation of* (3.6) *has just* m *simple poles, and their imaginary parts fulfil*

$$-\delta_j(g) = -\pi g^2 \frac{\mathrm{d}}{\mathrm{d}\lambda} (V\varphi_j, E_\lambda^0 E_\mathrm{d} V\varphi_j)\Big|_{\lambda=\lambda_0} + O(g^3), j = 1, \ldots, m, \tag{3.25a}$$

where the spectral decomposition $\{E_\lambda^0\}$ *refers to* H_0 *and* $\varphi_j = \varphi_j(0)$.

REMARK 3.3.7. If the Friedrichs condition holds, then the poles are at the same time poles of $R_\mathrm{u}^{\mathrm{II}}(\,.\,, H_g)$ by Theorem 3.5. In such a case, the decay rates corresponding to the second-sheet poles are

$$\Gamma_j(g) = 2\pi g^2 \frac{\mathrm{d}}{\mathrm{d}\lambda} (V\varphi_j, E_\lambda^0 P_\mathrm{c}(H_0) V\varphi_j)\Big|_{\lambda=\lambda_0} + O(g^3). \tag{3.25b}$$

Replacement of E_d by $P_\mathrm{c}(H_0)$ in (3.25b) is justified by Proposition 3.2, which implies that

$$\|E_\lambda^0 E_\mathrm{d} V\varphi_j\|^2 = \mathrm{const} + \|E_\lambda^0 P_\mathrm{c}(H_0) V\varphi_j\|^2$$

holds in some neighbourhood of λ_0. Notice further that the relation (3.25a) is independent of the used factorization $V = B^*A$.

Proof of the theorem. According to Theorems 3.3 and 3.4, the poles of $X^{\mathrm{II}}(\,.\,,g)$ exist and

$$z_j(g) = \lambda_0 + \beta_{j1} g + \beta_{j2} g^2 + \tilde{z}_j(g), \tag{3.24b}$$

where $\tilde{z}_j(g) = O(g^{2+\epsilon})$ with $\epsilon > 0$. Using (3.7), (3.8) and (3.8a), the equation $\gamma(z, g) = 0$ which determine the singularities may be rewritten as

$$\det\left\{ \mathrm{pr_u}\left(-z + T_g + gB^*A + g \sum_{n=1}^{\infty} B^*(-g\phi_\mathrm{d}^{\mathrm{II}}(z, g))^n A\right)\right\}$$

$$= \det\{\mathrm{pr_u}(-z + H_g - g^2 B^* \phi_\mathrm{d}^{\mathrm{II}}(z, g) A + O(g^3))\}$$

$$= \det\{(\lambda^{(k)}(g) - z)\delta_{kl} - g^2 Y_{kl}(z, g) + O_{kl}(g^3)\} = 0,$$

where

$$Y_{kl}(z, g) = (B\varphi_k(g), \phi_\mathrm{d}^{\mathrm{II}}(z, g) A\varphi_l(g)). \tag{3.26}$$

The functions defined by (3.26) are analytic in $Z_0 \times U_4$ because of (a) and the analycity of $\varphi_j(\,.\,)$, so $Y_{kl}(z_j(\,.\,)\,,\,.\,)$ is analytic in some neighbourhood of 0 and $Y_{kl}(z_j(g), g) = Y_{kl}(\lambda_0, 0) + O(g)$. Substituting this together with (3.24) to the last equation and dividing it by g^m, we get

$$\det\{(\lambda_{k1} - \beta_{j1})\delta_{kl} + g(\lambda_{k2} - \beta_{j2})\delta_{kl} - gY_{kl}(\lambda_0, 0) -$$
$$- g^{-1}\tilde{z}_j(g)\delta_{kl} + O_{kl}(g^2)\} = 0. \tag{3.27}$$

Up to the first order in g, this determinant equals $\Pi_{k=1}^{m}(\lambda_{k1} - \beta_{j1}) + O(g)$, but the numbers $\lambda_{11}, \ldots, \lambda_{m1}$ are mutually different due to the assumption, and therefore

$$\beta_{j1} = \lambda_{j1}, \qquad j = 1, \ldots, m. \tag{3.28}$$

In view of Theorem 3.4, each Puiseux cycle consists therefore of just one root, so $\tilde{z}_j(g) = O(g^3)$ may be included in the remainder term. Using this fact together with (3.28), we are able to rewrite equation (3.27) in the form

$$g \prod_{k \neq j} (\lambda_{k1} - \lambda_{j1})(\lambda_{j2} - \beta_{j2} - Y_{jj}(\lambda_0, 0)) + O(g^2) = 0,$$

which yields $\beta_{j2} = \lambda_{j2} - Y_{jj}(\lambda_0, 0)$. Further, all the coefficients λ_{jn} are real so the definition of $\phi_{\mathrm{d}}^{\mathrm{II}}$ together with hermiticity of V give

$$\begin{aligned}\operatorname{Im}\beta_{j2} &= -\operatorname{Im}(B\varphi_j, \phi_{\mathrm{d}}^{\mathrm{II}}(\lambda_0, 0)A\varphi_j)\\ &= -\operatorname{Im}(B\varphi_j, AR_{\mathrm{d}}(\lambda_0 + i0, H_0)B^*A\varphi_j)\\ &= -\operatorname{Im}(V\varphi_j, R_{\mathrm{d}}(\lambda_0 + i0, H_0)V\varphi_j).\end{aligned}$$

Finally, we may rewrite the last expression in terms of the spectral decomposition of H_0 projected to $\mathscr{H}_{\mathrm{d}}$ in analogy with the proof of Proposition 3.2; we obtain

$$\begin{aligned}\operatorname{Im}\beta_{j2} &= -\frac{1}{2i}(V\varphi_j, [R_{\mathrm{d}}(\lambda_0 + i0, H_0) - R_{\mathrm{d}}(\lambda_0 - i0, H_0)]V\varphi_j)\\ &= -\pi \frac{\mathrm{d}}{\mathrm{d}\lambda}(V\varphi_j, E_\lambda^0 E_{\mathrm{d}} V\varphi_j)\Big|_{\lambda=\lambda_0},\end{aligned}$$

i.e. the desired result. ∎

3.4. Symmetries and Broken Symmetries

The total energy characterized by the Hamiltonian H belongs to some complete set of compatible observables which is used to characterize states of the considered system. In general, there are many such complete sets but this arbitrariness is actually very limited, because we should choose among the 'standard' observables such as the components of momenta, spins, isospins, etc. Members of the complete set are frequently useful for simplification of the problem. An observable is called **dynamical symmetry** of the considered system (with respect to H) if the corresponding self-adjoint operator A on $\mathscr{H}$ commutes with H, or equivalently, with the propagator U_t for all real t. We may denote $U(\xi) = \exp(iA\xi)$, then another equivalent reformulation of the last statement says that $[U_t, U(\xi)] = 0$ for all $t, \xi \in \mathbb{R}$. This is useful because often the unitary group $\{U(\xi) : \xi \in \mathbb{R}\}$ has a direct physical meaning. Of course, all subspaces of the type $E_A(\Delta)\mathscr{H}$ with $\Delta \in \mathrm{Bor}(\mathbb{R})$ are preserved during the time evolution mediated by U_t; in particular, if ψ is an eigenvector of A corresponding to an eigenvalue a, then the same is true for the vectors $U_t\psi$.

Among these symmetries, some are important from the viewpoint of decay; namely, those for which the operator A (or equivalently, the operators $U(\xi)$) is reduced at the same time by the projection E_{u}. Such a symmetry we shall call **the symmetry of decay**. Clearly $U(\xi)$ commutes with $E_{\mathrm{u}}U_tE_{\mathrm{u}}$, and therefore

$$\mathrm{Tr}\{V_tU(\xi)\rho U(-\xi)V_t^*\} = \mathrm{Tr}\{U(\xi)V_t\rho V_t^*U(-\xi)\} = \mathrm{Tr}\{\rho V_t^*V_t\}.$$

Hence we obtain the following assertion:

PROPOSITION 3.4.1. *Let* $U(\xi) = \exp(iA\xi)$ *be a symmetry of the decay. Then the reduced propagator fulfils* $[V_t, U(\xi)] = 0$ *for all* $t, \xi \in \mathbb{R}$, *and the decay laws are invariant*, $P_\rho(t) = P_{\rho'}(t)$, where $\rho' = U(\xi)\rho U(\xi)^{-1}$.

REMARK 3.4.2. The above description is fashioned for continuous symmetry transformations. As for the discrete ones, let us recall that in view of the well-known theorem of Wigner (cf. Barut and Rączka (1977, §13.2)) the symmetry transformations are allowed to be either unitary or antiunitary. Since we have not used the linearity of $U(\xi)$ in its deduction, Proposition 4.1 is preserved when $U(\xi)$ is replaced by an operator representing the considered transformation. However, some caution is needed when dealing with the antiunitary transformations, because in such a case, commutativity of Θ with the Hamiltonian is equivalent to $\Theta U_t\Theta^{-1} = U_{-t}$, $t \in \mathbb{R}$.

Suppose that a symmetry of the decay is described by a self-adjoint A with purely point spectrum, $A = \Sigma_j\, a_jP_j$, then we may write

$$\mathscr{H} = \sum_j{}^{\oplus} \mathscr{H}_j, \qquad \mathscr{H}_{\mathrm{u}} = \sum_j{}^{\oplus} \mathscr{H}_{\mathrm{u}j}, \tag{4.1}$$

where $\mathscr{H}_j = P_j\mathscr{H}$ and $\mathscr{H}_{\mathrm{u}j} = P_j\mathscr{H}_{\mathrm{u}}$. According to the assumption, U_t commutes with all the projections P_j, so the propagator decomposes into the direct sum of operators $\mathrm{pr}_j\, U_t$. On the other hand, at least some of the subspaces $\mathscr{H}_{\mathrm{u}j} \subset \mathscr{H}_j$ must be non-invariant under $\{U_t\}$, otherwise the postulate (u2) would be violated. Thus the given symmetry allows us to simplify the study of the decay process via its decomposition to the mutually independent decay processes characterized by the eigenvalues a_j of the observable A. Moreover, it may happen that some of these partial problems are trivial because the respective decays are forbidden; this is usually due to some selection rule, i.e. another symmetry of the decay. Notice that we have already met decompositions of this type in connection with the inverse-decay problem – cf. Proposition 1.4.3 and Problem 4.1.

Examples of the described decomposition can be found easily, for instance, with A being an angular-momentum variable. Formally the same procedure is used if A is not only a symmetry but also a *superselection rule*. The only difference is that superpositions are not allowed now between different coherent subspaces $\mathscr{H}_j$, i.e. each state of the considered system is reduced by all the projections P_j. Hence, for example, every decay is supposed to preserve the electric charge or baryon number (though the last-named superselection rule has been challenged recently in connection with the possible non-stability of protons).

Orthogonal sum decompositions may also be performed in cases when the spectrum of the symmetry A is not purely point, but this is usually not very interesting from a practical point of view. There is another important particular case in which the character of $\sigma(A)$ is completely irrelevant; namely, when decomposes into the tensor product $\mathscr{H}^{(1)} \otimes \mathscr{H}^{(2)}$ and

$$H = \overline{H_1 \otimes I_2}\,, \qquad A = \overline{I_1 \otimes A_2} \tag{4.2a}$$

with self-adjoint H_1, A_2 on $\mathscr{H}^{(1)}$, $\mathscr{H}^{(2)}$, respectively. If we have simultaneously

$$\mathscr{H}_{\mathrm{u}} = \mathscr{H}_{\mathrm{u}}^{(1)} \otimes \mathscr{H}^{(2)}, \tag{4.2b}$$

i.e. $E_{\mathrm{u}} = E_{\mathrm{u}}^{(1)} \otimes I_2$ for some subspace $\mathscr{H}_{\mathrm{u}}^{(1)} \subset \mathscr{H}^{(1)}$, then A obviously represents symmetry of the decay and the reduced propagator equals $V_t = V_t^{(1)} \otimes I_2$, where $V_t^{(1)} = \mathrm{pr}_{\mathrm{u}1}\, U_t^{(1)} = \mathrm{pr}_{\mathrm{u}1} \exp(-iH_1 t)$. Conversely, if $\{\mathscr{H}^{(1)}, U_t^{(1)}, E_{\mathrm{u}}^{(1)}\}$ is a minimal unitary dilation of the function $t \mapsto V_t^{(1)}$, then the same is true for $\{\mathscr{H}, U_t, E_{\mathrm{u}}\}$ and $t \mapsto V_t$ (cf. Problem 2). Furthermore, these conclusions remain valid when we replace the Hamiltonian from (4.2a) by a more general one,

$$H = \overline{H_1 \otimes I_2 + I_1 \otimes H_2}\,, \tag{4.2c}$$

where H_2 is a self-adjoint operator on $\mathscr{H}^{(2)}$ which commutes with A_2.

These considerations suit the case when our system consists of two subsystems S_j referring to the spaces $\mathscr{H}^{(j)}$. It is not important here whether these subsystems can be actually separated from each other (as in the case of two particles forming the joint two-particle system) or whether they are fictional, representing merely a suitable grouping of the degrees of freedom. The subsystems S_1, S_2 are said to be *non-interacting* if the total Hamiltonian of the joint system is of the form (4.2c), where H_j is the Hamiltonian of S_j. The above results then say that each observable of the second system represents symmetry of the decay for the first one. The tensor-product decomposition of the system under consideration into non-interacting (actual or fictional) subsystems is one more way in which we can separate just a part of the problem which is relevant to the decay. One has to check only that the decay laws are not sensitive to the presence of the other subsystem. Since the reduced propagator is of the form $V_t = V_t^{(1)} \otimes I_2$, the last statement is obviously valid for states $\rho = \rho^{(1)} \otimes \rho^{(2)}$, where Ran $\rho^{(1)} \subset \mathscr{H}_u^{(1)}$ and $\rho^{(2)}$ is an arbitrary density matrix on $\mathscr{H}^{(2)}$. However, not all states of the joint system are of this form. Let ρ be an arbitrary density matrix on $\mathscr{H}$ (with possible restrictions imposed by the superselection rules), then there is a unique reduced state $\rho_1(\rho)$ of S_1 determined by the relation

$$\mathrm{Tr}((\overline{A_1 \otimes I_2})\rho) = \mathrm{Tr}(A_1 \rho_1(\rho)) \tag{4.3a}$$

for all bounded observables A_1 of S_1 – cf. Jauch (1969, §11–8), Blank and Exner (1977, 1980, §7.6.2). The decay operator $P_t^{(1)} = V_t^{(1)*} V_t^{(1)}$ is Hermitean so it may be regarded as an observable, and therefore for any ρ with Ran $\rho \subset \mathscr{H}_u$ the relation (1.2.3a) gives the equality

$$P_{\rho_1(\rho)}(t) = P_\rho(t), \qquad t \geqslant 0. \tag{4.3b}$$

It is illustrative to derive this formula using the explicit form of the reduced state $\rho_1(\rho)$ (Problem 16).

Let us now return to the theme discussed in the preceding sections and assume that the system is characterized by a pair of self-adjoint operators H_0, H and that $\mathscr{H}_u$ is an m-dimensional eigenspace of H_0 referring to an eigenvalue λ_0. If A is a dynamical symmetry with respect to both H_0 and H, then it commutes particularly with the projection on $\mathscr{H}_u$, so it represents symmetry of the decay. The operator $\mathrm{pr}_u A$ is Hermitean and finite-dimensional, so it has a purely point spectrum; further, the eigenspaces of A are preserved by U_t, hence one may use a decomposition of the type (4.1) as far as the minimal space relevant to the decay is concerned.

The easiest situation arises when the *spectrum of* $\mathrm{pr}_u A$ *is simple*; then we get an orthogonal sum of m partial problems, and each of them concerns the perturbation of λ_0 as a simple eigenvalue. The perturbation may or may not remove the energy degeneration of the full problem, but we are able to make some important conclusions: the positions of the second-sheet poles depend *analytically* on the coupling constant (provided the theorems derived in the preceding section may be applied) and the residuum operators corresponding to different poles are mutually *orthogonal* in the sense that they act in mutually orthogonal subspaces. In such a case, we say that the considered decay process *decomposes completely due to the dynamical symmetry under consideration*. In fact, the observable A here may stand for compatible observables $A_1, \dots, A_n$ such that the system $\mathscr{A} = \{\mathrm{pr}_u A_j : j = 1, \dots, n\}$ is maximal, i.e. that its commutant $\mathscr{A}'$ coincides with the bicommutant $\mathscr{A}''$; according to the well-known theorem of von Neumann (Akhiezer and Glazman, 1966, §VI.92), one can always find a single Hermitean operator which generates the algebra $\mathscr{A}''$.

Unfortunately, such a complete decomposition is not always possible. We often meet the situation when the unperturbed Hamiltonian H_0 possesses a symmetry A, but A does not commute with the total Hamiltonian H. Of course, A is not the symmetry of decay in this case; we say that the symmetry is **broken** by the interaction that is responsible for decay. Now the conclusions derived above need be no longer valid; in particular, the pole residua are not necessarily orthogonal, so the problem cannot be split in to a discussion of 'one-dimensional' decays. A fairly simple and illustrative example concerns the *decay of neutral kaons.*

EXAMPLE 3.4.3. As we shall discuss in the next section, the effect of the 'co-ordinate' part of the wave functions on kaon decay may be neglected in a reasonable approximation. Hence we may consider the two-dimensional state Hilbert space $\mathscr{H}_{K^0}$ only, which is spanned by

$$\varphi_{K^0} = \begin{pmatrix} 1 \\ 0 \end{pmatrix}, \qquad \varphi_{\bar{K}^0} = \begin{pmatrix} 0 \\ 1 \end{pmatrix} \tag{4.4a}$$

or by the pair of non-orthogonal vectors

$$\varphi_S = N_\epsilon \begin{pmatrix} 1+\epsilon \\ 1-\epsilon \end{pmatrix}, \qquad \varphi_L = N_\epsilon \begin{pmatrix} 1+\epsilon \\ -1+\epsilon \end{pmatrix}, \tag{4.4b}$$

where $N_\epsilon = (1 + |\epsilon|^2)^{-1/2}$ and ϵ is a complex parameter; its value is fixed experimentally to be

$$\epsilon = (2.3 \times 10^{-3}) \exp(\pi i/4) \tag{4.4c}$$

with an error of about few percent (see Bricman *et al.* (1978) or more recent editions of these tables). Let us introduce the operators P_S, P_L which are represented, on the basis of (4.4a), as follows

$$P_S = \tfrac{1}{2}\begin{pmatrix} 1 & b_\epsilon \\ b_\epsilon^{-1} & 1 \end{pmatrix}, \qquad P_L = \tfrac{1}{2}\begin{pmatrix} 1 & -b_\epsilon \\ -b_\epsilon^{-1} & 1 \end{pmatrix}, \tag{4.5}$$

where $b_\epsilon = (1 + \epsilon)/(1 - \epsilon)$. It is easy to see that these operators are non-Hermitean idempotents (i.e. non-orthogonal projections) which fulfil the relations

$$P_j P_k = \delta_{jk} P_k, \quad j, k = \text{S, L}, \tag{4.6a}$$

$$P_j \varphi_k = \delta_{jk} \varphi_k, \quad j, k = \text{S, L}. \tag{4.6b}$$

The time evolution in $\mathscr{H}_{K^0}$ is governed (in the rest system of the kaon) by the operator family

$$V_t = P_S \exp(-iz_S t) + P_L \exp(-iz_L t), \tag{4.7}$$

where $z_j = m_j - i\delta_j$ with $m_S \approx m_L \approx 497.67$ MeV and

$$m_L - m_S = 3.52 \times 10^{-12}\ \text{MeV}, \tag{4.8a}$$

$$\delta_S = \tfrac{1}{2}\Gamma_S = 3.69 \times 10^{-12}\ \text{MeV}, \tag{4.8b}$$

$$\delta_L = \tfrac{1}{2}\Gamma_L = 0.635 \times 10^{-14}\ \text{MeV} \tag{4.8c}$$

(the experimental errors are again listed in the tables of particle properties, but they are not substantial for the present considerations). Consequently, φ_S is the shortlived state of the neutral kaon with the mean life $T_S = \hbar\Gamma_S^{-1} = 0.892 \times 10^{-10}$ sec, while the longlived φ_L has $T_L = \hbar\Gamma_L^{-1} = 5.18 \times 10^{-8}$ sec. What is important is that the relation (4.6a) implies *the operators* (4.7) *to fulfil the semigroup condition* (1.1.1).

Let us now see how this phenomenological description can be reconciliated with the above considerations. Since we work in the rest system of the kaon, we are allowed to identify its Hamiltonian with the mass operator (cf. the next section). Suppose the kaon interacts only strongly. In such a case, the mass operator would have a single real eigenvalue (presumably differing no more than few MeV from $m_S \approx m_L$) of multiplicity 2. The weak interaction acts here as a perturbation which splits the eigenvalue pole and shifts its daughter poles to the points z_S, z_L of the second sheet given by (4.8). Then we get the reduced propagator (4.7) in the pole-and-real-axis approximation if only P_S, P_L given by (4.5) are the residuum operators of the respective poles.

Now the antiunitary operator Θ_{CP} of the combined charge and space parity is preserved in the strong interactions. The vectors

$$\varphi_\pm = 2^{-1/2}(\varphi_{K^0} \pm \varphi_{\bar{K}^0}) \tag{4.9}$$

are eigenvectors of Θ_{CP} corresponding to the eigenvalues ± 1. The weak interaction, however, breaks the CP-symmetry. Let us restrict ourselves, for simplicity, to non-leptonic decays; then if CP were the symmetry of the decay, only the processes $K_S \to 2\pi$ and $K_L \to 3\pi$ would be allowed because the 2π and 3π states have a CP-parity equal to +1 and −1, respectively. Actually a small fraction of the long-lived neutral kaons decays into two pions, either $\pi^+\pi^-$ or $\pi^0\pi^0$. The ratio of such decays is about 3×10^{-3}; it is related directly to the parameter ϵ given by (4.4c) which characterizes how much the CP-symmetry is violated.

Let us finally stress that in spite of the relation (4.6a), the pole residua are *not* orthogonal in the above-mentioned sense, since they project onto one-dimensional subspaces spanned by the vectors φ_S, φ_L which fulfil $(\varphi_S, \varphi_L) = 4\,|N_\epsilon|^2 \operatorname{Re}\epsilon$, and ϵ cannot be purely imaginary within the range of experimental errors. Hence the decay of neutral kaons offers the example of an actual unstable system, with semigroup evolution, which cannot be decomposed into an orthogonal sum of one-dimensional semigroups.

3.5. Relativistic Invariance

"The experience has shown to me that it is easy to lecture the theory of relativity to the students which are inclined to theoretical speculations and willing to 'believe in the theory'."

V. VOTRUBA

Among all symmetries of a given system, a distinguished role is played by those related to the transformations of space-time variables. For a major part of micro-physical systems, gravitational effects are inessential within the present quantum theory, so that the space-time symmetries are described by the Poincaré group

$\mathscr{P}$, or eventually by the Galilei group in the low-velocities limit. The transformation of coordinates between two inertial systems is given by the relation

$$x'_\mu = \Lambda_\mu{}^\nu x_\nu + a_\mu, \tag{5.1}$$

where the matrix Λ belongs to the group SO(3, 1) and a is a four-vector. For simplicity, we shall consider here the connected component of $\mathscr{P}$ only, thus avoiding a discussion of the space and time inversions. The composition law of two Poincaré transformations reads

$$(\Lambda, a)(\Lambda', a') = (\Lambda\Lambda', a + \Lambda a') \tag{5.2a}$$

so that $(\Lambda, a)^{-1} = (\Lambda^{-1}, -\Lambda^{-1}a)$ and the unit element is $(I, 0)$. The relation (5.2a) further shows that $(\Lambda, a)(I, b)(\Lambda, a)^{-1} = (I, \Lambda b)$, so the translations form an invariant subgroup and $\mathscr{P}$ decomposes into the semidirect product SO(3, 1) $\rtimes T_4$ of the six-parameter proper Lorentz group and the Abelian group of space-time translations. The Lie algebra $L_{\mathscr{P}}$ of $\mathscr{P}$ is generated by ten elements $l_{\mu\nu} = -l_{\nu\mu}$, $p_\mu, \mu, \nu = 0, 1, 2, 3$.

As mentioned in Section 1, 2, the postulate of (special) relativistic invariance claims that the the state Hilbert space $\mathscr{H}$ of any isolated system is a carrier space space of some (strongly continuous) unitary representation $U : \mathscr{P} \to \mathscr{B}(\mathscr{H})$ under which the dynamical variables of the system transform in a specific way. In particular, some important observables are identified directly with the generators $L_{\mu\nu}$, P_μ of the corresponding representation of the Lie algebra $L_{\mathscr{P}}$: the total energy with $H \equiv P_0$, the components of momentum with the generators P_j of space translations, etc. These operators fulfil the commutation relations

$$[L_{\mu\nu}, L_{\rho\sigma}] = i(g_{\nu\rho}L_{\mu\sigma} - g_{\mu\rho}L_{\nu\sigma} + g_{\nu\sigma}L_{\rho\mu} - g_{\mu\sigma}L_{\rho\nu}), \tag{5.3a}$$

$$[L_{\mu\nu}, P_\sigma] = i(g_{\nu\sigma}P_\mu - g_{\mu\sigma}P_\nu), \tag{5.3b}$$

$$[P_\mu, P_\nu] = 0 \tag{5.3c}$$

for $\mu, \nu, \rho, \sigma = 0, 1, 2, 3$, which may be derived from the relations

$$U(\Lambda, a)U(\Lambda', a') = U(\Lambda\Lambda', a + \Lambda a') \tag{5.2b}$$

valid for all $(\Lambda, a), (\Lambda', a') \in \mathscr{P}$. The Poincaré group has two Casimir operators, i.e. there are two elements in the universal enveloping algebra of $L_{\mathscr{P}}$ represented by the operators

$$P^2 = P_\mu P^\mu, \tag{5.4a}$$

$$W^2 = \tfrac{1}{2} L_{\mu\nu}L^{\mu\nu}P^2 - L_{\mu\sigma}L^{\nu\sigma}P^\mu P_\nu, \tag{5.4b}$$

which are left invariant by the Poincaré transformations, i.e. they commute with all the operators $L_{\mu\nu}$ and P_μ. If the representation U is irreducible, then the operators (5.4) are multiplies of unity in view of the Schur's lemma. There is a complete classification of the unitary irreducible representations of $\mathscr{P}$ (e.g. Barut and Rączka (1977, §17.2). For the physically most important class of these representations, values of the invariants are m^2 and $m^2 s(s+1)$, where $m \in \mathbb{R}_+$ and $s = 0, \frac{1}{2}, 1, \frac{3}{2}, \ldots$; to these parameters conventionally the meaning of *mass* and *spin* is ascribed. There is also a third invariant as far as the proper Poincaré group (without inversions) is considered, namely sgn P_0.

Let us now turn to the relativistic invariance of the decay processes. As usual, we shall associate with the decaying system, and with the larger isolated system, a pair of Hilbert spaces $\mathscr{H}_u$, $\mathscr{H}$. In analogy with the reduced propagator, we may define the operator-valued function $V: \mathscr{P} \to \mathscr{B}(\mathscr{H}_u)$ by

$$V(\Lambda, a) = \mathrm{pr_u}\, U(\Lambda, a), \tag{5.5}$$

where U is the unitary representation of $\mathscr{P}$ ascribed to the isolated system. According to Theorem 1.4.1, the function V is strongly continuous, positive define and fulfils $V(I, 0) = I_u$. If $\mathscr{H}_u$ is preserved by the transformations belonging to some subgroup of $\mathscr{P}$, then the corresponding operators $V(\Lambda, a)$ are unitary and we can simplify the problem by this symmetry as discussed in the preceding section. However, repeating the simple argument from Section 1.2, we see that the operators $V(\Lambda, a)$ cannot fulfil the group law analogous to (5.2b). In other words, *V cannot be a non-unitary representation of $\mathscr{P}$.*

This fact does not exclude the possibility that non-unitary representations of $\mathscr{P}$ may be used for a description of unstable systems in the following weaker sense: there is such a representation $\tilde{V}$, and $\tilde{V}(\Lambda, a)$ coincides with $V(\Lambda, a)$ for some reasonable subset of $\mathscr{P}$, say, for arbitrary Λ and all four-vectors a belonging to the forward light cone. Such representations were constructed (see Zwanziger (1963) and other references in notes to this section). Typically their Lorentz parts are unitary while the translations are realized as multiplication by $\exp(-iq \cdot a)$, where q is a complex four-momentum vector. This is a natural generalization of the complex-energy description of the decaying-system time evolution, of course, if we neglect the difficulty with the energy semiboundedness.

Unfortunately, even such a point of view cannot be retained. The reason is that it does not respect Euclidean invariance (the first and maybe the most important of the laws on which the physics is built – E. Wigner *dixit*). It is fairly reasonable to assume that two observers, whose reference frames are obtained one from the other by space translations and rotations, will determine by their instruments exactly the same decay law and other characteristics for a given unstable system. Hence, in particular, the operators $V(I, a)$ for $a = (0, \mathbf{a})$ should be unitary, which is not true for the above-mentioned representations. Furthermore, the translational invariance implies that *the operators $V(\Lambda, 0)$ referring to the pure Lorentz transformations (boosts) must not be unitary.* In order to see this, notice that the relation (5.2b) yields the identity

$$U(I, \Lambda a)U(\Lambda^{-1}, 0)U(I, -a)U(\Lambda, 0) = U(I, \Lambda a - \Lambda^{-1}a). \tag{5.6a}$$

Let Λ be a boost in some direction and $a = (0, \mathbf{a})$ with $\mathbf{a}$ having the same direction. As shown in Section 1,.2, the operator $V(\Lambda, a)$ is unitary for some (Λ, a) iff the corresponding $U(\Lambda, a)$ commutes with E_u. Then all the operators on the l.h.s. of

(5.6a) commute with E_u, and the same is naturally true for $U(I, \Lambda a - \Lambda^{-1}a)$. Without loss of generality, one may assume that the boost is paralel to the first axis and $a = (0, a_1, 0, 0)$ so that $\Lambda a = (-a_1 \operatorname{sh} \beta, a_1 \operatorname{ch} \beta, 0, 0)$ and

$$\Lambda a - \Lambda^{-1}a = (-2a_1 \operatorname{sh} \beta, \mathbf{0}). \tag{5.6b}$$

But a_1, β are arbitrary parameters from $\mathbb{R}$ and $(-1, 1)$, respectively, and therefore E_u should commute with U_t for all $t \in \mathbb{R}$. Of course, this contradicts to (u2).

We are now going to discuss how the representation U might appear in the simplest case of a free elementary particle. A stable elementary particle is naturally associated with some of the above-mentioned irreducible representations characterized by a mass m and spin s. An unstable particle is still determined by a definite value of the spin, but this is no longer true for the mass. The most simple ansatz is therefore to consider a *direct integral of the irreducible representations with a fixed spin*. The carrier space $\mathscr{H} = \mathscr{H}(m_0, s)$ of such a representation is given by

$$\mathscr{H} = L^2\left([m_0, \infty) \times \mathbb{R}^3, \mathrm{d}m \otimes \frac{\mathrm{d}^3p}{2(m^2 + \mathbf{p}^2)^{1/2}}\right) \otimes \mathbb{C}^{2s+1}, \tag{5.7a}$$

i.e. the scalar product of $\psi, \varphi \in \mathscr{H}$ is

$$(\psi, \varphi) = \sum_{j=-s}^{s} \int_{m_0}^{\infty} \mathrm{d}m \int_{\mathbb{R}^3} \frac{\mathrm{d}^3p}{2(m^2 + \mathbf{p}^2)^{1/2}} \overline{\psi_j(m, \mathbf{p})}\varphi_j(m, \mathbf{p}). \tag{5.7b}$$

It is useful sometimes to separate fully the kinematical variables from the mass. To this end, one has to employ the four-velocity $k = p/m$, i.e. to introduce the Hilbert space

$$\widetilde{\mathscr{H}} = L^2([m_0, \infty)) \otimes L^2\left(\mathbb{R}^3, \frac{\mathrm{d}^3k}{2(1 + \mathbf{k}^2)^{1/2}}\right) \otimes \mathbb{C}^{2s+1}, \tag{5.8a}$$

which is isomorphic to $\mathscr{H}$ by means of the relations

$$\tilde{\psi}_j(m, \mathbf{k}) = m\psi_j(m, m\mathbf{k}), \qquad \psi_j(m, \mathbf{p}) = m^{-1}\tilde{\psi}_j(m, m^{-1}\mathbf{p}) \tag{5.8b}$$

valid for all $j = -s, -s+1, \dots, s$, $m \in [m_0, \infty)$ and $\mathbf{p}, \mathbf{k} \in \mathbb{R}^3$.

The representation U acts on the space (5.7a) according to the following prescription

$$(U(\Lambda, a)\psi)(m, \mathbf{p}) = \mathrm{e}^{-ip \cdot a} S(m, s; \Lambda)\psi(m, \mathbf{p}_\Lambda), \tag{5.9}$$

where $p \cdot a = p_\mu a^\mu$; further, $\mathbf{p}_\Lambda$ is the three-vector part of $\Lambda^{-1}p$, and $S(m, s; \Lambda)$ is a $(2s + 1) \times (2s + 1)$ matrix the explicit form of which may be constructed using representations of the little group SU(2). The relation (5.9) is especially simple in the case of pure translations, because $S(m, s; I)$ is the unit matrix, so for $x = (t, \mathbf{x})$ we get

$$(\psi, U(I, x)\varphi) = \sum_{j=-s}^{s} \int_{m_0}^{\infty} \mathrm{d}m \int_{\mathbb{R}^3} \frac{\mathrm{d}^3p}{2(m^2 + \mathbf{p}^2)^{1/2}} \times$$

$$\times \exp\{-i(t(m^2 + \mathbf{p}^2)^{1/2} - \mathbf{x} \cdot \mathbf{p})\} \overline{\psi_j(m, \mathbf{p})}\varphi_j(m, \mathbf{p}). \tag{5.10}$$

In particular, for the pure time translations and $\varphi = \psi$ we have

$$(\psi, U_t\psi) = \sum_{j=-s}^{s} \int_{m_0}^{\infty} dm \int_{\mathbb{R}^3} \frac{d^3p}{2(m^3 + \mathbf{p}^2)^{1/2}} \times$$
$$\times \exp\{-it(m^2 + \mathbf{p}^2)^{1/2}\} | \psi_j(m, \mathbf{p})|^2. \tag{5.11a}$$

Changing the variables $(m, \mathbf{p})$ to $(\lambda, \mathbf{p})$ with $\lambda = p_0 = (m^2 + \mathbf{p}^2)^{1/2}$, we may rewrite the last expression in the form

$$(\psi, U_t\psi) = \int_{m_0}^{\infty} d\lambda\, e^{-i\lambda t} \times$$
$$\times \left\{ \sum_{j=-s}^{s} \int_{V_\lambda} \frac{d^3p}{2(\lambda^2 - \mathbf{p}^2)^{1/2}} \, |\psi_j((\lambda^2 - \mathbf{p}^2)^{1/2}, \mathbf{p})|^2 \right\}, \tag{5.11b}$$

where $V_\lambda = \{\mathbf{p} : |\mathbf{p}| \leqslant (\lambda^2 - m_0^2)^{1/2}\}$.

The crucial point lies in the *choice of the subspace* $\mathcal{H}_u$ which would be ascribed to the unstable particle alone. If this space was one-dimensional (spanned by some $\psi \in \mathcal{H}$), then (5.11) would yield the non-decay amplitude. However, we have argued above that $\mathcal{H}_u$ should be invariant, particularly with respect to the space translations. This is impossible for a one-dimensional $\mathcal{H}_u$, because the momentum operators P_j have purely continuous spectra so ψ cannot be their eigenvector. Nevertheless, we shall formulate below a heuristic argument which shows that in most cases the relations (5.11) may be accepted as expressions of the non-decay amplitude in a reasonable approximation.

We shall consider first the scalar particles, $s = 0$. Our most important hypothesis is that some state of the unstable particle describes by a wave function which factorizes

$$\psi(m, \mathbf{p}) = f(m)g(\mathbf{p}). \tag{5.12}$$

Next we adopt various simplifying assumptions. First, we set

$$\operatorname{supp} f = (M - \eta, M + \eta) \subset [m_0, \infty), \tag{5.13a}$$

$$\int_{m_0}^{\infty} |f(m)|^2 \, dm = \int_{M-\eta}^{M+\eta} |f(m)|^2 \, dm = 1. \tag{5.13b}$$

The parameter η is a positive number, assumed to be much less than M. Further, we assume

$$\operatorname{supp} g = B_\epsilon = \{\mathbf{p} : |\mathbf{p}| < \epsilon\} \tag{5.13c}$$

so the support of g is centred at $\mathbf{p} = \mathbf{0}$. For small enough ϵ, this is practically equivalent to the assumption that the particle dwells in its rest system. According to (5.9), the space translations give $\psi_{\mathbf{a}} : \psi_{\mathbf{a}}(m, \mathbf{p}) = e^{i\mathbf{a}\cdot\mathbf{p}}\psi(m, \mathbf{p})$ when acting on $\psi = \psi_0$. Since $\psi_{\mathbf{a}}$ should belong to $\mathcal{H}_u$ for all $\mathbf{a} \in \mathbb{R}^3$, and the exponentials form a complete set in $L^2(\mathbb{B}_\epsilon)$, we may put

$$\mathcal{H}_u = \{\psi : \psi(m, \mathbf{p}) = f(m)g(\mathbf{p}), g \in L^2(B_\epsilon)\}. \tag{5.14a}$$

As a set, this $\mathscr{H}_{\mathrm{u}}$ coincides with $\{f\}_{\mathrm{lin}} \otimes L^2(B_\epsilon)$. The scalar product is, however, different: the norm of $\psi \in \mathscr{H}_{\mathrm{u}}$ is given by

$$\|\psi\|^2 = \int_{M-\eta}^{M+\eta} \mathrm{d}m\, |f(m)|^2 \int_{B_\epsilon} \frac{\mathrm{d}^3 p}{2(m^2 + \mathbf{p}^2)^{1/2}} |g(\mathbf{p})|^2 .$$

Let $\|\,.\,\|_2$ denote the norm in $L^2(B_\epsilon)$, $\|g\|_2^2 = \int_{B_\epsilon} |g(\mathbf{p})|^2\, \mathrm{d}^3 p$. We may use it to estimate the norm $\|\psi\|$, or vice versa, to derive the inequalities

$$2(M-\eta)\, \|\psi\|^2 \leqslant \|g\|_2^2 \leqslant 2((M+\eta)^2 + \epsilon^2)^{1/2}\, \|\psi\|^2 . \tag{5.15}$$

This shows particularly that a sequence $\{\psi_n\} \subset \mathscr{H}_{\mathrm{u}}$ is Cauchy iff the same is true for the corresponding sequence $\{g_n\} \subset L^2(B_\epsilon)$; hence $\mathscr{H}_{\mathrm{u}}$ defined by (5.14a) is a (closed) subspace in $\mathscr{H}$. The inequality (5.20a) below shows that $\epsilon \ll M$ and the same restriction was imposed on η, so the function g corresponding to a unit $\psi \in \mathscr{H}_{\mathrm{u}}$ fulfils $\|g\|_2 \approx (2M)^{1/2}$.

Let us now inspect the action of the time translations on a unit ψ from the chosen subspace (5.14a). According to (5.9), they multiply $\psi(m, \mathbf{p})$ by $\exp(-it(m^2 + \mathbf{p}^2)^{1/2})$. This expression does not factorizes, but for ϵ small enough, one may try to approximate it by e^{-imt}. Since $\epsilon \ll M$, we may restrict ourselves to the first two terms of the expansion

$$\exp(-it(m^2 + \mathbf{p}^2)^{1/2}) = \mathrm{e}^{-imt} \left\{ 1 - i \frac{\mathbf{p}^2 t}{2m} + O(\mathbf{p}^4) \right\} .$$

The propagator is correspondingly written as $U_t = U_t^{(0)} + U_t^{(1)}$, neglecting the remainder. In order to estimate the influence of the second term, we take an arbitrary unit vector $\varphi \in \mathscr{H}_{\mathrm{u}}$, $\varphi(m, \mathbf{p}) = f(m)h(\mathbf{p})$, and express

$$(\varphi, U_t^{(1)}\psi) = \int_{M-\eta}^{M+\eta} \mathrm{d}m\, \mathrm{e}^{-imt}\, |f(m)|^2 \int_{B_\epsilon} \frac{\mathrm{d}^3 p}{2(m^2 + \mathbf{p}^2)^{1/2}} \left\{ -i \frac{\mathbf{p}^2 t}{2m} \right\} \overline{h(\mathbf{p})} g(\mathbf{p})$$

The relations (5.13), (5.15) yield the following inequalities

$$\begin{aligned} |(\varphi, U_t^{(1)}\psi)| &\leqslant \frac{1}{2(M-\eta)}\, \frac{\epsilon^2 t}{2(M-\eta)}\, \|h\|_2\, \|g\|_2 \\ &\leqslant \frac{((M+\eta)^2 + \epsilon^2)^{1/2}}{M-\eta}\, \frac{\epsilon^2 t}{2(M-\eta)} . \end{aligned}$$

Hence we may estimate $\|E_{\mathrm{u}} U_t^{(1)} \psi\| = \sup\{|(\varphi, U_t^{(1)}\psi) : \varphi \in \mathscr{H}_{\mathrm{u}}, \|\varphi\| = 1\}$. Since both η, ϵ are supposed to be much less than M, we find it to be $\lesssim \epsilon^2 t/2M$. The approximation mentioned above is therefore possible under the condition

$$\frac{\epsilon^2 t}{2M} \ll 1 . \tag{5.16}$$

In such a case, the norm of the difference between $E_{\mathrm{u}} U_t \psi$ and $E_{\mathrm{u}} U_t^{(0)} \psi$ is very small and we are allowed to write $V_t \psi \approx E_{\mathrm{u}} U_t^{(0)} \psi$.

In the next step, we shall verify that the last expression is close to $(\psi, U_t^{(0)}\psi)\psi$. To this end, we take an arbitrary unit vector $\varphi \in \mathscr{H}_{\mathrm{u}}$, $\varphi(m, \mathbf{p}) = f(m)h(\mathbf{p})$, which is orthogonal to ψ. This orthogonality, together with (5.13b), makes it possible to estimate $(h, g)_2$ from the identity

$$\frac{1}{2M}(h, g)_2 = \int_{M-\eta}^{M+\eta} \mathrm{d}m\, |f(m)|^2 \int_{B_\epsilon} \left\{\frac{1}{2M} - \frac{1}{2(m^2 + \mathbf{p}^2)^{1/2}}\right\} \overline{h(\mathbf{p})} g(\mathbf{p})\, \mathrm{d}^3 p.$$

Since η, ϵ are much less than M, we have the following estimate valid up to higher-order terms

$$\left|\frac{1}{2M} - \frac{1}{2(m^2 + \mathbf{p}^2)^{1/2}}\right| \lesssim \frac{1}{2M}\left(\frac{\eta}{M} + \frac{\epsilon^2}{2M^2}\right). \tag{5.17a}$$

Combining this with the Hölder inequality, we obtain

$$|(h, g)_2| \lesssim \left(\frac{\eta}{M} + \frac{\epsilon^2}{2M^2}\right) \|h\|_2\, \|g\|_2 . \tag{5.17b}$$

Now we are able to estimate the scalar product $(\varphi, U_t^{(0)}\psi)$:

$$\begin{aligned} |(\varphi, U_t^{(0)}\psi)| \leqslant{} & \left|\frac{1}{2M}\int_{M-\eta}^{M+\eta} \mathrm{d}m\, \mathrm{e}^{-imt}\, |f(m)|^2\, (h, g)_2\right| + \\ & + \int_{M-\eta}^{M+\eta} \mathrm{d}m\, |f(m)|^2 \int_{B_\epsilon} \left|\frac{1}{2(m^2 + \mathbf{p}^2)^{1/2}} - \right. \\ & \left. - \frac{1}{2M}\right| |h(\mathbf{p})|\, |g(\mathbf{p})|\, \mathrm{d}^3 p. \end{aligned}$$

Applying (5.17a) and (5.17b) to the second and first term on the r.h.s., respectively, and using the Hölder inequality again, we get

$$|(\varphi, U_t^{(0)}\psi)| \lesssim \frac{2}{2M}\left(\frac{\eta}{M} + \frac{\epsilon^2}{2M^2}\right) \|h\|_2\, \|g\|_2 \int_{M-\eta}^{M+\eta} |f(m)|^2\, \mathrm{d}m.$$

However, φ and ψ are assumed to be unit vectors so $\|h\|_2 \approx \|g\|_2 \approx (2M)^{1/2}$. Finally, the normalization condition (5.13b) yields

$$|(\varphi, U_t^{(0)}\psi)| \lesssim 2\, \frac{\eta}{M} + \frac{\epsilon^2}{M^2}. \tag{5.18}$$

Since φ is an arbitrary unit vector from $\mathscr{H}_{\mathrm{u}}$ orthogonal to ψ, we see that $U_t^{(0)}\psi$ stays nearly parallel to ψ. Hence we may write

$$(V_t\psi)(m, \mathbf{p}) \approx \psi(m, \mathbf{p}) \int_{m_0}^{\infty} \mathrm{d}\mu\, |f(\mu)|^2\, \mathrm{e}^{-i\mu t} \int_{B_\epsilon} \frac{\mathrm{d}^3\kappa}{2(m^2 + \kappa^2)^{1/2}}\, |g(\kappa)|^2 .$$

Moreover, the inequality (5.17a) allows us to replace the denominator in the last integral by $2M$; the corresponding error is again, at most, comparable with the r.h.s. of (5.18). Thus we also have

$$(V_t\psi)(m, \mathbf{p}) \approx \psi(m, \mathbf{p}) \int_{m_0}^{\infty} \mathrm{e}^{-i\mu t}\, |f(\mu)|^2\, \mathrm{d}\mu. \tag{5.19b}$$

Concluding the above discussion, we may say that *if the three-momentum spread of ψ is sufficiently narrow, the decay goes effectively as if $\mathscr{H}_u$ was one-dimensional.* In that case, the non-decay amplitude is given by (5.11), and it may be approximated by the integrals appearing in the relations (5.19). Of course, the approximation also needs $\eta \ll M$, but this can be achieved as we shall see later.

The argument presented is easily generalized for particles with non-zero spin. One has only to use the rotational invariance of $\mathscr{H}_u$, then the following choice is natural

$$\mathscr{H}_u = \{\psi : \psi(m, \mathbf{p}) = f(m)g(\mathbf{p}), g \in L^2(B_\epsilon) \otimes \mathbb{C}^{2s+1}\}. \tag{5.14b}$$

Mimicking the above reasoning, we arrive again at approximation (5.19b).

We must therefore ask the circumstances under which conditions (5.16) and $\eta \ll M$ are valid. In any realistic description of unstable particles, the function $|f(\,.\,)|^2$ should have a sharp peak of more or less Breit–Wigner shape. In such a case, its width Γ may be characterized by the inverse of the mean life, while M has to be associated with the position of the peak. For all real unstable particles, M is much larger than Γ: the ratio M/Γ varies from 1.06×10^5 for Σ^0 to 1.31×10^{27} for neutrons (with exception of π^0, η and Σ^0, its lower bound is 10^{11}). Hence we can choose η so that the inequalities $\Gamma \ll \eta \ll M$ hold. The first of these ensures that the truncation of the mass distribution $|f(\,.\,)|^2$ to the interval $(M - \eta, M + \eta)$ will cause negligible change in the decay law. In fact, simple estimates (similar to those performed in Example 1.5.15 and Section 2.3) can show that for all practical purposes it is sufficient to choose $\eta \approx 10^2\Gamma$.

Of course, the condition (5.16) cannot hold for all values of t, but it is reasonable to demand its validity in the region where the decay law is actually measured, i.e. up to few Γ^{-1}. Then the three-momentum spread $\Delta p \equiv \epsilon$ of ψ must obey $(\Delta p)^2 \ll M\Gamma$ or

$$\Delta p \ll c^{-1}(M\Gamma)^{1/2} \tag{5.20a}$$

when we return to the conventional system of units. In order to appreciate this restriction, let us rewrite it by the uncertainty relation to the form

$$\Delta q \gg \hbar c(M\Gamma)^{-1/2}. \tag{5.20b}$$

On this heuristic level, we have thus come to the following result: *the above conclusion about the effective one-dimensionality of $\mathscr{H}_u$ is practically applicable provided that the considered particle is not spatially localized so sharply that* (5.20b) *is violated.* However, this condition is fulfilled almost always in actual experimental arrangements, as Table 1 shows.

Finally, let us look how the decaying particle behaves when being not at rest. Let a reference frame S belong to the observer, and assume that the rest system S' of the particle moves with a velocity β respectively to S (Figure 3). Of course, we must not only sandwich the propagator between $U(\Lambda_{\pm\beta}, 0)$, similarly as a simple-minded look on the factor multiplying the time variable in the Lorentz transformation does not yield the time dilatation. From the viewpoint of the

TABLE 3.1

Particle	$\hbar c(M\Gamma)^{-1/2}$ [cm]	Particle	$\hbar c(M\Gamma)^{-1/2}$ [cm]
μ	1.11×10^{-4}	n	0.763
τ	1.2×10^{-8}	Λ	3.74×10^{-7}
$\pi^{\pm}$	1.05×10^{-5}	Σ^+	3.53×10^{-7}
π^0	6.03×10^{-10}	Σ^0	5.4×10^{-12}
η	1.3×10^{-11}	Σ^-	2.71×10^{-7}
$K^{\pm}$	3.85×10^{-6}	Ξ^0	3.6×10^{-7}
K_S^0	3.26×10^{-7}	Ξ^-	2.71×10^{-7}
K_L^0	7.85×10^{-6}	Ω^-	1.70×10^{-7}
$D^{\pm}$	1.7×10^{-8}	Λ_c^+	5.3×10^{-9}
D^0	1.2×10^{-8}		
F	8.0×10^{-9}		

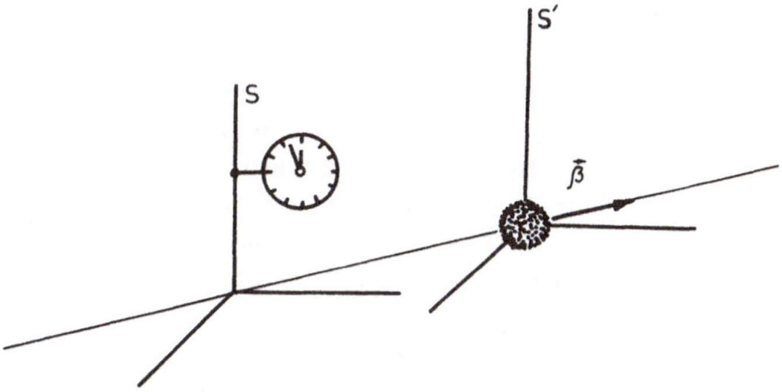

3.3. To the decay of a moving unstable particle.

reference frame S, we are interested in the space-time shift on $x = (t, \beta t)$. If the condition (5.20a) is valid – i.e. if we may describe the particle by a single vector $\psi \in \mathscr{H}$ which refers to its rest system – then the observer will associate with it the vector $U(\Lambda_\beta, 0)^{-1}\psi$. Hence the non-decay amplitude equals

$$v_\psi(t;\beta) = (U(\Lambda_\beta, 0)^{-1}\psi,\ U(I, x)U(\Lambda_\beta, 0)^{-1}\psi). \tag{5.21a}$$

Using the relations (5.2) and (5.9), we may now rewrite this expression as follows

$$\begin{aligned} v_\psi(t;\beta) &= (\psi,\ U(I, \Lambda_\beta x)\psi) \\ &= \sum_{j=-s}^{s} \int_{m_0}^{\infty} \mathrm{d}m \int_{\mathbb{R}^3} \frac{\mathrm{d}^3 p}{2(m^2 + \mathbf{p}^2)^{1/2}} \exp(-ip \cdot \Lambda_\beta x)\, |\psi_j(m, \mathbf{p})|^2 . \end{aligned}$$

However, the Lorentz transformation gives $\Lambda_\beta x = (t(1-\beta^2)^{1/2}, \mathbf{0})$, so

$$v_\psi(t;\beta) = v_\psi(t(1-\beta^2)^{1/2}; \mathbf{0}). \tag{5.21b}$$

This yields the relation

$$P_\psi(t;\beta) = P_\psi(t(1-\beta^2)^{1/2};\mathbf{0}), \tag{5.22}$$

which is valid as far as the approximation identifying the decay law $P_\psi(t;\beta)$ with the square of (5.12a) may be used.

REMARKS 3.5.1. (a) The relation (5.22) is proved by numerous experiments. It was even used thirty years ago for a direct check of the relativisitic time dilatation. The cosmic-ray muons appear as secondary particles in the stratosphere (more exactly, at a height of about 30 km). Since their mean life is 2.197×10^{-6} sec, almost no trace of them would reach the earth's surface if their decay was not slowed down due to their high velocities.

(b) As we have seen, the possibility of neglecting the **p**-spread of the wave function is conditioned by requirements of the type shown by (5.16); this particularly concerns the validity of (5.22). Sometimes these requirements are fulfilled even better than inequalities (5.20) and Table 1 show – namely, in the cases when the particle suffers repeated measurements, which we have studied in the preceding chapter. Since the decay starts anew after each measurement (which has given a positive result), we need not require (5.16) to hold for times comparable with Γ^{-1}, but merely with the mean time between two neighbouring measurements, and that is usually a few orders of magnitude shorter. As an example, consider the model treated in Section 2.3: there the mean time between measurements is $\sim 10^{-4}\,\Gamma^{-1}$. Instead of (5.21b), we obtain thus the condition $\Delta q \gg 10^{-8}$ cm for charged kaons, but actually they are localized within the range of bubble diameter, i.e. about 10^{-2} cm. Similar conclusions may also be obtained for other unstable particles and track-monitoring devices. Notice that they are particularly useful for *justifying the assumption* (r) from Section 2.1

Let us conclude the chapter with some comments concerning the *problem of identity* of unstable systems. According to the theory developed in the first chapter, the decay process is characterized fully by the operator-valued measure (1.4.5); in particular, by the function $\lambda \mapsto (\psi, E_\lambda\psi)$ when $\mathscr{H}_u$ is a one-dimensional subspace spanned by ψ and $\{E_\lambda\}$ is the spectral decomposition of the total Hamiltonian. We have seen that these functions exhibit the peculiar behaviour connected with the presence of poles in the analytically continued reduced resolvent, as far as we are concerned with the decays which allow description based on perturbations of embedded eigenvalues.

Knowledge of the poles (together with the corresponding residua) can substitute for the complete information about the measure (1.4.5) only when the pole-and-real-axis approximation can be justified. These matters were discussed in Sections 1, 1.6, 2.3 and in other places, and it is not necessary to repeat the conclusions here. It is a fact of life, however, that the approximation works quite well for unstable systems which really occur in quantum microphysics; so we are allowed to characterize the structure of the measure (1.4.5) by a mere few numbers within the range of experimental errors, e.g. by the mean energy and mean life in the simplest case when $\mathscr{H}_u$ is one-dimensional. If we combine this information with other quantum numbers (spins, parities, etc.) which arise eventually from symmetry considerations, we may identify the unstable system fully.

Let us recall the argument of Haag and Kastler mentioned in Section 1.6, according to which an experimentalist determines not a point but a certain subset in the state space. For a one-dimensional $\mathscr{H}_u$, the true (dynamically generated) function $\lambda \mapsto (\psi, E_\lambda \psi)$ as well as its possible modifications (e.g. by truncation due to energy filtering) do not coincide exactly with the Breit–Wigner distribution resulting from the pole-and-real-axis approximation, but it makes no difference provided that we stay within experimentally fixed limits. Generalization of this argument to higher-dimensional $\mathscr{H}_u$'s is straightforward.

Until now we have only spoken about the time evolution. Considerations of the present section show how to proceed when a relativistically invariant identification is needed. We shall restrict ourselves to unstable elementary particles and assume that some condition of the type (5.16), (5.20) is fulfilled. Then the Hamiltonian may be approximated by the mass operator in the rest system of the particle. The state of the undecayed particle is characterized by a vector ψ from the Hilbert space (5.7a). The theory should yield a shape of ψ, especially its mass dependence, as a solution to the dynamical problem. In practice, one matches this task in the following way: ψ is supposed to be of the form (5.12) with f being the Breit–Wigner distribution (possibly truncated). The width Γ of the peak is calculated, more or less on the basis of the Fermi golden rule. On the other hand, the (mean value of) mass is usually taken from experiment, because even finding the 'unperturbed' eigenvalues is a non-trivial problem for the present theory of strong interactions. This method, although crude, works reasonably well and it will probably be a long time before we have anything better.

Notice finally that the conclusion about the effective one-dimensionality of $\mathscr{H}_u$ for the unstable elementary particles derived above concerns the space-time part of the wave function only. If that part of the problem related to internal degrees of freedom cannot be decomposed completely, we have $\dim \mathscr{H}_u > 1$. Hence for the neutral kaons, for example, the space $\mathscr{H}_u$ is effectively two-dimensional provided that conditions (5.15) are valid; this justifies the assumption made in Example 4.3.

Notes to Chapter 3

SECTION 3.1. The description of photon emission from excited atoms, which we have used for illustrative purposes here, goes back to the classical paper of Weisskopf and Wigner (1930a). General 'decay systems' of this type were discussed by Horwitz and Sigal (1978). Notice that in order to describe the decay of the nth excited state, one does not need the whole subspace $(\Sigma_n^{\oplus} \mathscr{H}_a^{(n)} \otimes \{\Psi_0\}_{\text{lin}})^{\perp}$ as $\mathscr{H}_d$ because the latter should not contain the states to which the system must not pass because of either energy conservation or selection rules. Notice further that the splitting of the total Hamiltonian may not have any implications for the dynamical character of forces included to H_0, V. In the case of radiation damping, the interaction under consideration is entirely electromagnetic, and we divide it into the (mostly) Coulomb part included in H_0 and the 'radiation corrections' responsible for the decay. On the other hand, elementary-particle physics offers a lot of examples, in which the interactions are of a different nature, e.g. when the 'strongly stable' particles (like neutrons, neutral kaons, etc.) decay weakly.

• There are unstable systems which do not fit into the embedded-eigenvalue scheme. The well-known example concerns a particle which is confined initially inside a potential barrier. This case is of great practical interest: let us recall the theories of α-radioactivity by Gamow (1928), of heavy nuclei fission (cf. Malik and Sabatier (1973)) and many other 'tunneling-decay' effects. In the 'practical' quantum theory, situations of this type are usually handled by means of quasi-classical methods following the original Gamow's idea, while the embedded-eigenvalue problem rely on Fermi golden rule and the related machinery. Nevertheless, the barrier problem can also be viewed as a perturbation one. If the barrier is made infinitely thick or high, then we have a potential well whose energy levels may be found; one naturally expects that for large but finite barrier thickness (height) the spectrum of the Hamiltonian should be concentrated around the former energy eigenvalues. Hence we deal here with the perturbations of the *isolated* eigenvalues which disappear in the continuous spectrum as soon the perturbation is switched on. The intuitively expected spectral concentration can be illustrated on simple models; however, they cannot substitute a theory. The latter has been elaborated for the systems which are essentially one-dimensional (including spherically symmetric potential barriers). As mentioned above, the perturbation can be switched off in different ways: by removing a cut-off imposed on the potential (Howland, 1971) or by blowing up some strictly positive part of the potential (Emch and Sinha, 1979; Ashbaugh and Harrell, 1982).

• Another well-studied problem dealing with the spectral concentration and isolated-eigenvalue perturbations is the Stark-type effect; for its exposition and bibliography we refer to Reed and Simon (1978, §XII.5 and notes to this section); see also Herbst (1979) or Graffi and Grecchi (1981). Another closely related problem concerns the *Wannier ladder*, i.e. the systems described by a potential which is the sum of a linear term and a periodic one. A treatment of spectral concentration in this case together with further references can be found, e.g. in Avron (1982); the model is interesting in connection with the so-called Zener effect in solid-state physics.

• Notice also that the parameter which characterizes the spectral concentration can have no relation to the strength or range of interaction at all. An example is provided by the recent work of Gesztesy *et al.* (1982) who studied the non-relativistic limit of the systems described by Dirac Hamiltonian with an external field determined by a potential. In that case, the system possesses resonances which turn to the (isolated) eigenvalues of the corresponding Pauli Hamiltonian when the velocity of light $c \to \infty$.

• The idea of connecting decay phenomena with the poles of the analytically continued resolvent was implicitly presented in the paper of Weisskopf and Wigner (1930a). In the course of time, this connection has been revealed and refined gradually – see, e.g. Höhler (1958), Schwinger (1960), Grossman (1964), Horwitz and Marchand (1969b, 1971), Simon (1973), Baumgärtel (1975), and Horwitz and Sigal (1978). The pole-and-real-axis approximation, usually called simply 'pole approximation', represents a common tool for the description of decays; nonetheless, the rigorous results on its justification are scarce. For the Friedrichs model, this problem was discussed by Demuth (1976). Notice also that the pole-and-real-axis approximation can generally expected to be better the closer the pole is to the real axis. As we shall see in Sections 2 and 3 the imaginary part of the pole position is often proportional to g^2, up to higher-order terms in the coupling constant. This situation calls to mind the weak-coupling limit in the master equations mentioned briefly in the notes to Section 1.2: in both cases, the purely semigroup evolution is achieved in the limit $g \to 0$, with respect to the rescaled time $t' = g^2 t$.

SECTION 3.2. The model discussed here was formulated originally by Friedrichs (1948) with the intention to remedy defects of formal perturbation theory in application to the embedded-eigenvalue problems. Though this paper outlined clearly relations of the model to the Weisskopf–Wigner theory of the line breadth and to the Auger effect, it awoke at first no response among physicists. A new impetus came in the middle of the 1950s in connection with the so-called ***Lee model*** (Lee, 1954), a solvable field-theoretical model in which a heavy particle with two internal states (called N- and V-particle, $m_V > m_N$) interacts with a field of light bosons (Θ-particles). The model was further studied, e.g. by Glaser and Källén (1957), Araki *et al.* (1957), Höhler (1958), Lévy (1959) and many others; its description together with a more complete bibliography can be found in Schweber (1961, § 12b), Baz *et al.* (1966, chap. VIII). There are two characteristic situations here. If $m_V < m_N + m_\Theta$, then the eigenvalue m_V of the 'bare' mass operator is isolated and one studies its shift caused by the perturbation (mass renormalization). On the other hand, a large enough m_V is embedded into the continuous spectrum which starts at $m_N + m_\Theta$. In such a case, the V-particle becomes unstable and decays to N with emission of Θ. Description of this decay is essentially the same as in the Friedrichs model (with a simple eigenvalue embedded into non-simple continuous spectrum – see Theorem 3.1), so some authors prefer to speak of the Lee–Friedrichs model.

- Proposition 2.1 is taken from Horwitz and Marchand (1971). This paper contains a detailed discussion of the Friedrichs model, in both the original and multidimensional cases, which covers the essential physical aspects of the problem. The decay problem in the Friedrichs-type models is treated in numerous papers; we have chosen for its exposition a way which is in our opinion straightforward and at the same time illustrative. The obtained solution is, of course, local in the sense that it concerns small values of g only (weak-coupling case). With a particular choice of the function v, Höhler (1958) found the 'global' pole trajectories; it appears that if the coupling is sufficiently strong, the pole may eventually return to the first sheet (on negative real axis). Example 2.5, which illustrates the dangers hidden in the thresholds, is borrowed from Howland (1974).
- The 'Fermi golden rule' is one the basic parts of the quantum lore, and has been used successfully for many years despite its poor justification. The standard textbook derivation (Fermi, 1960, lecture 23) is discussed critically in Reed and Simon (1978, notes to § XII.6). The problem remained to disturb the mathematically inclined people, so gradually the rule was set on a more rigorous footing – see Howland (1972–1975), Simon (1973) and Baumgärtel (1975). These results are illustrated here in the Friedrichs-model framework; in Section 3 we shall establish the Fermi rule for more general perturbations.
- The connection between the decay and resonance phenomena is outlined only briefly within the scope of this book. General references to the quantum scattering theory from a 'physical' point of view are, e.g. the monographs by Newton (1966), Baz *et al.* (1966) or Taylor (1972); a mathematically rigorous treatment of the same subject can be found in Amrein *et al.* (1977), Reed and Simon (1979) and Amrein (1981). The bulk of this material concerns the quantum-mechanical scattering theory (especially, the potential scattering) which is more developed than its relativistic and field-theoretical counterparts. As for the example treated here, the scattering aspect of the problem has already been considered in the original Friedrichs paper (1948); a detailed discussion of the Friedrichs model as a scattering system was given by Horwitz and Marchand (1971).
- Various definitions of 'resonances' appear in the literature. Most frequently, this notion means the *scattering resonance* identified with a pole in the analytically continued scattering amplitude. The presence of such a pole has characteristic consequences, a bump (or gap) in the cross section and a sheer increase in the phase shift on π, especially if the pole is close to the real axis. It should be noted, however, that observation of these effects does not necessarily mean that the scattering amplitude possesses a pole – cf. Calucci and Ghirardi (1968) or Fonda *et al.* (1978). The number of resonances can be sometimes determined from a Levinson-type theorem (Horwitz and Marchand, 1971; Wollenberg, 1977). The perturbations of scattering resonances has been studied recently by Albeverio and Høegh-Krohn (1982). For one-dimensional Schrödinger operators whose potentials are of a compact support, the lower bounds on resonance widths were given by Harrell (1982).
- However, the poles of the scattering amplitude are not the only objects dubbed 'resonances':

this term is also used for the poles of the analytically continued reduced resolvent (*resolvent resonances*), for the *dilation-analytic resonances* mentioned below and others. Appearance of such a 'multivalued' notion is quite natural because these types of resonances correspond to various mathematical concepts aiming to describe essentially the same physical effect. In spite of this intuitive coincidence, the problem of establishing (for a given class of interactions) rigorous equivalence is, in general, difficult. In our example, the equivalence of scattering resonances and the second-sheet poles of the reduced resolvent is expressed by Proposition 2.6; a discussion of this problem for more general perturbations can be found in Howland (1975) or Baumgärtel *et al.* (1978). The equivalence between scattering resonances and dilation-analytic resonances for various classes of Schrödinger operators was established by Simon (1973), theorem 7.2) and Hagedorn (1979). Other related results are due to Jensen (1977, 1980) and Balslev (1978). For alternative approaches to the resonance problem see, e.g. Davies (1975a), Bailey and Schieve (1977), Sudarshan *et al.* (1978), Parravicini *et al.* (1980), Romo (1980), Böhm (1980, 1981), Gadella (1983) and Horwitz and Katznelson (1983b). Most of them seek some way in which the resonances could be treated as generalized eigenvalues of H (corresponding to the poles of the analytically continued resolvent). Since this is impossible in the Hilbert-space framework, the authors advocate the use of other structures such as rigged Hilbert spaces, partial-inner-product spaces, etc.

• The relation (2.20b) shows that the lifetime of the resonant state is proportional to the derivative of the phase shift with respect to energy at $\lambda = \lambda_p$. This quantity can be interpreted as the *time delay* of the scattered particle compared to the free one. Though too small to be measured directly for the really observed resonances, this quantity is nevertheless of interest. A more precise definition can be formulated as follows. We take a suitable region B_r in configuration space, say, a ball of radius r, and denote by P_r the corresponding projection on the state space $\mathscr{H}$. For a given $\psi \in \mathscr{H}_{ac}(H_0)$, the states $U_t^0\psi$ and $U_t\Omega_-\psi$ are asymptotically equal as $t \to -\infty$. Then

$$T_r(\psi) = \int_{\mathbb{R}} (\|P_rU_t\Omega_-\psi\|^2 - \|P_rU_t^0\psi\|^2)\, dt$$

is the difference between the times spent by these two states within B_r. This expression can have a finite limit $T(\psi)$ as $r \to \infty$, which is identified with the sought time delay, even if the limit of the sojourn times themselves eventually does not exist. Of course, for simple scattering systems – as the one connected with the Friedrichs model – one must choose some other appropriate one-parameter family of projections tending strongly to I for $\{P_r\}$.

• If $\mathscr{H}_u$ is one-dimensional, the corresponding decay law is an even function of time; thus the actual 'lifetime' of the resonance is $2\Gamma^{-1} = \delta_p^{-1}$. For a vector ψ whose energy distribution is given roughly by the corresponding Breit–Wigner function, $|\psi(\lambda)|^2 \approx (\delta_p/\pi)((\lambda - \lambda_p)^2 + \delta_p^2)^{-1}$ around $\lambda = \lambda_p$, we expect therefore $T(\psi) = \delta_p^{-1}$. Further, δ_p^{-1} can be expressed easily by means of the derivative of $\delta(\lambda) = -(i/2) \ln S(\lambda)$, and therefore we arrive at the so-called *Eisenbud–Wigner formula*:

$$T(\psi) = (\psi, T_{EW}\psi), \qquad T_{EW} = -i\overline{S(\lambda)}\,\frac{d}{d\lambda}S(\lambda).$$

It appears that this relation, which we have obtained by a heuristic argument for one particular ψ, can be proved exactly for many scattering systems, and that it holds usually on a dense subset of $\mathscr{H}_{ac}(H_0)$.

• The fact that the time delay is proportional to the derivative of the phase shift was recognized first in 1948 by Eisenbud in an unpublished dissertation, and later noted by Wigner (1955). The first rigorous treatment of the EW-formula can be found in the paper by Jauch *et al.* (1972). An alternate proof for simple scattering systems is due to Martin (1976). In the case of potential scattering, Gustafson and Sinha (1980) extended the results of Jauch *et al.* to the potentials which decrease at infinity as $r^{-5/2-\epsilon}$, and Jensen (1981), using the 'geometric' method, was able to cover those with the asymptotics $r^{-2-\epsilon}$. Recently Bollé *et al.* (1983)

have treated the case of long-range forces (including the Coulomb potential), in which the definition of time delay requires a modification. For N-body scattering, see Bollé and Osborn (1979). The dependence of time delay on the intensity of an electric field in the Stark effect was studied by Sinha (1978). There are also other definitions of time delay; e.g. that of Narnhofer (1980). For a more complete exposition of these problems, we refer to the recent review by Martin (1981).

SECTION 3.3. We often deal here with the analytic (meromorphic) operator-valued functions. We have already mentioned that an operator-valued function is strongly analytic iff it is weakly analytic. Moreover, it appears that also the norm-analycity is equivalent to both these notions (Hille and Phillips, 1957, §3.10), so we may speak about analytic operator-valued functions without further specification.

- The mathematical theory of embedded-eigenvalue perturbations was initiated by the work of Friedrichs discussed in the preceding section and Titchmarch (1951); for a more complete bibliography see Reed and Simon (1978, notes to §XII.6). The series of papers by Howland (1968–1975) and Baumgärtel with collaborators (1974–1978) are of particular importance. They gradually treat more general perturbations, starting from the finite-rank and compact operators, eventually reaching the unbounded ones. A substantial part of this work relies on the factorization technique sketched here, which belongs to Kato (1966b). Proposition 3.2 and Theorem 3.3 are adapted from Baumgärtel (1975). Theorem 3.4 was given first by Howland (1972) in the particular case when T_g is g-independent; the present modification of this result is due to Baumgärtel (1975) as well as Theorem 3.5. The operator $W(z, g)$ appearing in the relations (3.20) is known under different names; in particular, Howland (1975) calls it the Livsic matrix (after M. S. Lifshitz). Discussion of the Fermi rule (Theorem 3.6) follows essentially Howland (1974) and Baumgärtel (1975). Further generalization of the present considerations to the case when $H_g = T_g + B_g^* A_g$, and the last term too is not necessarily linear in the coupling constant, may be found in Baumgärtel and Demuth (1976). Notice also that the methods described here are sometimes useful for the study of isolated-eigenvalue perturbations; namely, if the perturbation can be divided into two parts, the first of which shifts the eigenvalue so that it becomes embedded – see Demuth (1974), and Rauch (1980) for a particular class of Schrödinger operators.
- Specific methods have been developed for the Schrödinger operators, $H = -\frac{1}{2}\Delta + gV$. In this approach, the potential V is assumed to fulfil certain analycity requirements, and one introduces the operators U_θ, which express particularly space-coordinates dilatation if θ is real, and study the operators $H(\theta) = U_\theta H U_\theta^{-1}$ for complex θ. It appears that this transformation preserves the eigenvalues and thresholds, and rotates the essential spectrum (i.e. cuts off the resolvent) so the resonance poles (identified with the eigenvalues of $H(\theta)$ for complex θ) can be revealed. For a description of this dilation-analytic method and a more complete bibliography we refer to Aguilar and Combes (1971), Simon (1973) and Reed and Simon (1978, §XII.6). There is also a related technique of local distortions – see Babbitt and Balslev (1976) and the references contained in their paper. The rotation of the essential spectrum has also been used for other operators – cf. Horwitz and Sigal (1978). Another modification of the dilation-analytic technique has been developed for the treatment of the Stark effect – see Herbst (1979) or Graffi and Grecchi (1981).

SECTION 3.4. Instead of one-parameter unitary groups, we could consider more general symmetries described by unitary representations of some topological group. In Section 5 we shall study such a problem for the Poincaré group. There is an enormous variety of problem connected with symmetries of physical systems and their treatment using representations of groups, Lie algebras, etc., and we can only touch this field briefly here. As for the references, we content ourselves with mentioning the monograph by Barut and Rączka (1977), in which an extensive bibliography and guide to problems may be found. An illustrative example of the angular-momentum decomposition in application to the embedded-eigenvalues perturbation, namely the Auger effect, is due to Simon (1973, §5).

• The decay of neutral kaons is described by the semigroup (4.7) within the range of experimental errors. Nevertheless, there are speculations about a possible breakdown of the semigroup condition. Horwitz and Marchand (1969b) found with the help of a Friedrichs-type model and the experimentally known non-orthogonality of φ_S, φ_L that the residuum operators must fulfil

$$P_S = \begin{pmatrix} O(1) & O(\epsilon) \\ O(\epsilon) & O(\epsilon^2) \end{pmatrix}, \qquad P_L = \begin{pmatrix} O(\epsilon^2) & O(\epsilon) \\ O(\epsilon) & O(1) \end{pmatrix};$$

they concluded that the CP-symmetry is violated as $P_S P_L = O(\epsilon)$. However, we may express the residuum operators (4.5) in the basis of the CP-eigenstates (4.9) obtaining

$$P_S = (1-\epsilon^2)^{-1} \begin{pmatrix} 1 & \epsilon \\ -\epsilon & -\epsilon^2 \end{pmatrix}, \qquad P_L = (1-\epsilon^2)^{-1} \begin{pmatrix} -\epsilon^2 & -\epsilon \\ \epsilon & 1 \end{pmatrix};$$

these relations yield exact validity of (4.6a) and show the weak point of the above argument. It is not excluded, of course, that more exact measurements will reveal a departure from the semigroup time evolution, but this has to be commented in the same way as the possible violations due to energy semiboundedness in Section 1.6: there is ample room for such cases in the quantum-kinematical scheme discussed here, and the semigroup evolution is merely a very useful approximation.

• Since we believe that the weak and electromagnetic forces are no more than different aspects of the same interaction, then after its switching off we would in fact have an eigenvalue of multiplicity 4 for the kaon mass operator, because the charged and neutral kaons are alike from the point of view of the strong interaction. However, charge conservation is a superselection rule, so the neutral kaon decay may be treated independently. Let us also mention that besides such standard and commonly accepted superselection rules, some other were proposed for a description of the decay and resonance processes. For instance, Lurçat (1968) postulated mass as a continuous superselection rule; but such a theory excludes *a priori* time characterization of decays. Even more non-standard is the model mentioned by Piron (1969) in which the fact of the decay at an instant τ represents a superselection rule.

SECTION 3.5. The exposition in this section follows essentially Exner (1983b). The first systematic study of the unitary irreducible representations of $\mathscr{P}$ can be found in the classical paper by Wigner (1939). The attempts to associate non-unitary irreducible representations of $\mathscr{P}$ with unstable particles were connected with the group-theoretical boom in elementary-particle physics during the 1960s and early 1970s. This trend was stimulated partly by a certain pessimism concerning perspectives of the quantum field theory; it was not strange, therefore, that some authors tried to accompany the constructed representations with modifications in postulates of the quantum theory, e.g. Simonius (1970) or Rączka (1973). As for the proposed non-unitary representations, Zwanziger (1963) treated the simplest possibility, when the representations are characterized by a complex four-momentum which equals $(m - i\delta)(1, 0)$ in the rest system of the particle. More general representations, in which the real and imaginary parts of the complex four-momentum are not necessarily parallel each other, was studied, e.g. by Beltrametti and Luzzatto (1965), Roffman (1967), Kawai and Gotō (1969), Simonius (1970) or Weldon (1976). Schulman treated the so-called Poincaré semigroup, i.e. the part of $\mathscr{P}$ which consists of the proper Lorentz group and translations to the forward light cone (absolute future); the obtained representations are closely related to the results of the above-named authors. He also studied representations which are reducible but not completely reducible. A more general treatment of such representations (dubbed usually as indecomposable) is due to Raczka (1973); see also Barut and Rączka (1977, §17.4). A particular aim of considering the indecomposable representations was to obtain the decay laws connected with the higher-order poles of the analytically continued reduced resolvent (see notes to Section 1.5).

• The approach discussed in the present section – namely, to choose a reducible unitary representation of $\mathscr{P}$ on the 'large' state space $\mathscr{H}$, and to project it on a suitably chosen subspace $\mathscr{H}_u$ – is opposite in a sense to the one based on non-unitary representations. Needless to say, it is physically much better motivated and more general. Direct integrals of unitary irreducible representations over mass were first used by Beskow and Nilsson (1967), and later independently by Jersák (1969b). Williams (1971) made an attempt to reconcile both the above-named approaches by constructing minimal unitary dilation of Zwanziger's representation. He obtained the Hilbert space (5.8a) with $m_0 = -\infty$ and a unitary representation $\tilde{U}$ of $\mathscr{P}$ on $\mathscr{H}$ which coincides with (5.9) when transformed by means of (5.8b). The Zwanziger's representation is recovered if one projects $\tilde{U}$ to the subspace

$$\mathscr{H}_u^Z = \{f\}_{\text{lin}} \otimes L^2(\mathbb{R}^3, d^3k/2k_0) \otimes \mathbb{C}^{2s+1},$$

where $f(m) = (\Gamma/2\pi)^{1/2}(m - M + (i/2)\Gamma)^{-1}$. Williams himself regarded the below-unbounded mass spectrum as the main defect, but we know already that it can be rectified by a mass filtering procedure without any observable consequences. The fact that $\mathscr{H}_u$ should be Euclidean invariant together with its implication for the non-unitarity of $V(\Lambda_\beta, 0)$ was pointed out by Havlíček and Exner (1973). For a description of the time-dilatation experiments with cosmic-ray muons, see Votruba (1969, §IV.4.3).

• Of course, the condition (5.20b) does not need its momentum (or velocity) distribution to be supported by some ball only. In order to take possible 'tails' of these distributions into account (simultaneously preserving the translational invariance of $\mathscr{H}_u$), a more careful mathematical treatment of the problem is needed. However, we are not going to pursue this task here. On the heuristic level, one might say that in the subspace referring to the Zwanziger's representation a lot of space is left unemployed. One actually needs only

$$\mathscr{H}'_u = \{f\}_{\text{lin}} \otimes L^2(B_\kappa, d^3k/2k_0) \otimes \mathbb{C}^{2s+1},$$

where $\kappa \approx \epsilon/M$. This subspace is in some sense 'intermediate' between the two mentioned above. One can derive for it a conclusion analogous to (5.19b) with more ease. On the other hand, $\mathscr{H}'_u$ is no longer translationally invariant, though the violation becomes essential only on large distances (Problem 17).

• As for the problem of identity, Schwinger (1960) pointed out that the concept of an unstable particle becomes meaningless when its decay law depends on details of the measuring procedure; in particular, on energy filtering. We have seen in the preceding chapters that this is unlikely for 'true' unstable (i.e. metastable) systems. On the other hand, for scattering resonances the attainable energy resolution allows a fine energy filtering, but the time evolution of the resonance as a separate object can hardly be studied experimentally, and even associating a subspace $\mathscr{H}_u$ with the resonance is a speculative matter. Nonetheless, resonances are identifiable objects in a well-understood sense: they have spin, parity, isospin, etc. and the characteristic behaviour of the cross section and phase shift connected with poles of the scattering amplitude. Hence we have two mutually exclusive effects (Beskow and Nilsson, 1967) which are generated essentially by the same dynamical mechanism, and each of them has an identity criteria of its own.

Chapter 4

Pseudo-Hamiltonians

"The virtue of the model is that it eliminates a large number of degrees of freedom and leaves one with a more analytically tractable problem."

E. B. DAVIES

After establishing the usefulness and plausibility of the semigroup approximation in the preceding chapters, we study here the semigroup time evolution from the point of view of its infinitesimal generator, which we call pseudo-Hamiltonian. First we discuss the relations of such an operator to the total Hamiltonian of the system under consideration. Section 2 is devoted to the mathematical characterization of pseudo-Hamiltonians, which appear to belong to the class of maximal dissipative operators; we deduce here various criteria under which a given operator is (essentially) maximal dissipative. In Section 3, an important particular class of pseudo-Hamiltonians is treated, namely the Schrödinger-type operators with complex absorptive potentials. Next we study one of the situations where the pseudo-Hamiltonian description often appears in practice: the optical approximation in the many-channel scattering processes. We derive some conditions under which this approximation may be expected to be good. Motivated by this problem, we present in the last section some fundamental notions of the non-unitary scattering theory.

4.1. Pseudo-Hamiltonians and Quasi-Hamiltonians

The preceding chapters have illustrated the usefulness of the semigroup approximation for a description of unstable systems. Here we are going to discuss the semigroup time evolution from the viewpoint of its infinitesimal generator.

Non-self-adjoint Hamiltonians were used as a tool for the phenomenological description of certain non-stationary processes from the early years of quantum theory. Some examples have already been mentioned, others will be discussed below. Such treatments usually start from the Schrödinger-type equation

$$i\,\frac{\mathrm{d}}{\mathrm{d}t}\,\psi_t = H_{\mathrm{p}}\psi_t, \tag{1.1}$$

where H_p is a phenomenological Hamiltonian on a suitable Hilbert space $\mathscr{H}_p$. On a formal level, the solution of this equation referring to an initial datum $\psi_0 = \varphi$ is $\psi_t = \exp(-iH_p t)\varphi$ and the evolution operator has a semigroup property. Since H_p is not self-adjoint, the norm of ψ_t need not be conserved. It is not allowed to grow, however, unless some 'sources' are introduced into the theory. The last restriction is often associated, again more or less formally, with the condition

$$\operatorname{Im}(\psi, H_p\psi) \leqslant 0 \quad \text{for all } \psi \in D(H_p). \tag{1.2}$$

The connection between equation (1.1) and semigroup time evolution permits a more rigorous formulation. Let us recall that to each continuous contractive semigroup $\{V_t : t \geqslant 0\}$ on a Hilbert space $\mathscr{H}$ there corresponds a unique infinitesimal generator A, which is defined by

$$A\psi := \lim_{t \to 0+} t^{-1}(I - V_t)\psi. \tag{1.3}$$

The domain $D(A)$ consists of all the vectors $\psi \in \mathscr{H}$ for which the limit (1.3) makes sense, and the operator A determined in this way is densely defined and closed. Usually the requirements of continuity and contractivity may be imposed on the semigroup evolution, in which case the operator iH_p should be the infinitesimal generator of some continuous contractive semigroup. Phenomenological Hamiltonians of this type will be dubbed **pseudo-Hamiltonians** here. It means particularly that each pseudo-Hamiltonian is supposed to be densely defined and closed.

The considerations presented in the preceding chapters allow us to understand easily how the phenomenological description may be incorporated into the standard quantum-theoretical framework. The object of investigation has to be regarded as a part of some larger isolated system, to which a Hamiltonian H on a state Hilbert space $\mathscr{H} \supset \mathscr{H}_p$ is ascribed, and the phenomenological evolution operator $\exp(-iH_p t)$ should be identified with projection of the propagator $U_t = \exp(-iHt)$ to the subspace $\mathscr{H}_p$. This interpretation, though naturally applicable to most of the phenomenological models, raises some problems of both a physical and a mathematical nature:

(i) The described scheme does not fully preserve the dual character of the Hamiltonian – namely, to be simultaneously the operator of the total energy and the infinitesimal generator of time translations. This is true, of course, for the operator H which refers to an isolated system. In the phenomenological description, however, the role of the time-translation generator is entrusted to the pseudo-Hamiltonian H_p, while the observable energy should be associated rather with $\mathrm{pr}_p\, H$. Thus the interrelations between H and H_p must be formulated thoroughly. In particular, the inevitably approximative character of the semigroup reduced evolution should be taken into account.

(ii) Another related problem is of a more mathematical nature but not less important. As we have seen, each pseudo-Hamiltonian is a multiple of the generator of a continuous contractive semigroup. It would be useful to have some equivalent expressions of this condition, in particular, various necessary and sufficient conditions under which concrete classes of operators may be accepted as pseudo-Hamiltonians.

Let us start with the approximative character of the semigroup reduced evolution. We have already treated this problem in Section 1.6, but there are two reasons which force us to choose a slightly different approach here. The first concerns the technical aspects of the results derived in Section 1.6. Proposition 1.6.3 may be applied without restrictions only if $\dim \mathscr{H}_p < \infty$. For an infinite-dimensional $\mathscr{H}_p$ one must start with checking the relation $\text{s-lim}_{b\to\infty} E_p^b = E_p$, and this task may not be easy. However, there is a more important reason. In Section 1.6 we assumed the Hamiltonian to be given, and we changed the state space by the addition of an energy filter. It is more natural in the present context to regard the phenomenological state space $\mathscr{H}_p$ as given, and to suppose that the 'large' Hilbert space $\mathscr{H}$ together with the corresponding Hamiltonian H are determined only approximatively.

In general, the accuracy of the semigroup description may depend on the state of the system and the time interval involved. Hence we take an arbitrary positive ϵ, T and a finite subset $M \subset \mathscr{H}_p$, and ask whether a triple $\{\mathscr{H}^\xi, H^\xi, P^\xi\}$ could be found to each $\xi \equiv (\epsilon, T, M)$ with the following properties:

(p1) $\mathscr{H}_p = P^\xi \mathscr{H}^\xi$,
(p2) H^ξ is self-adjoint and semibounded,
(p3) $\|(\text{pr}_p \exp(-iH^\xi t) - V_t)\psi\| < \epsilon$ for all $\psi \in M$ and $t \in [0, T]$.

Notice that the requirements of finiteness imposed on the time interval and the subset of states are quite natural from the viewpoint of any real experimental situation. It appears that the above conditions are easily fulfilled; one may even choose $\mathscr{H}$ and P independently of ξ and the operators H^ξ as functions of a single operator H.

PROPOSITION 4.1.1. *Let $\{V_t : t \geqslant 0\}$ be a continuous contrative semigroup on $\mathscr{H}_p$, and $\{\mathscr{H}, H, P\}$ be the triple resulting from its minimal unitary dilation. Then to any positive ϵ, T and a finite subset $M \subset \mathscr{H}_p$, there exists a sequence $\{f_n\}$ of real Borel functions such that*

(a) *the triple $\{\mathscr{H}, f_n(H), P\}$ fulfils* (p1)–(p3) *and* $\inf \sigma(f_n(H)) \geqslant -n$;
(b) *the sequence $\{f_n(H)\}$ converges to H as $n \to \infty$ in the strong resolvent sense, i.e.* $\text{s-lim}_{n\to\infty} (f_n(H) - z)^{-1} = (H - z)^{-1}$ *for each non-real z.*

Proof. Theorem 1.5.1 implies $\sigma(H) = \mathbb{R}$, with the exception of the trivial case when $\{V_t\}$ is unitary and $H_p = H$. Let $\{f_n\}$ be a sequence of real Borel functions such that $f_n(x) \geqslant -n$ and $\lim_{n\to\infty} f_n(x) = x$ for all $x \in \mathbb{R}$. One can take, for example, $f_n(x) = x\Theta(x + n)$, where Θ is the Heaviside function. Then the operators $f_n(H)$ are self-adjoint, $\inf \sigma(f_n(H)) \geqslant -n$, and the functional-calculus rules imply

$$\text{s-lim}_{n\to\infty} \exp(-if_n(H)t) = \exp(-iHt) \tag{1.4}$$

for each $t \in \mathbb{R}$. Further, using the Trotter theorem (cf. Reed and Simon (1972, theorem and problem VIII.21)), we find that $f_n(H)$ converges to H as $n \to \infty$ in the strong resolvent sense, and the convergence in (1.4) is uniform with respect to t in any finite interval. We have $V_t = \mathrm{pr_p}\, e^{-iHt}$, so

$$\|\mathrm{pr_p} \exp(-if_n(H)t) - V_t)\psi\| \leq \|(\exp(-if_n(H)t) - \exp(-iHt))\psi\|$$

and the condition (p3) with $H^\xi = f_n(H)$ is easily seen to be valid for n larger than some $n(\xi)$. ∎

REMARK 4.1.2. Since the condition $\lim_{n\to\infty} f_n(x) = x$ was only important for the application of the Trotter theorem, one can also carry out the approximation with *bounded* operators. There are various other modifications of the last result; we leave them to the reader.

Theorem 1.4.1 and the Stone theorem together imply that, to a given pseudo-Hamiltonian H_p on $\mathscr{H}_\mathrm{p}$, there is a (unique up to unitary equivalence) self-adjoint operator H which corresponds to $U_t = \exp(-iHt)$ from the minimal unitary dilation of $t \mapsto \exp(-iH_\mathrm{p}t)$. We shall call this H the **quasi-Hamiltonian** referring to H_p; it substitutes for the operator of total energy in the above-described sense.

Let us now examine in more detail the relations between a pseudo-Hamiltonian H_p and the corresponding quasi-Hamiltonian H. According to (1.3), we have

$$iH_\mathrm{p}\varphi = \lim_{t\to 0+} t^{-1}(\varphi - V_t\varphi) = \lim_{t\to 0+} P[t^{-1}(\varphi - U_t\varphi)]$$

for all $\varphi \in D(H_\mathrm{p})$. On the other hand, $\lim_{t\to 0+} t^{-1}(\varphi - U_t\varphi)$ exists for each $\varphi \in D(H)$ and equals $iH\varphi$ so

$$H_\mathrm{p}\varphi = (\mathrm{pr_p}\, H)\varphi \quad \text{for all } \varphi \in D(H_\mathrm{p}) \cap D(H). \tag{1.5}$$

The trouble is that the relation (1.5) might eventually hold for the zero vector only:

EXAMPLE 4.1.3. Let $\dim \mathscr{H}_\mathrm{p} = 1$ with H_p multiplying on $z = \lambda_0 - i\delta$, $\delta > 0$. Due to Example 1.4.8, the quasi-Hamiltonian H acts as $(H\psi)(\lambda) = \lambda\psi(\lambda)$ on $L^2(\mathbb{R})$, when $\mathscr{H}_\mathrm{p}$ is identified with the linear span of $\psi_0 : \psi_0(\lambda) = (\delta/\pi)^{1/2}(\lambda - z)^{-1}$. However, $\psi_0 \notin D(H)$, so $D(H_\mathrm{p}) \cap D(H) = \{0\}$.

It is clear that the unboundedness of H is essential here, and therefore the difficulty might be rectified by considering bounded functions of H only. To this end, we replace (1.5) by the corresponding relation between the resolvents.

PROPOSITION 4.1.4. *Let H_p be a pseudo-Hamiltonian on $\mathscr{H}_\mathrm{p}$ and H and corresponding quasi-Hamiltonian on $\mathscr{H} \supset \mathscr{H}_\mathrm{p}$. Then the equality*

$$R(z, H_\mathrm{p}) = R_\mathrm{p}(z, H) \tag{1.6}$$

holds for the reduced resolvent $R_{\mathrm{p}}(z, H) = \mathrm{pr}_{\mathrm{p}}(H - z)^{-1}$ *and all complex* z *with* $\mathrm{Im}\, z > 0$. *Furthermore, we have*

$$(\varphi, E_H([\lambda, \mu])\psi) + (\varphi, E_H((\lambda, \mu))\psi)$$
$$= \frac{1}{\pi i} \lim_{\eta \to 0+} \int_{\lambda}^{\mu} (\varphi, [R(\xi + i\eta, H_{\mathrm{p}}) - R(\xi - i\eta, H_{\mathrm{p}}^*)]\psi)\, \mathrm{d}\xi, \tag{1.7}$$

where $\varphi, \psi \in \mathscr{H}_{\mathrm{p}}$ *and* λ, μ *are real,* $\lambda < \mu$.

Proof. According to Theorem 2.1 below, the open upper halfplane belongs to the resolvent set $\rho(H_{\mathrm{p}})$, and the relations

$$(\varphi, R(z, H_{\mathrm{p}})\psi) = i \int_0^t \mathrm{e}^{izt} (\varphi, V_t\psi)\, \mathrm{d}t = i \int_0^t \mathrm{e}^{izt} (\varphi, U_t\psi)\, \mathrm{d}t$$
$$= (\varphi, R(z, H)\psi)$$

hold for all $\varphi, \psi \in \mathscr{H}_{\mathrm{p}}$ so (1.6) is proved. We have

$$R(z, H_{\mathrm{p}})^* = R(\bar{z}, H_{\mathrm{p}}^*), \qquad z \in \rho(H_{\mathrm{p}}), \tag{1.8}$$

according to definition of the reduced resolvent. Now the relations (1.6), (1.8) give $R_{\mathrm{p}}(\xi - i\eta, H) = R(\xi - i\eta, H_{\mathrm{p}}^*)$, so (1.7) follows from (3.1.1). ■

This result yields the following alternative for the eigenvalues of a pseudo-Hamiltonian H_{p}:

THEOREM 4.1.5. *Let* $z_0 = \lambda_0 - i\delta$ *be an eigenvalue of* H_{p} *corresponding to an eigenvector* $\psi_0 \in \mathscr{H}_{\mathrm{p}}$. *Then there are the following possibilities*:

either $\delta = 0$, *in which case* $z_0 = \lambda_0$ *is at the same time the eigenvalue of the quasi-Hamiltonian* H,

or $\delta > 0$, *then* $\psi_0 \in \mathscr{H}_{\mathrm{ac}}(H)$ *and*

$$\frac{\mathrm{d}}{\mathrm{d}\lambda}(\psi_0, E_\lambda\psi_0) = \frac{\delta}{\pi}[(\lambda - \lambda_0)^2 + \delta^2]^{-1}, \tag{1.9}$$

where $E_\lambda \equiv E_H((-\infty, \lambda])$. *In particular,* $\psi_0 \notin D(H)$.

Proof. The possibility $\delta < 0$ is excluded by Theorem 2.1 below. For any complex z with $\mathrm{Im}\, z > 0$, we have $R(z, H_{\mathrm{p}})\psi_0 = (z_0 - z)^{-1}\psi_0$. Thus we may put $\varphi = \psi = \psi_0$ in (1.7) and perform the integration, which yields

$$(\psi_0, E_H([\lambda, \mu])\psi_0) + (\psi_0, E_H((\lambda, \mu))\psi_0)$$
$$= \frac{2}{\pi} \lim_{\eta \to 0+} \left[\mathrm{arctg}\, \frac{\mu - \lambda_0}{\delta + \eta} - \mathrm{arctg}\, \frac{\lambda - \lambda_0}{\delta + \eta}\right]. \tag{*}$$

If $\delta = 0$, we set $\lambda = \lambda_0$ so (*) acquires the form

$$(\psi_0, E_H([\lambda_0, \mu])\psi_0) + (\psi_0, E_H((\lambda_0, \mu))\psi_0) = 1$$

for each $\mu > \lambda_0$. One can always find a decreasing sequence $\{\mu_n\}_{n=1}^{\infty}$ such that $\lim_{n\to\infty} \mu_n = \lambda_0$ and $E_H(\{\mu_n\}) = 0$ for all n. The last property together with additivity of the spectral measure give

$$(\psi_0, E_H(\{\lambda_0\})\psi_0) + 2(\psi_0, [E_{\mu_n} - E_{\lambda_0}]\,\psi_0) = 1,$$

and since $\lambda \mapsto E_\lambda$ is strongly continuous, we get $(\psi_0, E_H(\{\lambda_0\})\psi_0) = 1$ as $n \to \infty$. This proves the first part. On the other hand, a similar argument shows in the case $\delta > 0$ that $(\psi_0, E_H(\{\lambda\})\psi_0) = 0$ for each $\lambda \in \mathbb{R}$, so $\psi_0 \in \mathcal{H}_c(H)$. Then the limit $\lambda \to -\infty$ in (*) leads to the relation

$$(\psi_0, E_\mu\psi_0) = \frac{1}{\pi}\,\text{arctg}\,\frac{\mu - \lambda_0}{\delta} + \tfrac{1}{2},$$

which yields the remaining part of the assertion. ∎

Thus each eigenspace corresponding to a *real* eigenvalue of H_p reduces both the operators H_p and H. It implies, for example, that any two eigenvectors of H_p corresponding to different real eigenvalues are mutually orthogonal. Of course, this is only a particular case of the assertion expressed by Proposition 1.4.3. On the other hand, eigenvectors of H_p corresponding to different *non-real* eigenvalues are in no case obliged to be orthogonal. Let us recall, for instance, the decay of neutral kaons (for another illustrative example see Proposition 6.2.14 below):

EXAMPLE 4.1.6. The semigroup (3.4.7) is generated by the pseudo-Hamiltonian H_{K^0} on $\mathcal{H}_{K^0} = \mathbb{C}^2$: $H_{K^0} = z_S P_S + z_L P_L$. In view of (3.4.6), $z_j = m_j - i\delta_j$ are eigenvalues of H_{K^0} corresponding to the non-orthogonal eigenvectors φ_j, $j = \text{S, L}$.

Notice that the considerations presented above bear a direct resemblance to those of Section 3.1. The fact that the resolvents of a physical Hamiltonian and of the quasi-Hamiltonian H are mutually close in the strong resolvent sense, together with the relation (1.6), show that the description of an open system through a pseudo-Hamiltonian may be regarded as a *generalization to the pole-and-real-axis approximation*. This is particularly apparent when the non-real eigenvalues of a pseudo-Hamiltonian are considered.

Let us finally mention the *time-dependent pseudo-Hamiltonians*. We have not in mind here the fact that for a reduced propagator V_t which is semigroup only approximatively, the time derivative $\dot{V}_t$ cannot be a constant operator multiple of V_t. This time dependence is not important from the practical point of view and we neglect it in the well-defined approximative sense. There are cases, however, when the considered microsystem is not only dissipative itself (as a consequence of an internal dynamical mechanism), but in addition it is exposed to the influence of some external forces. Such a situation occurs, for instance, when an excited atom is placed into a non-constant (e.g. oscillating) electromagnetic field. It is clear that such a time dependence of the phenomenological Hamiltonian raises some specific problems. In particular, the construction of the solution to the evolution equation

$$i \frac{\mathrm{d}}{\mathrm{d}t} \psi_t = H(t)\psi_t \tag{1.10}$$

– where $H(t)$ represents a pseudo-Hamiltonian for each fixed t from the considered interval and $H(\,.\,)$ eventually fulfils suitable smoothness conditions – is in general much more difficult than the analogous task for equation (1.1). Below, we shall present various results concerning this problem.

4.2. Maximal Dissipative Operators

"His Majesty discovered not the least curiosity to enquire into the laws, government, history, religion, or manners of the countries where I had been but confined his questions to the state of mathematics."

J. SWIFT, *A Voyage to Laputa*

Since we are now going to discuss the pseudo-Hamiltonians alone, we shall mostly drop the subscript 'p' used in the preceding section. We have seen that each pseudo-Hamiltonian may be naturally required to be (modulo i) infinitesimal generator of a continuous contractive semigroup. The fundamental characterization of such operators is given by the following criterion:

THEOREM 4.2.1. *A closed operator H on $\mathscr{H}$ is of the form $H = -iA$, with A being infinitesimal generator of a continuous contractive semigroup, iff the following conditions hold simultaneously*:

(i) *H is densely defined and* $\mathrm{Ran}(H - i\lambda) = \mathscr{H}$ *for some* $\lambda > 0$;
(ii) $\mathrm{Im}(\psi, H\psi) \leqslant 0$ *for all* $\psi \in D(H)$.

Moreover, if H satisfies (i) *and* (ii), *then the whole open upper halfplane belongs to the resolvent set $\rho(H)$ and*

$$R(z, H)\psi = i \int_0^\infty \mathrm{e}^{izt}\, V_t\psi \,\mathrm{d}t, \tag{2.1}$$

where $V_t = \exp(-iHt)$, *holds for all* $\psi \in \mathscr{H}$ *and complex z with* $\mathrm{Im}\, z > 0$.

Proof. One has only to combine some well-known results concerning continuous contractive semigroups. The conditions (i), (ii) represent the Hilbert-space version of the basic criterion (Reed and Simon, 1975, theorem X.48), while the remaining assertion follows directly from the Hille–Yosida theorem. ∎

The condition (ii) in the above theorem is just the requirement expressed by (1.2). A densely defined operator with this property will be called **dissipative**; we shall also use the term **accretive** for the corresponding operator $A = iH$ which fulfils $\mathrm{Re}(\psi, A\psi) \geqslant 0$ for all $\psi \in D(A)$. We see that dissipativity is necessary but, in general, not sufficient for a (closed) operator H to be acceptable as a pseudo-Hamiltonian. The following simple consequences may be drawn from Theorem 2.1:

COROLLARY 4.2.2. *Let H be a closed operator on $\mathscr{H}$, then $A = iH$ generates a continuous contractive semigroup if the operators H and $-H^*$ are simultaneously dissipative.*

Proof. It holds that

$$\|(H - i\lambda)\psi\|^2 = \|H\psi\|^2 - 2\lambda\, \mathrm{Im}(\psi, H\psi) + \lambda^2 \|\psi\|^2 .$$

Thus for a positive λ, the requirement of dissipativity yields

$$\|(H - i\lambda)\psi\| \geqslant \lambda \|\psi\| \quad \text{for all } \psi \in D(H), \tag{2.2a}$$

and analogously

$$\|(H^* + i\lambda)\varphi\| \geqslant \lambda \|\varphi\| \quad \text{for all } \varphi \in D(H^*) \tag{2.2b}$$

The first of these inequalities implies $\mathrm{Ran}(H - i\lambda)$ to be closed, while the second shows that $\mathrm{Ker}(H^* + i\lambda) = \{0\}$. Together they give

$$\mathrm{Ran}(H - i\lambda) = \overline{\mathrm{Ran}(H - i\lambda)} = (\mathrm{Ker}(H^* + i\lambda))^{\perp} = \mathscr{H},$$

so Theorem 2.1 can be applied. ■

COROLLARY 4.2.3. *Let $H = H_0 - iW$ with H_0 self-adjoint and W bounded accretive, then $A = iH_0 + W$ generates a continuous contractive semigroup.*

Proof. One can verify easily hypotheses of the preceding corollary to be fulfilled. ■

Now we shall deduce an alternative criterion in which the condition (i) of Theorem 2.1 is replaced by a suitable maximality requirement. To this end, we employ the **Cayley transformation**, which assigns to an operator H on $\mathscr{H}$ – such that $H - i$ is injective – an operator V defined on $D(V) = \mathrm{Ran}(H - i)$ by the relation

$$V = (H + i)(H - i)^{-1}. \tag{2.3}$$

LEMMA 4.2.4. *The Cayley transformation defines a bijective mapping between the set of dissipative operators H on $\mathscr{H}$ and the set of contractions V on $\mathscr{H}$ such that* $\mathrm{Ran}(V - I)$ *is dense in $\mathscr{H}$. The operator H is closed iff the same is true for V.*

Proof. Let first H be dissipative and $\psi \in D(H)$. Then $\varphi \in D(V)$, where

$$\varphi = (H - i)\psi, \qquad V\varphi = (H + i)\psi. \tag{2.4}$$

Dissipativity of H implies contractivity of V:

$$\begin{aligned} \|V\varphi\|^2 &= \|H\psi\|^2 + \|\psi\|^2 + 2\,\mathrm{Im}(\psi, H\psi) \\ &\leqslant \|H\psi\|^2 + \|\psi\|^2 - 2\,\mathrm{Im}(\psi, H\psi) = \|\psi\|^2 \end{aligned}$$

for each $\varphi \in \mathrm{Ran}(H - i)$. Since V is bounded, it is closed iff $D(V) = \mathrm{Ran}(H - i)$ is closed, which is, further, equivalent to the closedness of H (Problem 18). The relations (2.4) may be inverted

$$\psi = \frac{1}{2i}(V\varphi - \varphi), \qquad H\psi = \tfrac{1}{2}(V\varphi + \varphi); \tag{2.5}$$

it shows that $V - I$ is injective and its range equals $D(H)$.

Conversely, let V be a contraction such that $V - I$ is injective. Then the relations (2.5) define a linear operator H on $\mathscr{H}$ with $D(H) = \mathrm{Ran}(V - I)$ and

$$\mathrm{Im}(\psi, H\psi) = \tfrac{1}{4}(\|V\psi\|^2 - \|\psi\|^2) \leqslant 0.$$

Until now, we have not used the fact that H is densely defined. Combined with the last inequality, it implies H to be dissipative. We shall further show that if $\mathrm{Ran}(V - I)$ is dense, then $V - I$ is injective. Suppose that $V\psi - \psi = 0$ for some $\psi \in D(V)$. For arbitrary complex α and $\chi \in D(V)$, the inequality $\|V(\chi + \alpha\psi)\|^2 \leqslant \|\chi + \alpha\psi\|^2$ holds, which gives

$$\begin{aligned}&\|V\chi\|^2 + \mathrm{Re}[\alpha(V\chi, V\psi)] + |\alpha|^2\, \|V\psi\|^2 \\ &\quad \leqslant \|\chi\|^2 + 2\,\mathrm{Re}[\alpha(\chi, \psi)] + |\alpha|^2\, \|\psi\|^2.\end{aligned}$$

Using the contractivity of V again, together with the assumption about ψ, we can bring the last relation to the form

$$2\,\mathrm{Re}[\alpha(V\chi - \psi, \psi)] \leqslant \|\chi\|^2 - \|V\chi\|^2.$$

However, α is arbitrary complex, so $(V\chi - \chi, \psi) = 0$. Finally, the set $\{V\chi - \chi : \chi \in D(V)\}$ is dense in $\mathscr{H}$ due to the assumption, thus $\psi = 0$ and $V - I$ is injective. ∎

This lemma suggests a way in which the condition replacing (i) from Theorem 2.1 may formulated. We need the following notion: a dissipative operator is called **maximal dissipative** if it has no proper dissipative extensions.

THEOREM 4.2.5. (Phillips). *An operator $A = iH$ generates a continuous contractive semigroup on $\mathscr{H}$ iff the operator H is maximal dissipative.*

Proof. The generator of a continuous contractive semigroup is always closed and densely defined (see Remark 2.6 below). According to Theorem 2.1, the operator H is then dissipative with $\mathrm{Ran}(H - i) = \mathscr{H}$. Lemma 2.4 establishes a one-to-one correspondence between the dissipative extensions of H and the contractive extensions of its Cayley transform V. However, $D(V) = \mathrm{Ran}(H - i)$, so no proper extension exists and H is maximal. On the other hand, if H is maximal dissipative, then the corresponding V is defined on the whole $\mathscr{H}$ and is therefore closed. Thus H is closed from Lemma 2.4, so iH generates a continuous contractive semigroup due to Theorem 2.1. ∎

REMARK 4.2.6. It is not difficult to verify that the generator of any continuous contractive semigroup on a Banach space X is densely defined and closed (Davies,

1980a, lemmas 1.1 and 1.5). This means particularly that the assumption of closedness is, in fact, not needed in Theorem 2.1. Notice, however, that Theorem 2.1 generalizes to the Banach-space case, while the same is not true for Theorem 2.5.

An important class among the operators treated above is formed by the *symmetric* operators: they are densely defined and fulfil $\text{Im}(\varphi, H\varphi) = 0$ for all $\varphi \in D(H)$. The Phillips' result calls to mind the Stone theorem. The generator of a continuous unitary group is self-adjoint, and therefore maximal symmetric. The converse is not true, of course, but one can prove a weaker assertion which reflects the fact that a maximal symmetric operator has at least one deficiency index that is zero (Problems 19 and 20). It is also clear from these considerations the place that is occupied by the true physical Hamiltonians in the class of all pseudo-Hamiltonians.

The particular case of self-adjoint operators also suggests the introduction of one more notion. It is often useful to get information about the maximal dissipativity of an operator H without having a complete knowledge of the domain $D(H)$. In particular, it concerns the case of differential operators, where one usually has to deal with a manifold $D \subset D(H)$ of sufficiently smooth functions. We are interested primarily in the situations when D is a core for H. A dissipative operator H is called **essentially maximal dissipative** (e.m.d.) if $\bar{H}$ is maximal dissipative. Due to Lemma 2.4, H has the last named property iff its Cayley transform is densely defined. Since $D(V) = \text{Ran}(H - i)$, we obtain the following criterion:

PROPOSITION 4.2.7. *A dissipative operator H on $\mathscr{H}$ is e.m.d. iff* $\text{Ran}(H - i)$ *is dense in $\mathscr{H}$.*

Each e.m.d. operator possesses, of course, a unique maximal dissipative extension, namely its closure. On the other hand, if $\text{Ran}(H - i)$ is not dense for a dissipative operator H, then different maximal dissipative extensions (in general, infinitely many) are possible. The idea of generalizing the von Neumann's theory of symmetric extensions arises naturally. Such a theory has been actually constructed but we are not going to discuss it here as it will not be necessary for our considerations.

Now we would like to derive some perturbative criterion for pseudo-Hamiltonians, i.e. a condition ensuring that the addition of an operator to a given pseudo-Hamiltonian will not spoil its maximal dissipativity. A particular result of this type is given by Corollary 2.3. It admits the following substantial generalization (cf. also Problems 32 and 33):

THEOREM 4.2.8. *Let H be e.m.d. and assume B to be a closed accretive operator with $D(B) \supset D(H)$. If there are non-negative numbers $a < 1$ and b such that*

$$\|B\varphi\| \leqslant a\,\|H\varphi\| + b\,\|\varphi\| \quad \text{for all } \varphi \in D(H), \tag{2.6a}$$

then $H - iB$ is e.m.d. Furthermore, $D(\bar{H}) \subset D(B)$ and the operator $\bar{H} - iB = \overline{H - iB}$ is maximal dissipative.

REMARK 4.2.9. The condition (2.6a) may be replaced by a slightly stronger requirement, namely

$$\|B\varphi\|^2 \leqslant a^2 \|H\varphi\|^2 + b \|\varphi\|^2 \qquad \text{for all} \qquad \varphi \in D(H). \tag{2.6b}$$

Obviously (2.6b) implies (2.6a). On the other hand, from (2.6a) follows the validity of (2.6b) with $a' = a(1+\eta)$, $b' = b(1+\eta^{-1})$ for an arbitrary positive η. Hence the two conditions are fully interchangeable with respect to Theorem 2.8.

Proof of the theorem. Let us check first that $D(\bar{H}) \subset D(B)$ and that the validity of (2.6a) may be extended to $D(\bar{H})$. To an arbitrary $\varphi \in D(\bar{H})$, we can choose a sequence $\{\varphi_n\} \subset D(H) \subset D(B)$ such that $\varphi_n \to \varphi$. Then $H\varphi_n \to \bar{H}\varphi$ and the inequality $\|B\varphi_n - B\varphi_m\| \leqslant a \|H\varphi_n - H\varphi_m\| + b \|\varphi_n - \varphi_m\|$ implies the sequence $\{B\varphi_n\}$ to be Cauchy. Thus it converges so $\varphi \in D(\bar{B})$, but B is assumed to be closed, which means $D(\bar{H}) \subset D(B)$. Moreover, (2.6a) holds for each φ_n and $B\varphi_n \to B\varphi$, so we obtain $\|B\varphi\| \leqslant a \|\bar{H}\varphi\| + b \|\varphi\|$ for all $\varphi \in D(\bar{H})$.

Further, we shall verify that $D(H)$ is a core for $\bar{H} - iB$, i.e. $\bar{H} - iB = \overline{H - iB}$. Let $\{\varphi_n\}$ be as above, then the relation $(H - iB)\varphi_n \to (\bar{H} - iB)\varphi$ shows that $\bar{H} - iB \subset \overline{H - iB}$. Combing (2.6a) with the triangle inequality, we easily obtain the estimates

$$\|H\psi\| \leqslant \frac{1}{1-a} \|(H - iB)\psi\| + \frac{b}{1-a} \|\psi\|,$$

$$\|B\psi\| \leqslant \frac{a}{1-a} \|(H - iB)\psi\| + \frac{b}{1-a} \|\psi\|$$

valid for all $\psi \in D(H)$. Consequently, if $\{(H - iB)\varphi_n\}$ is Cauchy so are $\{H\varphi_n\}$ and $\{B\varphi_n\}$; thus the inclusion $D(\overline{H - iB}) \subset D(\bar{H}) \cap D(B) = D(\bar{H} - iB)$ also holds.

Hence it is sufficient to prove maximal dissipativity of $H - iB$ in the case when H is closed. For a positive λ, the same argument which has yielded (2.2a) gives $\|(H - i\lambda)\psi\| \geqslant \|H\psi\|$. Since $\psi = (H - i\lambda)^{-1}\varphi$ belongs to $D(H)$ for each $\varphi \in \mathscr{H}$, the last inequality implies $\|H(H - i\lambda)^{-1}\| \leqslant 1$. In the same way, we obtain $\|(H - i\lambda)^{-1}\| \leqslant \lambda^{-1}$. Then the assumption (2.6a) gives

$$\|B(H - i\lambda)^{-1}\varphi\| \leqslant a \|H(H - i\lambda)^{-1}\varphi\| + b \|(H - i\lambda)^{-1}\varphi\| \leqslant$$
$$\leqslant (a + b\lambda^{-1}) \|\varphi\|$$

so $\|B(H - i\lambda)^{-1}\| < 1$ for all λ large enough. In such a case, the operator $I - iB(H - i\lambda)^{-1}$ maps bijectively $\mathscr{H}$ onto $\mathscr{H}$. Since $\text{Ran}(H - i\lambda) = \mathscr{H}$ and the equality

$$H - iB - i\lambda = (I - iB(H - i\lambda)^{-1})(H - i\lambda)$$

holds, we get $\text{Ran}(H - iB - i\lambda) = \mathscr{H}$. At the same time, $H - iB$ is densely defined and

$$\text{Im}(\varphi, (H - iB)\varphi) = \text{Im}(\varphi, H\varphi) - \text{Re}(\varphi, B\varphi) \leqslant 0$$

for all $\varphi \in D(H)$, so Theorems 2.1 and 2.5 yield the desired result. ∎

Let H be a pseudo-Hamiltonian on $\mathscr{H}$ and $\{V_t = e^{-iHt} : t \geqslant 0\}$ the corresponding continuous contractive semigroup. Proposition 1.4.3 shows that an eventual unitary part of the problem may be separated; further, each real eigenvalue of H is at the same time an eigenvalue of the respective quasi-Hamiltonian in view of Theorem 1.5. These facts appear to be particular cases of the structure theorem formulated below. First we introduce the necessary notions. A subspace $\mathscr{G} \subset \mathscr{H}$ is **invariant** under the semigroup $\{V_t\}$ if $V_t\mathscr{G} \subset \mathscr{G}$ for all $t \geqslant 0$. The semigroup $\{V_t\}$ is said to be **unitary on a subspace** $\mathscr{G}$, if $\mathscr{G}$ is invariant and each V_t maps it isometrically *onto* itself. In that case, the operator $V_t \upharpoonright \mathscr{G} = \mathrm{pr}_{\mathscr{G}}\, V_t$ is unitary for each $t \geqslant 0$ and the semigroup $\{\mathrm{pr}_{\mathscr{G}}\, V_t : t \geqslant 0\}$ may be extended to a one-parameter unitary group by setting $\mathrm{pr}_{\mathscr{G}}\, V_{-t} := \mathrm{pr}_{\mathscr{G}}\, V_t^*$. On the other hand, the semigroup $\{V_t\}$ is called **completely non-unitary** (c.n.u.) if the only subspace on which $\{V_t\}$ is unitary is $\mathscr{G} = \{0\}$. An operator H is called **completely dissipative** if $A = iH$ generates a c.n.u. semigroup.

THEOREM 4.2.10. *Let H be a maximal dissipative operator on $\mathscr{H}$, then there is a unique orthogonal decomposition $\mathscr{H} = \mathscr{H}_0 \oplus \mathscr{H}_1$ with the following properties: both the subspaces $\mathscr{H}_j$ reduce H so $H = H_0 \oplus H_1$, where $H_j = \mathrm{pr}_j\, H$; further, H_0 is self-adjoint and H_1 completely dissipative.*

Proof. We shall establish first the corresponding decomposition of the semigroup $\{V_t = e^{-iHt} : t \geqslant 0\}$. Let $\mathscr{G} = \{\psi : \|V_t\psi\| = \|\psi\| \text{ for all } t \geqslant 0\}$, then the parallelogram identity implies

$$\begin{aligned} 2\|\psi\|^2 + 2\|\varphi\|^2 &= 2\|V_t\psi\|^2 + 2\|V_t\varphi\|^2 \\ &= \|V_t(\psi+\varphi)\|^2 + \|V_t(\psi-\varphi)\|^2 \leqslant \|\psi+\varphi\|^2 + \|\psi-\varphi\|^2 \\ &= 2\|\psi\|^2 + 2\|\varphi\|^2 \end{aligned}$$

for any $\psi, \varphi \in \mathscr{G}$ so $\|V_t(\psi \pm \varphi)\| = \|\psi \pm \varphi\|$ and $\mathscr{G}$ is a linear manifold. It is easy to see that $\mathscr{G}$ is closed, i.e. a subspace in $\mathscr{H}$. For arbitrary $\psi \in \mathscr{G}$, positive t and $s \geqslant 0$ we have $\|V_sV_t\psi\| = \|V_{s+t}\psi\| = \|\psi\| = \|V_t\psi\|$, which means that $\mathscr{G}$ is invariant under $\{V_t\}$. Consider further the subspace

$$\mathscr{H}_0 = \bigcap_{t \geqslant 0} V_t\mathscr{G}$$

in $\mathscr{H}$, which is clearly also invariant under $\{V_t\}$. A vector $\psi \in \mathscr{H}_0$ belongs to $V_s\mathscr{G}$ for each $s \geqslant 0$ and V_s is isometric on $\mathscr{G}$, thus there is a unique $\varphi_s \in \mathscr{G}$ such that $\psi = V_s\varphi_s$. Then we have

$$\psi = V_s\varphi_s = V_{s+t}\varphi_{s+t} = V_s(V_t\varphi_{s+t})$$

for arbitrary $s, t \geqslant 0$, which gives $\varphi_s = V_t\varphi_{s+t}$, but $\varphi_{s+t} \in \mathscr{G}$ so $\varphi_s \in V_t\mathscr{G}$ for each $t \geqslant 0$. Hence $\varphi_s \in \mathscr{H}_0$, i.e. the inverse $(V_s \upharpoonright \mathscr{H}_0)^{-1}$ exists for any $s \geqslant 0$.

Consider now a subspace $\mathscr{F} \subset \mathscr{H}$, on which $\{V_t\}$ is unitary. We shall show its complement $\mathscr{F}^\perp$ to be invariant, $V_t(\mathscr{F}^\perp) \subset \mathscr{F}^\perp$ for all $t \geqslant 0$. Suppose this is not true, then one can find unit vectors $\psi \in \mathscr{F}$, $\varphi \in \mathscr{F}^\perp$ and a positive t such that $(V_t\varphi, \psi)$ is real and *positive*. Since $\psi \in \mathscr{F}$, there is a unit $\chi \in \mathscr{F}$ such that $\psi = V_t\chi$. Then we take $\alpha \in \mathbb{R}$ and express

$$\| V_t(\alpha\varphi + \chi)\|^2 = \|\alpha V_t\varphi + \psi\|^2 = 1 + 2\alpha(V_t\varphi, \psi) + \alpha^2 \| V_t\varphi\|^2 .$$

Contractivity of V_t demands

$$g(\alpha) \equiv \frac{1 + 2\alpha(V_t\varphi, \psi) + \alpha^2 \| V_t\varphi\|^2}{1 + \alpha^2} \leqslant 1,$$

but $g(0) = 1$ and $g'(0) = 2(V_t\varphi, \psi) > 0$. In order to avoid contradiction, $\mathscr{F}^\perp$ must be invariant. Recall that $\mathscr{F}$ is an arbitrary subspace on which $\{V_t\}$ is unitary. All the operators V_t map $\mathscr{F}$ isometrically onto $\mathscr{F}$ so $\mathscr{F} \subset \mathscr{H}_0$, and therefore $\{V_t \restriction \mathscr{H}_1 : t \geqslant 0\}$, where $\mathscr{H}_1 = \mathscr{H}_0^\perp$, is completely non-unitary. Moreover, $\mathscr{F}^\perp = (\mathscr{F}^\perp \cap \mathscr{H}_0) \oplus (\mathscr{F}^\perp \cap \mathscr{H}_1)$ and $\{V_t\}$ is unitary on $\mathscr{F}^\perp \cap \mathscr{H}_0$, so the decomposition is unique.

Now H commutes with a bounded operator iff the same is true for all V_t, $t \geqslant 0$ (Davies, 1980a, theorem 1.15). We have seen that both the subspaces $\mathscr{H}_j, j = 0, 1$, are invariant under $\{V_t\}$; hence the semigroup is reduced by the corresponding projections E_j, and the same is true for H. Since the projection of $\{V_t\}$ on $\mathscr{H}_1$ was proved to be c.n.u., the operator H_1 is completely dissipative. On the other hand, we have mentioned that the projection of $\{V_t : t \geqslant 0\}$ to $\mathscr{H}_0$ may be extended to a one-parameter unitary group by $\mathrm{pr}_0\, V_{-t} := \mathrm{pr}_0\, V_t^*$. Its strong continuity for all $t \in \mathbb{R}$ is easily verified so H_0 is self-adjoint due to the Stone theorem. ∎

In particular, if $\mathscr{H}$ is finite-dimensional, we have the following criteria of complete dissipativity:

COROLLARY 4.2.11. *Let H be dissipative and* $\dim \mathscr{H} < \infty$, *then the following assertions are equivalent*:

(a) *H is completely dissipative*;
(b) *H has no real eigenvalue*;
(c) $\lim_{t\to\infty} e^{-iHt}\psi = 0$ *holds for each* $\psi \in \mathscr{H}$.

Proof. Since $\mathscr{H}$ is finite-dimensional, H is automatically maximal. We shall show (a) ⇒ (b) ⇒ (c) ⇒ (a) using the notation introduced in the above proof. If there is a real eigenvalue λ_0, $H\psi_0 = \lambda_0\psi_0$, then Theorem 1.5 gives $\psi_0 \in \mathscr{H}_0$ so $\{V_t\}$ is not completely non-unitary. The implication (b) ⇒ (c) can be check using the Jordan form of the matrix representing H (Problem 21). Finally, if (c) holds, then no non-zero vector ψ can fulfil $\| V_t\psi \| = \| \psi \|$ for all $t \geqslant 0$ so $\mathscr{H}_0 = \{0\}$. ∎

Only a part of the above conclusions extends to infinite-dimensional Hilbert spaces. For instance, it is clear that a self-adjoint operator with empty point spectrum is not completely dissipative, though it has no real eigenvalues. On the other hand, it may happen that H is completely dissipative and the corresponding semigroup is isometric (Problem 23) so $\text{s-lim}_{t\to\infty} V_t$ does not exist. Some sufficient conditions of complete dissipativity may be extracted easily from proof of Theorem 2.10 (Problem 22). Another one is given by the following assertion:

PROPOSITION 4.2.12. *Let H be a maximal dissipative operator of H. Let, further, B be bounded and 'strictly accretive', i.e.* $\mathrm{Re}(\psi, B\psi) > 0$ *for each non-zero* $\psi \in \mathscr{H}$. *Then $H - iB$ is completely dissipative.*

Proof. According to Theorem 2.8, $H - iB$ is maximal dissipative. We denote $V_t = \mathrm{e}^{-i(H-iB)t}$. The function $t \mapsto V_t\psi$ is continuously differentiable for each ψ from $D(H - iB) = D(H)$, and

$$\frac{\mathrm{d}}{\mathrm{d}t} V_t\psi = -i(H - iB)\,V_t\psi$$

(Davies, 1980a, Lemma 1.3). Then $t \mapsto \|V_t\psi\|^2$ is continuously differentiable too, and

$$\begin{aligned}\frac{\mathrm{d}}{\mathrm{d}t}\|V_t\psi\|^2 &= 2\,\mathrm{Im}(V_t\psi, HV_t\psi) - 2\,\mathrm{Re}(V_t\psi, BV_t\psi)\\ &\leqslant -2\,\mathrm{Re}(V_t\psi, BV_t\psi)\end{aligned}$$

so we obtain

$$\|\psi\|^2 - \|V_t\psi\|^2 \geqslant 2\int_0^t \mathrm{Re}(V_s\psi, BV_s\psi)\,\mathrm{d}s$$

for each $\psi \in D(H)$. However, the integrated function is bounded by $\|B\|\,\|\psi\|^2$ and $D(H)$ is dense so the last inequality also holds for each $\psi \in \mathscr{H}$. A strong continuity of $\{V_t\}$ implies the integrated function to be continuous; further, $\mathrm{Re}(V_0\psi, BV_0\psi) = \mathrm{Re}(\psi, B\psi)$. Hence we have $\|\psi\| > \|V_t\psi\|$ for each non-zero $\psi \in \mathscr{H}$ and all positive t, and the complete dissipativity follows (cf. Problem 22). ∎

Concluding this section, we return to the time-dependent pseudo-Hamiltonians. The first complication related to solution of equation (1.10) stems from the fact that the instant when time evolution starts is no longer irrelevant. Consequently, the **propagator** should be defined in this case as a two-parameter family of contractions $V(s, t) \in \mathscr{B}(\mathscr{H})$ which obeys the following conditions:

(i) $V(t, s)V(s, r) = V(t, r)$ for all relevant $r \leqslant s \leqslant t$,
(ii) $V(t, t) = I$,
(iii) $(t, s) \mapsto V(t, s)$ is jointly continuous in the strong operator topology of $\mathscr{B}(\mathscr{H})$.

A propagator is called **unitary** if all the operators $V(t, s)$ are unitary. If the pseudo-Hamiltonians $H(t)$ are bounded, the following well-known assertion is applicable:

THEOREM 4.2.13 (Dyson expansion). *Let $J \subset \mathbb{R}$ be a closed interval and let $H : J \to \mathscr{B}(\mathscr{H})$ be a strongly continuous function whose values are bounded* *dissipative operators. Then there is a propagator V on J such that for all $s, t \in J$, $s \leq t$, and an arbitrary $\varphi \in \mathscr{H}$, the function $t \mapsto \psi_t$, $\psi_t := V(t, s)\varphi$ solves the equation*

$$i \frac{\mathrm{d}}{\mathrm{d}t} \psi_t = H(t)\psi_t \tag{2.7}$$

with an initial datum $\psi_s = \varphi$. The propagator V is unitary if the operators $H(t)$ are Hermitean.

Proof. We set

$$V(t, s)\varphi = \varphi + \sum_{n=1}^{\infty} V_n(t, s), \tag{2.8a}$$

$$V_n(t, s)\varphi = (-i)^n \int_s^t \mathrm{d}t_1 \int_s^{t_1} \mathrm{d}t_2 \ldots \int_s^{t_{n-1}} \mathrm{d}t_n H(t_1)H(t_2)\ldots H(t_n)\varphi. \tag{2.8b}$$

Let first $\{u_n\} \subset J$ be a convergent sequence, $u_n \to u$. We apply the uniform boundedness principle (for convex functionals, cf. (A.1)) to $g_n : g_n(\psi) = \|H(u_n)\psi\|$. Strong continuity of $H(\,.\,)$ implies that $\|g_n(\psi)\|$ is bounded for each $\psi \in \mathscr{H}$, hence there is a positive M such that

$$\|H(u_n)\psi\| = \sup_{\|\psi\|=1} \|H(u_n)\psi\| \leq M.$$

It is sufficient to consider a finite interval J only, because otherwise we could divide J into finite subintervals and construct the global solution using property (i). If J is compact, the last inequality implies $\|H(\,.\,)\|$ to be bounded, $M_{st} = \sup\{\|H(u)\| : s \leq u \leq t\} < \infty$. Thus the integrand in (2.8b) is bounded,

$$\|H(t_1)\ldots H(t_n)\varphi\| \leq M_{st}^n \|\varphi\|. \tag{2.9a}$$

It is also continuous, because the multiplication of operators is sequentially continuous with respect to the strong operator topology, and therefore the r.h.s. of (2.8b) makes sense and defines the operator $V_n(s, t)$. The inequality (2.9a) yields

$$\|V_n(t, s)\varphi\| \leq \frac{1}{n!} M_{st}^n |t-s|^n \|\varphi\| \tag{2.9b}$$

for all $\varphi \in \mathscr{H}$, hence

$$\|V^{(n)}(t, s) - V^{(m)}(t, s)\| \leq \sum_{k=n+1}^{m} \frac{1}{n!} M_{st}^n |t-s|^n$$

holds for $V^{(k)}(t, s) = \Sigma_{n=1}^k V_n(t, s)$ and $n < m$, so the series in (2.8a) converges with respect to the operator norm. Property (ii) evidently holds, and we shall prove further the relation

$$\operatorname*{u-lim}_{k \to \infty} V^{(k)}(t, s) V^{(k)}(s, r) = V(t, r), \tag{2.10a}$$

which yields (i). Let $K = \max\{M_{st}\,|t-s|,\, M_{rs}\,|s-r|\}$, then the inequality (2.9b) implies

$$\left\| \sum_{n=k+1}^{2k} \sum_{j=n-k}^{k} V_j(t,s)V_{n-j}(s,r) \right\|$$

$$\leqslant \sum_{n=k+1}^{2k} \frac{1}{n!} K^n \sum_{j=0}^{n} \binom{n}{j} \leqslant \sum_{n=k+1}^{\infty} \frac{1}{n!} (2K)^n .$$

The r.h.s. tends to zero as $k \to \infty$ so it is sufficient to check the equality

$$\sum_{j=0}^{n} V_j(t,s)V_{n-j}(s,r)\varphi = V_n(t,r)\varphi \tag{2.10b}$$

for each $\varphi \in \mathcal{H}$ and natural n. Since bounded operators may be interchanged with an integral from a vector-valued function (Dunford and Schwartz, 1958, theorem III.2.19), we have

$$V_j(t,s)V_l(s,r)\varphi = (-i)^{j+l} \int_s^t \mathrm{d}t_1 \ldots \int_s^{t_{j-1}} \mathrm{d}t_j \int_r^s \mathrm{d}s_1 \ldots$$

$$\ldots \int_r^{s_{l-1}} \mathrm{d}s_l H(t_1)\ldots H(t_j)H(s_1)\ldots H(s_l)\varphi$$

and (2.10b) verifies easily by induction. Moreover, if $H(\,.\,)$ is Hermitean-valued, one has $V_n(t,s)^* = V_n(s,t)$ for all n. Combining it with the operator-norm continuity of the map $B \mapsto B^*$, we find the propagator to be unitary. Let us further check (iii). Since the Lebesgue measure of the sets

$$N_{st}^n = \{(t_1,\ldots,t_n) \in \mathbb{R}^n : s \leqslant t_j \leqslant t_{j-1}, j = 0,1,\ldots,n,\ t_0 = t\}$$

is continuous with respect to (s, t) and $\|H(\,.\,)\|$ is bounded, each $V_n(\,.\,,.\,)\varphi$ is jointly continuous. It is easy to see that the series in (2.8a) converges uniformly in $J \times J$ so $V(\,.\,,.\,)$ is also jointly continuous. Furthermore, the series

$$\sum_n \frac{\mathrm{d}}{\mathrm{d}t} V_n(t,s)\varphi = -i \sum_n H(t)V_n(t,s)\varphi$$

also converges uniformly in $J \times J$; this fact allows us to verify that (2.8a) defines the desired solution to equation (2.7). Finally, one finds similarly, as in the proof of Proposition 2.12,

$$\|\psi_t\|^2 = \|\psi_s\|^2 + 2\int_s^t \operatorname{Im}(\psi_r, H(r)\psi_r)\,\mathrm{d}r$$

so contractivity of $V(t, s)$ follows from the assumed dissipativity of the operators $H(r)$. ∎

The pseudo-Hamiltonians which are practically interesting are mostly unbounded, so one cannot apply to them the proved theorem directly. There is a particular case, however, in which the Dyson expansion may be used if we employ the *interaction picture*. Assume $H(t)$ to be of the form

$$H(t) = H_0 + U(t), \tag{2.11}$$

where H_0 is self-adjoint and time-independent, while $U : J \to \mathscr{B}(\mathscr{H})$ is a strongly continuous function whose values are dissipative operators. Assume, further, the existence of a propagator $V : J \times J \to \mathscr{B}(\mathscr{H})$, which determines the solution to equation (2.7) with the pseudo-Hamiltonian (2.11). The function $\tilde{U}$: $\tilde{U}(t) = \exp(iH_0 t)U(t)\exp(-iH_0 t)$ then fulfils the hypotheses of the theorem, and therefore solution to the equation

$$i\frac{\mathrm{d}}{\mathrm{d}t}\tilde{V}(t, s)\varphi = \tilde{U}(t)\tilde{V}(t, s)\varphi \tag{2.12a}$$

exists. The connection between $\tilde{V}(\,.\,,\,.\,)$ and the solution to the original equation is easily seen to be

$$V(t, s) = \mathrm{e}^{-iH_0 t}\,\tilde{V}(t, s)\,\mathrm{e}^{iH_0 s}. \tag{2.12b}$$

The relations (2.8) and (2.12b) together yield an explicit form of the series expressing $V(t, s)\varphi$. Let us stress that in order to apply the Dyson expansion to the pseudo-Hamiltonians (2.11), we need to assume additionally the existence of the solution $t \mapsto V(t, s)\varphi$, or equivalently, that the operators $V(t, s)$ map the domain $D(H_0)$ onto itself.

To this end, a more general existence criterion is required than the one given by Theorem 2.13. The expansion (2.8) was obtained by iterations of the integral equation

$$V(t, s)\varphi = \varphi - i\int_s^t H(u)\,V(u, s)\varphi\,\mathrm{d}u, \tag{2.13}$$

which is formally equivalent to the original equation (2.7). Another possible trick is based on approximating the operator-valued function $H(\,.\,)$ by a sequence of 'step functions'. One typical result is quoted in the Appendix (A.12); a more detailed discussion of such methods is left to the following chapters (see Sections 5.5 and 6.1).

4.3. Schrödinger Pseudo-Hamiltonians

Now we are going to study an important special case of pseudo-Hamiltonians on $\mathscr{H} = L^2(\mathbb{R}^d)$, which consists of generalized Schrödinger operators. We set

$$H = H_0 + U, \tag{3.1}$$

where H_0 is the standard *free Hamiltonian* on $L^2(\mathbb{R}^d)$ and U is a multiplication operator corresponding to some complex-valued potential $u(\,.\,)$.

The free Hamiltonian is determined through a multiple of the Laplacian: for sufficiently smooth functions $\varphi \in L^2(\mathbb{R}^d)$ one has

$$H_0\varphi = -\tfrac{1}{2}\Delta\varphi \equiv -\tfrac{1}{2}\sum_{j=1}^{d}\frac{\partial^2\varphi}{\partial x_j^2}. \tag{3.2}$$

It is known that the operator defined by (3.2) on $C_0^\infty(\mathbb{R}^d)$ is essentially self-adjoint; its closure is therefore identified naturally with H_0. Furthermore, the unitary equivalence

$$H_0 = F_d^{-1}SF_d \tag{3.3}$$

holds, where F_d is the Fourier–Plancherel operator on $L^2(\mathbb{R}^d)$ and S is the self-adjoint operator of multiplication by $s : s(k_1, \ldots, k_d) = \frac{1}{2}\Sigma_{j=1}^{d}\, k_j^2$.

Now we must specify the class of considered complex potentials. Let $u : \mathbb{R}^d \to \mathbb{C}$ be a Borel functions, then the corresponding operator $U : (U\varphi)(x) = u(x)\varphi(x)$ with a natural domain is closed and densely defined. Its adjoint U is the operator of multiplication by $\bar{u}$, etc. It is reasonable to allow u to assume infinite values on some Borel set M_u^∞ since otherwise singular potentials would be excluded. However, we restrict our attention to the **almost regular** functions only, which are those with M_u^∞ being a Borel set of Lebesgue measure zero.

PROPOSITION 4.3.1. *A Borel function $u : \mathbb{R}^d \to \mathbb{C}$ is almost regular iff to any non-zero Borel $M \subset \mathbb{R}^d$ there is a non-zero compact Borel $N \subset M$ such that* $\int_N |u(x)|\,dx < \infty$.

Proof. If u is not almost regular, then $M = M_u^\infty$ has no such subset. For an almost regular u, the decomposition $M = (M \cap M_u^\infty) \cup (M \cap M_u^{\text{fin}})$ may be used, where

$$M_u^{\text{fin}} = \bigcup_n M_u^{(n)}, \qquad M_u^{(n)} = \{x \in \mathbb{R}^d : |u(x)| < n\}.$$

It shows that $M \cap M_u^{\text{fin}}$ is non-zero so $M \cap M_u^{(n)}$ is non-zero for some n; the assertion then follows from the fact that Lebesgue measure is σ-finite and regular. ∎

Let u be an almost regular complex potential. If the operator (3.1) with $D(H) = D(H_0) \cap D(U)$ is densely defined, we call it the **generalized Schrödinger operator** (related to u). The same term might eventually be used for the closure $\bar{H}$ whose existence is guaranteed: as $H^* \supset H_0 + U^*$ and $D(U^*) = D(U)$, it holds that $D(H^*) \supset D(H)$ so H^* is densely defined. Furthermore, we have $H^* = \overline{H^*} \supset H_0 + U^*$, i.e.

$$H^* \supset \overline{H_0 + U^*}\,. \tag{3.4a}$$

It is suitable to introduce the following notions. Let J be a *conjugation* on a Hilbert space $\mathscr{H}$, i.e. an antiunitary and idempotent operator, then a densely defined operator H on $\mathscr{H}$ is called ***J*-symmetric** if

$$JHJ \subset H^*. \tag{3.5a}$$

A J-symmetric H clearly fulfils $\overline{JHJ} \subset H^*$. Moreover, if equalities hold in these inclusions,

$$JHJ = H^*, \tag{3.5b}$$

$$\overline{JHJ} = H^*, \tag{3.5c}$$

then H is called **J-self-adjoint** and **essentially J-self-adjoint**, respectively. In the particular case of $\mathscr{H} = L^2(\mathbb{R}^d)$, we shall always have in mind the *complex conjugation* when speaking about J, $J\psi = \overline{\psi}$. The relation (3.4a) is then a consequence of J-symmetry, because $JHJ = H_0 + U^*$, and similarly the equalities

$$H^* = H_0 + U^*, \tag{3.4b}$$

$$H^* = \overline{H_0 + U^*} \tag{3.4c}$$

express the J-self-adjointness and essential J-self-adjointness of $H = H_0 + U$, respectively.

THEOREM 4.3.2. *Let $u : \mathbb{R}^d \to \mathbb{C}$ be almost regular, $u = v - iw$. The corresponding generalized Schrödinger operator* (3.1) *on $L^2(\mathbb{R}^d)$ is J-symmetric; it is dissipative if the condition*

$$-w(x) = \operatorname{Im} u(x) \leqslant 0 \quad \text{a.e. in } \mathbb{R}^d \tag{3.6}$$

holds. Moreover, if H is essentially J-self-adjoint, then validity of the condition (3.6) *is sufficient for it to be essentially maximal dissipative.*

Proof. We have already seen that H is J-symmetric. The restriction to almost regular potentials has the following consequence: to any non-zero Borel set $M \subset \mathbb{R}^d$ there exists a compact Borel subset N, whose Lebesgue measure fulfils $0 < m(N) < \infty$, such that $C_0^\infty(N) \subset D(H)$. This is obvious from the inclusion $C_0^\infty(N) \subset D(H_0)$ and from Proposition 3.1 which yields $C_0^\infty(N) \subset D(U)$. Suppose now the Borel set $M_- = \{x \in \mathbb{R}^d : w(x) < 0\}$ to be non-zero. Due to the presented argument, there is a non-zero compact Borel $N_- \subset M_-$ such that $C_0^\infty(N_-) \subset D(H)$, and any non-zero $\varphi \in C_0^\infty(N_-)$ satisfies

$$\operatorname{Im}(\varphi, H\varphi) = -(\varphi, W\varphi) = -\int_{N_-} w(x)\, |\varphi(x)|^2 \, dx > 0$$

so H cannot be dissipative. Since H is densely defined, the condition (3.6) implies it to be dissipative.

We have $D(U) = D(U^*)$ so the domains of H and $H_0 + U^*$ coincide. The condition (3.6) yields

$$\operatorname{Im}(\varphi, (H_0 + U^*)\varphi) = (\varphi, W\varphi) \geqslant 0$$

for all $\varphi \in D(H_0 + U^*)$. If H is essentially J-self-adjoint, then the last inequality in combination with (3.4c) gives

$$\operatorname{Im}(\varphi, H^*\varphi) \geqslant 0$$

for all $\varphi \in D(H^*)$ so $-H^* = -(\bar{H})^*$ is dissipative. Consequently, $\bar{H}$ is maximal dissipative in view of the Phillips' theorem and Corollary 2.2. ∎

For the sake of brevity, it is useful to introduce the notion of the **dissipative Schrödinger operator** on $L^2(\mathbb{R}^d)$ as the one specified by the following requirements:

(d1) *it is a densely defined operator of the form* (3.1),
(d2) *the potential* $u : \mathbb{R}^d \to \mathbb{C}$ *is almost regular and fulfils the dissipativity condition* (3.6).

Considerations of the preceding section show that we need criteria for such operators to be e.m.d. One sufficient condition is given by the above-proved theorem, but it is not very convenient from a practical point of view, because for unbounded potentials a check of J-self-adjointness may not be easy. In addition, the condition is not necessary at the same time: there are dissipative Schrödinger operators which are e.m.d. but not essentially J-self-adjoint or even J-symmetric (see Remark 3.10 below).

It would be useful to have some sufficient conditions formulated in terms of the potential alone. To this purpose, perturbation-theory methods can be used (cf. also Problem 34):

THEOREM 4.3.3. *Let H be a dissipative Schrödinger operator on $L^2(\mathbb{R}^d)$ with a potential $u \in L^p(\mathbb{R}^d) + L^\infty(\mathbb{R}^d)$, where*

$$p = 2 \qquad \text{for } d \leqslant 3 \tag{3.7a}$$

$$p > \tfrac{1}{2}d \quad \text{for } d \geqslant 4 \tag{3.7b}$$

then H is maximal dissipative. Moreover, $C_0^\infty(\mathbb{R}^d)$ is a core for H.

Proof. The operator H_0 is essentially self-adjoint (e.s.a.) on $C_0^\infty(\mathbb{R}^d)$ and fulfils there are following relations (Reed and Simon, 1975, theorem IX.28): if $d \leqslant 3$, then a positive b (independent of ψ) corresponds to every $a > 0$ such that

$$\|\psi\|_\infty \leqslant a\,\|H_0\psi\| + b\,\|\psi\| \tag{3.8a}$$

for each $\psi \in C_0^\infty(\mathbb{R}^d)$. On the other hand, if $d > 4$ and $2 \leqslant q < 2d/(d-4)$, or $d = 4$ and $q > 2$, then to any positive a there is a constant b (depending only on a, d, q) such that the inequality

$$\|\psi\|_q \leqslant a\,\|H_0\psi\| + b\,\|\psi\| \tag{3.8b}$$

holds for each $\psi \in C_0^\infty(\mathbb{R}^d)$.

The operator U is closed and dissipative with $D(U) \supset C_0^\infty(\mathbb{R}^d)$. According to the assumption, $u = u_1 + u_2$, where $u_1 \in L^p(\mathbb{R}^d)$ and $u_2 \in L^\infty(\mathbb{R}^d)$. Each function $\psi \in C_0^\infty(\mathbb{R}^d)$ is bounded so

$$\|U\psi\| \leqslant \|U_1\psi\| + \|U_2\psi\| \leqslant \|u_1\|\,\|\psi\|_\infty + \|u_2\|_\infty\,\|\psi\| \tag{3.9a}$$

holds for $d \leqslant 3$, where $\| . \|$ is the L^2-norm as usually. If $d \geqslant 4$, one has to employ the Hölder inequality to get

$$\|U\psi\| \leqslant \|u_1\|_p \, \|\psi\|_q + \|u_2\|_\infty \, \|\psi\| \tag{3.9b}$$

for each $\psi \in C_0^\infty(\mathbb{R}^d)$, where $p^{-1} + q^{-1} = \frac{1}{2}$. This relation, together with (3.7b), gives $2 < q < 2d/(d-4)$ for $d \geqslant 5$ and $q > 2$ for $d = 4$. One may therefore combine the inequalities (3.8) and (3.9) to get

$$\|U\psi\| \leqslant a \|u_1\|_p \, \|H_0\psi\| + (b \|u_1\|_p + \|u_2\|_\infty) \, \|\psi\| \tag{3.10}$$

for all $\psi \in C_0^\infty(\mathbb{R}^d)$. Now it is sufficient to choose a so that $a \|u_1\|_p < 1$, then Theorem 2.8 may be applied to the operators $H_0 \upharpoonright C_0^\infty(\mathbb{R}^d)$ and iU; since $C_0^\infty(\mathbb{R}^d)$ is a core for H_0, the theorem is proved. ∎

Thus we have obtained a class of complex potentials for which the corresponding generalized Schrödinger operator is not only e.m.d. but also maximal dissipative. This class is reasonably large; it contains, e.g., potentials with isolated singularities of the type $|x - x_0|^{-\alpha}$, where $\alpha < \frac{1}{2}d$ if $d \leqslant 3$, and $\alpha < 2$ if $d \geqslant 4$.

Let us further mention a modification of the preceding result, which is particularly interesting from the viewpoint of n-particle open systems with u corresponding to an external field plus translationally invariant *two-particle forces*. This assumption we formulate as follows:

(a) *Let $d = 3n$, and let $\{u_j : \mathbb{R}^d \to \mathbb{C}, j = 1, \dots, N\}$ be a (finite) set of almost regular functions with the following property: there is a three-dimensional projection P_j on $\mathbb{R}^d$ for each j such that $u_j(x) = u_j(P_j x)$ for all $x \in \mathbb{R}^d$.*

In other words, each u_j depends on one three-dimensional vector argument only. For a fixed j, one may choose a function $\tilde{u}_j : \mathbb{R}^3 \to \mathbb{C}$ and a basis in $\mathbb{R}^d$ so that

$$u_j(x_1, \dots, x_{3n}) = \tilde{u}_j(x_1, x_2, x_3) \tag{3.11}$$

for all $x \in \mathbb{R}^d$.

THEOREM 4.3.4. *Let H be a dissipative Schrödinger operator on $L^2(\mathbb{R}^{3n})$ related to a potential $u = \sum_{j=1}^N u_j$. Suppose that $\{u_j\}_{j=1}^N$ fulfils the assumption* (a) *and $\tilde{u}_j \in L^2(\mathbb{R}^3) + L^\infty(\mathbb{R}^3)$ for $j = 1, \dots, N$. Then H is maximal dissipative and $C_0^\infty(\mathbb{R}^{3n})$ represents a core for it.*

Proof. Let us first notice that the inequality (3.10) has its 'quadratic counterpart', similarly as in Remark 2.9: to $\varphi \in C_0^\infty(\mathbb{R}^3)$ and $a_j > 0$, one can find a positive b_j such that

$$\|\tilde{U}_j\varphi\|^2_{L^2(\mathbb{R}^3)} \leqslant a_j^2 \, \|H_0^{(j)}\varphi\|^2_{L^2(\mathbb{R}^3)} + b_j^2 \, \|\varphi\|^2_{L^2(\mathbb{R}^3)}, \tag{3.12}$$

where $H_0^{(j)}$ refers to $-\frac{1}{2}\sum_{i=1}^3 \partial^2/\partial x_i^2$ in the corresponding basis. For an arbitrary $\psi \in C_0^\infty(\mathbb{R}^{3n})$, we have

$$\|U_j\psi\|^2 = \int_{\mathbb{R}^{3n-3}} \mathrm{d}x_4 \dots \mathrm{d}x_{3n} \int_{\mathbb{R}^3} |\tilde{u}_j(x_1, x_2, x_3)\psi(x_1, \dots, x_{3n})|^2 \, \mathrm{d}x_1 \, \mathrm{d}x_2 \, \mathrm{d}x_3$$

In order to estimate this norm, we employ the relations (3.12) and (3.3) together with the decomposition $F_d = F_{d-3} \otimes F_3$ of the Fourier–Plancherel operator,

$$\|U_j\psi\|^2 \leqslant \int_{\mathbb{R}^{3n-3}} dx_4 \ldots dx_{3n} \left\{ a_j^2 \int_{\mathbb{R}^3} |(H_0^{(j)}\psi)(x_1, \ldots, x_{3n})|^2 \, dx_1 \, dx_2 \, dx_3 + \right.$$

$$\left. + b_j^2 \int_{\mathbb{R}^3} |\psi(x_1, \ldots, x_{3n})|^2 \right\} dx_1 \, dx_2 \, dx_3$$

$$= a_j^2 \int_{\mathbb{R}^{3n}} \left(\sum_{l=1}^{3} \tfrac{1}{2} k_l^2 \right) |(F_{3n}\psi)(k_1, \ldots, k_{3n})|^2 \, dk_1 \ldots dk_{3n} + b_j^2 \, \|\psi\|^2$$

$$\leqslant a_j^2 \int_{\mathbb{R}^{3n}} \left(\sum_{l=1}^{3n} \tfrac{1}{2} k_l^2 \right) |(F_{3n}\psi)(k_1, \ldots, k_{3n})|^2 \, dk_1 \ldots dk_{3n} + b_j^2 \, \|\psi\|^2$$

$$= a_j^2 \, \|H_0\psi\|^2 + b_j^2 \, \|\psi\|^2.$$

It is important to note that the deduction relies on the basis used in (3.11), but the result is coordinate-independent, because the norms of $U_j\psi$ and $H_0\psi$ do not change when the coordinate system in $\mathbb{R}^{3n}$ is rotated. Then the Schwartz inequality gives

$$\|U\psi\|^2 = \left\| \sum_{j=1}^{N} U_j\psi \right\|^2 \leqslant \sum_{j,m=1}^{N} \|U_j\psi\| \, \|U_m\psi\|;$$

introducing $a = \max_{1 \leqslant j \leqslant N} a_j$, $b = \max_{1 \leqslant j \leqslant N} b_j$, we obtain the inequality

$$\|U\psi\|^2 \leqslant N^2 a^2 \|H_0\psi\|^2 + N^2 b^2 \|\psi\|^2$$

for each $\psi \in C_0^\infty(\mathbb{R}^d)$. Thus one must choose the numbers a_j in such a way that $N^2a^2 < 1$, then the assertion follows from Theorem 2.8. ■

The potentials covered by Theorems 3.3 and 3.4 are not allowed to grow at infinity. Now we shall show that this restriction may be relaxed *when the real part of the potential is semibounded*. We denote by $L^2_{loc}(\mathbb{R}^d)$ the set of all locally square-integrable functions $f : \mathbb{R}^d \to \mathbb{C}$, i.e. such that to each $x \in \mathbb{R}^d$ a neighbourhood $\Omega_x \subset \mathbb{R}^d$ can be found on which f is L^2-integrable. The following assertion is valid:

THEOREM 4.3.5. *Let H be a dissipative Schrödinger operator on $L^2(\mathbb{R}^d)$ related to a potential $u = v - iw$. Assume that $u \in L^2_{loc}(\mathbb{R}^d)$ and $\gamma \equiv \operatorname{inf\,ess} \{v(x) : x \in \mathbb{R}^d\} > -\infty$, then H is e.m.d. and the same is true for $H \upharpoonright C_0^\infty(\mathbb{R}^d)$. Moreover, the halfline $(-\infty, \gamma)$ belongs to the resolvent set $\rho(\bar{H})$.*

Proof. If H is e.m.d. so is $H - \gamma$, i.e. without loss of generality we may set $\gamma = 0$, or equivalently

$$v(x) \geqslant 0 \quad \text{a.e. in } \mathbb{R}^d. \tag{3.13}$$

The operators H and $H_c = H \upharpoonright C_0^\infty(\mathbb{R}^d)$ are dissipative as well as their closures; we need to prove that $\overline{H_c}$ is maximal dissipative, then $H_c \subset H$ would give $\overline{H_c} = \overline{H}$. Let $\mu > 0$, then

$$\mathrm{Re}(\varphi, (H+\mu)\varphi) = (\varphi, H_0\varphi) + (\varphi, V\varphi) + \mu \|\varphi\|^2 \geqslant 0$$

holds for each $\varphi \in C_0^\infty(\mathbb{R}^d)$ so we have $\mathrm{Re}(\varphi, (\overline{H_c}+\mu)\varphi) \geqslant 0$ for all $\varphi \in D(\overline{H_c})$ and the inequality

$$\|(\overline{H_c}+\mu)\varphi\| \geqslant \mu \|\varphi\|, \quad \varphi \in D(\overline{H_c}) \tag{3.14}$$

is proved in the same way as in Corollary 2.2. Thus $-\mu \in \rho(\overline{H_c})$ if $\mathrm{Ran}(\overline{H_c}+\mu)$ is dense in $L^2(\mathbb{R}^d)$. We have $(\overline{H_c})^* = H_c^*$ so the last assertion is also equivalent to $\mathrm{Ker}(H_c^* + \mu) = \{0\}$. We must check, therefore, that for any $\psi \in D(H_c^*)$, the relation

$$(H_c^* + \mu)\psi = 0 \tag{*}$$

implies $\psi = 0$. We shall use a distribution-theoretic argument. Since the domain of H_c is $C_0^\infty(\mathbb{R}^d)$, the standard test-function space, the equality (*) is equivalent to

$$\Delta\psi = 2(v + iw + \mu)\psi, \tag{**}$$

where the Laplacian is understood in the weak (distributional) sense. We have $\psi \in L^2(\mathbb{R}^d) \subset L^2_{\mathrm{loc}}(\mathbb{R}^d)$ and $(u+\mu) \in L^2_{\mathrm{loc}}(\mathbb{R}^d)$ so (**) implies $\Delta\psi \in L^1_{\mathrm{loc}}(\mathbb{R}^d)$. Then the Kato inequality may be applied (Reed and Simon, 1975, theorem X.27): it holds that

$$\Delta|\psi| \geqslant \mathrm{Re}((\mathrm{sgn}\,\psi)\Delta\psi), \tag{3.15a}$$

where

$$(\mathrm{sgn}\,\psi)(x) = \begin{cases} 0 & \ldots \quad \psi(x) = 0 \\ \overline{\psi}(x)\,|\psi(x)|^{-1} & \ldots \quad \psi(x) \neq 0 \end{cases} \tag{3.15b}$$

Since the functions v and w are real-valued, we get

$$\Delta|\psi| \geqslant 2\,\mathrm{Re}((\mathrm{sgn}\,\psi)(v\psi + iw\psi + \mu\psi))$$

$$= 2\,\mathrm{Re}(v|\psi| + iw|\psi| + \mu|\psi|) = 2(v+\mu)|\psi|.$$

In view of (3.13), it implies particularly $\Delta|\psi| \geqslant 0$.

Now the idea is to combine this inequality with positivity of the operator $-\Delta$. The trouble is that $\chi = |\psi|$ may not be contained in its domain. Thus we choose an approximative unity, i.e. a family $\{j_\epsilon\}$, $j_\epsilon(x) = \epsilon^{-d}j(x/\epsilon)$, where j is a non-negative function from $C_0^\infty(\mathbb{R}^d)$ such that $\int_{\mathbb{R}^d} j(x)\,dx = 1$, and consider the convolution $\chi_\epsilon := \chi * j_\epsilon$. We have $\Delta\chi_\epsilon = \chi * \Delta j_\epsilon$ and $\Delta j_\epsilon \in C_0^\infty(\mathbb{R}^d)$ so $\Delta\chi_\epsilon \in L^2(\mathbb{R}^d)$ in view of the Young inequality, $\|\Delta\chi_\epsilon\|_2 \leqslant \|\chi\|_2 \|\Delta j_\epsilon\|_1$. Then we get $(\chi_\epsilon, \Delta\chi_\epsilon) \leqslant 0$, where the equality holds only if $\chi_\epsilon = 0$, because the self-adjoint operator $-\Delta$ is positive and the spectrum of $(-\Delta)^{1/2}$ is purely continuous.

On the other hand, $\Delta\chi_\epsilon = \Delta\chi * j_\epsilon$. According to the definition of the convolution, we then have $(\Delta\chi_\epsilon)(\varphi) = (\Delta\chi)(\tilde{j}_\epsilon * \varphi)$ for each $\varphi \in \mathscr{S}(\mathbb{R}^d)$, where $\tilde{j}_\epsilon(x) = j_\epsilon(-x)$. Thus if $\Delta\chi \geqslant 0$ in the sense of distributions, $\Delta\chi_\epsilon \geqslant 0$ also holds. Further, $\Delta\chi_\epsilon \in L^2(\mathbb{R}^d)$ yields $(\Delta\chi_\epsilon)(x) \geqslant 0$ a.e. in $\mathbb{R}^d$ so $(\chi_\epsilon, \Delta\chi_\epsilon) \geqslant 0$. Combining the two inequalities, we get $\chi_\epsilon = 0$. Finally, $\{j_\epsilon\}$ is an approximative unity so χ_ϵ converges weakly to χ as $\epsilon \to 0$; this gives $\chi = 0$ and therefore $\psi = 0$.

Hence the halfline $(-\infty, 0)$ belongs to the resolvent set of $\overline{H_{\mathrm{C}}}$, which implies particularly that this operator is maximal dissipative (Problem 25), so the proof is finished. ■

Besides the sought criterion of maximal dissipativity, we have obtained at the same time the result concerning the spectra of such pseudo-Hamiltonians. It can be further strengthened provided that the potential assumes values from some special sector of the complex plane only (Problem 26). The obtained criteria of maximal dissipativity were deduced by an adaptation of the methods used for the self-adjointness proofs of Schrödinger operators with real-valued potentials. It is clear that this methodological receipt leaves other open possibilities, but we are not going to pursue this task further.

In addition to the operators considered above, one often studies *generalized Schrödinger operators on a halfline*, i.e. operators on $L^2(\mathbb{R}_+)$ obtained by a suitable extension of the operator H:

$$(H\varphi)(x) = -\tfrac{1}{2}\varphi''(x) + u(x)\varphi(x) \tag{3.16}$$

defined on a set of sufficiently smooth functions from $L^2(\mathbb{R}_+)$. Such operators are not only interesting in themselves, but they can be used to simplify the treatment of those d-dimensional problems in which the potential $\tilde{u} : \mathbb{R}^d \to \mathbb{C}$ is *spherically symmetric*, i.e. there is a function $u : \mathbb{R}_+ \to \mathbb{C}$ such that $\tilde{u}(x) = u(r)$ holds for each $x \in \mathbb{R}^d$, where $r = (\Sigma_{j=1}^d x_j^2)^{1/2}$.

Since the reduction of a d-dimensional spherically symmetric problem goes essentially in the same way as for the real-valued potentials, we limit ourselves with a brief sketch. The Hilbert space $L^2(\mathbb{R}^d)$ decomposes into the tensor product $L^2(\mathbb{R}_+, r^{d-1}\,\mathrm{d}r) \otimes L^2(S^{d-1}, \mathrm{d}\Omega)$, where $\mathrm{d}\Omega$ is the standard (rotationally invariant) measure on the $(d-1)$-dimensional sphere. Each vector $x \in \mathbb{R}^d$ may be written as (x, ϑ), where ϑ is an appropriate set of angular variables. For the sake of simplicity, we shall consider only the potentials u which are *continuous* on $\mathbb{R}^d\backslash\{0\}$. In that case, the domain of the generalized Schrödinger operator $H_0 + \widetilde{U}$ clearly contains the set $D = C_0^\infty(\mathbb{R}^d\backslash\{0\})$ of the infinitely differentiable functions with compact supports separated from the origin. In addition to D, we shall use the following subsets of D:

$$D_{\mathrm{p}} = \{\psi \in L^2(\mathbb{R}^d) : \psi(x) = f(r)g(\vartheta), f \in C_0(\mathbb{R}_+\backslash\{0\}), g \in C^\infty(S^{d-1})\},$$

$$D_{\min} = C_0^\infty(\mathbb{R}_+\backslash\{0\}) \circ C^\infty(S^{d-1}) \equiv (D_{\mathrm{p}})_{\mathrm{lin}}.$$

The last one is clearly dense in $L^2(\mathbb{R}^d)$. The operator $H_0 + \widetilde{U}$ acts on $\psi \in D_p$ as follows:

$$((H_0 + \widetilde{U})\psi)(r, \vartheta) = \left(-\tfrac{1}{2}\,\frac{d^2f}{dr^2} - \frac{d-1}{2r}\,\frac{df}{dr}\right)(r)g(\vartheta) +$$
$$+ u(r)f(r)g(\vartheta) + \frac{1}{2r^2}\,f(r)(\Delta_S g)(\vartheta), \tag{3.17}$$

where Δ_S is Laplace–Beltrami operator on $L^2(S^{d-1}, d\Omega)$. This operator is e.s.a. on $C^\infty(S^{d-1})$ and has a purely point spectrum, which consists of non-negative eigenvalues $\kappa_l = l(l + d - 2)$, $l = 0, 1, 2, \ldots$ (notice that within the standard sign convention, the LB-operator in $\mathbb{R}^d$ coincides with $-\Delta$). The corresponding eigenspaces $\mathscr{G}_l$, $l = 0, 1, 2, \ldots$, are finite-dimensional and contained in $C^\infty(S^{d-1})$. Now we employ the decomposition

$$L^2(\mathbb{R}^d) = \sum_{l=0}^{\infty}{}^{\oplus}\, \mathscr{H}_l, \tag{3.18a}$$

where

$$\mathscr{H}_l = L^2(\mathbb{R}_+, r^{d-1}\,dr) \otimes \mathscr{G}_l, \qquad l = 0, 1, 2, \ldots \tag{3.18b}$$

In order to check that $H_0 + \widetilde{U}$ is e.m.d. on D, it is sufficient to show that the same holds for its restriction to $D_{\min}$. The operator $(H_0 + \widetilde{U}) \upharpoonright D_{\min}$ is easily seen to be reduced by projections to the subspaces $\mathscr{H}_l$ and its lth component equals

$$(H_0 + \widetilde{U}) \upharpoonright D_l = \widetilde{H}_l \otimes I_{\mathscr{G}_l}, \tag{3.19a}$$

where $D_l \equiv D_{\min} \cap \mathscr{H}_l$ and the operators $\widetilde{H}_l$ act on $f \in C_0(\mathbb{R}^d \setminus \{0\})$ as follows

$$(\widetilde{H}_l f)(r) = \left[-\tfrac{1}{2}\,\frac{d^2}{dr^2} - \frac{d-1}{2r}\,\frac{d}{dr} + u(r) + \frac{\kappa_l}{2r^2}\right] f(r). \tag{3.19b}$$

Hence we have to verify that all the operators $\widetilde{H}_l$ are e.m.d. (cf. Problem 27 and 28). Further, we pass to the operators $H^{(l)}$ on $L^2(\mathbb{R}_+)$ by means of the unitary operator $S : L^2(\mathbb{R}_+, r^{d-1}\,dr) \to L^2(\mathbb{R}_+)$ defined by $(Sf)(r) = r^{(d-1)/2} f(r)$. Obviously S maps $C_0(\mathbb{R}_+ \setminus \{0\})$ onto itself and $H^{(l)} = S\widetilde{H}_l S^{-1}$ acts as

$$(H^{(l)}f)(r) = \left[-\tfrac{1}{2}\,\frac{d^2}{dr^2} + u(r) + \frac{1}{2r^2}\left(\kappa_l + \frac{(d-1)(d-3)}{4}\right)\right] f(r), \tag{3.20}$$

so the problem is now to establish the essentially maximal dissipativity of the operators (3.20) (cf. Problem 29). Concluding the above considerations, we can formulate the following assertion:

PROPOSITION 4.3.6. *Let H be a dissipative Schrödinger operator on $L^2(\mathbb{R}^d)$ corresponding to a spherically symmetric potential $\tilde{u}$, which is continuous on $\mathbb{R}^d \setminus \{0\}$. If the operators $H^{(l)}$ given by* (3.20) *with $\kappa_l = l(l + d - 2)$, $l = 0, 1, 2, \ldots$, are e.m.d. on $C_0^\infty(\mathbb{R}_+ \setminus \{0\})$, then H is e.m.d. on $C_0^\infty(\mathbb{R}^d \setminus \{0\})$.*

REMARK 4.3.7. (a) The assumption of continuity played no essential role in the above deduction; we have used it only to ensure that $D \subset D(H)$. Most of the known results about the generalized Schrödinger operators on a halfline, however, concern just the case when $u(\,.\,)$ is continuous on $(0, \infty)$.

(b) For some operators, the domain D might appear to be too small. Consider, for instance, the case of a single free particle, $d = 3$ and $u = 0$, then $H = -\frac{1}{2}\Delta$ is *not* essentially self-adjoint on D (Reed and Simon, 1975), theorem X.11), though it extends to the self-adjoint (and therefore maximal dissipative) operator. The same must be true for at least one of the operators $H^{(l)}$; it is a well-known fact that $H^{(0)} = -\frac{1}{2}(d^2/dr^2)$ is not e.s.a. on $C_0(\mathbb{R}_+\backslash\{0\})$. Of course, it does not disqualify the method. The described 'partial-wave decomposition' may also be carried out for a large domain than D, say C^∞-functions with some weaker restriction imposed on the behaviour around the origin. We only need some caution when passing from $L^2(\mathbb{R}_+, r^{d-1}\,dr)$ to $L^2(\mathbb{R}_+)$ because the operator S does not necessarily map the 'radial part' of such a domain onto itself.

Before proceeding further, we want to illustrate the meaning of the fact that an operator is e.m.d. To this end, we formulate below two examples concerning Schrödinger operators with *real-valued* potentials. Such operators commute with the complex conjugation J so they have equal deficiency indices due to the well-known theorem of von Neumann. Consequently, they are either e.s.a., or they are not e.s.a. but possess self-adjoint extensions. In the last case, however, the operator can have at the same time a properly dissipative extension with a quite reasonable physical meaning. Such a situation occurs, e.g. for *strongly attractive potentials*:

EXAMPLE 4.3.8. Consider the Schrödinger operator $H = H_0 + V$ on $L^2(\mathbb{R}^3)$ corresponding to the spherically symmetric potential v:

$$v(r) = gr^{-2}. \tag{3.21}$$

Proposition 3.6 allows us to reduce the problem to a discussion of the operators

$$T_\alpha : (T_\alpha f)(r) = -f''(r) + \alpha r^{-2} f(r) \tag{3.22a}$$

defined on an appropriate domain in $L^2(\mathbb{R}_+)$, where the parameter α assumes the values $\alpha_l = 2g + l(l+1)$. The operator T_α with the domain $C_0(\mathbb{R}_+\backslash\{0\})$ is known to be e.s.a. iff $\alpha \geqslant 3/4$ (Reed and Simon, 1975), appendix to §X.1). For an arbitrary g, we have $\alpha_l \geqslant 3/4$ if l is large enough; this corresponds to the fact that the centre of attraction is screened by the centrifugal barrier if the particle has a sufficiently large angular momentum. On the other hand, if $\alpha < 3/4$, then the potential corresponds to the limit-circle case at $r = 0$ so the deficiency indices are (1,1). It is easy to see that the adjoint operator acts again as

$$(T_\alpha^* f)(r) = -f''(r) + v_\alpha(r) f(r), \tag{3.23}$$

where $v_\alpha(r) = \alpha r^{-2}$, and its domain consists of all $f \in L^2(\mathbb{R}_+)$ which are continuously differentiable with f' absolutely continuous, f'' locally square-integrable in $(0, \infty)$ and such that $-f'' + v_\alpha f \in L^2(\mathbb{R}_+)$. Without loss of generality, we may assume T_α to be closed, i.e. to choose

$$D(T_\alpha) = \{f \in D(T_\alpha^*) : f(0) = f'(0) = 0\}. \tag{3.22b}$$

Let λ be a complex number with $\operatorname{Re}\lambda > 0$. We can find the vectors $f_\lambda^\pm$ which span the deficiency subspaces $\mathscr{G}_\lambda^\pm = \operatorname{Ran}(T_\alpha \pm i\lambda)^\perp = \operatorname{Ker}(T_\alpha^* \mp i\lambda)$. The relation (3.23) gives an equation for $f_\lambda^\pm$ which may be further transformed into the Bessel equation with

$$\nu^2 = \alpha + \tfrac{1}{4}. \tag{3.24}$$

Since we look for $f_\lambda^\pm$ within $L^2(\mathbb{R}_+)$, the solutions are unique up to a multiplicative constant:

$$f_\lambda^+(r) = r^{1/2} H_\nu^{(1)}(\epsilon\kappa r), \tag{3.25a}$$

$$f_\lambda^-(r) = r^{1/2} H_\nu^{(2)}(\bar{\epsilon}\kappa r), \tag{3.25b}$$

where $\kappa = \lambda^{1/2}$, $\epsilon = e^{\pi i/4}$, the square root is chosen to be positive on $\mathbb{R}_+$ and with a cut along $\mathbb{R}_-$, and $\nu = -i\beta$, $\beta > 0$, for $\nu^2 < 0$.

With the knowledge of the vectors (3.25), one can construct self-adjoint extensions of T_α in the standard way (Problem 30). We shall exhibit a dissipative extension. Let $\xi : \mathbb{R}_+ \to \mathbb{R}$ be a C^∞-function such that $\xi(r) = 1$ in $[0, 1]$ and $\xi(r) = 0$ in $[2, \infty)$, then we set

$$f_0(r) = r^{1/2} J_\nu(\epsilon\kappa r)\xi(r).$$

Now we may define $\tilde{T}_\alpha$ as the restriction of T_α^*:

$$\tilde{T}_\alpha \subset T_\alpha^*, \tag{3.26a}$$

$$D(\tilde{T}_\alpha) = \{f = g + cf_0 : g \in D(T_\alpha), c \in \mathbb{C}\}. \tag{3.26b}$$

We have $f_0 \in D(T_\alpha^*)$, then $\tilde{T}_\alpha$ is densely defined and $\operatorname{Im}(f, \tilde{T}_\alpha f) = |c|^2 \operatorname{Im}(f_0, \tilde{T}_\alpha f_0)$. A simple *per partes* integration yields

$$\operatorname{Im}(f_0, \tilde{T}_\alpha f_0) = \lim_{r \to 0+} \operatorname{Im} f_0'(r)\overline{f_0(r)};$$

which shows that $\tilde{T}_\alpha$ is symmetric when $0 \leqslant \nu < 1$, i.e. $-1/4 \leqslant \alpha < 3/4$. On the other hand, if $\alpha < -1/4$, we have $f_0'(r)\overline{f_0(r)} = (\nu + \tfrac{1}{2})\,|\Gamma(\nu + 1)|^{-2} + O(r)$ for small r, and therefore

$$\operatorname{Im}(f_0, \tilde{T}_\alpha f_0) = -\frac{\beta}{|\Gamma(1 - i\beta)|^2} < 0.$$

Since $\beta = i\nu > 0$, the operator $\tilde{T}_\alpha$ is properly dissipative for $\alpha < -1/4$.

The existence of various extensions stems from the fact that the knowledge of the potential (3.21) alone does not provide us with information on what happens to a particle that has reached the centre of attraction. It may be either reflected in different ways or absorbed, and the extension $\tilde{T}_\alpha$ obviously should be associated with the last possibility (cf. also Problem 31).

The next example, which is even more simple, shows that if an operator is not e.m.d., it may possess plenty of properly dissipative extensions with a reasonable physical meaning. To illustrate this, we shall discuss the case of a particle moving freely on a *halfline, which is bounded by a partially absorbing barrier*:

EXAMPLE 4.3.9. By $AC^p[\mathbb{R}_+]$ we denote the set of all $\psi \in L^2(\mathbb{R}_+)$ such that $\psi^{(p-1)}$ is absolutely continuous in (any finite interval of) $\mathbb{R}_+$ and $\psi^{(p)} \in L^2(\mathbb{R}_+)$. We drop the inessential factor 1/2 (or $1/2m$) and consider the operator H:

$$H\psi = -\psi'', \qquad D(H) = \{\psi \in AC^2[\mathbb{R}_+] : \psi(0) = \psi'(0) = 0\}, \tag{3.27}$$

which is closed and symmetric with the deficiency indices (1,1), its adjoint being

$$H^*\psi = -\psi'', \qquad D(H^*) = AC^2[\mathbb{R}_+]. \tag{3.28}$$

Now we define the operator H_a as the following restriction of H^*:

$$H_a\psi = -\psi'', \qquad D(H_a) = \{\psi \in AC^2[\mathbb{R}_+] : \psi'(0) + a\psi(0) = 0\}. \tag{3.29a}$$

This operator is self-adjoint if $a \in \mathbb{R}$, and moreover, the set $\{H_a : a \in \mathbb{R}\}$ completed with H_∞ (in which case the boundary condition is $\psi(0) = 0$) covers all possible self-adjoint extensions of the operator H. Let us now consider the case when a lies in the upper halfplane of $\mathbb{C}$, i.e.

$$a = \alpha + i\beta, \quad \beta > 0. \tag{3.29b}$$

It is easy to see that H_a is again closed. Let us look for its adjoint: we take $\varphi \in D(H_a^*)$ and an arbitrary vector $\psi \in D(H_a)$ whose support is bounded. We have $H_a^* \subset H^*$ so a simple integration *per partes* yields

$$(\varphi, H_a\psi) = \bar{\varphi}(0)\psi'(0) - \bar{\varphi}'(0)\psi(0) + (H_a^*\varphi, \psi), \tag{3.30a}$$

and since ψ obeys the boundary condition given by (3.29a), we get

$$H_a^*\varphi = -\varphi'', \qquad D(H_a^*) = \{\varphi \in AC^2[\mathbb{R}_+] : \varphi'(0) + \bar{a}\varphi(0) = 0\}. \tag{3.31}$$

It is apparent that $D(H_a^*) \not\supset D(H_a)$ for a non-real a, so *the operator H_a is not symmetric*. In the same way as (3.30a), we get

$$\operatorname{Im}(\psi, H_a\psi) = \operatorname{Im} \bar{\psi}(0)\psi'(0) \tag{3.30b}$$

for ψ with a bounded support; using, further, the closedness of H_a, one checks easily that (3.30b) holds for all $\psi \in D(H_a)$. Then the definition (3.29) yields

$$\operatorname{Im}(\psi, H_a\psi) = -\beta\,|\psi(0)|^2 \leqslant 0 \quad \text{for } \psi \in D(H_a) \tag{3.32}$$

so *the operator H_a is closed and dissipative*. It is also *maximal*: the equation $-\psi'' + i\psi = 0$ is solved by $\psi_\pm(x) = \exp(\pm\epsilon x)$, where $\epsilon = \exp(\pi i/4)$. The solution

ψ_+ does not belong to $L^2(\mathbb{R}_+)$, while ψ_- is square-integrable but $\mathrm{Im}(\psi'_-(0) + \bar{a}\psi_-(0)) = -2^{-1/2} - \beta < 0$ so $\psi_- \notin D(H_a)$. Hence we obtain $\mathrm{Ker}(H_a^* + i) = \{0\}$ and the assertion is proved (it also follows from Corollary 2.2).

In order to illustrate the physical meaning of the extensions H_a, consider the scattering on the barrier in the time-independent picture, i.e. the reflection of plane waves. Of course, the latter do not belong to $L^2(\mathbb{R}_+)$, so one should rather treat the reflection of wave packets which represent the true physical states of the particle. Such a reformulation, however, does not alter the conclusion presented below, and it is therefore left to the reader. Let

$$\psi(x) = \mathrm{e}^{-ikx} + R\,\mathrm{e}^{ikx}, \tag{3.33a}$$

i.e. superposition of the incident plane wave travelling to the left and the reflected one. If ψ has to obey the boundary condition from (3.29a), then the reflection coefficient equals

$$R = -\frac{a - ik}{a + ik}\,. \tag{3.33b}$$

Thus the reflected wave changes its phase on a factor depending on a, k as in the case of real a. In addition, however, it is partially absorbed, because

$$|R|^2 = \frac{\alpha^2 + (\beta - k)^2}{\alpha^2 + (\beta + k)^2} < 1 \tag{3.33c}$$

holds for all positive k, if only $\beta > 0$.

REMARK 4.3.10. The above examples show at the same time that the condition of essentially maximal dissipativity given by Theorem 3.2 is sufficient but in general not necessary. For instance, let H' be a densely defined operator, $H' \subset H^*$, where the r.h.s. is given by (3.28). It is easy to see that $JH'J = H'$ so H' is J-symmetric. Thus the operators H_a given by (3.29) are maximal dissipative but not J-symmetric, let alone J-self-adjoint. The same argument applies to the operator $\tilde{T}_\alpha$, $\alpha < -1/4$, from Example 3.8.

It is clear now which kind of problems may arise in the case that an operator H, intended as pseudo-Hamiltonian for some theory or model, is not essentially maximal dissipative. Different maximal dissipative extensions of H refer to *different physical situations*, and an additional information is needed to specify the right one. Of course, this conclusion just repeats in a wider context what is well known for self-adjoint extensions of the symmetric operators which are not e.s.a. In the particular case of dissipative Schrödinger operators, the fact that $H = H_0 + U$ is not e.m.d. means that the problem under consideration is not described fully by the potential u alone. The additional specification is usually given here by some sort of boundary conditions as in the above two examples.

In the preceding section, we have defined the completely dissipative operators as those related to completely non-unitary contractive semigroups. Let us now turn to the question of the conditions under which (the closure of) a given dissipative Schrödinger operator is completely dissipative.

PROPOSITION 4.3.11. *Let a dissipative Schrödinger operator H on $L^2(\mathbb{R}^d)$ referring to a potential $u = v - iw$ be e.m.d. and 'strictly dissipative', i.e.*

$$-w(x) = \operatorname{Im} u(x) < 0, \quad \textit{a.e. in } \mathbb{R}^d, \tag{3.34}$$

then $\bar{H}$ is completely dissipative.

Proof. If $w \in L^\infty(\mathbb{R}^d)$, the assertion follows directly from Proposition 2.12. If w is not essentially bounded, we can always express it as a sum of almost regular functions w_1, w_2 on $\mathbb{R}^d$ such that w_1 fulfils the condition (3.6), while $w_2 \in L^\infty(\mathbb{R}^d)$ and fulfils (3.34). A possible choice is $w_2(x) = \max\{1, w(x)\}$ and $w_1 = w - w_2$. Since the operator H is assumed to be e.m.d. and W_2 is bounded, the symmetric form of Theorem 2.8 (cf. Problem 33) implies that $H_1 = H_0 + V - iW_1$ is also e.m.d. and $\bar{H} = \overline{H_1} - iW_2$. Applying Proposition 2.12 again, this time to the operators $\overline{H_1}$, W_2, we get the desired result. ∎

COROLLARY 4.3.12. *An operator H which obeys the hypotheses of Proposition 3.11 has no real eigenvalues.*

In this way the complete dissipativity of a generalized Schrödinger operator can be ensured, if the function w (which is often called the *absorptive part* of the potential by physicists) is strictly positive, with the possible exception of a zero subset of the configuration space. However, this condition is in no case necessary as the following assertion illustrates:

THEOREM 4.3.13 (Davies). *Let H be a dissipative Schrödinger operator on $L^2(\mathbb{R}^d)$ with a purely imaginary bounded potential $u = -iw \in L^\infty(\mathbb{R}^d)$. If the function w assumes positive values on some non-zero subset of $\mathbb{R}^d$, then the operator H is completely dissipative.*

Proof. A continuous contractive semigroup $\{V_t = e^{-iHt} : t \geq 0\}$ corresponds to H in view of Corollary 2.3. Suppose that $\{V_t\}$ is unitary on a subspace $\mathcal{H}_0 \subset L^2(\mathbb{R}^d)$. Then $\|V_t\psi\| = \|\psi\|$ for $\psi \in \mathcal{H}_0$ and $t \geq 0$ so $(V_t\psi, WV_t\psi) = 0$ (cf. the proof of Proposition 2.12). Since $W \geq 0$, the inequality

$$\|WV_t\psi\|^2 \leq \|W^{1/2}\|^2 \, \|W^{1/2} V_t\psi\|^2 = \|W^{1/2}\|^2 \, (V_t\psi, WV_t\psi) = 0$$

gives $WV_t\psi = 0$ for all $\psi \in \mathcal{H}_0$ and $t \geq 0$. Now $D(H)$ is invariant under $\{V_t\}$, and therefore $HV_t\psi = H_0 V_t\psi$ for each $\psi \in D(H) \cap \mathcal{H}_0$. Uniqueness of the generator then yields $V_t\psi = \exp(-iH_0t)\psi$ within $\mathcal{H}_0$. Since $\{V_t\}$ is assumed to be unitary on $\mathcal{H}_0$, $\exp(-iH_0t)$ maps $\mathcal{H}_0$ onto $\mathcal{H}_0$, and is therefore reduced by the projection E_0 referring to $\mathcal{H}_0$. Then the projections $P_n \equiv E_{H_0}([0, n])$ commute with E_0, so $P_n\psi$ lies in $\mathcal{H}_0$ for an arbitrary $\psi \in \mathcal{H}_0$.

The relation (3.3) is essential for the proof; it implies $P_n = F_d^{-1}\tilde{P}_nF_d$, where $\tilde{P}_n = E_S([0, n])$. Hence $F_dP_n\psi$ is a function of compact support and its inverse Fourier transform $P_n\psi$ is an entire analytic function. For $\psi \in \mathcal{H}_0$, the above considerations give $HP_n\psi = H_0P_n\psi$ so $(P_n\psi)(x) = 0$ must hold on the support of W. Consequently, $P_n\psi = 0$ and $\psi = \text{s-lim}_{n\to\infty} P_n\psi = 0$, i.e. $\mathscr{H}_0 = \{0\}$. ∎

Of course, the discussion of a given Schrödinger pseudo-Hamiltonian H is in no case exhausted by establishing that it is e.m.d.; it is rather the starting point, from which one can tread various ways. One of them, which will be followed in the next chapters, consists of constructing a solution of the Schrödinger-type equation referring to H, i.e. the operators $V_t = e^{-iHt}$. Another possible way, which is in some sense complementary to the first, relies on spectral analysis of the operator H; it is clear that neither of them can be universally preferred.

Many results are known concerning the spectral properties of the generalized Schrödinger operators, and even a brief exposition of the subject goes beyond the scope of the present text. We limit ourselves to quoting two theorems; references to their proofs and further information may be found in the notes. First we mention that for many important potentials, the essential spectrum is concentrated on the real axis, while the open lower halfplane contains isolated points of $\sigma(H)$ only. Consider again the set $L^p(\mathbb{R}^d) + L^\infty(\mathbb{R}^d)$. Its subset $L^p(\mathbb{R}^d) + L^\infty(\mathbb{R}^d)_\epsilon$ is formed by the functions u with the following property: the decomposition $u = u_1 + u_2$ exists to any $\epsilon > 0$ such that the L^∞-component u_2 fulfils $\|u_2\|_\infty < \epsilon$. As an example, consider a Coulomb potential which belongs to $L^2(\mathbb{R}^3) + L^\infty(\mathbb{R}^3)_\epsilon$ though it is not square-integrable. It appears that for a suitably chosen p, the potentials of the described class represent relatively compact perturbations to the free Hamiltonian, and therefore leave its essential spectrum invariant in view of the Weyl theorem:

THEOREM 4.3.14. *Let H be a dissipative Schrödinger operator related to a potential $u \in L^p(\mathbb{R}^d) + L^\infty(\mathbb{R}^d)_\epsilon$, where $p = \max\{d/2, 2\}$ if $d \neq 4$, and $p > 2$ if $d = 4$, then $\sigma_{ess}(H) = \mathbb{R}_+$.*

It is clear that the behaviour of the potential at infinity is decisive here. Modifying the present result, one can find the essential spectrum of the dissipative Schrödinger operators referring to the potentials which are sufficiently regular (locally) and have a finite limit as $|x| \to \infty$ (Problem 35).

Recall that the *essential spectrum* is defined for a closed operator H as the complement of the *discrete spectrum*, $\sigma_{ess}(H) = \sigma(H)\backslash\sigma_{disc}(H)$. The latter consist of all *isolated* points $\lambda \in \sigma(H)$ such that the corresponding (in general, non-orthogonal) projection $P_H(\{\lambda\})$, which is defined by

$$P_H(\{\lambda\})\psi = \frac{1}{2\pi i}\oint_{|\xi-\lambda|=\epsilon} (H-\xi)^{-1}\psi \, d\xi \tag{3.35}$$

(clockwise) for all $\psi \in \mathscr{H}$ and sufficiently small positive ϵ, is *finite-dimensional*. Now the mentioned result suggests that the spectrum might be purely discrete if the real and/or negative imaginary part of the potential grows at infinity. In the one-dimensional case, the following stronger assertion may be obtained by generalization of the Molchanov criterion for real-valued potentials:

THEOREM 4.3.15. (Lidskii). *Let H be a dissipative Schrödinger operator on $L^2(\mathbb{R})$ related to a potential $u \in L^1_{\mathrm{loc}}(\mathbb{R})$, $u = v - iw$, such that v fulfils the condition* (3.13). *Let further $\{J_n\}$ be a sequence of mutually disjoint intervals of equal length approaching $\pm\infty$. The spectrum of $\bar{H}$ is purely discrete iff*

$$\lim_{n\to\infty} \int_{J_n} (v(x) + w(x))\,\mathrm{d}x = \infty \tag{3.36}$$

holds for all such sequences $\{J_n\}$.

4.4. The Optical Approximation

There are various ways in which the pseudo-Hamiltonians are applied in quantum microphysics. The most straightforward of them is tied with the existence of non-real eigenvalues, which can be associated with decay modes of the considered object. In that case, the use of a pseudo-Hamiltonian has a clear physical meaning as discussed in Section 1, namely as the generalized pole-and-real-axis approximation to a solution of the full dynamical problem.

We are now going to illustrate another important application of the pseudo-Hamiltonians. It is related to various scattering processes – especially in nuclear physics – which involve a large number of particles. In such a case again, one is rarely able to solve the corresponding many-body problem exactly. In order to get some information about the process, we must therefore select a suitable part of the whole state Hilbert space from which the needed quantities can be extracted, while the influence of the rest is taken into account in some phenomenological way. Among these methods, a distinguished role is played by the optical model of nuclear reactions. For the sake of simplicity, we shall consider only the simplest situations when fast neutrons (with energies from units to hundreds of MeV) are scattered elastically by nuclei. The experiments tell us that two processes are combined here. A part of the neutrons is scattered directly (in the so-called shape-elastic scattering), while the other neutrons are captured in the target forming compound nuclei, which decay after some time, emitting the neutron again in a direction that is independent of its initial velocity.

The existence of two strictly distinguished scattering processes suggests that the state Hilbert space $\mathscr{H}$ of the whole system (consisting of the target nucleus and the bombarding neutron) decomposes naturally into an orthogonal sum $\mathscr{H}_0 \oplus \mathscr{H}_1$, where the subspaces $\mathscr{H}_0$, $\mathscr{H}_1$ are related to the shape- and compound-elastic scattering, respectively. For brevity, we shall refer to them as to channels, though especially $\mathscr{H}_1$ can contain many 'true' scattering channels characterized by internal quantum numbers.

Next we must see how decomposition of the state space reflects in the structure of the Hamiltonian. We suppose it to be of the following form

$$H = H_0 + H_1 + gV, \tag{4.1a}$$

where the operators H_j are reduced by the projections P_j corresponding to $\mathscr{H}_j$, $j = 0, 1$, the part of H_0 in $\mathscr{H}_1$ is zero and vice versa. For the sake of simplicity, the symbol H_j will denote in this section simultaneously the operator $H_j \upharpoonright \mathscr{H}_j = \mathrm{pr}_j H_j$. Selection of the operators H_j is arbitrary up to a certain extent. For instance, assume that V is bounded and define $V_{jk} = P_j V P_k$; then for any fixed value of the coupling constant g, the operator gV_{jj} or part of it may be included in H_j. As for the second channel ($j = 1$), we remove this arbitrariness by setting

$$V_{11} = 0; \tag{4.1b}$$

this is possible since we shall not be interested in the detailed structure of H_1. On the other hand, the operator H_0 is fixed by the requirement that it is the free Hamiltonian of the first ($j = 0$) channel. In the specific physical situation described above, H_0 would therefore govern the motion of the neutron if the interaction with the target nucleus was switched off. In addition, we accept the plausible assumption that the spectrum of H_0 is absolutely continuous.

In order to get an idea of what these operators might look like, consider the simplest optical-model situation:

EXAMPLE 4.4.1. The neutron is scattered on a nucleus which consists of A nucleons. If we neglect their spins as well as the Pauli principle, the state space $\mathscr{H}$ is $\mathscr{H}_n \otimes \mathscr{H}_N$, or more explicitly

$$\mathscr{H} = L^2(\mathbb{R}^3) \otimes L^2(\mathbb{R}^{3A}). \tag{4.2}$$

The Hamiltonian may be chosen in the following form

$$H = -\frac{1}{2m}\Delta_n + H_N + gV, \tag{4.3a}$$

where H_N expresses as $\overline{I_n \otimes h_N}$ by means of the nucleus Hamiltonian h_N on $\mathscr{H}_N$, and $H_n = -(1/2m)\Delta_n$ is similarly related to the operator (3.2), up to the neutron mass m. The interaction part is described, e.g. by a potential v:

$$(V\psi)(x_n, x_N) = v(x_n, x_N)\psi(x_n, x_N). \tag{4.3b}$$

Let the ground state of the nucleus be described by an eigenvector $u_0 \in L^2(\mathbb{R}^{3A})$ of h_N referring to an eigenvalue λ_0. Then we choose

$$\mathscr{H}_0 = \{\psi \in \mathscr{H} : \psi(x_n, x_N) = f(x_n)u_0(x_N), f \in L^2(\mathbb{R}^3)\}, \tag{4.4}$$

i.e. $P_0 = I - P_1 = I_n \otimes E_{h_N}(\{\lambda_0\})$. It is an easy exercise to write the Hamiltonian (4.3) in the form (4.1) with

$$H_0 = \left(-\frac{1}{2m}\Delta_n + \lambda_0\right) P_0, \tag{4.5a}$$

$$H_1 = \left(-\frac{1}{2m}\Delta_n + H_N\right) P_1 + gV_{11}, \tag{4.5b}$$

and to find the explicit expressions for V_{jk}. In particular,

$$(V_{00}\psi)(x_n, x_N) = u_0(x_N)\,(u_0, \psi(x_n, .\,))_N\,(u_0, v(x_n, .\,)u_0)_N, \tag{4.6}$$

where $(.\,, .\,)_N$ denotes the inner product in $\mathscr{H}_N = L^2(\mathbb{R}^{3A})$. The last relation shows that we cannot, generally, expect V_{00} to be zero; in fact, it is likely to be non-zero for all the potentials that are of physical interest.

Before proceeding further, let us collect the assumptions made up to this point. The Hamiltonian acts on $\mathscr{H} = \mathscr{H}_0 \oplus \mathscr{H}_1$ and is of the form (4.1), where:

(s) *the operators H_0, H_1 are self-adjoint, reduced by P_j, and their parts in $\mathscr{H}_1$, $\mathscr{H}_0$, respectively, are zero;*

(b) *V is Hermitean,* i.e. *bounded and self-adjoint, with $V_{11} = 0$;*

(f) *H_0 is the free Hamiltonian of the first channel and $P_{ac}(H_0) = P_0$.*

It is clear that these assumptions are not very restrictive and apply to other physical situations besides the one described above.

The S-matrix is determined by the formulae (3.2.14) and (3.2.16). Notice that its first-channel part $S_{00} = P_0 S P_0$, in which we are primarily interested, can be expressed even if we do not know the full free Hamiltonian. Since $P_{ac}(H_0) = P_0$, it holds that

$$S_{00} = \operatorname*{s-lim}_{t\to\infty} S_{00}^t, \tag{4.7a}$$

$$S_{00}^t = P_0 \exp(iH_0 t)\exp(-2iHt)\exp(iH_0 t)P_0. \tag{4.7b}$$

In the particular physical situation considered above, the knowledge of S_{00} allows us to determine the shape-elastic cross-section. Now the **optical approximation** consists of choosing a suitable operator $U_{opt} \in \mathscr{B}(\mathscr{H}_0)$ such that

$$S_{opt} = \operatorname*{s-lim}_{t\to\infty} S_{opt}^t, \tag{4.8a}$$

$$S_{opt}^t = \exp(iH_0 t)\exp(-2iH_{opt}t)\exp(iH_0 t), \tag{4.8b}$$

where $H_{opt} = H_0 + U_{opt}$, is in some sense near to S_{00}.

Our main goal in this section is to exhibit some conditions under which the optical approximation can be justified. We shall concentrate our attention on the difference between S_{00}^t and S_{opt}^t, leaving aside for the moment the problem of the existence of the limits in (4.7a) and (4.8a).

We start with a discussion of the possible choices for the operator U_{opt}. First we notice that for a self-adjoint H and bounded V, the perturbation theory result quoted in the Appendix (A.13) gives

$$\begin{aligned} e^{-i(H+V)t}\,\psi &= e^{-iHt}\,\psi - i\int_0^t e^{-i(H+V)s}\,B\,e^{-iH(t-s)}\,\psi\,ds \\ &= e^{-iHt}\,\psi - i\int_0^t e^{-iH(t-s)}\,B\,e^{-i(H+V)s}\,\psi\,ds. \end{aligned} \tag{4.9}$$

This relation can be used to express the operators we are going to compare. Applying it repeatedly, we obtain the formal expansion

$$S_{00}^{t} = P_0 - ig\int_{-t}^{t} ds\, e^{iH_0 s}\, V\, e^{-iH_0 s} +$$

$$+ (-ig)^2 \int_{-t}^{t} ds \int_{-t}^{s} du\, e^{iH_0 s}\, V\, e^{-i(H_0 + H_1)(s-u)}\, V\, e^{-iH_0 u} + O(g^3).$$

Further, we perform, again formally, the limit $t \to \infty$. Since $S_{00} = P_0 S P_0$ and $V = V_{00} + V_{01} + V_{10}$, we get

$$S_{00} = P_0 - ig\int_{\mathbb{R}} ds\, e^{iH_0 s}\, V_{00}\, e^{-iH_0 s} +$$

$$+ (-ig)^2 \int_{\mathbb{R}} ds \int_{-\infty}^{s} du\, e^{iH_0 s}\, V_{00}\, e^{-iH_0(s-u)}\, V_{00}\, e^{-iH_0 u} +$$

$$+ (-ig)^2 \int_{\mathbb{R}} ds \int_{-\infty}^{s} du\, e^{iH_0 s}\, V_{01}\, e^{-iH_1(s-u)}\, V_{10}\, e^{-iH_0 u} +$$

$$+ O(g^3). \tag{4.10a}$$

In the same way, one finds

$$S_{\text{opt}} = I_{\mathcal{H}_0} - i\int_{\mathbb{R}} ds\, e^{iH_0 s}\, U_{\text{opt}}\, e^{-iH_0 s} +$$

$$+ (-i)^2 \int_{\mathbb{R}} ds \int_{-\infty}^{s} du\, e^{iH_0 s}\, U_{\text{opt}}\, e^{-iH_0(s-u)}\, U_{\text{opt}}\, e^{-iH_0 u} +$$

$$+ O(U_{\text{opt}}^3). \tag{4.10b}$$

The idea is to guess a suitable choice of U_{opt} from the comparison of the expansions (4.10a) and (4.10b). Two candidates arise at once:

$$U_{\text{opt}}^{(+)} = gV_{00} - ig^2 \int_0^{\infty} e^{iH_0 t}\, V_{01}\, e^{-iH_1 t}\, V_{10}\, dt, \tag{4.11a}$$

$$U_{\text{opt}}^{(-)} = gV_{00} - ig^2 \int_0^{\infty} V_{01}\, e^{-iH_1 t}\, V_{10}\, e^{iH_0 t}\, dt; \tag{4.11b}$$

for both of them, the expansions coincide up to the second order in the coupling constant.

Unfortunately, neither of the operators (4.11a, b) can be accepted. Besides the ambiguity in choosing the right one, they have a more serious defect. For obvious reasons, H_{opt} should be a pseudo-Hamiltonian, i.e. U_{opt} is expected to be dissipative. This may not be true for the operators (4.11a, b) even if V is of finite rank:

EXAMPLE 4.4.2. Consider the following Friedrichs-type model (cf. Section 3.2): $\mathcal{H} = L^2(\mathbb{R}) \oplus \mathbb{C}$ and

$$(H_0 + H_1)\{f, \alpha\} = \{Qf, \lambda_0\alpha\},$$

$$V_{01}\{f, \alpha\} = \{\alpha v, 0\}, \qquad V_{10}\{f, \alpha\} = \{0, (v, f)\}.$$

We take, e.g. the operator (4.11a). For any $\psi = \{f, \alpha\}$, we have

$$(\psi, U_{\mathrm{opt}}^{(+)}\psi) = -ig^2(v, f)\int_0^\infty \mathrm{d}t \int_{\mathbb{R}} \bar{f}(\lambda)v(\lambda)\,\mathrm{e}^{i(\lambda-\lambda_0)t}\,\mathrm{d}\lambda.$$

Now we choose $v, f \in \mathcal{S}(\mathbb{R})$ such that $(v, f) \neq 0$, the function $\bar{f}(\,.\,)v(\,.\,)$ is real and assumes both positive and negative values. In that case, one obtains easily

$$\mathrm{Im}(\psi, U_{\mathrm{opt}}^{(+)}\psi) = -\pi g^2(v, f)\bar{f}(\lambda_0)v(\lambda_0).$$

Since the parameter λ_0 can be chosen arbitrarily, $U_{\mathrm{opt}}^{(+)}$ may not be dissipative; a similar argument applies to $U_{\mathrm{opt}}^{(-)}$.

The well-known formal argument about a solution of the coupled time-independent Schrödinger equations related to the two channels offers another possibility, namely

$$U_{\mathrm{opt}}(E) = gV_{00} - g^2 \lim_{\epsilon\to 0+} V_{01}R(E + i\epsilon, H_1)V_{10}, \tag{4.12a}$$

where $R(z, H_1) = (H_1 - z)^{-1}$ as usual. In fact, it is a one-parameter family of operators indexed by the number E which has the dimension of energy. We shall use the term **Feshbach optical potential** for any of the operators (4.12).

It appears, however, that there is a close connection between the operators (4.11) and (4.12a). This can be illustrated by the following heuristic argument. Suppose that the initial state of the system (for very large negative t) is a wave packet φ_E whose energy distribution is sharply peaked around a mean value E. In other words, in the total-energy measurement nearly all values fall into a narrow interval centered at E. Notice further that the energy distribution is specified mainly by the beam of the incident particles. It is especially apparent in the case of neutron scattering on nuclei. In the initial stage, the neutron and the nucleus are spatially well separated and behave as if they were described by the free Hamiltonian $H_n + H_N$ (cf. (4.3a)). This approximation is very good: the chief electromagnetic contribution, the Coulomb force, is absent for the neutrons, and the strong interaction is known to have the range of about 10^{-12} cm. The target nuclei may be supposed to be at rest (we neglect their thermal fluctuations) so the contribution to the total energy from H_N is given by the binding energy of the nucleus (e.g. by λ_0 in the particular situation considered in Example 4.1). Then the energy distribution of φ_E has the same shape as that of the neutron beam, and E differs from the mean energy of this beam by an additive constant.

Since φ_E lies entirely in $\mathscr{H}_0$ and H_0 is assumed to be the first-channel part of the free Hamiltonian, we can try to replace $e^{iH_0t}\varphi_E$ approximatively by $e^{iEt}\varphi_E$. Then the operator (4.11b) acts as

$$U_{opt}^{(-)}\varphi_E \approx gV_{00}\varphi_E - ig^2\int_0^\infty V_{01}\,e^{-iH_1t}\,V_{10}\,e^{iEt}\,\varphi_E\,dt$$

$$= gV_{00}\varphi_E - ig^2 \lim_{\epsilon\to 0+}\int_0^\infty V_{01}\,e^{-i(H_1-E-i\epsilon)t}\,V_{10}\varphi_E\,dt$$

$$= gV_{00}\varphi_E - g^2 \lim_{\epsilon\to 0+} V_{01}(H_1-E-i\epsilon)^{-1}V_{10}\varphi_E,$$

i.e. $U_{opt}^{(-)}\varphi_E \approx U_{opt}(E)\varphi_E$. An analogous argument applied to the adjoint of the operator (4.11a) shows that $U_{opt}^{(+)}\varphi_E \approx U_{opt}(E)\varphi_E$ also holds.

Moreover, the operators $U_{opt}(E)$ fulfil the dissipativity requirement. Let $\{E_\lambda^{(1)}\}$ denote the spectral decomposition of H_1. The assumption (b) implies V_{00} to be Hermitean and $V_{01}^* = V_{10}$ so

$$\mathrm{Im}(\psi, U_{opt}(E)\varphi) = -g^2 \lim_{\epsilon\to 0+} \mathrm{Im}(V_{10}\psi, R(E+i\epsilon, H_1)V_{10}\psi)$$

$$= -g^2 \lim_{\epsilon\to 0+} \mathrm{Im}\int_{\mathbb{R}} (\lambda-E-i\epsilon)^{-1}\,d(V_{10}\psi, E_\lambda^{(1)}V_{10}\psi)$$

holds for all $\psi\in\mathscr{H}_0$ and $E\in\mathbb{R}$, i.e.

$$\mathrm{Im}(\psi, U_{opt}(E)\psi) = -g^2 \lim_{\epsilon\to 0+}\int_{\mathbb{R}} \frac{\epsilon}{(\lambda-E)^2+\epsilon^2}\,d\,\|E_\lambda^{(1)}V_{10}\psi\|^2 \leqslant 0. \quad (4.13)$$

Thus the operators (4.12a) represent a reasonable candidate for an optimal optical approximation in the situation when the target nuclei are exposed to the beam of neutrons which is *nearly monochromatic* in a suitable energy scale. Notice, however, that the Feshbach potential need not be bounded even if it is true for V. Consequently, in order to use

$$H_{opt}(E) = H_0 + U_{opt}(E) \quad (4.12b)$$

as the pseudo-Hamiltonian related to the first channel, one must check that it is densely defined and maximal dissipative.

Now we are going to specify a class of the operators V for which the difference between S_{00}^t and S_{opt}^t referring to (4.12) can be estimated. The idea is to use expansions of the type (4.10) and to build the estimate on the *smooth-perturbation theory* of Kato. The central trick of the latter consists of factorizing the perturbation by means of a pair of operators which are smooth with respect to the unperturbed one (cf. (A.14)). Hence we shall assume the existence of some bounded operators C_0, C_1 such that $C_j = P_0C_jP_j$, $j = 0, 1$, and

$$V_{10} = C_1^*C_0, \qquad V_{01} = C_0^*C_1. \quad (4.14a)$$

The 'diagonal' term V_{00} can be factorized, e.g., as follows

$$V_{00} = B^*A \tag{4.14b}$$

with $A = |V_{00}|^{1/2}$ and $B = AW^*$, where $W : \overline{\text{Ran}\,|V_{00}|} \to \overline{\text{Ran}\, V_{00}}$ is a partial isometry according to the polar-decomposition theorem; since V_{00} is Hermitean, so is W and $W^2 = I_{\mathcal{H}_0}$.

REMARK 4.4.3. A concise formulation of the following theorem, as well as its proof, will need some notation which is collected below. First we introduce the vector-valued functions $\varphi_C^t : \mathbb{R} \to \mathcal{H}_0$ by $\varphi_C^t(s) = \chi_{[-t,\,t]}(s) C_0 \exp(-iH_0 s)\varphi$ for each $\varphi \in \mathcal{H}_0$ and all $t \geqslant 0$. Analogously one defines $\varphi_G^t(\,.\,)$ for $G = A, B$. Next we define the operator-valued functions $x_{jk}(\,.\,)$ and $z_{11}(\,.\,)$ from $\mathbb{R}$ to $\mathcal{B}(\mathcal{H}_0)$ by the relations

$$\begin{aligned} x_{00}(t) &= A\ \mathrm{e}^{-iH_0 t}\, B^*\Theta(t), & x_{01}(t) &= A\ \mathrm{e}^{-iH_0 t}\, C_0^*\Theta(t),\\ x_{10}(t) &= C_0\ \mathrm{e}^{-iH_0 t}\, B^*\Theta(t), & x_{11}(t) &= C_0\ \mathrm{e}^{-iH_0 t}\, C_0^*\Theta(t),\\ z_{11}(t) &= C_1\ \mathrm{e}^{-iH_1 t}\, C_1^*\Theta(t), \end{aligned} \tag{4.15}$$

where $\Theta = \chi_{\mathbb{R}_+}$ is the Heaviside function. Beside the standard norm $\|\,.\,\|$ and the inner product $(\,.\,,\,.\,)$ in $\mathcal{H}$, $\mathcal{H}_0$, we shall use the Hilbert space norm $\|\,.\,\|_2$ in $L^2(\mathbb{R};\ \mathcal{H}_0)$ corresponding to the inner product $\langle\,.\,,\,.\,\rangle$, and the operator norms $\|\,.\,\|_{\mathrm{u}}$, $\|\,.\,\|_{\mathrm{uf}}$ in $\mathcal{B}(\mathcal{H}_0)$ and $\mathcal{B}(L^2(\mathbb{R};\ \mathcal{H}_0))$, respectively. Finally, let a measurable operator-valued function $f : \mathbb{R} \to \mathcal{B}(\mathcal{H}_0)$ be given. We set $\|f\|_1 = \int_{\mathbb{R}} \|f(x)\|_{\mathrm{u}}\, \mathrm{d}x$ and $\|f\|_\infty = \sup\operatorname{ess}\{\|f(x)\|_{\mathrm{u}} : x \in \mathbb{R}\}$, and $\hat{f}(\,.\,)$ denotes the Fourier transform of f, $\hat{f}(x) = \int_{\mathbb{R}} \mathrm{e}^{ixy} f(y)\, \mathrm{d}y$. The symbol F stands for the operator of the 'left multiplication' by f on $L^2(\mathbb{R};\ \mathcal{H}_0)$, $(F\psi)(x) = f(x)\psi(x)$, and $\hat{F}$ is similarly related to $\hat{f}$; clearly $\|F\|_{\mathrm{uf}} = \|f\|_\infty$. The operator $\tilde{F}$ on $L^2(\mathbb{R};\ \mathcal{H}_0)$ is defined by means of the convolution, $(\tilde{F}\psi)(x) = \int_{\mathbb{R}} f(y)\psi(x-y)\, \mathrm{d}y$. Most of this notation will not be used in the following sections.

In order to make use of the smooth-perturbation technique, we adopt one more assumption, namely that the $\|\,.\,\|_1$-norms of the operator-valued functions (4.15) are *finite*. In this way five constants enter into the estimation, but they are not completely independent. The operators C_0, C_1 are given up to a multiplicative constant, $C_1^* C_0 = (aC_1)^*(a^{-1}C_0)$ for any positive a, and the same is true for the product gV. Then the last assumption can be reformulated as the following normalization condition:

(n) *The operator* V *factorizes according to* (4.14) *with the functions* (4.15) *fulfilling* $\|x_{00}\|_1 = 1, \max\{\|x_{10}\|_1, \|x_{01}\|_1, \|x_{11}\|_1\} = 1$ *and* $\|z_{11}\|_1 \equiv \eta < \infty$.

With these prerequisites, we can formulate the main result as follows:

THEOREM 4.4.4. *Assume* (s), (b), (f) *and* (n), *and denote* $\lambda = \max\{1, \eta\}$. *For each* $\psi \in \mathcal{H}_0$ *and* $|g| < (4\lambda)^{-1/2}$, *the inequality*

$$\|S_{00}^t\psi - S_{\mathrm{opt}}^t(E)\psi\| \leqslant g^2\, \frac{2^{-1/2}\alpha(E,\psi) + (2\beta_0(E) + \beta_1(E))\,\|\psi\|}{1 - (4\lambda)^{1/2}\,|g|} \tag{4.16a}$$

holds, where $S^t_{\mathrm{opt}}(E)$ *refers to the optical potential* (4.12a), *and*

$$\alpha(E, \psi) = \|(\tilde{Z}_{11} - \hat{Z}_{11}(E))\psi^t_C\|_2, \tag{4.16b}$$

$$\beta_j(E) = \|(\tilde{Z}_{11} - \hat{Z}_{11}(E))\tilde{X}_{1j}\|_{\mathrm{uf}}, \qquad j = 0, 1. \tag{4.16c}$$

The proof is divided into several steps:

LEMMA 4.4.5. *Assume* (s) *and* (n), *then*

$$\|\varphi^t_G\|_2^2 \leqslant 2\,\|\varphi\|^2, \qquad G = C, A, B \tag{4.17}$$

holds for each $t \geqslant 0$ *and all* $\varphi \in \mathscr{H}_0$.

Proof. Consider first φ^t_C. According to its definition, we have

$$\|\varphi^t_C\|_2^2 \leqslant \int_{\mathbb{R}} \|C_0\, \mathrm{e}^{-iH_0 s}\,\varphi\|^2\, \mathrm{d}s,$$

and the r.h.s. is $\leqslant 2\pi\,\|C_0\|^2_{H_0}\,\|\varphi\|^2$ in view of (A.14). Further, we employ the third of the equivalent expressions of this norm quoted in the Appendix. The relation (2.1) implies

$$[R(z, H_0) - R(\bar{z}, H_0)]\varphi = i \int_{\mathbb{R}} \exp(i\xi s - \eta\,|s| - iH_0 s)\, \mathrm{d}s,$$

where $z = \xi + i\eta$, $\eta > 0$, so we obtain

$$\begin{aligned}\|C_0\|^2_{H_0} &= \frac{1}{2\pi} \sup_{z,\psi} \left| \int_{\mathbb{R}} (\mathrm{e}^{i\xi t - \eta|t|}\,\mathrm{e}^{-iH_0 t}\, C_0^*\psi, C_0^*\psi)\, \mathrm{d}t \right| \\ &\leqslant \frac{1}{2\pi} \sup_{z,\psi} \int_{\mathbb{R}} \mathrm{e}^{-\eta|t|}\, |(C_0\,\mathrm{e}^{-iH_0 t}\, C_0^*\psi, \psi)|\, \mathrm{d}t \\ &\leqslant \frac{1}{2\pi} \int_{\mathbb{R}} \|C_0\,\mathrm{e}^{-iH_0 t}\, C_0^*\|_{\mathrm{u}}\, \mathrm{d}t.\end{aligned}$$

Since the function $t \mapsto \|C_0\,\mathrm{e}^{-iH_0 t}\,C_0^*\|_{\mathrm{u}}$ is even, we get

$$2\pi\,\|C_0\|^2_{H_0} \leqslant 2 \int_0^\infty \|C_0\,\mathrm{e}^{-iH_0 t}\, C_0^*\|_{\mathrm{u}}\, \mathrm{d}t = 2\,\|x_{11}\|_1$$

and (4.17) follows from (n). The other two cases are treated similarly. We have $A^* = B^*W$, so

$$2\pi\,\|A\|^2_{H_0} \leqslant 2 \int_0^\infty \|A\,\mathrm{e}^{-iH_0 t}\, A^*\|_{\mathrm{u}}\, \mathrm{d}t \leqslant 2\,\|x_{00}\|_1,$$

and an analogous applies to φ^t_B. ∎

LEMMA 4.4.6. *The condition* (n) *implies*

$$\|\tilde{X}_{jk}\|_{uf} \leqslant 1, \qquad j, k = 0, 1, \tag{4.18a}$$

$$\|\tilde{Z}_{11}\|_{uf} \leqslant \eta, \tag{4.18b}$$

$$\|\hat{Z}_{11}(E)\|_{uf} \leqslant \eta \qquad \textit{for all} \qquad E \in \mathbb{R}. \tag{4.18c}$$

Proof. Consider a function $f : \mathbb{R} \to \mathscr{B}(\mathscr{H}_0)$ with $\|f\|_1 < \infty$ and $\psi \in L^2(\mathbb{R}; \mathscr{H}_0)$. It holds

$$\|\tilde{F}\psi\|_2^2 = \int_{\mathbb{R}} \left\| \int_{\mathbb{R}} f(y)\psi(x-y)\,\mathrm{d}y \right\|^2 \mathrm{d}x$$

$$\leqslant \int_{\mathbb{R}} \mathrm{d}y\, \|f(y)\|_u \int_{\mathbb{R}} \mathrm{d}z\, \|f(z)\|_u \int_{\mathbb{R}} \mathrm{d}x\, \|\psi(x-y)\|\, \|\psi(x-z)\|$$

so the Hölder inequality applied to the last integral gives $\|\tilde{F}\psi\|_2^2 \leqslant \|f\|_1^2\, \|\psi\|_2^2$, i.e.

$$\|\tilde{F}\|_{uf} \leqslant \|f\|_1,$$

which proves (4.18a, b). The remaining inequality follows from $\|\hat{Z}_{11}(E)\|_{uf} = \|\hat{z}_{11}(E)\|_u \leqslant \|z_{11}\|_1$. ∎

Proof of Theorem 4.4. The adopted notation allows us to rewrite the optical potential (4.12a) in the form

$$U_{\text{opt}}(E) = gB^*A - ig^2 C_0^* \hat{Z}_{11}(E) C_0, \tag{4.19}$$

because

$$[U_{\text{opt}}(E) - gV_{00}]\varphi = \lim_{\epsilon \to 0+} \int_0^\infty \mathrm{e}^{i(E+i\epsilon)t}\, V_{01}\, \mathrm{e}^{-iH_1 t}\, V_{10}\varphi\, \mathrm{d}t$$

$$= \lim_{\epsilon \to 0+} \int_0^\infty \mathrm{e}^{i(E+i\epsilon)t}\, C_0^* C_1\, \mathrm{e}^{-iH_1 t}\, C_1^* C_0 \varphi\, \mathrm{d}t$$

and the dominated-convergence theorem is applicable since $\|z_{11}\|_1 < \infty$.

Now the idea is to combine the estimates collected in the above two lemmas with the expansions of S_{00}^t and S_{opt}^t analogous to those used in the derivation of the optical potential. We notice first that they are simply the series obtained by application of the Dyson expansion (in the interaction picture – see the text following Theorem 2.13) to $\exp(-2iHt)$ and $\exp(-2iH_{\text{opt}}(E)t)$. The operators V and $U_{\text{opt}}(E)$ are bounded due to the assumption, so both the exponentials exist and their Dyson expansions make sense.

Consider first S_{00}^t. For arbitrary $\varphi, \psi \in \mathscr{H}_0$, we have

$$(\varphi, S_{00}^t \psi) = (\varphi, \psi) + \sum_{k=1}^{\infty} (-ig)^k \int_{-t}^{t} \mathrm{d}s_1 \int_{-t}^{s_1} \mathrm{d}s_2 \ldots$$

$$\ldots \int_{-t}^{s_{k-1}} \mathrm{d}s_k\, (\varphi, \mathrm{e}^{iH_0 s_1}\, V\, \mathrm{e}^{-i(H_0+H_1)(s_1-s_2)}\, V \ldots V\, \mathrm{e}^{-iH_0 s_k}\, \psi)$$

Now one has to substitute $V = V_{00} + V_{01} + V_{10}$ and to use the factorization (4.14); it yields

$$(\varphi, S^t_{00}\psi) = (\varphi, \psi) + \sum_{k=1}^{\infty} (-ig)^k s^t_k(\varphi, \psi) \tag{4.20a}$$

where the sesquilinear forms $s^t_k(\varphi, \psi)$ are given by

$$\begin{aligned}
s^t_1(\varphi, \psi) &= \langle \varphi^t_B, \psi^t_A \rangle,\\
s^t_2(\varphi, \psi) &= \langle \varphi^t_B, \tilde{X}_{00}\psi^t_A \rangle,\\
s^t_3(\varphi, \psi) &= \langle \varphi^t_B, \tilde{X}_{00}\tilde{X}_{00}\psi^t_A \rangle + \langle \varphi^t_B, \tilde{X}_{01}\tilde{Z}_{11}\psi^t_C \rangle + \langle \varphi^t_C, \tilde{Z}_{11}\tilde{X}_{10}\psi^t_A \rangle,\\
s^t_4(\varphi, \psi) &= \langle \varphi^t_B, \tilde{X}_{00}\tilde{X}_{00}\tilde{X}_{00}\psi^t_A \rangle + \langle \varphi^t_B, \tilde{X}_{00}\tilde{X}_{01}\tilde{Z}_{11}\psi^t_C \rangle +\\
&\quad + \langle \varphi^t_B, \tilde{X}_{01}\tilde{Z}_{11}\tilde{X}_{10}\psi^t_A \rangle + \langle \varphi^t_C, \tilde{Z}_{11}\tilde{X}_{10}\tilde{X}_{00}\psi^t_A \rangle + \langle \varphi^t_C, \tilde{Z}_{11}\tilde{X}_{11}\tilde{Z}_{11}\psi^t_C \rangle,
\end{aligned} \tag{4.20b}$$

etc. The structure of the general s^t_k is easily found: neighbouring indices of the neighbouring operators must coincide, and the following pairings are *forbidden*:

$$\tilde{X}_{01}\tilde{X}_{11},\ \tilde{X}_{11}\tilde{X}_{10},\ \tilde{X}_{01}\tilde{X}_{10},\ \tilde{Z}_{11}\tilde{Z}_{11}. \tag{4.21}$$

The number n_k of the terms contained in s^t_k will be useful in the following. In order to determine it, notice that each operator product in s^t_k is due to (4.21) uniquely determined by the positions of the operators $\tilde{Z}_{11}$. Let c^m_{k-1} denote in how many ways m operators Z_{11} can be distributed over $k-1$ places. It is easy to see that $c^m_{k-1} = c^{m-1}_{k-3} + c^{m-1}_{k-4} + \dots$ so

$$n_k = \sum_{m=0}^{[k/2]} c^m_{k-1}, \qquad c^m_{k-1} = \binom{k-m}{m}, \tag{4.22}$$

where [.] denotes the integer part.

Next we pass to S^t_{opt}. Combining the Dyson expansion with the relation (4.19), we get

$$(\varphi, S^t_{\text{opt}}\psi) = (\varphi, \psi) + \sum_{k=1}^{\infty} b^t_k(\varphi, \psi), \tag{4.23a}$$

where

$$\begin{aligned}
b^t_1(\varphi, \psi) &= -ig\langle \varphi^t_B, \psi^t_A \rangle + (-ig)^2 \langle \varphi^t_C, \hat{Z}_{11}(E)\psi^t_C \rangle,\\
b^t_2(\varphi, \psi) &= (-ig)^2 \langle \varphi^t_B, \tilde{X}_{00}\psi^t_A \rangle + (-ig)^3 \langle \varphi^t_B, \tilde{X}_{01}\hat{Z}_{11}(E)\psi^t_C \rangle +\\
&\quad + (-ig)^3 \langle \varphi^t_C, \hat{Z}_{11}(E)\tilde{X}_{10}\psi^t_A \rangle + (-ig)^4 \langle \varphi^t_C, \hat{Z}_{11}(E)\tilde{X}_{11}\hat{Z}_{11}(E)\psi^t_C \rangle
\end{aligned} \tag{4.23a}$$

etc. In order to make the comparison of the expansions (4.20) and (4.23) possible, one must rearrange the latter according to the powers of $-ig$. It is possible provided that the series (4.23a) converges absolutely. Without loss of generality, one may assume $\|\varphi\| = \|\psi\| = 1$; then Lemmas 4.5 and 4.6 yield

$$|b_1^t(\varphi, \psi)| \leq |g| \, \|\varphi_B^t\|_2 \, \|\psi_A^t\|_2 + |g|^2 \, \|\varphi_C^t\|_2 \, \|\psi_C^t\|_2 \, \|\hat{Z}_{11}(E)\|_{\mathrm{uf}}$$
$$\leq 2(|g| + \eta g^2),$$
$$|b_2^t(\varphi, \psi)| \leq 2(g^2 + 2\eta \, |g|^3 + \eta^2 g^4)$$

etc. so the kth term is estimated by

$$|b_k^t(\varphi, \psi)| \leq 2 \, |g|^k (1 + \eta \, |g|)^k.$$

Hence the absolute convergence is guaranteed if only

$$|g| < \frac{1}{2\eta} \, [(1 + 4\eta)^{1/2} - 1]. \tag{4.24}$$

It is easy to see that after the rearrangement, the expansion (4.23) differs from (4.20) just by replacement of all the operators $\tilde{Z}_{11}$ by $\hat{Z}_{11}(E)$.

Now we are able to complete the proof comparing term-by-term the two expansions,

$$(\varphi, [S_{00}^t - S_{\mathrm{opt}}^t(E)] \, \psi) = \sum_{k=2}^{\infty} (-ig)^k d_k^t(\varphi, \psi). \tag{4.25}$$

Lemmas 4.5 and 4.6 together with (4.16b, c) give the following estimates for d_k^t:

$$|d_2^t(\varphi, \psi)| \leq \|\varphi_C^t\|_2 \, \alpha(E, \psi) \leq 2^{1/2} \, \|\varphi\| \, \alpha(E, \psi),$$
$$|d_3^t(\varphi, \psi)| \leq \|\varphi_B^t\|_2 \, \|\tilde{X}_{01}\|_{\mathrm{uf}} \, \alpha(E, \psi) + \|\varphi_C^t\|_2 \, \beta_0(E) \, \|\psi_A^t\|_2$$
$$\leq 2^{1/2} \, \|\varphi\| \, [\alpha(E, \psi) + 2^{1/2} \, \|\psi\| \, \beta_0(E)],$$
$$|d_4^t(\varphi, \psi)| \leq 2^{1/2} \, \|\varphi\| \, \alpha(E, \psi) + 2 \cdot 2^{1/2} \, \|\varphi\| \, \beta_0(E) \, 2^{1/2} \, \|\psi\| +$$
$$+ |\langle \varphi_C^t, [\tilde{Z}_{11} - \hat{Z}_{11}(E)] \tilde{X}_{11} \tilde{Z}_{11} \psi_C^t \rangle| +$$
$$+ |\langle \varphi_C^t, \hat{Z}_{11}(E) \tilde{X}_{11} [\tilde{Z}_{11} - \hat{Z}_{11}(E)] \, \psi_C^t \rangle|$$
$$\leq 2^{1/2} \, \|\varphi\| \{(1 + \eta) \alpha(E, \psi) + 2^{1/2} \, \|\psi\| \, (2\beta_0(E) + \beta_1(E))\}.$$

In order to estimate the general d_k^t, we rewrite in the form

$$d_k^t(\varphi, \psi) = \sum_{m=1}^{[k/2]} d_k^t(\varphi, \psi; m), \tag{4.26}$$

where $d_k^t(\varphi, \psi; m)$ contains the terms with just m operators $\tilde{Z}_{11}$ or $\hat{Z}_{11}(E)$. Their number equals mc_{k-1}^m, because similarly as above one has to add and subtract

$(m-1)c^m_{k-1}$ terms to single out the differences $D_{11} \equiv \tilde{Z}_{11} - \hat{Z}_{11}(E)$. We divide them into three groups in which D_{11} is followed by ψ^t_C and $\tilde{X}_{1j}$; numbers of their elements are denoted as $c^m_{k-1}(\alpha)$ and $c^m_{k-1}(\beta_j), j = 0, 1$, respectively.

LEMMA 4.4.7. *It holds that* $c^m_{k-1}(\alpha) = \binom{k-m-1}{m-1}$ *and* $c^m_{k-1}(\beta_0) = c^{m+1}_k(\beta_1) = m\binom{k-m-1}{m}$.

Proof. Consider first $c^m_{k-1}(\alpha)$. In the terms of this type, D_{11} must stand first from the right, and its left neighbour is $\tilde{X}_{11}$ or $\tilde{X}_{01}$ due to (4.22). Thus we have $k-3$ places for the arrangement of $m-1$ operators $\tilde{Z}_{11}$ or $\hat{Z}_{11}(E)$, symbolically

$$\vdash\!\frac{m-1}{k-3}\!\!-\tilde{X}D_{11},$$

which gives $c^m_{k-1}(\alpha) = c^{m-1}_{k-3} = \binom{k-m-1}{m-1}$. As for $c^m_{k-1}(\beta_1)$, we have the following possibilities

$$D_{11}\tilde{X}_{11}Z_{11}\tilde{X}\,\frac{m-2}{k-5}\!\dashv,$$

$$\vdash\!\frac{m-s-2}{k-l-6}\!-\tilde{X}D_{11}\tilde{X}_{11}Z_{11}\tilde{X}\!-\!\frac{s}{l}\!\dashv, \qquad l = 0, 1, \ldots, k-6$$

$$\max\left\{0, m-2-\left[\!\left[\frac{k-l-5}{2}\right]\!\right]\right\} \leqslant s \leqslant \min\left\{m-2, \left[\!\left[\frac{l+1}{2}\right]\!\right]\right\},$$

$$\vdash\!\frac{m-2}{k-5}\,\tilde{X}D_{11}\tilde{X}_{11}Z_{11},$$

which give

$$c^m_{k-1}(\beta_1) = 2\binom{k-m-2}{m-2} + \sum_{l=0}^{k-6}\ \sum_{s=0}^{[(l+1)/2]} \binom{k-l-m+s-3}{m-s-2}\binom{l+1-s}{s}$$

$$= \sum_{r=0}^{k-4}\ \sum_{s=0}^{[r/2]} \binom{k-r-m+s-2}{m-s-2}\binom{r-s}{s}.$$

We set here $\binom{a}{b} = 0$ unless $0 \leqslant b \leqslant a$; it is easy to see that this convention allows us to extend the summation over both r, s arbitrarily. In the same way, we obtain

$$c^m_{k-1}(\beta_0) = \sum_{r=0}^{m-1}\ \sum_{s=0}^{[r/2]} \binom{k-r-m+s-2}{m-s-1}\binom{r-s}{s} = c^{m+1}_k(\beta_1).$$

In order the evaluate this expression, we change the summation indices to $p = r - s$ and s. The limitations imposed on s imply $0 \leqslant s \leqslant p \leqslant s+k-2m-1 \leqslant k-m-2$ and $s \leqslant m-1$, so

$$c^m_{k-1}(\beta_0) = \sum_{s=0}^{m-1}\ \sum_{p=s}^{s+k-2m-1} \binom{k-m-2-p}{m-1-s}\binom{p}{s} = \sum_{s=0}^{m-1}\binom{k-m-1}{m}. \ \blacksquare$$

Proof of Theorem 4.4, *continued*. The proved lemma yields the following relations

$$|d_k^t(\varphi,\psi;m)| \leqslant 2^{1/2}\,\|\varphi\|\,\eta^{m-1}\left\{\binom{k-m-1}{m-1}\alpha(E,\psi)+\right.$$

$$+2^{1/2}\,\|\psi\|\left[(m-1)\binom{k-m-1}{m-1}\beta_1(E)+\right.$$

$$\left.\left.+m\binom{k-m-1}{m}\beta_0(E)\right]\right\}. \tag{4.27}$$

Thus one has to estimate the expressions

$$e_k(\xi,\eta)=\sum_{m=1}^{[k/2]}\eta^{m-1}c_{k-1}^m(\xi),\qquad \xi=\alpha,\beta_0,\beta_1. \tag{4.28}$$

Consider first $e_k(\xi)=e_k(\xi,1)$. The relations

$$e_k(\alpha)=e_{k-1}(\alpha)+e_{k-2}(\alpha),$$

$$e_k(\beta_1)=e_{k-2}(\alpha)+e_{k-2}(\beta_1)+e_{k-1}(\beta_1)$$

follow easily from Lemma 4.7; they imply by induction $e_k(\alpha)\leqslant 2^{k-3}$ and $e_k(\beta_1)=e_{k-1}(\beta_0)\leqslant 2^{k-3}$. Further, we have $e_k(\xi,\eta)\leqslant\lambda^{[k/2]-1}e_k(\xi)$ so

$$e_k(\alpha,\eta)\leqslant\lambda^{k/2-1}2^{k-3},\qquad e_k(\beta_j,\eta)\leqslant\lambda^{k/2-1}2^{k-2-j},\quad j=0,1. \tag{4.29}$$

The relations (4.25)–(4.29) together imply

$$|(\varphi,[S_{00}^t-S_{\rm opt}^t(E)]\psi)|\leqslant\sum_{k=2}^{\infty}|g|^k\sum_{m=1}^{[k/2]}|d_k^t(\varphi,\psi;m)|$$

$$\leqslant 2^{1/2}\,\|\varphi\|\sum_{k=2}^{\infty}|g|^k\left\{e_k(\alpha,\eta)\alpha(E,\eta)+\right.$$

$$\left.+2^{1/2}\,\|\psi\|\sum_{j=0,1}e_k(\beta_j,\eta)\beta_j(E)\right.$$

$$\leqslant g^2\,\|\varphi\|\sum_{n=0}^{\infty}(2\,|g|\,\lambda^{1/2})^n\times$$

$$\times\{2^{-1/2}\alpha(E,\psi)+(2\beta_0(E)+\beta_1(E))\}$$

for all $\varphi,\psi\in\mathscr{H}_0$; summing the geometric series, we arrive at the relation (4.16a). Finally, $(4\lambda)^{-1/2}\leqslant((1+4\eta)^{1/2}-1)/2\eta$ so the condition (4.24) is fulfilled, and therefore the proof is finished. ∎

Let us turn to a discussion of the result obtained. The inequality (4.16a) can be used to estimate the difference between S_{00} and $S_{\mathrm{opt}}(E)$, once their existence is established. We are naturally interested in the conditions under which this estimate might be expected to be good. To this end, consider the following one-parameter family of functions

$$z_{11}^{\omega} : z_{11}^{\omega}(t) = \omega^{-1} z_{11}(\omega^{-1} t), \quad \omega > 0, \tag{4.30}$$

whose members have the same normalization, $\|z_{11}^{\omega}\|_1 = \eta$ for all $\omega > 0$. We shall inspect how the r.h.s. of (4.16a) behaves for small ω. First we adopt a slightly stronger hypothesis about the operator-valued functions (4.15). Each of them is strongly continuous in view of (s); further, (n) shows that the functions $t \mapsto \|x_{1j}(t - \omega s) - x_{1j}(t)\|_u$, $j = 0, 1$, belong to $L(\mathbb{R})$ for each $s \in \mathbb{R}$, $\omega > 0$. We shall assume that

(u) *the functions* $x_{1j}(\,.\,)$, $j = 0, 1$, *are continuous with respect to* $\|\,.\,\|_u$, *and* $t \mapsto \|x_{1j}(t - y) - x_{1j}(t)\|_u$ *is majorized by an integrable function independently of* y.

Suppose now that the Hamiltonian depends on the parameter ω in such a way that the z_{11}-function is given by (4.30), while the remaining functions (4.15) do not change; this can be achieved, for example, with

$$H = H_0 + gV_{00} + g\omega^{-1/2}(V_{01} + V_{10}) + \omega^{-1} H_1 .$$

The corresponding operators (4.6b) and (4.8b) we denote by $S_{00}^{t,\omega}$ and $S_{\mathrm{opt}}^{t,\omega}(E)$, respectively.

THEOREM 4.4.8. *Let the hypotheses of Theorem* 4.4 *together with* (u) *be fulfilled, then*

$$\lim_{\omega \to 0} \|S_{00}^{t,\omega}\psi - S_{\mathrm{opt}}^{t,\omega}(E)\psi\| = 0. \tag{4.31}$$

Proof. In view of (4.16), we have to prove the relations

$$\lim_{\omega \to 0} \|[\tilde{Z}_{11}^{\omega} - \hat{Z}_{11}^{\omega}(E)]\,\psi_C^t\|_2 = 0, \tag{4.32a}$$

$$\lim_{\omega \to 0} \|[\tilde{Z}_{11}^{\omega} - \hat{Z}_{11}^{\omega}(E)]\,\tilde{X}_{1j}\|_{uf} = 0, \quad j = 0, 1. \tag{4.32b}$$

First we notice that

$$\begin{aligned}\|\hat{Z}_{11}^{\omega}(E) - \hat{Z}_{11}^{\omega}(0)\|_{uf} &= \|\hat{z}_{11}^{\omega}(E) - \hat{z}_{11}^{\omega}(0)\|_u \\ &= \left\| \int_{\mathbb{R}} z_{11}(s)\,(e^{i\omega Es} - 1)\,ds \right\|_u \leqslant \\ &\leqslant \int_{\mathbb{R}} \|z_{11}(s)\|_u\,|e^{i\omega Es} - 1|\,ds\end{aligned}$$

and $\|z_{11}\|_1 < \infty$ so the dominated-convergence theorem yields

$$\lim_{\omega \to 0} \|\hat{Z}_{11}^{\omega}(E) - \hat{Z}_{11}^{\omega}(0)\|_{\mathrm{uf}} = 0. \tag{4.33}$$

Hence it is sufficient to check the relations (4.32) for $E = 0$ only. Mimicking the proof of Lemma 4.6, we obtain the estimates

$$\|[\tilde{Z}_{11}^{\omega} - \hat{Z}_{11}^{\omega}(0)]\,\psi_C^r\|_2$$

$$\leqslant \int_{\mathbb{R}} \|z_{11}(s)\|_{\mathrm{u}} \left(\int_{\mathbb{R}} \|\psi_C^r(t - \omega s) - \psi_C^r(t)\|^2 \,\mathrm{d}t \right)^{1/2} \mathrm{d}s, \tag{4.34a}$$

$$\|[\tilde{Z}_{11}^{\omega} - \hat{Z}_{11}^{\omega}(0)]\,\tilde{X}_{1j}\|_{\mathrm{uf}}$$

$$\leqslant \int_{\mathbb{R}} \|z_{11}(s)\|_{\mathrm{u}} \int_{\mathbb{R}} \|x_{1j}(t - \omega s) - x_{1j}(t)\|_{\mathrm{u}} \,\mathrm{d}u\,\mathrm{d}s. \tag{4.34b}$$

The limit $\omega \to 0$ is performed through repeated use of the dominated-convergence theorem. One can justify it easily for the 'outer' integrals. As for the 'inner' ones, the assumption (u) applies to (4.34b). In the case of (4.34a), we use the fact that $\psi_C^r(\,.\,)$ is bounded (by $\|C_0\|_{\mathrm{u}}\,\|\psi\|$) and of a bounded support, so an integrable majorizing function is easily constructed for any fixed s. Finally,

$$\|\psi_C^r(t - \omega s) - \psi_C^r(t)\| \leqslant \|C_0\|_{\mathrm{u}}\,\|e^{i\omega H_0 s}\,\psi - \psi\|$$

and $t \mapsto \exp(iH_0 t)$ is strongly continuous. ■

What is the physical meaning of this result? The obtained assertion shows that the optical potential (4.12a) should yield a good approximation, when $z_{11}(\,.\,)$ was sharply peaked around $t = 0$. In the above used expansion of S_{00}^t, this function is responsible for the interaction with the second channel; in its lowest order, this is clearly seen from (4.10a). It is common practice in perturbative methods of quantum physics to understand the process under consideration as a (coherent) sum of partial processes related to the single terms of the perturbative expansion. In our example of neutron scattering on nuclei, the processes connected with the function $z_{11}(\,.\,)$ can be interpreted as *virtual excitations* of the target nucleus. Theorem 4.8 asserts that the optical approximation is plausible if these processes are *sufficiently rapid*. Moreover, in that case the true excitations (i.e. the process of formation of a compound nucleus) are likely to be rapid too; it is suggested heuristically by the expansion of S_{10}^t analogous to the one performed above. We can even make some roughly quantitative conclusions. The natural time scale of the scattering process is associated with the duration of interaction between the nucleus and the directly scattered neutron. Its order of magnitude may be characterized from the behaviour of functions containing $\exp(-iH_0 t)$. Now the proof of Theorem 4.8 shows that the functions $\psi_C^t(\,.\,)$ and $x_{1j}(\,.\,)$ must be varying slowly when compared with the time interval peculiar for $z_{11}(\,.\,)$. Thus one can say the applicability of the optical approximation demands the partial processes which involve (virtual) transitions to the compound nucleus to be very rapid in the natural time scale described above.

There is another problem here which is related to the spectrum of H_1. Theorem 4.4 can be applied if the latter is continuous, because otherwise $\|z_{11}\|_1$ need not be finite as the condition (n) requires. In practice, however, one often meets the situation when the part of $\sigma(H_1)$ related to the considered region of energy is purely discrete. Then the above-described theory can be used only after the replacement of H_1 by an operator H_1' which is near in some sense to H_1 but has a continuous spectrum. There are various ways of choosing H_1', one of which is to introduce a more realistic model of the compound nucleus, which would take into account decay modes other than release of the captured neutron; in particular, the possibility of γ-decay. As a result of the interaction with the electromagnetic field, the energy levels of the compound nucleus become diffuse, just as the atomic levels in the problem of radiation damping (cf. Section 3.1).

On the other hand, H_1' may be chosen as some 'smoothing' approximation to H_1. From the physical point of view, such a procedure can be defended on the grounds of the following argument. Variation of the total elastic cross section for the neutron–nucleus scattering is highly irregular; however, these irregularities disappear if the energy resolution is sufficiently poor. In fact, in the early experiments, which inspired formulation of the optical model, smooth curves were seen; only later was it recognized that they represented the smoothing of numerous narrow peaks and gaps. Let us construct a simple mathematical model appropriate to this situation. We choose a number $\alpha \in (0, 1]$ and consider the linear manifold $\mathscr{S}_\alpha \subset \mathscr{H}_0$, which consists of all $\varphi \in \mathscr{H}_0$ such that

$$|\varphi|_\alpha = \int_{\mathbb{R}} \|V\,\mathrm{e}^{-iH_0 t}\varphi\|\,(1 + |t|^\alpha)\,\mathrm{d}t \tag{4.35}$$

is finite. It is easy to see that the integrated function in (4.35) is continuous provided that V is bounded, and that $|\,.\,|_\alpha$ is a seminorm on $\mathscr{S}_\alpha$. Let us denote $S_t = \exp(iH_0 t)\exp(-2iHt)\exp(iH_0 t)$ and S_t' the corresponding operator obtained by replacement of H_1 by H_1'; then the following assertion holds:

PROPOSITION 4.4.9. *Assume* (s), (b), (f) *and* $\|H_1 - H_1'\|_u < \epsilon$, *then for each* $\varphi, \psi \in \mathscr{S}_\alpha$ *and all* $t \geqslant 0$ *we have*

$$|(\varphi, S_t\psi) - (\varphi, S_t'\psi)| \leqslant 2g^2\epsilon^\alpha\,|\varphi|_\alpha\,|\psi|_\alpha. \tag{4.36}$$

Proof. Since $H_0 + H_1$ is self-adjoint and V is bounded, we can express e^{-2iHt} applying twice the formula (4.9). Hence, we obtain

$$(\varphi, S_t\psi) = (\varphi, \psi) - ig\int_{-t}^{t} (V\,\mathrm{e}^{-iH_0 s}\varphi, \mathrm{e}^{-iH_0 s}\psi)\,\mathrm{d}s -$$

$$- g^2\int_{-t}^{t}\mathrm{d}s\int_{-t}^{s}\mathrm{d}u\,(V\,\mathrm{e}^{-iH_0 s}\varphi, \mathrm{e}^{-iH(s-u)}\,V\,\mathrm{e}^{-iH_0 u}\psi)$$

and the analogous expression for $(\varphi, S'_t\psi)$. The second term on the r.h.s. does not depend on H_1 so it cancels in the subtraction. Consequently,

$$|(\varphi, S_t\psi) - (\varphi, S'_t\psi)| = g^2 \left| \int_{-t}^{t} \mathrm{d}s \int_{-t}^{s} \mathrm{d}u \, (V \mathrm{e}^{-iH_0 s} \varphi, [\mathrm{e}^{-iH(s-u)} - \right.$$

$$\left. - \mathrm{e}^{-iH'(s-u)}] \, V \mathrm{e}^{-iH_0 u} \psi) \right|$$

$$\leqslant g^2 \int_{-t}^{t} \mathrm{d}s \int_{-t}^{s} \mathrm{d}u \, \|\mathrm{e}^{-iH(s-u)} -$$

$$- \mathrm{e}^{-iH'(s-u)}\|_{\mathrm{u}} \, \|V \mathrm{e}^{-iH_0 s} \varphi\| \, \|V \mathrm{e}^{-iH_0 u} \psi\|$$

Formula (4.9) yields the estimate $\|\mathrm{e}^{-iHx} - \mathrm{e}^{-iH'x}\|_{\mathrm{u}} \leqslant \|H - H'\|_{\mathrm{u}} \, |x|$. At the same time, this norm is $\leqslant 2$ for all x so we get

$$\|\mathrm{e}^{-iH(s-u)} - \mathrm{e}^{-iH'(s-u)}\|_{\mathrm{u}} \leqslant 2^{1-\alpha} \epsilon^{\alpha} \, |s-u|^{\alpha} \leqslant 2\epsilon^{\alpha} (|s|^{\alpha} + |u|^{\alpha})$$

$$\leqslant 2\epsilon^{\alpha} (1 + |s|^{\alpha}) \, (1 + |u|^{\alpha}).$$

Combining the last two inequalities with (4.35), we arrive at (4.36). ■

Let us now see which lesson can be drawn from this result. We shall not trouble ourselves with the convergence of the integral in (4.35), because it is likely to be of no physical importance (compare this with the problem discussed in Section 1.6). It is more interesting to see whether the value of $|\varphi|_{\alpha}$ may be very large for some φ, thus devaluing the estimate (4.36). According to the assumption (f), the spectrum of H_0 is absolutely continuous. However, if φ has a small enough energy spread – i.e. if the vector-valued measure $E_{H_0}(\,.\,)$ is concentrated essentially in some narrow interval – then the function $t \mapsto \exp(-iH_0 t)\varphi$ may be expected to vary slowly. Of course, the behaviour of the integrated function in (4.35) depends substantially on the operator V. In the physically interesting cases (as that considered in Example 4.1) $t \mapsto \|V \mathrm{e}^{-iH_0 t} \varphi\|$ can be expected to be slowly varying again, so $|\varphi|_{\alpha}$ eventually may be very large. In order to prevent this effect, *the chosen energy resolution should not, therefore, be too good*. Let us stress, however, that the value of the presented consideration is rather heuristic. The nuclear Hamiltonians mostly have finite discrete spectra so one can hardly achieve a reasonable *operator-norm* approximation of H_1 by an operator H'_1 with a purely continuous spectrum.

The problem can be also viewed from another angle. The isolated eigenvalues of H_1 give rise to embedded eigenvalues of the full 'free' Hamiltonian $H_0 + H_1$. According to the theory developed in Sections 3.2 and 3.3, the perturbation gV turns these eigenvalues into resonances. Of course, they are the 'resolvent' resonances which are not *a priori* identical with the scattering resonances, but we have mentioned that the presence of the latter is confirmed experimentally. The desirable smoothing is such that will suppress the detailed resonance structure; one usually wants an optical potential which would reproduce the measured cross-section averaged over suitable energy intervals. It shows that *the energy spread of the incident beam should be much larger than the spacing of the reasonances* (or of the energy levels of H_1, which has the same order of magnitude).

Hence, the application of the described optical approximation is conditioned by various requirements, which eventually may contradict each other. It concerns particularly the energy resolution which should be good enough to allow use of the optical potential (4.12a), and at the same time poor enough to admit a reasonable smoothing of the compound-nucleus energy spectrum. A more careful analysis of these requirements is needed for each concrete optical model of scattering. The only general conclusion which can be made is the following: the optical approximation needs *the energy levels of the compound nucleus to be numerous and close together*, so it presumably works better for scattering on heavy nuclei.

Until now we have discussed the optical approximation based on the potential (4.12a). Unfortunately, the latter is not very suitable for practical calculations, in which the state spaces involved are usually of the form $L^2(\mathbb{R}^d; \mathbb{C}^n)$, or a subspace of it. One would invite the optical potential to be *local*, i.e. to act as the operator of multiplication by some (matrix) function. In particular, if we neglect spins and other quantum numbers which eventually give rise to multicomponent wave functions, as well as the effects caused by the identity of particles, etc., the operators H_{opt} of this type are just the Schrödinger pseudo-Hamiltonians treated in the previous section. Notice also that from the practitioner's point of view, the solution of the problem reduces essentially, in this case, to the solution of a (system of) partial differential equation(s), so one has elaborate computer machinery at one's disposal. The local optical potentials apply successfully not only to the elastic neutron scattering, which we have chosen as a motivating example here, but also to many other nuclear reactions. Moreover, it is known that reasonable agreement with the experimental data is frequently achieved with the help of the optical potentials which are complex linear combinations of the spherically symmetric functions

$$V(r) = V_0\,[1 + \exp((r - r_0)/a_0)]^{-1}$$

(Saxon–Woods potentials), their derivatives (surface terms) and other similarly nice functions which have few adjustable parameters. It should be noted, however, that such optical potentials often suffer from various ambiguities.

These facts show that it is desirable to extend the results formulated in the present section to cover other optical potentials, primarily the local ones. The problem could be approached either directly, or by combining Theorem 4.4 with a suitable approximation of S_{opt}. In any case, more detailed information about the operator H_1 would be needed. Generally speaking, the task of finding a satisfactory rigorous justification of the optical approximation still lies ahead.

4.5. Non-unitary Scattering Theory

In order to make use of the scattering models discussed in the previous section, one must be able to establish the existence of the S-matrix. This suggests the idea of generalizing the scattering theory to the situations when the 'interacting' dynamics of the scattering system under consideration are described by a pseudo-Hamiltonian H. In the present section we are going to prove a few existence results and to quote some others. We shall not attempt to give a complete exposition with the weakest possible assumptions, mainly because the subject is, at present, a far from fully developed theory.

Assume the 'free' dynamics to be governed by a Hamiltonian H_0 on the state Hilbert space $\mathscr{H}$ of the problem. We adopt the following assumptions:

(a) *It holds that $H = H_0 + U$, where H_0 is self-adjoint and semibounded, and $U = V - iW$. Both the operators V, W are assumed to be self-adjoint, $W \geqslant 0$, and relatively compact with respect to H_0.*

(d) *The spectrum of H_0 is absolutely continuous. There is a dense subset $D \subset D(H_0)$ such that the function $t \mapsto \|U \exp(iH_0 t)\varphi\|$ belongs to $L(\mathbb{R}_+)$ for each $\varphi \in D$.*

(s) *The singularly continuous spectrum of the self-adjoint part $H_1 = H_0 + V$ of H is empty.*

The relative compactness of V in (a) means that $D(V) \supset D(H_0)$ and for some (and consequently, for all) $z \in \rho(H_0)$ the operator $V(H_0 - z)^{-1}$ is compact (Reed and Simon, 1978, §XIII.4), and similarly for W. Another equivalent expression of the relative compactness is: V is compact as the mapping from $\langle D(H_0), \|\,.\,\|_{H_0}\rangle$ to $\langle \mathscr{H}, \|\,.\,\|\rangle$, where the norm $\|\,.\,\|_{H_0}$ is defined by $\|\varphi\|_{H_0}^2 = \|H_0\varphi\|^2 + \|\varphi\|^2$. It is not difficult to see that the relative compactness implies the relative bound of U with respect to H_0 to be zero (Problem 38). Combining this fact with Theorem 2.8, we see that the operator H specified by (a) is maximal dissipative.

As a useful preliminary, let us discuss some important subspaces in $\mathscr{H}$ (if they will not be specified otherwise, they always refer to the operator H). By a *bound state*, we understand conventionally any eigenvector φ of H corresponding to a *real* eigenvalue. The closed linear span of all bound states will be denoted by $\mathscr{H}_{\text{bound}}$. Since H is non-self-adjoint, it may also have non-real eigenvalues in the lower complex halfplane. To each of them, there corresponds a subspace spanned by the respective (generalized) eigenvectors – cf. (3.35). Notice that $\sigma_{\text{ess}}(H)$ is, in view of (a), confined to $\mathbb{R}$ so each of these subspaces is finite-dimensional. We denote by $\mathscr{H}_{\text{nre}}$ the closed linear span of the subspaces $P_H(\{\lambda\})$, where λ runs over all non-real eigenvalues of H. Another subspace of interest consists of the *decaying states*, $\mathscr{H}_{\text{dec}} = \{\varphi : \lim_{t\to\infty} \|e^{-iHt}\varphi\| = 0\}$.

In the standard scattering theory, an important role is played by the *absolutely continuous* subspaces – cf. (3.2.14), (3.2.15). Unfortunately, this concept relies on the existence of spectral decomposition, so it cannot be directly adapted for the non-self-adjoint operator H. Instead, we introduce $\mathscr{H}_{\text{ac}}(H)$ as the closure of the following linear manifold in $\mathscr{H}$:

$$\mathscr{M}(H) = \{\varphi : \text{there is } c_\varphi \text{ such that}$$

$$\int_0^\infty |(\psi, e^{-iHt}\varphi)|^2\, dt \leqslant c_\varphi \|\psi\|^2 \text{ for all } \psi \in \mathscr{H}\}. \tag{5.1}$$

The definition is plausible, because for a self-adjoint operator H_0, the set $\mathscr{M}(H_0)$ is dense in $\mathscr{H}_{ac}(H_0)$ determined in the standard way (Problem 39). In particular, if (d) is valid, then $\mathscr{H}_{ac}(H_0) = \mathscr{H}$ irrespective of the definition used.

For the sake of simplicity, we denote $V_t = \exp(-iHt)$ and $V_t^0 = \exp(-iH_0 t)$. Let us collect some simple properties of the above-listed subspaces:

PROPOSITION 4.5.1. *Assume* (a). *Each bound state* φ *of* H *referring to an eigenvalue* $\lambda \in \mathbb{R}$ *belongs to* Ker W *and* $H_1\varphi = \lambda\varphi$. *The subspaces* $\mathscr{H}_{nre}$, $\mathscr{H}_{dec}$, $\mathscr{H}_{ac}$, $\mathscr{H}_{bound}$ *and its complement* $\mathscr{H}_{bound}^{\perp}$ *are all invariant under* $\{V_t : t \geqslant 0\}$. *Moreover, the inclusions*

$$\mathscr{H}_{nre} \subset \mathscr{H}_{dec} \subset \mathscr{H}_{bound}^{\perp}, \tag{5.2a}$$

$$\mathscr{H}_{nre} \subset \mathscr{H}_{ac} \subset \mathscr{H}_{bound}^{\perp} \tag{5.2b}$$

hold.

Proof. A bound state φ fulfils $(\varphi, H\varphi) = (\varphi, H_1\varphi) - i(\varphi, W\varphi) = \lambda\|\varphi\|^2$ so $(\varphi, W\varphi) = 0$. Since $W \geqslant 0$, it yields $W\varphi = 0$. In fact, this is just a particular case of Theorem 2.10: the operators H, H^* and H_1 coincide on $\mathscr{H}_{bound} \subset \mathscr{H}_0$, because the part of H in $\mathscr{H}_0$ is self-adjoint. The invariance check for $\mathscr{H}_{bound}$ is therefore easy, and the same is true for $\mathscr{H}_{nre}$, $\mathscr{H}_{dec}$ and $\mathscr{H}_{ac}$. Consider, further, $\varphi \in \mathscr{H}_{bound}^{\perp}$ and an arbitrary bound state ψ, $H\psi = \lambda\psi$. We shall employ the fact that $-iH^*$ generates the continuous contractive semigroup $\{V_t^* : t \geqslant 0\}$ (Davies, 1980a, theorem 1.34). It holds

$$(V_t\varphi, \psi) = (\varphi, V_t^*\psi) = e^{i\lambda t}(\varphi, \psi) = 0$$

so $\mathscr{H}_{bound}^{\perp}$ is also invariant. According to the assumption, each subspace $P_H(\{\lambda\})\mathscr{H}$ related to a non-real λ is invariant and finite-dimensional. The corresponding functions $t \mapsto \|V_t\varphi\|^2$ are then easily seen to have an exponentially decreasing bound (cf. Problem 21), which yields the first inclusions in (5.2a, b). Let, further, ψ be a bound state, $H\psi = \lambda\psi$, and $\varphi \in \mathscr{H}_{dec}$, then

$$0 = \lim_{t\to\infty} (\psi, V_t\varphi) = \lim_{t\to\infty} (V_t^*\psi, \varphi) = \lim_{t\to\infty} e^{-i\lambda t}(\psi, \varphi)$$

so $(\psi, \varphi) = 0$, and therefore $\mathscr{H}_{dec} \perp \mathscr{H}_{bound}$. Similarly, let ψ be a bound state and $\varphi \in \mathscr{M}(H)$, then

$$\int_0^\infty |(\psi, V_t\varphi)|^2 \, dt = \int_0^\infty |e^{-i\lambda t}(\psi, \varphi)|^2 \, dt \leqslant c_\varphi \|\psi\|^2$$

gives $(\psi, \varphi) = 0$, i.e. $\mathscr{M}(H) \subset \mathscr{H}_{bound}^{\perp}$ and (5.2b) holds. ∎

One is naturally interested whether equality may hold in some of the above inclusions. This is true for a finite-dimensional $\mathscr{H}$, where all the subspaces appearing in (5.2) coincide (Problem 40). In the general case, however, the problem is not easy and the full answer is not known. We shall formulate below one sufficient condition under which the equality

$$\mathscr{H}_{ac}(H) = \mathscr{H}_{bound}(H)^{\perp} \tag{5.3}$$

holds. First we prove the following auxiliary assertion:

LEMMA 4.5.2. *For a completely dissipative operator H on $\mathscr{H}$, we have $\mathscr{H}_{ac}(H) = \mathscr{H}$.*

Proof. Due to the assumption, the semigroup $\{V_t : t \geq 0\}$ is completely non-unitary. First we shall treat the particular case, when it is a semigroup of isometries. In view of (A.15), the space $\mathscr{H}$ may then be identified (up to an isomorphism) with $L^2(\mathbb{R}_+; \mathscr{G})$, where $\mathscr{G}$ is some auxiliary Hilbert space and $(V_t\varphi)(x) = \varphi(x - t)\Theta(x - t)$ for all $t \geq 0$. Let F denote the Fourier–Plancherel operator on $L^2(\mathbb{R}; \mathscr{G})$: if $\|\varphi(\,.\,)\|_{\mathscr{G}}$ decays rapidly enough at infinity, it acts as

$$(F\varphi)(k) = (2\pi)^{-1} \int_{\mathbb{R}} e^{-ikx}\, \varphi(x)\, dx.$$

One easily finds that

$$(FV_t\varphi)(k) = e^{-ikt}(F\varphi)(k), \qquad t \geq 0. \tag{$*$}$$

Consider now arbitary $\psi \in \mathscr{H}$ and $\varphi \in C_0^\infty(\mathbb{R}_+; \mathscr{G})$. Unitarity of F together with $(*)$ give

$$(\psi, V_t\varphi) = \int_0^\infty e^{-ikt}((F\psi)(k), (F\varphi)(k))_{\mathscr{G}}\, dk.$$

Using further the Parseval relation and Schwartz inequality, we get

$$\int_0^\infty |(\psi, V_t\varphi)|^2\, dt \leq \int_{\mathbb{R}} \|(F\psi)(k)\|_{\mathscr{G}}^2\, \|(F\varphi)(k)\|_{\mathscr{G}}^2\, dk.$$

However, φ is assumed to be a smooth function with a compact support, so

$$\|(F\varphi)(k)\|_{\mathscr{G}} \leq (2\pi)^{-1/2} \int_{\mathbb{R}} \|\varphi(x)\|_{\mathscr{G}}\, dx \equiv c_\varphi^{1/2} < \infty.$$

Hence $\mathscr{M}(H)$ contains the set $C_0^\infty(\mathbb{R}_+; \mathscr{G})$ which is dense in $L^2(\mathbb{R}_+; \mathscr{G})$, and the assertion holds.

In the general case, let $\{\mathscr{K}, U_t, P\}$ denote the minimal unitary dilation of $\{V_t\}$. The subspace $\mathscr{L} = \overline{\{U_t\mathscr{H} : t \geq 0\}_{\mathrm{lin}}}$ in $\mathscr{K}$ fulfils $\mathscr{H} \subset \mathscr{L}$ and $U_t\mathscr{L} \subset \mathscr{L}$ for all $t \leq 0$. Thus for $\varphi, \psi \in \mathscr{H}$ and $s, t \geq 0$, we have

$$\begin{aligned}(U_{-s}\psi, (I - P)U_t\varphi) &= (\psi, PU_s(I - P)U_tP\varphi)\\ &= (\psi, V_{s+t}\varphi) - (\psi, V_sV_t\varphi) = 0,\end{aligned}$$

i.e. $(I - P)U_t\varphi \in \mathscr{L}^\perp$. The subspace $\mathscr{H} = \mathscr{M} \oplus \mathscr{L}^\perp$ then fulfils $U_t\mathscr{M} \subset \mathscr{M}$ for each $t \geqslant 0$. This means that $\{U_t\}$ acts on $\mathscr{M}$ as a one-parameter semigroup of isometries; we may write $\mathrm{pr}_{\mathscr{M}}\, U_t = e^{-iBt}$ with some maximal dissipative operator B. According to Theorem 2.10, this operator decomposes into an orthogonal sum of two operators, one of which is self-adjoint and the other completely dissipative. To the latter we can apply the result of the first part of the proof. On the other hand, the unitary part of $\{e^{-iBt}\}$ represents a restriction of the semigroup $\{U_t : t \geqslant 0\}$ whose generator has an absolutely continuous spectrum in view of the complete non-unitarity of $\{V_t\}$ and (A.16). Combining these facts with the result of Problem 39, we see that $\mathscr{M}(B)$ is dense in $\mathscr{M}$. Finally, consider arbitrary $\varphi \in \mathscr{M}(B)$ and $\psi \in \mathscr{H}$. It holds that

$$\int_0^\infty |(\psi, V_t\varphi)|^2\, dt = \int_0^\infty |(\psi, PU_tP\varphi)|^2\, dt$$

$$= \int_0^\infty |(\psi, e^{-iBt}\, P\varphi)|^2\, dt \leqslant c_{P\varphi}\, \|\psi\|^2$$

so $P\varphi \in \mathscr{M}(H)$. Since $\mathscr{M}(B)$ is dense in $\mathscr{H}$ and P is the projection to $\mathscr{H}$, $\mathscr{M}(H)$ is also dense and the result follows. ∎

THEOREM 4.5.3 (Davies). *Under the assumptions* (a) *and* (s), *the relation* (5.3) *holds.*

Proof. According to Theorem 2.10, there are orthogonal decompositions $\mathscr{H} = \mathscr{H}_u \oplus \mathscr{H}_{cnu}$ and $H = H_u \oplus H_{cnu}$, with H_u self-adjoint and H_{cnu} completely dissipative. For $\varphi \in \mathscr{H}_u$, we have $(\varphi, H\varphi) = (\varphi, H_1\varphi) - i(\varphi, W\varphi)$. The operator H_1 is self-adjoint and $W \geqslant 0$ so $W\varphi = 0$, i.e. $\mathrm{pr}_u H = H_u = \mathrm{pr}_u H_1$. Further, we can decompose $\mathscr{H}_u$ with respect to H_1 as $\mathscr{H}_u = \mathscr{H}_{u,\mathrm{bound}} \oplus \mathscr{H}_{u,\mathrm{ac}}$, because the singularly continuous spectrum of H_1 is void due to (s). Here $\mathscr{H}_{u,\mathrm{bound}}$ is the closed linear span of the bound states of H_1 contained in $\mathscr{H}_u$ (which are at the same time bound states of H). However, H has no bound states in $\mathscr{H}_{cnu}$, otherwise H_{cnu} would not be completely dissipative. Hence $\mathscr{H}_{u,\mathrm{bound}} = \mathscr{H}_{\mathrm{bound}}(H)$. Further, $\mathscr{H}_{u,\mathrm{ac}} \subset \mathscr{H}_{\mathrm{ac}}(H)$ and the above lemma yields $\mathscr{H}_{cnu} \subset \mathscr{H}_{\mathrm{ac}}(H)$ so

$$\mathscr{H}_{\mathrm{ac}}(H) \supset \mathscr{H}_{u,\mathrm{ac}} \oplus \mathscr{H}_{cnu} = \mathscr{H}_{\mathrm{bound}}(H)^\perp.$$

Since the opposite inclusion follows from (5.2b), the proof is finished. ∎

After these preliminaries, we now turn to the definition of the *wave operators*. Recall that we have denoted $V_t = \exp(-iHt)$ and $V_t^0 = \exp(-iH_0t)$. The operator Ω_- is defined in the standard way as

$$\Omega\ (H, H_0) = \operatorname*{s-lim}_{t\to\infty} V_t V^0{}_t, \tag{5.4a}$$

where we have taken the assumption (d) into account; otherwise we should add the projection on $\mathscr{H}_{\mathrm{ac}}(H_0)$ to the r.h.s. of (5.4a) as in (3.2.14). Instead of Ω_+, we define rather directly the operator which is needed for the expression of the S-matrix. It will nevertheless be denoted by Ω_+^* to keep the correspondence with the unitary case; we set

$$\Omega_+^*(H, H_0)\varphi = \lim_{t\to\infty} V_{-t}^0 V_t \varphi \tag{5.4b}$$

for all φ for which the limit makes sense.

The assumption (d) allows us to handle the existence problem for Ω_- by means of the well-known Cook argument:

THEOREM 4.5.4. *Assume* (a) *and* (d), *then the operator* $\Omega_-(H, H_0)$ *exists. It is a contraction, which maps* $\mathscr{H}$ *into* $\mathscr{H}_{ac}(H)$ *and fulfils* $V_t\Omega_- = \Omega_- V_t^0$ *for all* $t \geqslant 0$. *If* U *is bounded, then the operator* Ω_- *is injective.*

Proof. We take $\varphi \in D$ and define $f(t) = V_t V_{-t}^0 \varphi$. Since φ belongs to the set $D(H) = D(H_0)$, which is invariant under V_{-t}^0, the function f is differentiable and $f'(t) = -iV_t U V_{-t}^0 \varphi$. Thus for any $0 \leqslant s < t$ we have

$$\|f(t) - f(s)\| = \left\| \int_s^t f'(x)\,dx \right\| \leqslant \int_s^t \|f'(x)\|\,dx$$
$$\leqslant \int_s^t \|UV_{-x}^0\varphi\|\,dx$$

so the net $\{f(t) : t \geqslant 0\}$ is Cauchy and $\lim_{t\to\infty} f(t) \equiv \Omega_-\varphi$ exists for all $\varphi \in D$. Further, we employ the $\frac{1}{3}\epsilon$-trick: D is dense in $\mathscr{H}$ so to every $\varphi \in \mathscr{H}$ and $\epsilon > 0$, there is $\varphi_\epsilon \in D$ such that $\|\varphi - \varphi_\epsilon\| < \frac{1}{3}\epsilon$. To this φ_ϵ, we can find t_ϵ such that $\|f_\epsilon(t) - f_\epsilon(s)\| < \frac{1}{3}\epsilon$ for all $t, s \geqslant t_\epsilon$, where $f_\epsilon(t) = V_t V_{-t}^0 \varphi_\epsilon$. Since $\|f(t) - f_\epsilon(t)\| \leqslant \|\varphi - \varphi_\epsilon\| < \frac{1}{3}\epsilon$, one finds easily $\|f(t) - f(s)\| < \epsilon$ for all $t, s \geqslant t_\epsilon$, hence the existence of Ω_- is established.

Contractivity of Ω_- is evident. The relation $\Omega_-\varphi = \lim_{s\to\infty} V_{t+s}V_{-t-s}^0\varphi = V_t\Omega_- V_{-t}^0\varphi$, which is valid for all $\varphi \in \mathscr{H}$ and $t \geqslant 0$, implies the intertwining property of Ω_-. Consider now arbitrary $\varphi \in \mathscr{M}(H_0)$ and $\psi \in \mathscr{H}$. We have

$$\int_0^\infty |(\psi, V_t\Omega_-\varphi)|^2\,dt = \int_0^\infty |(\psi, \Omega_- V_t^0\varphi)|^2\,dt$$
$$= \int_0^\infty |(\Omega_-^*\psi, V_t^0\varphi)|^2\,dt \leqslant c_\varphi \|\Omega_-^*\psi\|^2.$$

Since Ω_-^* is also contractive, $\Omega_-\varphi \in \mathscr{M}(H)$. Thus we have $\Omega_-\mathscr{M}(H_0) \subset \mathscr{M}(H)$, but Ω_- is bounded and $\overline{\mathscr{M}(H_0)} = \mathscr{H}_{ac}(H_0) = \mathscr{H}$ in view of (d) and Problem 39, so $\operatorname{Ran}\Omega_- \subset \overline{\mathscr{M}(H)} = \mathscr{H}_{ac}(H)$.

Finally, let us check the injectivity of Ω_- for a bounded U. If $\Omega_-\varphi = 0$, then $\Omega_- V_t^0\varphi = V_t\Omega_-\varphi = 0$ so

$$V_t^0 \operatorname{Ker}\Omega_- \subset \operatorname{Ker}\Omega_- \tag{5.5}$$

holds for all $t \geqslant 0$. The operator H_0 is semibounded due to the assumption, and therefore the functional-calculus rules imply the function $t \mapsto V_t^0$ to be strongly analytic in the open lower complex halfplane of t and strongly continuous in $\{t : \text{Im } t \leqslant 0\}$. Moreover, $R(\,.\,, H_0)$ is strongly analytic with exception of the part of $\mathbb{R}$ containing $\sigma(H_0)$. We have the relations

$$R(z, H_0)\varphi = i \int_0^\infty e^{izt}\, V_t^0 \varphi \, dt, \tag{5.6a}$$

$$\exp(-H_0 s) = \frac{1}{2\pi i} \int_C e^{-zs} R(z, H_0)\varphi \, dz, \tag{5.6b}$$

which are implied again by the functional calculus (cf. also Theorem 2.1). The integration contour in (5.6b) may be chosen, e.g. as $C = \{z = \gamma + \xi^2 + i\xi : \xi \in \mathbb{R}\}$ followed clockwise for some $\gamma < \inf \sigma(H_0)$. If $\Omega_-\varphi = 0$, then (5.5) together with (5.6a) give $\Omega_- R(z, H_0)\varphi = 0$ so $\exp(-H_0 s)\varphi = 0$ holds for all $s \geqslant 0$ in view of (5.6b). Combining this fact with the analycity of $t \mapsto V_t\varphi$, we see that (5.5) also holds for $t \leqslant 0$.

Next we derive a useful integral representation for Ω_-. Since U is bounded, (A.13) yields the equality

$$V_t\psi = V_t^0\psi - i \int_0^t V_s U V_{t-s}^0 \psi \, ds \tag{5.7a}$$

valid for all $\psi \in \mathscr{H}$ and $t \geqslant 0$. Setting now $\psi = V_{-t}^0\varphi$, we obtain the relation

$$\Omega_-\varphi = \varphi - i \int_0^\infty V_t U V_{-t}^0 \varphi \, dt \tag{5.7b}$$

valid for all φ for which the r.h.s. makes sense. The assumption (d) shows that (5.7b) holds at least for all $\varphi \in D$. We use it to express

$$\lim_{t \to \infty} \|\Omega_- V_{-t}^0 \varphi\| = \lim_{t \to \infty} \left\| V_{-t}^0 \varphi - i \int_t^\infty V_{t-s} U V_{-s}^0 \varphi \, ds \right\|.$$

Since the norm of the second term tends to zero as $t \to \infty$ and V_{-t}^0 is unitary, we get

$$\lim_{t \to \infty} \|\Omega_- V_{-t}^0 \varphi\| = \|\varphi\| \tag{5.8}$$

for all $\varphi \in D$. But D is dense in $\mathscr{H}$ and the operators $\Omega_- V_{-t}^0$ are uniformly bounded, so we may use the $\frac{1}{3}\epsilon$-trick again to show that (5.8) holds for each $\varphi \in \mathscr{H}$.

Consider now an arbitrary vector $\varphi \in \text{Ker}\, \Omega_-$. We have seen that $V_t^0\varphi \in \text{Ker}\, \Omega_-$ holds for all $t \in \mathbb{R}$, so the relation (5.8) yields $\|\varphi\| = 0$ and the injectivity of Ω_- is established. ■

The existence problem for $\Omega_{\mp}^{*}$ is more complicated. If U belongs to the trace class, one can adapt for this purpose the Kato–Birman theory. Let us prove first the following lemma:

LEMMA 4.5.5. *Let C be a compact operator and $\varphi \in \mathscr{H}_{\mathrm{ac}}(H)$, then* $\lim_{t\to\infty} \|CV_t\varphi\| = 0$.

Proof. The function $f : f(t) = |(\psi, V_t\varphi)|$ with arbitrary $\psi, \varphi \in \mathscr{H}$ fulfils

$$|f(t+\eta) - f(t)| = |(\psi, V_t(V_\eta\varphi - \varphi))| \leqslant \|\psi\|\,\|V_\eta\varphi - \varphi\|$$

so it is uniformly continuous in $\mathbb{R}_+$. If especially $\varphi \in \mathscr{H}(H)$, then f belongs at the same time to $L^2(\mathbb{R}_+)$, i.e. $\lim_{t\to\infty} |(\psi, V_t\varphi)| = 0$. This further implies $\lim_{t\to\infty} CV_t\varphi = 0$ for any finite-rank operator C. By the density argument, $\lim_{t\to\infty} \|CV_t\varphi\| = 0$ holds for all $\varphi \in \mathscr{H}_{\mathrm{ac}}(H)$. Finally, any compact operator C can be approximated by an operator-norm convergent sequence $\{C_n\}$ of finite-rank operators; it is elementary to get the desired result from $\|CV_t\varphi\| \leqslant \|C - C_n\|\,\|\varphi\| + \|C_nV_t\varphi\|$. ■

THEOREM 4.5.6 (Davies). *Assume that* (a) *is valid and U belongs to the trace class, then $\Omega_{\mp}^{*}\varphi$ exists for each $\varphi \in \mathscr{H}_{\mathrm{ac}}(H)$, and $\|\Omega_{\mp}^{*}\varphi\| \leqslant \|\varphi\|$.*

Proof. We denote $W(t) = V_{-t}^{0}V_t$; our aim is to prove that

$$\lim_{t,s\to\infty} \|[W(t) - W(s)]\varphi\| = 0 \tag{5.9a}$$

holds for any $\varphi \in \mathscr{H}_{\mathrm{ac}}(H)$; once the limit exists, the contractivity is obvious. Let $\psi \in D(H_0)$, then $W(\,.\,)\psi$ is differentiable and $W'(t)\psi = \; iV_{-t}^{0}UV_t\psi$ so

$$[W(t) - W(s)]\,\psi = -i\int_s^t V^0{}_xUV_x\psi\,\mathrm{d}x \tag{5.9b}$$

holds for all $s \leqslant t$. The r.h.s. makes sense for each $\psi \in \mathscr{H}$ and its norm is $\leqslant \int_s^t \|UV_x\psi\|\,\mathrm{d}x \leqslant \|U\|\,|t-s|\,\|\psi\|$. Thus it defines a bounded operator which coincides with $W(t) - W(s)$ on $D(H_0)$, i.e. the relation (5.9b) holds for all $\psi \in \mathscr{H}$.

Next we show that $B \equiv W(t) - W(s)$ is compact. Let $\{\varphi_n\} \subset \mathscr{H}$ be a weakly convergent sequence. The operator U is of the trace class, and therefore compact, so $-iV_{-x}^{0}UV_x$ is also compact. Consequently, the sequence $\{-iV_{-x}^{0}UV_x\varphi_n\}$ is norm convergent for each $x \in [s, t]$. The relation (5.9b) gives

$$B\varphi_n = -i\int_s^t V_{-x}^{0}UV_x\varphi_n\,\mathrm{d}x.$$

Due to the uniform boundedness principle, there is a positive K such that $\|\varphi_n\| \leqslant K$ for all n. Hence the dominated-convergence theorem may be used, which shows $\{B\varphi_n\}$ to be norm convergent as required.

Let $\varphi \in \mathscr{M}(H)$. The idea is to express the square of the norm appearing in

(5.9a) as an integral of its derivative. Since $W(t) - W(s)$ is compact, the above lemma gives

$$\lim_{x\to\infty} \|W(t+x)\varphi - W(s+x)\varphi\| = \lim_{x\to\infty} \|V^0_{-x}[W(t) - W(s)]\, V_x\varphi\| \leqslant$$

$$\leqslant \lim_{x\to\infty} \|[W(t) - W(s)]\, V_x\varphi\| = 0.$$

Then we have

$$\|[W(t) - W(s)]\varphi\|^2 = -\int_0^\infty \frac{\partial}{\partial x} \|W(t+x)\varphi - W(s+x)\varphi\|^2 \, dx$$

$$= -\int_0^\infty \frac{\partial}{\partial x} \{\|V^0_{-t-x}V_{t+x}\varphi\|^2 + \|V^0_{-s-x}V_{s+x}\varphi\|^2 -$$

$$- 2\,\mathrm{Re}(V^0_{-t-x}V_{t+x}\varphi, V^0_{-s-x}V_{s+x}\varphi)\} \, dx$$

$$= 2\int_0^\infty \{\mathrm{Im}(V^0_{-t-x}UV_{t+x}\varphi, V^0_{-t-x}V_{t+x}\varphi) -$$

$$- \mathrm{Im}(V^0_{-s-x}V_{s+x}\varphi, V^0_{-s-x}UV_{s+x}\varphi) +$$

$$+ i\,\mathrm{Re}(V^0_{-t-x}UV_{t+x}\varphi, V^0_{-s-x}V_{s+x}\varphi) -$$

$$- i\,\mathrm{Re}(V^0_{-t-x}V_{t+x}\varphi, V^0_{-s-x}UV_{s+x}\varphi) \, dx$$

$$\leqslant 2\int_0^\infty |(UV_{t+x}\varphi, V_{t+x}\varphi)| \, dx +$$

$$+ 2\int_0^\infty |(V_{s+x}\varphi, UV_{s+x}\varphi)| \, dx +$$

$$+ 2\int_0^\infty |(UV_{t+x}\varphi, V^0_{t-s}V_{s+x}\varphi)| \, dx +$$

$$+ 2\int_0^\infty |(V^0_{s-t}V_{t+x}\varphi, UV_{s+x}\varphi)| \, dx$$

(since U is bounded, the derivatives make sense for any $\varphi \in \mathcal{H}$).

Let us estimate, for example, the third integral. The operator U can be expressed in canonical form: there are orthonormal sets $\{\psi_n\}$, $\{\chi_n\} \subset \mathcal{H}$ and positive numbers λ_n such that $U\varphi = \Sigma_n \lambda_n(\chi_n, \varphi)\psi_n$ holds for all $\varphi \in \mathcal{H}$. Here λ_n are the singular values of U; since this operator belongs to the trace class, we have $\|U\|_1 = \Sigma_n \lambda_n < \infty$. Substituting for U, we get

$$I_3 \leqslant \int_0^\infty \sum_n \lambda_n |(V_{t+x}\varphi, \chi_n)(\psi_n, V^0_{t-s}V_{s+x}\varphi)| \, dx.$$

Using the monotone-convergence theorem, one can easily justify the interchange of the sum with the integral. Then we apply the Hölder inequality to the latter, obtaining thus

$$I_3 \leqslant \sum_n \lambda_n \left(\int_0^\infty |(\chi_n, V_{t+x}\varphi)|^2 \, dx \right)^{1/2} \left(\int_0^\infty |(V^0_{s-t}\psi_n, V_{s+x}\varphi)|^2 \, dx \right)^{1/2}$$

Since $\varphi \in \mathcal{M}(H)$, the second term of the product is $\leqslant c_\varphi^{1/2}$. The same estimation procedure may be applied to the other three integrals; together they give

$$\|[W(t) - W(s)]\varphi\|^2 \leqslant 8\, c_\varphi^{1/2} \sum_n \lambda_n \left(\int_s^\infty |(\chi_n, V_y\varphi)|^2 \, dy \right)^{1/2}$$

for all $\varphi \in \mathcal{M}(H)$ and $s \leqslant t$. This proves the relation (5.9a) for $\varphi \in \mathcal{M}(H)$. Finally, $\mathcal{M}(H)$ is dense in $\mathcal{H}_{ac}(H)$ and the operators $W(t)$ are uniformly bounded, so the proof is completed by the $\frac{1}{3}\epsilon$-trick. ∎

The existence criteria we have obtained are not very strong. In particular, the last proved theorem can never be applied when H is a Schrödinger pseudo-Hamiltonian. Fortunately, other results are applicable to that situation, though their number is much less than in the standard (unitary) scattering theory. Below, without proofs, we quote two of them; for references and further comments see the notes. The smooth-perturbation theory yields the following sufficient condition:

THEOREM 4.5.7 (Kato). *Let $H_0 + U$ be a Schrödinger pseudo-Hamiltonian on $\mathcal{H} = L^2(\mathbb{R}^3)$ referring to a potential u which fulfils*

$$\int_{\mathbb{R}^3} \int_{\mathbb{R}^3} \frac{|u(x)|\,|u(y)|}{|x-y|^2} \, dx \, dy < 4\pi^2, \tag{5.10a}$$

then there is a maximal dissipative extension H of $H_0 + U$ such that the wave operators $\Omega_-(H, H_0)$ and $\Omega_+^*(H, H_0)$ are defined on the whole $\mathcal{H}$. Moreover, each of them has a bounded inverse and the similarity relations

$$H = \Omega_- H_0 (\Omega_-)^{-1} = (\Omega_+^*)^{-1} H_0 \Omega_+^* \tag{5.11}$$

hold. Finally, H is a spectral operator.

REMARK 4.5.8. (a) Using this theorem, one must be sure that H is the 'right' extension. This difficulty does not arise, when we are able to show that $H_0 + U$ is e.m.d. Notice that the operator U specified by (5.10a) is H_0-bounded (with an arbitrarily small relative bound) in the quadratic-form sense (Reed and Simon, 1975, theorem X.19). However, this fact alone is not sufficient for Theorem 2.8.

(b) The l.h.s. of (5.10a) is nothing more than the squared Rollnik norm (Simon, 1971, chap. 1). One often meets other numerical factor on the r.h.s. due to a different value of mass. For $H_0 = (-\hbar^2/2m)\Delta$, the relation (5.10a) acquires the form

$$\|u\|_R < 2\pi\hbar^2 m^{-1}. \tag{5.10b}$$

Recently a 'geometrical' method has been applied to the problem. It provides us with the following results:

THEOREM 5.8 (Simon). *Let $H = H_0 + U$ be a pseudo-Hamiltonian on $\mathcal{H} = L^2(\mathbb{R}^d)$ which is of the following form: H_0 is given by (3.2), while U is dissipative and H_0-bounded with a relative bound < 1. Let further the 'Enss condition' be valid:*

$$\int_0^\infty h(r)\,\mathrm{d}r < \infty, \qquad h(0) < \infty, \tag{5.12a}$$

$$h(r) = \|U(H_0 - i)^{-1} F(|x| > r)\|, \tag{5.12b}$$

where $F(|x| > r)$ is the operator of multiplication by characteristic function of the indicated set. Then for the pair H, H_0,

(i) *the operator Ω_- exists,*
(ii) *the limit $\Omega_+^*\varphi$ exists for all $\varphi \in \mathcal{H}_{\mathrm{bound}}^\perp$.*

The class of operators considered here is, of course, also not very general; but, particularly for Ω_+^*, this theorem yields a stronger existence result than the one derived above. The direct comparison is made possible by Theorem 5.3, which exhibits conditions under which there is a unique natural domain for Ω_+^*.

The construction of the scattering theory, however, is not completed by proving the existence of the wave operators. One must also check that they yield a meaningful S-matrix, which would map $\mathcal{H}$ *bijectively* on $\mathcal{H}$. In the standard scattering theory, it is realized through verification of completeness of the wave operators – cf. (3.2.15); this task is usually not less difficult than the existence proofs themselves. The situation is even more complicated in non-unitary theory. There are few cases where the complete answer is known, notably the one considered in Theorem 5.7, but they are all rather special.

Within the framework of the hypotheses made in this section, we have in particular Theorems 5.4 and 5.6 – which may be used to establish the existence of the S-maxtrix – and Theorem 5.8 (cf. Problem 42). The most important trouble induced by the non-unitarity is that the existence of the S-matrix and its invertibility are now separate problems, because the wave operators are no longer partial isometries, and therefore automatically injective. As a simple example (which violates, however, the assumption (a)), consider $H = H_0 - iI$, where $\Omega_+^* = \Omega_- = 0$, so the S-matrix exists but does not give a reasonable scattering theory. Such a situation is prevented by the stated hypotheses, but many questions remain unanswered. For example, though Ω_- is injective by Theorem 5.4, its range need not be the whole $\mathcal{H}_{ac}(H)$:

EXAMPLE 4.5.9. Suppose H has an eigenvalue $\lambda = \lambda_0 - i\delta$, $\delta > 0$, of a finite multiplicity. Let $P = P_H(\{\lambda\})$ be the corresponding projection – cf. (3.35); it commutes with H, and therefore with V_t for each $t \geqslant 0$ (Reed and Simon, 1978, theorem XII.5; Davies, 1980a, theorem 1.15). Then we have

$$\|P\Omega_-\varphi\| = \lim_{t\to\infty} \|PV_tV^0_{-t}\varphi\| \leqslant \lim_{t\to\infty} \|PV^0_{-t}\varphi\|$$

for all $\varphi \in \mathscr{H}$, but P is of finite rank and $\mathscr{H}_{ac}(H_0) = \mathscr{H}$, so the limit is zero by an argument analogous to Lemma 5.5. Hence we obtain

$$P_H(\{\lambda\}) \cap \text{Ran}\ \Omega_- = \{0\}; \tag{5.13}$$

however, the subspace $P_H(\{\lambda\})$ is contained in $\mathscr{H}_{ac}(H)$ by Proposition 5.1.

We shall limit ourselves with formulating one simple criterion for invertibility of the S-matrix. We shall need some more hypotheses:

(b) *The operator U is bounded.*
(c) *There is a conjugation operator J on $\mathscr{H}$ such that $JH_0 = H_0J$, $JV = VJ$ and $JW = WJ$.*
(e) *The limit $\Omega_+^*\varphi$ exists for all $\varphi \in \mathscr{H}^{\perp}_{\text{bound}}$.*

The first of these ensures injectivity of Ω_- through Theorem 5.4; the third can be checked, e.g. by means of Theorem 5.3 and 5.6, or Theorem 5.8. Assumption (c) is often fulfilled in practice; it is the new ingredient which makes the proof of Theorem 5.11 possible.

Let us introduce the reverse wave operators by the relations

$$V_+\varphi = \lim_{t\to\infty} V_t^* V_t^0 \varphi, \tag{5.14a}$$

$$V_-\varphi = \lim_{t\to\infty} V_t^0 V_t^* \varphi \tag{5.14b}$$

for those φ for which the limits make sense. It is an easy exercise to see that the operators $JV_tJ = V_t^*$, $t \geqslant 0$, form the semigroup which is generated by $-iH^* = -iJHJ$, if only the assumptions (a) and (c) are valid.

LEMMA 4.5.10. *Assume* (a), (c)–(e) *and* (s), *then the operators V_+, V_- exist with $D(V_+) = \mathscr{H}$ and $D(V_-) = \mathscr{H}^{\perp}_{\text{bound}}$. Moreover, if we set $V_-\varphi = \Omega_+^*\varphi = 0$ for all $\varphi \in \mathscr{H}_{\text{bound}}$, then the following relations hold*:

$$J\Omega_-J = V_+ = (\Omega_+^*)^* = JV_-^*J. \tag{5.15}$$

Proof. Since (a) and (d) ensure the existence of Ω_-, we have

$$V_+\varphi = \lim_{t\to\infty} JV_t(V_t^0)^*J\varphi = J \lim_{t\to\infty} V_tV^0_{-t}J\varphi = J\Omega_-J\varphi$$

for all $\varphi \in \mathscr{H}$. In the same way, one proves $V_-\varphi = J\Omega_+^*J\varphi$ for $J\varphi \in \mathscr{H}^{\perp}_{\text{bound}}$, and Proposition 5.1 together with (c) imply easily that J maps $\mathscr{H}_{\text{bound}}$ and $\mathscr{H}^{\perp}_{\text{bound}}$

onto themselves. After extending V_- and Ω_+^* to the whole $\mathscr{H}$, we get $V_- = J\Omega_+^* J$, i.e. $JV_-^* J = (\Omega_+^*)^*$.

The remaining equality is proved as follows: If $\varphi \in \mathscr{H}_{\text{bound}}^\perp$ and $\psi \in \mathscr{H}$, we have

$$(\varphi, V_+\psi) = \lim_{t\to\infty} (\varphi, V_t^* V_t^0 \psi) = \lim_{t\to\infty} (V_{-t}^0 V_t \varphi, \psi) = (\Omega_+^* \varphi, \psi).$$

If $\varphi \in \mathscr{H}_{\text{bound}}$, then $(\Omega_+^*\varphi, \psi) = 0$ according to the definition. On the other hand, $V_+ = \Omega_-(-H^*, -H_0)$ so application of Theorems 5.4 and 5.3 gives $V_+\mathscr{H} \subset \mathscr{H}_{\text{ac}}(H^*) = \mathscr{H}_{\text{bound}}(H^*)^\perp$ and $(\varphi, V_+\psi) = 0$. Consequently, $V_+ = (\Omega_+^*)^*$. ■

Notice that one may also use the notation suggested by the standard scattering theory,

$$V_+ = \Omega_+, \qquad V_- = \Omega_-^*; \tag{5.14c}$$

the relation (5.15) shows that the operators $\Omega_\pm^*$ extended in the described way are actually adjoints of $\Omega_\pm$.

THEOREM 4.5.11 (Davies). *Under the assumptions* (a)–(e) *and* (s), *the following assertions are equivalent*:

(i) $S = \Omega_+^*\Omega_-$ *maps bijectively* $\mathscr{H}$ *onto* $\mathscr{H}$;

(ii) Ran Ω_- *is closed*;

(iii) $\mathscr{H}$ *decomposes into a Banach-space direct sum*, $\mathscr{H} = (\text{Ran}\ \Omega_-) \oplus \mathscr{H}_{\text{dec}} \oplus \mathscr{H}_{\text{bound}}$; *here the first two subspaces are both orthogonal to the last one, though they are not necessarily orthogonal to each other.*

Proof. The existence of S is ensured by (a), (d) and (e).

(i) ⇒ (ii) According to the inverse-mapping theorem, S^{-1} is bounded, so there is a positive a such that $\|S\psi\| \geqslant a\,\|\psi\|$ for all $\psi \in \mathscr{H}$. Since the operators Ω_+^*, Ω_- are contractive, we have $\|\psi\| \geqslant \|\Omega_-\psi\| \geqslant a\,\|\psi\|$, and therefore Ran Ω_- is closed.

(ii) ⇒ (i) Ran Ω_- is closed and Ω_- is injective by Theorem 5.4, so we may again use the inverse-mapping theorem, which implies the existence of a positive b such that $\|\Omega_-\varphi\| \geqslant b\,\|\varphi\|$ for all $\varphi \in \mathscr{H}$. Then the relation

$$\|S\varphi\| = \lim_{t\to\infty} \|V_{-t}^0 V_t \Omega_-\varphi\| = \lim_{t\to\infty} \|V_{-t}^0 \Omega_- V_t^0 \varphi\| \geqslant b \lim_{t\to\infty} \|V_t^0\varphi\| = b\,\|\varphi\|$$

shows that S is injective and has a closed range. We have $V_+ = J\Omega_- J$ due to (5.15), so Ran $V_+ = J$ Ran Ω_- is also closed. Since $V_+ = \Omega_-(-H^*, -H_0)$ is injective, the same argument as above implies that the operator V_-V_+ is injective with a closed range; we notice that $V_-V_+ = (\Omega_+^*\Omega_-)^* = S^*$. Let $\varphi \in (\text{Ran}\ S)^\perp$, then $0 = (\varphi, S\psi) = (S^*\varphi, \psi)$ holds for all $\psi \in \mathscr{H}$. Thus $S^*\varphi = 0$, but S^* is injective, and therefore Ran $S = \mathscr{H}$.

(iii) ⇒ (i) Closedness of Ran Ω_- is contained trivially in (iii).

(i), (ii) ⇒ (iii) Proposition 5.1 and Theorem 5.4 give $\mathscr{H}_{\text{dec}} \subset \mathscr{H}^{\perp}_{\text{bound}}$ and $\text{Ran}\,\Omega_- \subset \mathscr{H}^{\perp}_{\text{bound}}$. Suppose $\Omega_-\varphi \in \mathscr{H}_{\text{dec}}$, then

$$\|S\varphi\| = \lim_{t\to\infty} \|V^0_{-t} V_t \Omega_-\varphi\| = \lim_{t\to\infty} \|V_t\Omega_-\varphi\| = 0,$$

but S is bijective so $\varphi = 0$, i.e.

$$\mathscr{H}_{\text{dec}} \cap \text{Ran}\,\Omega_- = \{0\} \tag{5.16}$$

(notice that in view of (5.2a), this relation strengthens (5.13)). For an arbitrary $\varphi \in \mathscr{H}$, we have a unique decomposition $\varphi = \varphi_0 + \varphi_1$ with $\varphi_0 \in \mathscr{H}^{\perp}_{\text{bound}}$ and $\varphi_1 \in \mathscr{H}_{\text{bound}}$. According to (e), $\Omega_+^*\varphi_0 = \lim_{t\to\infty} V^0_{-t} V_t\varphi_0$ exists and equals some $\psi \in \mathscr{H}$. Further, the range of $S = \Omega_+^*\Omega_-$ is $\mathscr{H}$ so there is some $\varphi_2 \in \text{Ran}\,\Omega_-$ such that $\Omega_+^*\varphi_2 = \psi$. Let us denote $\varphi_3 = \varphi_0 - \varphi_2$, then

$$\lim_{t\to\infty} V^0_{-t} V_t\varphi_3 = \Omega_+^*\varphi_0 - \Omega_+^*\varphi_2 = 0,$$

but this means

$$\lim_{t\to\infty} \|V^0_{-t} V_t\varphi_3\| = \lim_{t\to\infty} \|V_t\varphi_3\| = 0$$

so $\varphi_3 \in \mathscr{H}_{\text{dec}}$. Hence we obtain $\varphi = \varphi_1 + \varphi_2 + \varphi_3$ with $\varphi_2 \in \text{Ran}\,\Omega_-$ and $\varphi_3 \in \mathscr{H}_{\text{dec}}$. Finally, the uniqueness of the orthogonal decomposition together with (5.16) imply that the decomposition is unique. ∎

Notes to Chapter 4

SECTION 4.1. The use of complex quantities for the description of some non-stationary processes (mostly damped oscillations) originates in classical physics from which it was inherited by the new-born quantum mechanics – see, for instance, Gamow (1928). A large review of older works on 'non-self-adjoint' problems in both classical and quantum physics is due to Dolph (1961); see also Brodsky and Lifshitz (1958). The standard reference to operator semigroups is the monograph by Hille and Phillips (1957); an updated exposition of this theory is due to Davies (1980a). The bulk of the material can be found, however, in most of the large monographs on functional analysis and its applications, such as Kato (1966a, chap. IX), Yosida (1966, chaps. IX, XIII and XIV) or Reed and Simon (1957, §§X.8, X.9).

• Every true physical Hamiltonian H belongs to the class of pseudo-Hamiltonians according to the definition, and it is also clear that in this case H represents at the same time the corresponding quasi-Hamiltonian. The conventions concerning the generators are different: one mostly uses $e^{\mp iHt}$ for unitary groups but e^{-At} (or sometimes e^{At}) for contractive semigroups. We shall be concerned with the Schrödinger-type equation (1.1) so we prefer to write the corresponding contractive semigroup as e^{-iHt}.

• Though the phenomenological non-self-adjoint Hamiltonians are widely used, rigorous studies of the subject are not so numerous. The assertions derived in this section are mostly taken from Blank *et al.* (1979), where the notions of pseudo-Hamiltonian and quasi-Hamiltonian were introduced. In the next section we shall show that maximal dissipativity is the substantial property which makes an operator acceptable as a pseudo-Hamiltonian. It does not mean, however, that all maximal dissipative operators are pseudo-Hamiltonians. The situation is exactly the same as for the true Hamiltonians which must obey various physical requirements (depending on the considered theory or model) besides self-adjointness.

SECTION 4.2. Most of the material in this section is standard and may be derived from the references quoted in the notes to the preceding section. The term 'dissipative' is used by various authors with ambiguity given by a power of the imaginary unit; we choose the definition which suits the concept of pseudo-Hamiltonian. Also, in the mathematical literature the dissipative operators are often not required to be densely defined, but merely to obey the condition (ii) of Theorem 2.1. Then a caution is needed; for instance, there are operators which are maximal dissipative in that sense but not closed (Phillips, 1959). We prefer to include the existence of a dense domain into the definition, so our (maximal) dissipative operators represent a direct generalization to the (maximal) symmetric operators.

• Theorem 2.5 is due to Phillips (1959). The Cayley transform V of the generator H is sometimes called cogenerator – cf. Sz.-Nagy and Foias (1970, §III.8). The notions of accretivity and dissipativity extend to the Banach-space operators with the help of so-called normed tangent functionals – see, e.g. Reed and Simon (1975, §X.8), where the Banach-space version of Theorem 2.1 may be found. An example showing the invalidity of Theorem 2.5 for continuous contractive semigroups on Banach spaces is due to Lumer and Phillips (1961). The notion of essentially maximal dissipative operator comes from Crandall and Phillips (1968), who constructed the mentioned extension theory for dissipative operators. The following assertion is sometimes useful (Reed and Simon, 1975, theorem X.49; Davies, 1980a, theorem 1.9): *let A generate a continuous contractive semigroup, then a dense $D \subset D(A)$ is core for A if it is invariant under the semigroup.*

• Theorem 2.8 generalizes the fundamental perturbative result for self-adjoint operators, known as the Kato–Rellich theorem (Reed and Simon, 1975, theorem X.12). Its validity may also be extended to generators of continuous contractive semigroups on Banach spaces; however, the presented proof works there only if $a < \frac{1}{2}$ (Nelson, 1964). The proof for $a < 1$ was given by Gustafson (1966); a somewhat weaker result concerning the relative bound case is due to Chernoff (1972). The structural result contained in Theorem 2.10 is due to Sz.-Nagy and Foias (1970, prop. III.8.3). The proof is adopted (with a small improvement) from Davies (1980a, §6.2) as well as Corollary 2.11 and Proposition 2.12. The formula (2.8) which is essential for Theorem 2.13 is named after F. Dyson, who used the 'time-ordered exponentials' in his pioneer work (1949) on quantum electrodynamics.

SECTION 4.3. In the next section, we shall show one possible way in which the non-self-adjoint interaction terms in pseudo-Hamiltonians may arise. The introductory considerations of this section are taken from Blank *et al.* (1979), where Theorem 3.2 as well as Theorem 3.3 for $d \leqslant 3$ were proved. The notions of J-symmetry and J-self-adjointness are due to Glazman (1957). In distinction to the ordinary symmetric operators, each J-symmetric operator can be shown to have a J-self-adjoint extension; for more details and further references see Glazman (1963, §I.22). The properties of the free Hamiltonian H_0 may be found in nearly all monographs on linear operators – see, e.g. Reed and Simon (1975, §IX.7). Let us remark also that the semigroup techniques were applied to solving parabolic differential equations long ago (see Reed and Simon (1975, notes to §X.8) for references), but the interest was concentrated primarily on the diffusion (heat) equation rather than on the Schrödinger-type equation treated here.

• The derived sufficient conditions of the (essentially) maximal dissipativity in terms of the potential are straightforward generalizations of the well-known criteria of (essential) self-adjointness. Historically the first of these 'self-adjoint' results (corresponding to Theorem 3.4) was obtained by Kato (1951; 1966, §5.5.3), for the generalization to the d-dimensional case, see Nelson (1964) and further references quoted by Reed and Simon (1975, notes to §X.2). The self-adjoint version of Theorem 3.5 is again due to Kato (1973); see also Reed and

Simon (1975, theorem X.28 and notes to §X.4). As for the properly dissipative case, some weaker results have been known for some time. Glazman (1963, theorem I.46) proved for the continuous complex-valued potentials u with a below-bounded real part, that the corresponding generalized Schrödinger operator H is essentially J-self-adjoint on the domain which consists of all functions from $C^2(\mathbb{R}^d)$ with compact supports. Hence, if u fulfils the dissipativity condition (3.6), H is e.m.d. on this domain by virtue of Theorem 3.2. In the one-dimensional case, some weaker sufficient conditions for J-self-adjointness (not requiring continuity of the potential) were derived earlier by Lidskii (1960).

• The generalized Schrödinger operators on a halfline were studied by many authors; a detailed exposition with numerous references was given by Liance (1969). The reduction of the spherically symmetric problem to the treatment of a family of such operators (Proposition 3.6) is standard; a review of the tensor product formalism for Hilbert-space operators can be found, e.g. in Blank and Exner (1976, 1977). Example 3.8 is due to Nelson (1964) who also proved the extension T_α to be maximal. Let us mention that the limit point–limit circle method of Weyl, so useful in the self-adjoint case, may also be extended to the generalized one-dimensional Schrödinger operators – see Lidskii (1960).

• Proposition 3.11 follows easily from the results of the preceding section. One would invite, however, some weaker sufficient conditions for complete dissipativity. Theorem 3.13, which is due to Davies (1980a, §6.2), suggests that such conditions are likely to exist. There is also the following closely related result (Davies, 1975b): *let H be a dissipative Schrödinger operator related to a purely imaginary $u = -iw \in L^\infty$, then* $\lim_{|x|\to\infty} w(x) = \gamma > 0$ *implies* $\lim_{t\to\infty} \|e^{-iHt}\| = 0$. In the last mentioned paper also, other limiting cases for Schrödinger pseudo-Hamiltonians with purely imaginary potentials are discussed. In particular, Davies found (under some assumptions on support of the potential – cf. also his monograph on open systems (1976, §7.4)) the limit $\text{s-lim}_{\lambda\to\infty} \exp(-i(H_0 - i\lambda W)t)$ to equal $\exp(-(i/2)\Delta_D^M t)$, where Δ_D^M is the Dirichlet Laplacian related to $M = \text{supp}\, W$. It is interesting to compare this result with the assertions proved in Example 2.4.11 and Section 6.3.

• Proof of Theorem 3.14 relies on the relative compactness of the potential with respect to H_0; it is not important whether u assumes real or complex values. We refer to Reed and Simon (1978, §XIII.4), where also an extension of the result to the potentials $u \in R + (L^\infty)_\epsilon$, R being the Rollnik class, may be found. Various particular cases and modifications are known: see, e.g. Glazman (1963, theorems II.31, 32) for the continuous potentials and $d = 1$. It should be noted, however, that the essential spectrum of a generalized Schrödinger operator can behave pathologically in comparison with the self-adjoint case: it can contain the so-called *spectral singularities* (Naimark, 1954; Liance, 1969).

• The necessary and sufficient condition under which the essential spectrum of a Schrödinger operator is void was given by Molchanov (1953). The generalization to the case of complex potentials is due to Lidskii (1957, 1960); another proof for the particular case of continuous potentials can be found in Glazman (1963, theorem II.30). Let us make few comments on this result. Its validity is not limited by the dissipativity condition (3.6); it holds for other generalized Schrödinger operators as well, but in that case, one must replace $w(x)$ by $|w(x)|$ in (3.36). The original Molchanov criterion concerns Schrödinger operators on $L^2(\mathbb{R}^d)$, $d \geqslant 1$, One can expect the corresponding extension of Theorem 3.15 to hold, with $\{J_n\}$ replaced by a sequence of cubes approaching infinity, but this seems not to be done. Notice also that the potential is required only to be of L^1_{loc} (compare to Theorem 3.5). Such potentials can also be considered for $d > 1$ if the corresponding Schrödinger operator is understood in the sense of quadratic forms – cf. Reed and Simon (1975, theorem X.33).

• The principal role played by the spectral measure in the theorem of self-adjoint operators is well known. One is therefore interested whether an analogous object could be ascribed to the non-self-adjoint operators under consideration. If such a measure exists, its values are not longer orthogonal projections, because in that case the corresponding operator would be normal. In general, however, this is not true for dissipative operators as the example of generalized Schrödinger operators shows. There is a wider class of operators which admit a sort of spectral decomposition, the so-called *spectral operators*, whose extensive discussion with a rich bibliography may be found in Dunford and Schwartz (1971). With a spectral operator T on $\mathscr{H}$, one can associate a σ-additive spectral measure $P_T(\,.\,)$ defined on Bor($\mathbb{C}$) whose values are (in general, non-orthogonal) projections from $\mathscr{B}(\mathscr{H})$. The correspondence is as follows: for an arbitrary Borel $\Delta \subset \mathbb{C}$, the projection $P_T(\Delta)$ reduces the operator T and $\sigma(T \upharpoonright P_T(\Delta)\mathscr{H}) \subset \bar{\Delta}$. One is tempted to write the spectral operator as $T = \int_{\mathbb{C}} \lambda \, dP_T(\lambda)$ (with the obvious meaning of the integral); but this is not the most general form even in the finite-dimensional case. A bounded spectral operator can be always expressed as the sum of the above written operator (which is said to be of the *scalar type*) and an operator S which is *quasi-nilpotent*, i.e. such that $\lim_n \|S^n\|^{1/n} = 0$. Notice that the limit is just the expression of the spectral radius: S is quasi-nilpotent iff $\sigma(S) = \{0\}$. The described decomposition of a bounded spectral operator represents a direct generalization to the Jordan normal form of a matrix. Unfortunately, knowledge of the conditions under which a generalized Schrödinger operator is spectral is relatively poor. For the operators on a halfline, this is true if the potential is infinitely differentiable and falls quickly enough at infinity (Dunford and Schwartz, 1971, theorem XX.1.12). Of course, the condition is far from necessary. Generalizations of this result for the operators on $L^2(\mathbb{R})$ and for matrix potentials are due to Siroid (1981). The correspondence between the spectral operators and the above-mentioned singularities should also be noted: a generalized Schrödinger operator is spectral if the latter are absent (Liance, 1969).

• The discrete spectrum of a Schrödinger pseudo-Hamiltonian H is of immediate importance in view of the considerations presented in Section 1. In particular, one is interested in the conditions which must be imposed on the potential in order to ensure finiteness of $\sigma_{\mathrm{disc}}(H)$, absence of its accumulation points, etc. Various results of that type are scattered in the mathematical literature, but we are not going to gather them here; some classical ones may be found, e.g. in Dolph (1961) or Glazman (1963, §VI.66). Some indirect information about eigenvalues for a special type of complex potentials can be derived from the studies of coupling-constant analycity for Schrödinger operators with real potentials – besides the standard literature on perturbation theory one can mention, e.g. Blankenbecler *et al.* (1977) or Klaus and Simon (1980). The stability of eigenvalues with respect to the coupling constant for generalized Schrödinger operators has been studied recently by Vock and Hunziker (1982).

• In this connection, let us stress that, in the non-self-adjoint case, the vectors referring to a point $\lambda \in \sigma_{\mathrm{disc}}(H)$ need not be eigenvectors of H as simple matrix examples show. The subspace determined by the projection (3.35) may contain at the same time the *generalized eigenvectors* which fulfil $(H - \lambda)^n \psi = 0$ for some integer $n > 1$. Though their appearance is mathematically quite natural, they did not find use in (the considered branch of) physics until now: remember that, in view of Proposition 1.4, the generalized eigenvectors are directly related to the higher-order poles in the reduced resolvent (cf. Section 3.3 and the notes to Section 1.5).

• Besides the present context, there are other situations in which we are looking for complex eigenvalues of a generalized Schrödinger operator. Let us recall the dilation-analytic theory of resonances mentioned in the notes to Section 3.3. There is a difference here which amounts to complex conjugation. Consider the simplest case of a Schrödinger operator H on a halfline (Reed and Simon, 1978, §XII.6). The operators $H(\theta)$, Im $\theta > 0$, have their essential spectrum rotated into the lower complex halfplane, so that for Im θ large enough the sought second-sheet poles can be revealed. At the same time, one has $H(\theta) = e^{-2\theta}\, G(\theta)$, where $G(\theta)$ is a generalized Schrödinger operator. Hence the resonance problem is reduced to finding eigenvalues of $G(\theta)$ in the upper halfplane.

SECTION 4.4. The term 'optical' comes from the analogy with the scattering of light, where a partially absorbing medium is described conventionally by a complex refraction index. An early attempt to introduce an imaginary term into the potential responsible for the capture of the incident particles was undertaken by Bethe (1940); a more realistic optical model using a complex square-well potential was elaborated by Feshbach *et al.* (1954). The idea was pursued by many authors who treated various nuclear reactions, energies and complex potentials; a partial summary can be found in Úlehla *et al.* (1964). Gradually the description based on a suitable (local) optical potential became the standard tool of practical nuclear physics. It applies today not only to the scattering of elementary particles on nuclei, but also to collisions between the nuclei (Wildermuth and Tang, 1977, chap. 11), nuclear heavy-ion reactions (Hodgson, 1978, especially chap. 3), etc. An introductory reading to these problems, together with numerous references, can be found, e.g. in Hodgson (1980, chap. 1; 1981, chap. 2).

• Considerations of the present section show that there are (at least) two ways of applying pseudo-Hamiltonians to the description of scattering, which are complementary in a sense. This can be illustrated on the example of the elastic scattering of particles on nuclei, where the possibility of resonance effects is taken into account through describing the particle motion by a pseudo-Hamiltonian H. If we are interested primarily in the resonances, we seek non-real eigenvalues of H. On the other hand, the optical approximation treated here is concerned with the shape-elastic scattering, and the non-self-adjointness of H reflects in this case neglection of the (averaged) resonance scattering. We should remark, however, that the terminology is not strictly fixed. Sometimes the adjective 'optical' means simply a complex potential, as, for example, in the recent study by Gal *et al.* (1981), where such a potential is used for modelling of the resonance scattering of Σ-hyperons on light nuclei.

• In this section, we follow mainly the papers by Davies (1978) and Exner and Úlehla (1983). In the first of these, the problem was formulated and a modification of Theorem 4.4 was proved under the simplifying assumptions $E = 0$ and $V_{00} = 0$; notice that the latter means a serious restriction, as shown by Example 4.1. The present generalization comes from the last-mentioned paper. The smooth-perturbation technique used in the proof was introduced by Kato (1966b); the necessary facts are collected in the Appendix (A.14); for more information see Reed and Simon (1978, §XIII.7). The formal 'time-independent' derivation of the optical potential (4.12a) is due to Feshbach (1958, 1963) and may be found in textbooks (e.g. Taylor (1972, §19.4); see also Dolph (1961), Bencze and Chandler (1982)). As we have mentioned, the Feshbach optical potential is generally non-local which makes its practical applications troublesome (for a pedagogical example, see Doyle *et al.* (1975)).

• Proposition 4.9 is essentially due to Davies (1978), as well as the physical conclusion that the optical approximation is characterized by the existence of different time scales. He also pointed out the following fact. There is a 'strong version' of Proposition 4.9 (cf. Problem 37), but it gives an estimate which is not applicable for large times. A possible interpretation is that the 'smoothing procedure' suppresses part of the wave function which is much delayed in the course of the scattering process, though not small in norm.

SECTION 4.5. A non-unitary scattering matrix, obtained by introducing complex wave shifts into the partial-wave decomposition, is often used for phenomenological analysis of inelastic nuclear reactions (cf. Landau and Lifshitz (1974, §142) and the optical-model literature quoted in the notes to the preceding section). In spite of this fact, the rigorous non-unitary scattering theory has been treated by few authors only, especially by Kato (1966b), Goldstein (1970, 1971), Martin (1975), Simon (1979b) and Davies (1978, 1980b). Our exposition follows essentially the two last-mentioned papers. There are various notational conventions for the wave operators and related quantities; we choose the notation to correspond with the unitary case mentioned in Section 3.2.

• Theorem 5.3 is adopted from Davies (1980b); its proof and the definition of $\mathscr{H}_{ac}(H)$ itself are based on the concept of square-integrable vectors (Davies, 1980a, §6.5). The idea of using Cook's argument to prove the existence of Ω_-, as well as the injectivity check in Theorem 5.4, are due to Martin (1975); the presented formulation comes from Davies (1980b). Notice that the boundedness of U is used exclusively to establish the relations (5.7). Thus this assumption may be eventually replaced by some weaker one, e.g. that U is a class $\mathscr{P}$ perturbation (Davies, 1980a, §3.1). Theorem 5.6 was obtained by Davies (1978) who adapted to this purpose the Kato–Birman theory (cf. Reed and Simon (1979, §XI.3)). In fact, Davies proved somewhat stronger results than the one presented here: he considered the generalized wave operators (cf. Problem 41), and he also showed $\Omega_{\pm}^{*}$ to exist if $(H - i)^{-m}U(H - i)^{-n}$ is of the trace class for some m, n. Theorem 5.7 is due to Kato (1966b), a detailed study of the Rollnik potentials can be found in Simon (1971). In Kato's paper other sufficient conditions are given for the existence and invertibility of the wave operators; in particular, for the cases when H is a Schrödinger pseudo-Hamiltonian on $L^2(\mathbb{R}^d)$, $d \geqslant 3$, and on $L^2(\mathbb{R}_+)$. It should be noted that for real potentials, the inequality (5.10) is one of the sufficient conditions for absence of the singularly continuous spectrum (Reed and Simon, 1978, theorem XIII.21).

• Theorem 5.8 was proved by Simon (1979b) by the 'geometrical' method suggested originally by Enss (1978). He also found that under hypotheses of the theorem, each non-zero real eigenvalue of H had a finite multiplicity and the only possible limit of such eigenvalues was zero. Example 5.9 is essentially due to Martin (1975); see also Rimon (1979b). Theorem 5.11 comes from Davies' (1980b) paper, where alternative sufficient conditions for the bijectivity of S can be found.

Chapter 5

Feynman Path Integrals

"... one feels as Cavalieri must have felt calculating the volume of a pyramid before the invention of calculus."

R. P. FEYNMAN

The present chapter is devoted to a rigorous treatment of the 'sums over paths' proposed by Feynman as an alternative calculus for quantum mechanics. We begin with a formulation of the problem and a brief survey of some approaches to it. In Section 2 we study a theory in which the path integrals under consideration are defined by a Parseval-type equality, and refer to the functions on an abstract Hilbert space of paths. The latter is specified in Section 3: we discuss here various ways in which the path spaces referring to a quantum-mechanical system may be characterized. The most popular, at least among physicists, are the methods which follow Feynman's original suggestion and define the path integral through the approximations based on piecewise linear paths. Section 4 is devoted to a detailed discussion of these methods. In particular, they allow us to extend the definition given in Section 2. Another 'sequential' definition of the Feynman path integral based on the product formulae is given in Section 5. The final section of this chapter contains more information about the other F-integral theories and their mutual relations.

5.1. The Integrals that are Not Integrals: a Brief Survey

"I believe that the standpoints should be put off in the ante-room like hats and walking-sticks; once you permit a man with a standpoint to enter, he would certainly make some damage or at least he would quarrel with the others."

K. ČAPEK, *The Second-Pocket Stories*

Mathematical concepts born in physics often prove their fertility twice: first as a suitable but more or less formal calculus, and after that as a theory based on an appropriately rigorous treatment of the original idea. So the delta function represents an extremely useful computational tool, but the full power of this concept, which certainly goes beyond Dirac's intention, was not revealed before formulation of the distribution theory. One can therefore easily understand why there is so much temptation in the attempts to construct a mathematically sound theory of Feynman path integrals.

The primary physical aim was (and remained to be) to find a suitable calculus for treating the dynamics of particles and fields, which could eventually substitute the standard operator methods of the quantum theory in those situations where they are cumbersome or inapplicable. With the same intention, algebraical methods were introduced later into the quanum field theory; however, both problems appeared to be hard enough, and neither of the proposed formalisms could be claimed to be satisfactorily elaborated at the present time.

For the sake of simplicity, let us consider a single spinless particle with the free Hamiltonian $H_0 = (-\hbar^2/2m)\Delta$ which interacts with an external field described by a potential v. Feynman's main idea has two basic physical ingredients. The first of them is the superposition principle, or more explicitly, the fact that for any quantum process in the absence of measurements, it is just the transition amplitude (and not directly the probability) which is expressed as the sum of contributions from the partial processes. Feynman specified it to the space–time propagation of the particle: he proposed to calculate the probability amplitude of the transition between the points y and x (i.e. the kernel of the integral operator e^{-iHt}, where H is a suitable self-adjoint extension of $H_0 + V$) as a 'sum' of contributions from all possible paths connecting the two points. As a motivating example, one may consider the situation when the particle passes through a sequence of many-slit screens (and we do not register its 'trajectory').

The other basic assumption concerns the weight with which the contributions from particular paths are counted. Feynman postulated that each path γ, which connects the points $\gamma(0) = y$ and $\gamma(t) = x$, enters into the 'sum' with the weight factor $\exp\{(i/\hbar)S(\gamma)\}$, where

$$S(\gamma) = \int_0^t \left[\tfrac{1}{2}m\,|\dot{\gamma}(\tau)|^2 - v(\gamma(\tau))\right] \mathrm{d}\tau \tag{1.1}$$

is the action along the path γ, well known from classical mechanics. Since the paths are infinitely many, the sum over them may be written by the same right as a formal integral. Feynman's principal conclusion is then expressed as

$$(e^{-iHt/\hbar}\psi)(x) = \int_{\Gamma_x} \exp\left\{\frac{i}{\hbar}S(\gamma)\right\}\psi(\gamma(0))\,\mathrm{D}\gamma; \tag{1.2}$$

here ψ is the wave function at the initial instant $t = 0$, and Γ_x denotes a suitable set of paths ending at the point x. To be just, we should mention that this result has already been contained in an earlier paper by Dirac, but Feynman was the first who grasped fully its importance and gave a heuristic argument showing that the function determined by the r.h.s. of (1.2) should obey the Schrödinger equation with the Hamiltonian H. He also devised numerous applications of the new formalism, in which he was successfully followed by many other physicists.

The weakest point of the formalism rests with the definition of the path sums or path integrals. Feynman himself proposed a straightforward method which calls to mind the definition of the Riemann integral. In this approach, the set of all paths is replaced by its subset consisting of the polygonal paths; the particle velocity is assumed constant in the intervals $(\tau_j, \tau_{j+1}), j = 0, 1, \ldots$ (cf. Figure 1). The value of the 'integrated' function on the r.h.s. of (1.2) is now determined by the values $\gamma(0), \gamma(\tau_1), \ldots$ only, i.e. by a finite set of variables. It is then easy to write an integral which may be understood as a finite approximation to the sought object, and to define the latter as a limit with $\tau_{j+1} - \tau_j \to 0$. However, Feynman gave no conditions under which such a limit might exist. In fact, he did not even try to do so; by an expression of F. J. Dyson, "mathematical rigor is the last thing that Feynman was ever concerned about".

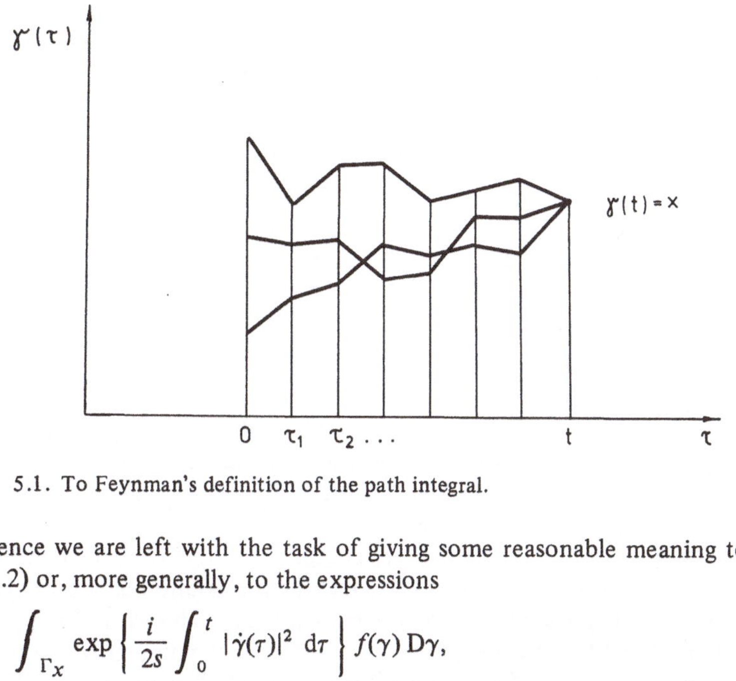

5.1. To Feynman's definition of the path integral.

Hence we are left with the task of giving some reasonable meaning to the r.h.s. of (1.2) or, more generally, to the expressions

$$\int_{\Gamma_x} \exp\left\{\frac{i}{2s}\int_0^t |\dot{\gamma}(\tau)|^2 \, d\tau\right\} f(\gamma)\, D\gamma, \tag{1.3}$$

where f is a complex function on the path space Γ_x and s is a real parameter. It is useful to recall that such expressions with $s = -ic$, $c > 0$, were considered much earlier with the object of describing Brownian motion and related processes. These path integrals were introduced by N. Wiener who used the ideas previously formulated by Einstein and Smoluchowski, and they had been later extensively studied. The main difference between the two cases is that the Wiener integrals represent a mathematically more manageable object. The expression (1.3) is always formal, however, since the path spaces of interest are presumably infinite-dimensional; in that case no Lebesgue-type measure exists which could stand for $D\gamma$

(cf. Problem 43). For a purely imaginary s, say $s = -i$, it does not matter. Loosely speaking, the singularities of the exponential term and of $\mathrm{D}\gamma$ cancel each other, and we are able to replace the formal expression $\exp\{-\frac{1}{2}\int_0^t |\dot\gamma(\tau)|^2\, d\tau\}\,\mathrm{D}\gamma$ by $dw(\gamma)$, where w is the **Wiener measure.**

We can present here neither the beautiful mathematical theory concerning the Wiener measure and Wiener integral nor its applications in quantum physics; it is not allowed by both the purpose and size of this book. This field is very wide and intensively studied, and there are a few excellent monographs to consult. References are given in the notes; we shall use them occasionally if information about the Wiener integral is needed.

For the present, recall that the Wiener measure is usually defined first on special sets of paths called *cylindrical*, and after that it is extended on a suitable σ-algebra of subsets in the path space under consideration. For the first step, one may equivalently use the cylindrical functions. Their definition will be given in the next section; roughly speaking, we mean those functions of a path which depend actually on a 'finite number of variables' only. One postulates for them the relation

$$\int_{\Gamma_x} \exp\left\{\frac{i}{2s}\int_0^t |\dot\gamma(\tau)|^2\, d\tau\right\} f(\gamma(\tau_0), \dots, \gamma(\tau_{n-1}))\,\mathrm{D}\gamma$$

$$= \prod_{j=0}^{n-1} (2\pi i s\delta_j)^{-d/2} \int_{\mathbb{R}^{nd}} \exp\left\{\frac{i}{2s}\sum_{j=0}^{n-1} |\gamma_{j+1}-\gamma_j|^2\,\delta_j^{-1}\right\} f(\gamma_0, \dots$$

$$\dots, \gamma_{n-1})\, d\gamma_0 \dots d\gamma_{n-1}, \tag{1.4}$$

which defines the formal path integral on the l.h.s.; the positive parameter $c = is$ is named *dispersion* and the corresponding Wiener measure is denoted w_c. In the relation (1.4), τ_j are fixed points in $[0, t]$ with $\delta_j = \tau_{j+1} - \tau_j$, and d is the dimension of the underlying configuration space.

The relation (1.4) can be used for other complex s too, as far as its r.h.s. makes sense. In particular, for a real positive s it is just the prescription used by Feynman to define the finite approximations to his path integral. This is nothing strange, because formula (1.4) for the integration of cylindrical functions is quite natural; it seems reasonable *to demand its validity for each possible definition of the path integral* (1.3). By this requirement, we also specify a minimal set on which the path integrals discussed below should coincide.

Formula (1.4) also illustrates well the main troubles connected with defining the expressions (1.3). For a real s, the integral on the r.h.s. of (1.4) often makes sense only as a principal value or not at all (remember that the physical interest concerns primarily the cases when modulus of the 'integrated' function equals 1 – cf. (1.2)). For example, consider the function

$$f : f(\gamma) = \exp\{-i\,|\gamma(0) - x|^2/2st\},$$

which is certainly nice enough. The r.h.s. of (1.4) is proportional in this case to the Lebesgue measure of $\mathbb{R}^d$, i.e. it is infinite. It follows that any straightforward Feynman-like approach to the definition of (1.3) is burdened by the fact that the approximating expressions are necessarily given by improper integrals. Needless to say, this is connected with considerable mathematical inconvenience. Not only is the result of such a method dependent on the limiting prescription used for calculating the integrals, but at the same time, we are bereft of the possibility of applying the powerful machinery of the standard theory of integration.

One could try to overcome this difficulty by a popular trick: to add a small negative imaginary part to s and define a Wiener-type measure for this complex parameter. After that the path integral of interest might be defined as a limit when s returns (non-tangentially) to the real axis. At a glance, the proposal seems reasonable, because the non-zero Im s adds an exponentially decaying factor which could suppress the divergences. Unfortunately, it fails on the infinite dimensionality of the path space as the following assertion shows:

THEOREM 5.1.1 (Cameron). *A (finite, complex) measure μ_s such that for each cylindrical function $f : f(\gamma) = f(\gamma(\tau_0), \ldots, \gamma(\tau_{n-1}))$, the integral $\int_{\Gamma_x} f(\gamma)\, d\mu_s(\gamma)$ is given by the r.h.s. of the relation* (1.4), *exists iff $s = -ic$ with some positive c.*

Proof. The existence proof of $\mu_s = w_{is}$ for a purely imaginary s may be adopted from theory of the Wiener measure (see, e.g. Kuo (1975, §I.3), Reed and Simon (1975, §X.11)). On the other hand, if Re $s \neq 0$, we may choose a partition $\sigma = \{\tau_j : 0 = \tau_0 < \tau_1 < \cdots < \tau_n = t\}$ of $[0, t]$ and construct the function

$$g_\sigma : g_\sigma(\gamma) = \exp\left\{ -\frac{a}{2} \sum_{j=0}^{n-1} |\gamma(\tau_{j+1}) - \gamma(\tau_j)|^2\, \delta_j^{-1} \right\}, \tag{1.5}$$

where a is a complex number with Re $a \geqslant 0$. The r.h.s. of (1.4) is easily calculated for g_σ: the substitution $y_j = (\gamma_{j+1} - \gamma_j)\delta_j^{-1/2}$; $j = 0, 1, \ldots, n-1$, where $\gamma_n \equiv x$, brings it to the form

$$(2\pi i s)^{-nd/2} \left(\int_{\mathbb{R}} \exp\left\{ -\left(\frac{a}{2} - \frac{i}{2s}\right) y^2 \right\} dy \right)^{nd}$$

so

$$\int_{\Gamma_x} g_\sigma(\gamma)\, d\mu_s(\gamma) = (1 + ias)^{-nd/2} \tag{1.6}$$

holds if only the measure μ_s exists. In that case, however, the total variation $|\mu_s|$ of μ_s is a finite positive measure on Γ_x; since $|g_\sigma(\gamma)| \leqslant 1$ for all γ, the relation (1.6) gives

$$|\mu_s|(\Gamma_x) \geqslant |1 + ias|^{-nd/2}. \tag{1.7a}$$

Suppose now $s = s_1 - is_2$ with a non-zero s_1 and $s_2 \geqslant 0$. Choosing $a = (\epsilon + is_2)|s|^{-2}$ with $\epsilon > 0$, we easily get $\bar{s}(1 + ias) = s_2 + \epsilon$ so

$$|1 + ias| = \frac{s_2 + \epsilon}{(s_1^2 + s_2^2)^{1/2}} < 1 \tag{1.7b}$$

holds for all sufficiently small ϵ. Since n may be chosen arbitrarily large, the relations (1.7) contradict the assumed existence of μ_s. ∎

Formulation of the theorem is somewhat vague, since we have not specified fully the set of cylindrical functions which should be inserted into the relation (1.4). The functions (1.5) used in the proof are, however, nice enough to be contained in any reasonable choice of this set. The main lesson from the above considerations is that *one cannot hope to treat the 'integrals'* (1.3) *within the framework of measure theory*, of course, unless the parameter s is purely imaginary. In other words, the path integrals under consideration are not integrals in the same way that the generalized functions (distributions) are not functions; we must look for alternative mathematical ways how to define and use them.

One of the possible ways can be built on *analytical continuation*. Suppose we know the Wiener integrals

$$I_s(f) = \int_{\Gamma_x} f(\gamma)\, dw_{is}(\gamma) \tag{1.8}$$

for a given function $f : \Gamma_x \to \mathbb{C}$ with dispersions belonging to some interval, e.g. $1 - \delta \leqslant is \leqslant 1$. Assume further that these $I_s(f)$ represent boundary values of a function, which is analytic in a suitable domain of the lower complex halfplane of s, say, the one sketched on Figure 2. In that case, one may define naturally the Feynman integral of the function f as the analytical continuation of $I_s(f)$ to the point $s = 1$. Notice that the described procedure is, in fact, analytical continuation

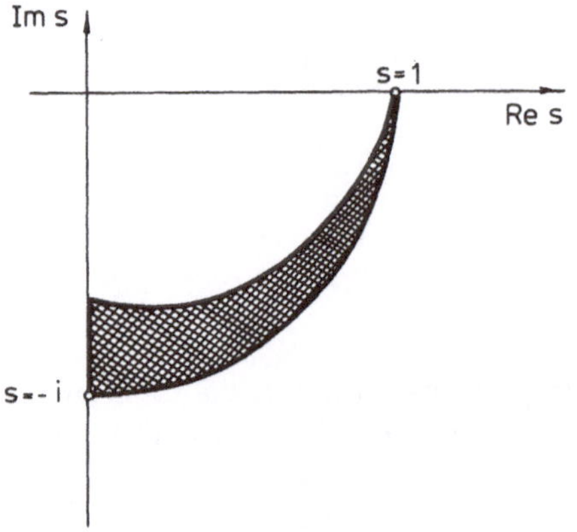

5.2. Analytic continuation in s.

in mass, because the parameter s is, in view of (1.1)–(1.3), proportional to the inverse mass.

There is also another way in which analytical continuation is used, namely as continuation *in time*. It concerns a special class of 'integrated' functions, which is, however, physically the most interesting one. For definiteness, consider again the Schrödinger operator $H = -\frac{1}{2}\Delta + V$ on $L^2(\mathbb{R}^3)$, and assume that the potential $v \in L^2(\mathbb{R}^3) + L^\infty(\mathbb{R}^3)$. Instead of e^{-iHt}, one may try to get a path-integral expression for e^{-Ht}. It is given by the **Feynman–Kac formula**,

$$(e^{-Ht}\psi)(x) = \int_{\Gamma_x} \exp\left\{-\int_0^t v(\gamma(\tau))\,d\tau\right\} \psi(\gamma(0))\,dw(\gamma), \tag{1.9}$$

which is the imaginary-time analogue of the relation (1.2), but in distinction to the latter, with a well-defined r.h.s. The Kato–Rellich theory implies the operator H to be semibounded, thus the function $t \mapsto e^{-iHt}$ is strongly analytic in the open lower halfplane (similarly as in the proof of Lemma 2.4.6 or Theorem 4.5.4), and we can find $e^{-iHt}\psi$ from (1.9) by analytical continuation.

Another group of definitions, which is again suitable primarily for the 'integrated' functions of the type

$$f(\gamma) = \exp\left\{-i\int_0^t v(\gamma(\tau))\,d\tau\right\} \psi(\gamma(0)), \tag{1.10a}$$

is based on various modifications of the Trotter product formula; it was originated by E. Nelson. This method is close to Feynman's original idea in the sense that it defines the path integral under consideration as a limit of some sequence of 'finite-dimensional' integrals. The difference is confined in the choice of this approximating sequence. Its members are again the integrals appearing on the r.h.s. of (1.4); however, instead of $f(\gamma)$ given by (1.10a) for a particular polygonal path γ (referring to a partition σ), they now contain the expressions

$$f^{(\sigma)}(\gamma) = \exp\left\{-i\sum_{j=0}^{n-1} v(\gamma(\tau_j))\,\delta_j\right\} \psi(\gamma(0)) \tag{1.10b}$$

in which the integral is replaced by the corresponding Darboux–Riemann sum. Intuitively one expects that the two approaches could yield the same values of path integrals, because (1.10a) and (1.10b) are close to each other for sufficiently fine partitions σ, but the strict convergence proofs are rather delicate.

Let us mention some other 'non-sequential' methods. The first of them is based on the observation made by K. Itô, which makes a modified 'limiting' approach possible. One can bypass Theorem 1.1 by choosing a more sophisticated exponential factor, which converts the formal Feynman integral into a well-defined integral with respect to a certain Gaussian measure. Having a family of such integrals corresponding to different Gaussian measures, one is able to define the sought path integral as the limit along the directed set of their covariance operators.

Next there is a group of definitions which employ Fourier transformation in various ways. Among them, particularly simple and elegant is the method of S. Albeverio and R. Høegh-Krohn. It is inspired by the following Parseval-type equality

$$(2\pi i)^{-n/2} \int_{\mathbb{R}^n} \exp\left\{\frac{i}{2}|x|^2\right\} f(x)\,\mathrm{d}x$$

$$= \int_{\mathbb{R}^n} \exp\left\{-\frac{i}{2}|y|^2\right\} \tilde{f}(y)\,\mathrm{d}y, \tag{1.11a}$$

where

$$f(x) = \int_{\mathbb{R}^n} \mathrm{e}^{-ix\cdot y} \tilde{f}(y)\,\mathrm{d}y, \tag{1.11b}$$

which is valid, for example, for each $f \in \mathscr{S}(\mathbb{R}^n)$. One may notice that the l.h.s. of (1.11a) bears a close resemblance with the formal path integral (1.3). The idea is therefore to use an analogous equality as the definition relation in the case when $\mathbb{R}^n$ is replaced by an infinite-dimensional path space and the counterpart of the l.h.s. makes no sense.

This survey could be continued, but we prefer to stop here. Our aim is to give an introduction to the treatment of the next sections rather than review all rigorous approaches to Feynman path integrals; some methods and problems, which we can mention only epigrammatically or not at all, are listed below and in the notes. As we shall see in the following, the actual number of the Feynman-integral theories is greatly enlarged due to the fact that various approaches can be combined. It is possible, for example, to use first the method of Albeverio and Høegh-Krohn and then extend this definition by polygonal-path approximations, or to define the path integral (1.3) sequentially for all s from the open lower halfplane and then let the parameter approach the real axis, etc. We are firmly convinced that such a diversity of methods is rather promising for a future theory. It stresses, however, the need for unification, that is, for establishing equivalence relations between various definitions on particular classes of functions, which should be as large as possible.

The last remark calls to mind another problem. Until now we were concerned with different ways of defining the path integral (1.3) for a given function $f(\,.\,)$. As for the class of functions for which a particular method works, we have argued that it should contain all cylindrical functions, with possible continuity or smoothness restrictions. But this is only a minimal requirement and one would like to know how many functions are actually allowed. Each of these classes certainly forms a linear space, so various Feynman integrals may be interpreted as linear functionals by analogy with the standard integration theory. However, the task of determining their domains is difficult, and in many cases they can be described only implicitly. Aside from purely mathematical reasons, a better understanding of this problem would be useful in the following sense. The core of the assertion contained in Theorem 1.1 is that the Feynman-type measures cannot exist because the exponential term in (1.3) is wildly oscillating unless s is purely imaginary. All the mentioned methods try to bypass this difficulty by choosing a suitable class of functions whose behaviour would be smooth enough in some sense to cancel the influence of the oscillations. With a detailed knowledge of the definition domains for different Feynman integrals, we would probably have a better intuition in deciding which of the 'smoothing' procedures is more natural, powerful, etc.

Before we proceed further, it is useful to set some *notational conventions.* A Feynman path integral corresponding to the formal expression (1.3) will be denoted most frequently by $I_s^{\alpha}(f)$, where the superscript specifies the definition used. There are situations, however, where this notation is not very convenient. Then we shall use either (1.3) with the specification added over the integral sign, or alternatively

$$\int_{\Gamma_x}^{\alpha} f(\gamma)\, D\phi_s(\gamma).$$

For the sake of brevity, we shall often replace the terms Feynman (path) integral and Wiener integral by F- and W-integral, respectively.

Concluding this section, let us list briefly some important problems which are left out of the scope.

(i) We are concerned exclusively with the 'configuration space' F-integrals. The related 'phase space' (or Hamiltonian) F-integrals, which are particularly useful in various quantization problems, will not be treated here.

(ii) With a few brief exceptions, we are not going to discuss the relations between F-integrals and the classical limit of quantum theory. This connection is one of the most beautiful features of Feynman's theory: on the heuristic level, one can see immediately that the less the value of Planck's constant, the more the probability of finding the particle concentrated around its classical trajectory. However, we intend to apply F-integral methods to open systems, for which the relations to their classical counterparts (if the latter exist: quantization, classical limit, etc.) are much more complicated and less important than for the isolated systems.

(iii) We have in mind mostly the systems which have (actually or effectively) a finite number of degrees of freedom, and therefore we study the essentially quantum-mechanical path integrals, and not the more general 'integrals over histories' needed for applications in quantum field theories. One should notice, however, that no such assumption is used in the theory exposed in the next section.

(iv) The configuration space considered will be always $\mathbb{R}^d$. Hence we shall not be concerned with the systems of identical particles, systems with constraints, as well as with quantum mechanics in a curved space–time.

5.2. Feynman Maps on the Algebra $\mathscr{F}(\mathscr{H})$

In this section we are going to treat the F-integral theory which is inspired by the relations (1.11). We shall begin by formulating hypotheses about the structure of the path space. The set of all paths ending at x is affine rather than linear space unless $x = 0$, but the formal expression (1.3) suggests that the equality

$$\int_{\Gamma_x} f(\gamma)\, \mathrm{D}\phi_s(\gamma) = \int_{\Gamma_0} f(\gamma + x)\, \mathrm{D}\phi_s(\gamma) \tag{2.1}$$

might be valid. In fact, this relation is not difficult to prove for specific F-integrals considered in the following sections: notice that it holds for the cylindrical functions in view of (1.4), as well as for the W-integrals (Kuo, 1975, theorem II.1.1). In the framework of this section, however, the relation (2.1) will be postulated; then we may assume without loss of generality that the paths form a vector space.

As for its metric structure, we assume it to be a *real separable Hilbert space*, which will in the following be denoted by $\mathscr{H}$. Its concrete realizations and further comments on structure of the path spaces will be given in the next section. To a given $\mathscr{H}$, we define the following two sets. $\mathscr{M}(\mathscr{H})$ denotes the set of all finite complex Borel measures μ on $\mathscr{H}$, i.e. with $|\mu|(\mathscr{H}) < \infty$; the Borel sets are, of course, related to the norm topology in $\mathscr{H}$. The set $\mathscr{F}(\mathscr{H})$ is formed by the Fourier transforms of these measures:

$$\mathscr{F}(\mathscr{H}) = \left\{ f : f(\gamma) = \int_{\mathscr{H}} e^{i(\gamma,\gamma')}\, d\mu(\gamma'),\, \mu \in \mathscr{M}(\mathscr{H}) \right\}, \tag{2.2}$$

where $(.\,,.)$ is the inner product in $\mathscr{H}$. The elements of $\mathscr{F}(\mathscr{H})$ will play the role of 'integrable' functions.

Let us start with some technical preliminaries. The set $\mathscr{M}(\mathscr{H})$ is a vector space: any linear combination of $\mu, \nu \in \mathscr{M}(\mathscr{H})$ belongs again to $\mathscr{M}(\mathscr{H})$, because the definition of the total variation implies $|\alpha\mu + \nu|(M) \leqslant |\alpha|\,|\mu|(M) + |\nu|(M)$ for all $\alpha \in \mathbb{C}$ and $M \in \mathrm{Bor}(\mathscr{H})$. Let, further, $\{\mu_n\} \subset \mathscr{M}(\mathscr{H})$ be a sequence such that $|\mu_n - \mu_m|(\mathscr{H}) \to 0$ as $n, m \to \infty$. It is easy to see that $\mu(M) := \lim_{n\to\infty} \mu_n(M)$ exists for each $M \in \mathrm{Bor}(\mathscr{H})$ and that μ defined in this way belongs to $\mathscr{M}(\mathscr{H})$. In other words, $\mathscr{M}(\mathscr{H})$ becomes a Banach space when equipped with the norm $|.|(\mathscr{H})$.

Next we shall show that there is a natural algebraical structure inherent to $\mathscr{M}(\mathscr{H})$. To this end, we introduce *convolution* of the measures $\mu, \nu \in \mathscr{M}(\mathscr{H})$: we denote $M - \gamma \equiv \{\gamma' - \gamma : \gamma' \in M\}$ and set

$$(\mu * \nu)(M) = \int_{\mathscr{H}} \mu(M - \gamma)\, d\nu(\gamma) \tag{2.3a}$$

for each $M \in \mathrm{Bor}(\mathscr{H})$. Since $\mu(M - \gamma)$ can be expressed as the integral from the characteristic function of the set $M - \gamma$, we may employ the Fubini theorem (for complex measures, cf. (A.6)) to rewrite (2.3a) in the symmetric form

$$(\mu * \nu)(M) = \int_{\mathcal{H}\times\mathcal{H}} \chi_M(\gamma + \gamma')\, d(\nu \otimes \mu)(\gamma, \gamma'), \tag{2.3b}$$

where $\nu \otimes \mu$ is the product measure of ν and μ. The set $\{\{\gamma, \gamma'\} : \gamma + \gamma' \in M\}$ is mapped homeomorphically to $M \times \mathcal{H}$, so it is Borel in $\mathcal{H} \times \mathcal{H}$ for each $M \in \mathrm{Bor}(\mathcal{H})$. Then the last integral makes sense and defines a complex Borel measure on $\mathcal{H}$. The mapping $(\mu, \nu) \mapsto \mu * \nu$ is obviously bilinear and commutative: the equality $(\mu * \nu)(M) = (\nu * \mu)(M)$ follows from (2.3b). The last-mentioned relation, together with the image-measure theorem (cf. (A.4)), imply

$$(\mu * \nu)(M) = \int_M d(\nu \otimes \nu)\,(\varphi^{(-1)}(\gamma'')),$$

where $\varphi : \mathcal{H} \times \mathcal{H} \to \mathcal{H}$ is defined by $\varphi(\gamma, \gamma') = \gamma + \gamma'$. Then for any $f \in L(\mathcal{H}, \mu * \nu)$, the last equality in combination with (A.5) and the Fubini theorem gives

$$\begin{aligned} f(\gamma)\, d(\mu * \nu)(\gamma) &= \int_{\mathcal{H}\times\mathcal{H}} f(\gamma + \gamma')\, d(\mu \otimes \nu)(\gamma, \gamma') \\ &= \int_{\mathcal{H}} d\mu(\gamma) \int_{\mathcal{H}} d\nu(\gamma')\, f(\gamma + \gamma'). \end{aligned} \tag{2.4}$$

In particular, this relation yields $(\mu * (\nu * \rho))(M) = ((\mu * \nu) * \rho)(M)$ for each $M \in \mathrm{Bor}(\mathcal{H})$, i.e. associativity of the convolution. Finally, let g be an arbitrary Borel function on $\mathcal{H}$ such that $|g(\gamma)| \leqslant 1$, then we may use (2.4) again to get the inequality

$$\begin{aligned} \left| \int_{\mathcal{H}} g(\gamma)\, d(\mu * \nu)(\gamma) \right| &\leqslant \int_{\mathcal{H}} \left| \int_{\mathcal{H}} g(\gamma + \gamma')\, d\nu(\gamma') \right| d|\mu|(\gamma) \\ &\leqslant \int_{\mathcal{H}} d|\mu|(\gamma) \int_{\mathcal{H}} d|\nu|(\gamma')\, |g(\gamma + \gamma')| \\ &\leqslant |\mu|(\mathcal{H})\, |\nu|(\mathcal{H}), \end{aligned}$$

which yields

$$|\mu * \nu|(\mathcal{H}) \leqslant |\mu|(\mathcal{H})\, |\nu|(\mathcal{H}). \tag{2.5}$$

The above discussion can be summed as follows:

PROPOSITION 5.2.1. *The space $\mathcal{M}(\mathcal{H})$ equipped with the norm $|\,.\,|(\mathcal{H})$ and the product $*$ is a commutative Banach algebra.*

Let us now turn to the set $\mathcal{F}(\mathcal{H})$. Continuity of $(\gamma, .)$ implies continuity of $e^{i(\gamma, .)}$ so the latter is Borel measurable and the Fourier transform makes sense for each $\mu \in \mathcal{M}(\mathcal{H})$. It is also clear that $\mathcal{F}(\mathcal{H})$ is a vector space with respect to pointwise addition and scalar multiplication. We want to show that the Banach-algebra structure of $\mathcal{M}(\mathcal{H})$ transfers isomorphically to $\mathcal{F}(\mathcal{H})$. The crucial point is to prove that the correspondence between the measures from $\mathcal{M}(\mathcal{H})$ and their Fourier transforms is *bijective*; in view of linearity it is sufficient to check that $f = 0$ implies $\mu_f = 0$. If $f = 0$, then $f(\gamma) \pm f(-\gamma) = 0$ holds for all $\gamma \in \mathcal{H}$ so

$$\int_{\mathcal{H}} \cos(\gamma, \gamma')\, d\mu_f(\gamma') = \int_{\mathcal{H}} \sin(\gamma, \gamma')\, d\mu_f(\gamma') = 0.$$

Since the integrated functions here are real-valued, the integrals with respect to $\operatorname{Re} \mu_f$ and $\operatorname{Im} \mu_f$ must also be zero, and therefore

$$\int_{\mathcal{H}} e^{i(\gamma,\gamma')}\, d \operatorname{Re} \mu_f(\gamma') = \int_{\mathcal{H}} e^{i(\gamma,\gamma')}\, d \operatorname{Im} \mu_f(\gamma') = 0 \tag{$*$}$$

holds for all $\gamma \in \mathcal{H}$. Consider now, for example, the signed measure $\rho \equiv \operatorname{Re} \mu_f$. Let $\rho = \rho_1 - i\rho_2$ be its Jordan decomposition. The positive measures ρ_1, ρ_2 have disjoint supports so $\rho_1 \neq \rho_2$ unless both of them are zero. However, Fourier transformation (denoted by F) is known to be injective for the positive measures on a (real, separable) Hilbert space (Skorokhod, 1975, §1.3) so $F\rho_1 \neq F\rho_2$ unless $\rho = 0$. Combining this fact with the relation $(*)$, we get $\operatorname{Re} \mu_f = 0$; the same argument applies to $\operatorname{Im} \mu_f$.

The abbreviation μ_f for the measure corresponding to a function $f \in \mathcal{F}(\mathcal{H})$ therefore makes sense and we shall use it whenever it proves convenient. There is a natural norm on $\mathcal{F}(\mathcal{H})$ defined by

$$\|\cdot\|_0 : \|f\|_0 = |\mu_f|\,(\mathcal{H}). \tag{2.6}$$

Let us collect some simple properties of $\mathcal{F}(\mathcal{H})$:

PROPOSITION 5.2.2. *The space $\mathcal{F}(\mathcal{H})$ with the norm (2.6) is a functional Banach algebra with unity. Each $f \in \mathcal{F}(\mathcal{H})$ is norm continuous and bounded, $\|f\|_\infty \leqslant \|f\|_0$. If $h : \mathbb{C} \to \mathbb{C}$ is an entire analytic function and $f \in \mathcal{F}(\mathcal{H})$, then the composed mapping $h \circ f$ belongs to $\mathcal{F}(\mathcal{H})$ as well.*

Proof. We have shown that the Fourier transformation F maps $\mathcal{M}(\mathcal{H})$ bijectively on $\mathcal{F}(\mathcal{H})$, so the latter with the norm (2.6) is a Banach space isometric to $\mathcal{M}(\mathcal{H})$. The convolution is transformed by F into pointwise multiplication: the relation (2.4) together with Fubini theorem give

$$(F(\mu * \nu))(\gamma) = \int_{\mathcal{H}} e^{i(\gamma,\gamma')}\, d(\mu * \nu)(\gamma')$$

$$= \int_{\mathcal{H}\times\mathcal{H}} e^{i(\gamma,\gamma'+\gamma'')}\, d(\mu \otimes \nu)(\gamma', \gamma'') = (F\mu)(\gamma)\,(F\nu)(\gamma)$$

for all μ, $\nu \in \mathcal{M}(\mathcal{H})$ and $\gamma \in \mathcal{H}$. The Dirac measure μ_e supported by zero, $\mu_e(\{0\}) = 1$ and $\mu_e(\mathcal{H}\setminus\{0\}) = 0$, belongs certainly to $\mathcal{M}(\mathcal{H})$; then $e = F\mu_e$ is the unit element of $\mathcal{F}(\mathcal{H})$, $e(\gamma) = 1$ for all $\gamma \in \mathcal{H}$. Since $\mathcal{H}$ is first countable with respect to the norm topology (even second countable), a function $f : \mathcal{H} \to \mathbb{C}$ is continuous if it is sequentially continuous. Consider a norm-convergent sequence $\{\gamma_n\} \subset \mathcal{H}$ with a limit γ. For an arbitrary $f \in \mathcal{F}(\mathcal{H})$, the dominated-convergence theorem yields

$$\lim_{n\to\infty} f(\gamma_n) = \lim_{n\to\infty} \int_{\mathcal{H}} \exp(i(\gamma_n, \gamma'))\, d\mu_f(\gamma')$$

$$= \int_{\mathcal{H}} \exp(i(\gamma, \gamma'))\, d\mu_f(\gamma') = f(\gamma)$$

so f is norm continuous. Further, the inequality $|f(\gamma)| \leqslant \int_{\mathcal{H}} d\,|\mu_f|\,(\gamma')$ gives

$$\|f\|_\infty = \sup_{\gamma\in\mathcal{H}} |f(\gamma)| \leqslant |\mu_f|\,(\mathcal{H}) = \|f\|_0 .$$

Finally, let us check that $\mathcal{F}(\mathcal{H})$ is closed under composition with entire functions. Let h be expressed by the series $h(z) = \Sigma_{n=0}^{\infty} a_n z^n$ with the infinite radius of convergence, then $h \circ f = \Sigma_{n=0}^{\infty} a_n f^n$. The sequence $\{a_n f^n\}$ is absolutely summable, because $\Sigma_{n=0}^{\infty} \|a_n f^n\|_0 \leqslant \Sigma_{n=0}^{\infty} |a_n|\, \|f\|_0^n < \infty$. Since $\mathcal{F}(\mathcal{H})$ is a Banach space, it is at the same time summable (Reed and Simon, 1972, theorem III.3), i.e. $\Sigma_{n=0}^{N} a_n f^n$ converges with $N \to \infty$ in the $\|\,.\,\|_0$-norm to some element of $\mathcal{F}(\mathcal{H})$. ■

Now we are able to formulate the main definition. The following subsets of $\mathbb{C}$ will be used: $\mathbb{C}_F = \{z : z \neq 0, \text{Im}\, z \leqslant 0\}$ and its interior, i.e. the open lower halfplane $\mathbb{C}_F^0 = \{z : \text{Im}\, z < 0\}$. To any $s \in \mathbb{C}_F$, we define the mapping $I_s^{\text{f}} : \mathcal{F}(\mathcal{H}) \to \mathbb{C}$ by

$$I_s^{\text{f}} : I_s^{\text{f}}(f) = \int_{\mathcal{H}} \exp\left\{ -\frac{is}{2}\|\gamma\|^2 \right\} d\mu_f(\gamma). \tag{2.7}$$

The definition makes sense: continuity of the norm implies the function $\exp\{-(is/2)\|\,.\,\|^2\}$ to be continuous; hence it is Borel measurable and at the same time bounded. If $s = 1$, then (2.7) is just the definition of the F-integral given by Albeverio and Høegh-Krohn. With the classical oscillatory integrals in mind, they then proposed the name Fresnel integral for I_1^{f}. Adopting this convention, we call I_s^{f} **Fresnelian Feynman map** (or complex F-integral) with the parameter s. It will be specified by the superscript 'f'; for brevity we shall speak mostly about the F_s-*map*. In the particular case when s is real and positive, I_s^{f} is called the **Fresnelian Feynman integral** (or Fresnel integral) with the parameter s; the subscript s may be omitted for $s = 1$.

PROPOSITION 5.2.3. *For each $s \in \mathbb{C}_F$, the F_s-map is a normalized linear functional and $\|I_s^{\text{f}}\| = 1$. For a fixed $f \in \mathcal{F}(\mathcal{H})$, the function $s \mapsto I_s^{\text{f}}(f)$ is single-valued analytic in $\mathbb{C}_F^0$ and continuous in $\mathbb{C}_F$.*

Proof. Linearity and boundedness are obvious, the normalization follows from

$$I_s^{\mathrm{f}}(e) = \int_{\mathcal{H}} \exp\left\{ -\frac{is}{2}\|\gamma\|^2 \right\} \mathrm{d}\mu_e(\gamma) = 1 .$$

The conditions under which the integral in (2.7) may be differentiated with respect to s verify easily: it holds that

$$\left| -\frac{i}{2}\|\gamma\|^2 \exp\left\{ -\frac{is}{2}\|\gamma\|^2 \right\} \right| \leqslant \tfrac{1}{2}\|\gamma\|^2 \exp\{\tfrac{1}{2}\|\gamma\|^2 \operatorname{Im} s\}$$

and the r.h.s. belongs to $L(\mathcal{H}, \mu_f)$ for $s \in \mathbb{C}_F^0$. Thus the function $s \mapsto I_s^{\mathrm{f}}(f)$ is differentiable in each open halfplane $\{s : \operatorname{Im} s < s_0 < 0\}$ and

$$\frac{\mathrm{d}}{\mathrm{d}s} I_s^{\mathrm{f}}(f) = -\frac{i}{2}\int_{\mathcal{H}} \|\gamma\|^2 \exp\left\{ -\frac{is}{2}\|\gamma\|^2 \right\} \mathrm{d}\mu_f(\gamma). \tag{2.8}$$

Consequently, $s \mapsto I_s(f)$ is single-valued analytic in $\mathbb{C}_F^0$. Its continuity in $\mathbb{C}_F$ follows from the dominated-convergence theorem. ■

On the other hand, the functional I_s^{f} is not positive with respect to the natural involution in $\mathcal{F}(\mathcal{H})$ unless s is purely imaginary, as the relation (2.11) below illustrates. It is nothing strange: had the Feynman measure existed, it would be complex. Notice also that $I_s^{\mathrm{f}}(f)$ may be extended by continuity to $I_0^{\mathrm{f}}(f) = f(0)$, but we shall not need this relation (see the notes). More important for us is the fact that in some cases the 'integration' may be performed in successive steps.

THEOREM 5.2.4 (Fubini property). *Let $f \in \mathcal{F}(\mathcal{H})$, where $\mathcal{H}$ decomposes into an orthogonal sum $\mathcal{H}_1 \oplus \mathcal{H}_2$; we denote $f(\gamma) = f(\gamma_1, \gamma_2)$, where $\gamma = \gamma_1 + \gamma_2$ with $\gamma_i \in \mathcal{H}_i$. The function $f(\,.\,, \gamma_2)$ belongs to $\mathcal{F}(\mathcal{H}_1)$ for each $\gamma_2 \in \mathcal{H}_2$; further, $\int_{\mathcal{H}_1}^{\mathrm{f}} f(\gamma_1, .\,)\, \mathrm{D}\phi_s(\gamma_1)$ belongs to $\mathcal{F}(\mathcal{H}_2)$ and the relation*

$$\int_{\mathcal{H}}^{\mathrm{f}} f(\gamma)\, \mathrm{D}\phi_s(\gamma) = \int_{\mathcal{H}_1}^{\mathrm{f}} \left(\int_{\mathcal{H}_2}^{\mathrm{f}} f(\gamma_1, \gamma_2)\, \mathrm{D}\phi_s(\gamma_1) \right) \mathrm{D}\phi_s(\gamma_2) \tag{2.9}$$

holds; the same is true if the roles of $\mathcal{H}_1$ and $\mathcal{H}_2$ are interchanged.

Proof. Suppose that f corresponds to a measure $\mu \in \mathcal{M}(\mathcal{H})$, then we write $\mathrm{d}\mu(\gamma) = \mathrm{d}\mu(\gamma_1, \gamma_2)$ and define

$$\mu_{\gamma_2}(M_1) = \tilde{\mu}_{\gamma_2}(M_1 \times \mathcal{H}_2),$$

$$\tilde{\mu}_{\gamma_2}(M) = \int_M \exp\{i(\gamma_2, \gamma_2')\}\, \mathrm{d}\mu(\gamma_1, \gamma_2)$$

for all $M_1 \in \operatorname{Bor}(\mathcal{H}_1)$ and $\gamma_2 \in \mathcal{H}_2$. It is easy to see that $\mu_{\gamma_2} \in \mathcal{M}(\mathcal{H}_1)$; the image-measure theorem gives

$$\int_{\mathcal{H}_1} g(\gamma_1)\, \mathrm{d}\mu_{\gamma_2}(\gamma_1) = \int_{\mathcal{H}} g(\gamma_1)\, \mathrm{d}\tilde{\mu}_{\gamma_2}(\gamma_1, \gamma_2)$$

for each $g \in L(\mathscr{H}_1, \mu_{\gamma_2})$. The Borel measure $\tilde{\mu}_{\gamma_2}$ on $\mathscr{H}$ is absolutely continuous with respect to μ so the last integral can be expressed by means of μ and the Radon–Nikodým derivative (cf. (A.5)); it holds that

$$\int_{\mathscr{H}_1} g(\gamma_1)\, \mathrm{d}\mu_{\gamma_2}(\gamma_1) = \int_{\mathscr{H}} g(\gamma_1')\exp\{i(\gamma_2, \gamma_2')\}\, \mathrm{d}\mu(\gamma_1', \gamma_2') \tag{2.10}$$

for each $g \in L(\mathscr{H}_1, \mu_{\gamma_2})$. Since $(\gamma, \gamma') = (\gamma_1, \gamma_1') + (\gamma_2, \gamma_2')$, the equality (2.10) for $g(\gamma_1') = \exp\{i(\gamma_1, \gamma_1')\}$ shows that $f(\,.\,, \gamma_2)$ belongs to $\mathscr{F}(\mathscr{H}_1)$ for any fixed γ_2 and

$$\int_{\mathscr{H}_1}^{\mathrm{f}} f(\gamma_1, \gamma_2)\, \mathrm{D}\phi_s(\gamma_1) = \int_{\mathscr{H}_1} \exp\left\{-\frac{is}{2}\|\gamma_1\|^2\right\} \mathrm{d}\mu_{\gamma_2}(\gamma_1).$$

The r.h.s. of the last relation can be expressed using (2.10) with $g(\gamma_1) = \exp\{-(is/2)\,\|\gamma_1\|^2\}$: we obtain

$$\int_{\mathscr{H}_1}^{\mathrm{f}} f(\gamma_1, \gamma_2)\, \mathrm{D}\phi_s(\gamma_1) = \int_{\mathscr{H}} \exp\left\{-\frac{is}{2}\|\gamma_1'\|^2 + i(\gamma_2, \gamma_2')\right\} \mathrm{d}\mu(\gamma_1', \gamma_2')$$

and the same argument as above gives

$$\int_{\mathscr{H}_1}^{\mathrm{f}} f(\gamma_1, \gamma_2)\, \mathrm{D}\phi_s(\gamma_1) = \int_{\mathscr{H}} \exp\{i(\gamma_2, \gamma_2')\}\, \mathrm{d}\nu_s(\gamma_2'),$$

where ν_s is the complex Borel measure from $\mathscr{M}(\mathscr{H}_2)$ defined by the relation

$$\nu_s(M_2) = \int_{\mathscr{H}_1 \times M_2} \exp\left\{-\frac{is}{2}\|\gamma_1\|^2\right\} \mathrm{d}\mu(\gamma_1, \gamma_2)$$

for each $M_2 \in \mathrm{Bor}(\mathscr{H}_2)$. Hence $\int_{\mathscr{H}_1}^{\mathrm{f}} f(\gamma_1, .\,)\, \mathrm{D}\phi_s(\gamma_1)$ belongs to $\mathscr{F}(\mathscr{H}_2)$ and

$$\begin{aligned}
&\int_{\mathscr{H}_2}^{\mathrm{f}} \left(\int_{\mathscr{H}_1}^{\mathrm{f}} f(\gamma_1, \gamma_2)\, \mathrm{D}\phi_s(\gamma_1)\right) \mathrm{D}\phi_s(\gamma_2) \\
&\quad = \int_{\mathscr{H}_2} \exp\left\{-\frac{is}{2}\|\gamma_2\|^2\right\} \mathrm{d}\nu_s(\gamma_2) \\
&\quad = \int_{\mathscr{H}} \exp\left\{-\frac{is}{2}\|\gamma_1\|^2 - \frac{is}{2}\|\gamma_2\|^2\right\} \mathrm{d}\mu(\gamma_1, \gamma_2) = I_s^{\mathrm{f}}(f).
\end{aligned}$$

It is clear that the same argument works when the roles of $\mathscr{H}_1$ and $\mathscr{H}_2$ are interchanged. ■

A function $f : \mathscr{H} \to \mathbb{C}$ is called **cylindrical** (or **tame**) if there is a finite-dimensional projection P on $\mathscr{H}$ such that $f(\gamma) = f(P\gamma)$ for all $\gamma \in \mathscr{H}$. In that case, f is said to have **basis** (to be based) in the subspace $P\mathscr{H}$. The set of all cylindrical functions in $\mathscr{F}(\mathscr{H})$ is denoted by $\mathscr{F}^t(\mathscr{H})$.

THEOREM 5.2.5. *Let $f \in \mathcal{F}(\mathcal{H})$ be a cylindrical function based in $P\mathcal{H}$ with $\dim P = N$ and $f_P := f \upharpoonright P\mathcal{H}$, then the relation*

$$I_s^{\mathrm{f}}(f) = (2\pi i s)^{-N/2} \int_{P\mathcal{H}} \exp\left\{\frac{i}{2s} \|\gamma_P\|^2\right\} f_P(\gamma_P)\, dm(\gamma_P) \tag{2.11}$$

holds for each $s \in \mathbb{C}_F^0$, where m is the Lebesgue measure on $P\mathcal{H}$. It holds for real s too, if $f_P \in L(P\mathcal{H}, m)$.

For the proof, we need the following auxiliary assertion:

LEMMA 5.2.6. *For a non-zero real number s, consider the integral*

$$K_s^N(y, \alpha) = (2\pi i s)^{-N/2} \int_{C_\alpha} \exp\left\{\frac{i}{2s} |x + sy|^2\right\} dx, \tag{2.12a}$$

over the cube $C_\alpha = \{x : |x_j| \leqslant \alpha, j = 1, \ldots, N\}$ in $\mathbb{R}^N$. There is a positive K_s^N such that

$$|K_s^N(y, \alpha)| \leqslant K_s^N \tag{2.12b}$$

holds for all $y \in \mathbb{R}^N$ and $\alpha \geqslant 0$. Moreover, for each $y \in \mathbb{R}^N$ we have

$$\lim_{\alpha \to \infty} K_s^N(y, \alpha) = 1. \tag{2.12c}$$

Proof. Consider first the integral $C(a, b) = \int_0^a \exp(ibt^2)\, dt$ with some positive a and b. Since the function $z \mapsto \exp(ibz^2)$ is entire, $C(a, b)$ may be evaluated by contour integration. Choosing the contour sketched on Figure 3, we get

$$C(a, b) = -ia \int_0^{\pi/4} \exp(iba^2\, \mathrm{e}^{2i\varphi})\, \mathrm{e}^{i\varphi}\, \mathrm{d}\varphi + \exp\left(\frac{\pi i}{4}\right) \int_0^a \exp(-bt^2)\, \mathrm{d}t \tag{2.13a}$$

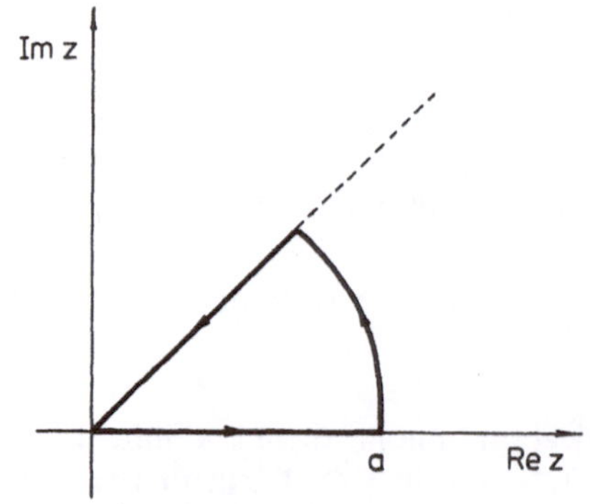

5.3. The integration contour used in proof of Lemma 5.2.6.

The first integral on the r.h.s. (call it I_1) can be estimated as follows

$$|I_1| \leqslant \frac{a}{2}\left(\int_0^{a^{-3/2}} + \int_{a^{-3/2}}^{\pi/2}\right) \exp(-ba^2 \sin \xi)\, d\xi$$

provided that $a \geqslant (\pi/2)^{-2/3}$. For $\xi \in [a^{-3/2}, \pi/2]$, we may use the inequalities $\sin \xi \geqslant \frac{1}{2}\xi \geqslant \frac{1}{2}a^{-3/2}$, which give

$$|I_1| \leqslant \tfrac{1}{2}a^{-1/2} + \frac{\pi a}{4}\exp(-\tfrac{1}{2}ba^{1/2}). \tag{$*$}$$

The second integral on the r.h.s. of (2.13a) is estimated easily as well as the whole of $C(a, b)$ for small values of a: together we obtain

$$|C(a,b)| \leqslant \begin{cases} a & \ldots a \leqslant (\pi/2)^{-2/3} \\ \frac{1}{2}\left(\frac{\pi}{b}\right)^{1/2} + \frac{1}{2}a^{-1/2} + \frac{\pi a}{4}\exp(-\frac{1}{2}ba^{1/2}) & \\ & \ldots a \geqslant (\pi/2)^{-2/3} \end{cases} \tag{2.13b}$$

so $|C(a, b)|$ is for each b majorized by a constant independent of a. The relations (2.13a) and ($*$) further imply

$$\lim_{a\to\infty} C(a,b) = \exp\left(\frac{\pi i}{4}\right)\int_0^\infty \exp(-bt^2)\, dt = \left(\frac{\pi i}{4b}\right)^{1/2}. \tag{2.13c}$$

Since $C(a, -b) = \overline{C(a, b)}$, the relations (2.13b) and (2.13c) also hold for each negative b. Now we employ the equality

$$K_s^1(y, \alpha) = (2\pi i s)^{-1/2} \int_{-\alpha}^{\alpha} \exp\left\{\frac{i}{2s}(x+sy)^2\right\} dx$$

$$= (2\pi i s)^{-1/2}\, [C(\alpha + sy, 1/2s) + C(\alpha - sy, 1/2s)]\,.$$

Combining it with (2.13b), we see that $|K_s^1(y, \alpha)|$ is majorized by a constant K_s^1 which depends on s only; further, (2.13c) yields (2.12c) for $N = 1$. Finally, the Fubini theorem allows us to express (2.12a) in the form

$$K_s^N(y, \alpha) = \prod_{j=1}^{N} K_s^1(y_j, \alpha)$$

so $K_s^N = (K_s^1)^N$ and the relation (2.12c) holds. ∎

Proof of Theorem 2.5. We start by considering the case of a finite-dimensional $\mathscr{H} = \mathbb{R}^N$ when all functions are cylindrical. We take $f \in \mathscr{F}(\mathbb{R}^N)$, $f(x) = \int_{\mathbb{R}^N} e^{ix\cdot y}\, d\mu_f(y)$, and define

$$I_s'(f) = (2\pi i s)^{-N/2} \int_{\mathbb{R}^N} \exp\left\{\frac{i}{2s}|x|^2\right\} f(x)\, dx. \tag{2.14a}$$

The integral exists for each $s \in \mathbb{C}_F$ due to the assumption. Let us substitute for $f(x)$. If $s \in \mathbb{C}_F^0$, then we have

$$I_s'(f) = (2\pi i s)^{-N/2} \int_{\mathbb{R}^N} d\mu_f(y) \prod_{j=1}^{N} \int_{\mathbb{R}} \exp\left\{ \frac{i}{2s} x_j^2 + i x_j y_j \right\} dx_j$$

by the Fubini theorem; evaluating the last integral, we get

$$I_s'(f) = \int_{\mathbb{R}^N} \exp\left\{ -\frac{is}{2} |y|^2 \right\} d\mu_f(y) = I_s^{\mathrm{f}}(f). \tag{2.15}$$

For real s one cannot apply the Fubini theorem directly, because $(x, y) \mapsto \exp\{(i/2s)|x|^2 + ix \cdot y\}$ does not belong to $L(\mathbb{R}^N \times \mathbb{R}^N, m \otimes \mu_f)$. However, we have assumed $f \in L(\mathbb{R}^N)$ so the dominated-convergence theorem allows us to express $I_s'(f)$ as follows

$$I_s'(f) = \lim_{\alpha \to \infty} (2\pi i s)^{-N/2} \int_{C_\alpha} \exp\left\{ \frac{i}{2s} |x|^2 \right\} f(x)\, dx, \tag{2.14b}$$

where C_α is the cube introduced in Lemma 2.6. Now the integrations may be interchanged, so after a short manipulation we get

$$I_s'(f) = \lim_{\alpha \to \infty} \int_{C_\alpha} \exp\left\{ -\frac{is}{2} |y|^2 \right\} K_s^N(y, \alpha)\, d\mu_f(y).$$

Finally, using Lemma 2.6 together with the dominated-convergence theorem, we arrive again at the relation (2.15).

In the general case, we use the decomposition $\mathscr{H} = P\mathscr{H} \oplus P'\mathscr{H}$, $P' = I - P$. The function f can be written as $f(\gamma) = e(P'\gamma) f_P(P\gamma)$ so the Fubini property in combination with linearity and normalization of I_s^{f} yields

$$\begin{aligned} I_s^{\mathrm{f}}(f) &= \int_{P\mathscr{H}}^{\mathrm{f}} f_P(P\gamma) \left(\int_{P'\mathscr{H}}^{\mathrm{f}} e(P'\gamma)\, D\phi_s(P'\gamma) \right) D\phi_s(P\gamma) \\ &= \int_{P\mathscr{H}}^{\mathrm{f}} f_P(P\gamma)\, D\phi_s(\gamma). \end{aligned} \tag{2.16}$$

Now it is sufficient to apply the result from the first part of the proof to the r.h.s., and the relation (2.11) follows. ∎

Later we shall use Theorem 2.5 to extend the definition (2.7) by polygonal-path approximations. For this purpose, the following property is crucial:

PROPOSITION 5.2.7. *Let* $\{E_n\}_{n=1}^{\infty}$ *be a sequence of orthogonal projections on* $\mathscr{H}$ *such that* s-$\lim_{n\to\infty} E_n = I$, *then for each* $f \in \mathscr{F}(\mathscr{H})$ *we have*

$$\lim_{n \to \infty} \int_{\mathscr{H}}^{\mathrm{f}} f(E_n\gamma)\, D\phi_s(\gamma) = \int_{\mathscr{H}}^{\mathrm{f}} f(\gamma)\, D\phi_s(\gamma). \tag{2.17}$$

Proof. For a given n, we decompose $\mathscr{H}$ into the orthogonal sum $E_n'\mathscr{H} \oplus E_n\mathscr{H}$. The restriction $f_n = f \upharpoonright E_n\mathscr{H}$ can be expressed as

$$f(E_n\gamma) = \int_{\mathscr{H}} \exp\{i(E_n\gamma, E_n\gamma')\}\, \mathrm{d}\mu_f(E_n'\gamma', E_n\gamma').$$

By analogy with the proof of Theorem 2.4, we define the measure μ_n on $E_n\mathscr{H}$ by $\mu_n(M) = \mu_f(E_n'\mathscr{H} \times M)$ for all $M \in \mathrm{Bor}(E_n\mathscr{H})$, and check the relation

$$\int_{E_n\mathscr{H}} g(\gamma_2')\, \mathrm{d}\mu_n(\gamma_2') = \int_{\mathscr{H}} g(E_n\gamma')\, \mathrm{d}\mu_f(E_n'\gamma', E_n\gamma')$$

for any $g \in L(E_n\mathscr{H}, \mu_n)$. Applying it to $g(\gamma_2') = \exp\{i(E_n\gamma, \gamma_2')\}$, we get $f_n \in \mathscr{F}(E_n\mathscr{H})$. Applying it further to $g(\gamma_2') = \exp\{-(is/2)\|\gamma_2'\|^2\}$, we obtain

$$I_s^{\mathrm{f}}(f_n) = \int_{\mathscr{H}} \exp\left\{-\frac{is}{2}\|E_n\gamma'\|^2\right\} \mathrm{d}\mu_f(\gamma');$$

the l.h.s. is equal to $\int_{\mathscr{H}}^{\mathrm{f}} f(E_n\gamma)\, \mathrm{D}\phi_s(\gamma)$ similarly as in (2.16). Finally, the relation

$$\lim_{n\to\infty} \exp\left\{-\frac{is}{2}\|E_n\gamma'\|^2\right\} = \exp\left\{-\frac{is}{2}\|\gamma'\|^2\right\}$$

holds due to the assumption, so (2.17) follows from the dominated-convergence theorem. ∎

Let us now see how $\int_{\mathscr{H}}^{\mathrm{f}} f(\gamma)\mathrm{D}\phi_s(\gamma)$ behaves when the 'integration variable' is changed. Consider first the translations:

PROPOSITION 5.2.8. *Let α be an arbitrary vector from $\mathscr{H}$ and $f \in \mathscr{F}(\mathscr{H})$. For $s \in \mathbb{C}_F$, we define*

$$f_\alpha : f_\alpha(\gamma) = \exp\left\{\frac{i}{s}(\gamma, \alpha)\right\} f(\gamma + \alpha) \tag{2.18a}$$

and assume $f_\alpha \in \mathscr{F}(\mathscr{H})$ (this is fulfilled automatically if s is real), then the following relation is valid

$$I_s^{\mathrm{f}}(f_\alpha) = \exp\left\{-\frac{i}{2s}\|\alpha\|^2\right\} I_s^{\mathrm{f}}(f). \tag{2.18}$$

Proof. First, assume s to be real non-zero. The image-measure theorem together with (A.5) show that f_α is a Fourier transform of the measure $\mu_\alpha \in \mathscr{F}(\mathscr{H})$ defined by

$$\mu_\alpha(M) = \int_{M+\alpha/s} \mathrm{e}^{i(\alpha,\gamma')}\, \mathrm{d}\mu_f(\gamma')$$

for all $M \in \mathrm{Bor}(\mathscr{H})$. Further, we have $\mathscr{H} - \alpha/s = \mathscr{H}$ so

$$I_s^{\mathrm{f}}(f_\alpha) = \int_{\mathscr{H}} \exp\left\{ -\frac{is}{2} \|\gamma' + \alpha/s\|^2 + i(\alpha, \gamma') \right\} \mathrm{d}\mu_f(\gamma')$$

$$= \exp\left\{ -\frac{i}{2s} \|\alpha\|^2 \right\} I_s^{\mathrm{f}}(f).$$

If $s \in \mathbb{C}_F^0$, we choose a sequence $\{E_n\}_{n=1}^\infty$ of projections, $\dim E_n = N$, such that $\text{s-lim}_{n\to\infty} E_n = I$. Since f_α is assumed to belong to $\mathscr{F}(\mathscr{H})$, we may use Theorem 2.5 and Proposition 2.7, which give

$$I_s^{\mathrm{f}}(f_\alpha) = \lim_{n\to\infty} \int_{\mathscr{H}}^{\mathrm{f}} f_\alpha(E_n\gamma)\, \mathrm{D}\phi_s(\gamma)$$

$$= \lim_{n\to\infty} (2\pi i s)^{-N/2} \int_{E_n\mathscr{H}} \exp\left\{ \frac{i}{2s} \|E_n\gamma\|^2 \right\} f_\alpha(E_n\gamma)\, \mathrm{d}m(E_n\gamma)$$

$$= \lim_{n\to\infty} (2\pi i s)^{-N/2} \exp\left\{ -\frac{i}{2s} \|E_n\alpha\|^2 \right\} \times$$

$$\times \int_{E_n\mathscr{H}} \exp\left\{ \frac{i}{2s} \|E_n\gamma'\|^2 \right\} f(E_n\gamma')\, \mathrm{d}m(E_n\gamma')$$

$$= \lim_{n\to\infty} \exp\left\{ -\frac{i}{2s} \|E_n\alpha\|^2 \right\} \int_{\mathscr{H}}^{\mathrm{f}} f(E_n\gamma')\, \mathrm{D}\phi_s(\gamma')$$

$$= \exp\left\{ -\frac{i}{2s} \|\alpha\|^2 \right\} I_s^{\mathrm{f}}(f). \qquad \blacksquare$$

PROPOSITION 5.2.9. *Let R be a regular isometric operator on $\mathscr{H}$ and $f \in \mathscr{F}(\mathscr{H})$, then*

$$\int_{\mathscr{H}}^{\mathrm{f}} f(R\gamma)\, \mathrm{D}\phi_s(\gamma) = \int_{\mathscr{H}}^{\mathrm{f}} f(\gamma)\, \mathrm{D}\phi_s(\gamma). \tag{2.19}$$

The proof is based again on the image-measure theorem: the function f_R : $f_R(\gamma) = f(R\gamma)$ expresses as

$$f_R(\gamma) = \int_{\mathscr{H}} \exp\{i(\gamma, R^{-1}\gamma')\}\, \mathrm{d}\mu_f(\gamma') = \int_{\mathscr{H}} \mathrm{e}^{i(\gamma,\gamma'')}\, \mathrm{d}\mu_f(R\gamma'')$$

so

$$I_s^{\mathrm{f}}(f_R) = \int_{\mathscr{H}} \exp\left\{ -\frac{is}{2} \|R^{-1}\gamma'\|^2 \right\} \mathrm{d}\mu_f(\gamma')$$

$$= \int_{\mathscr{H}} \exp\left\{ -\frac{is}{2} \|\gamma'\|^2 \right\} \mathrm{d}\mu_f(\gamma') = I_s^{\mathrm{f}}(f). \qquad \blacksquare$$

The above two assertions exhibit the transformation properties of the F_s-maps under *Euclidean motions*, $\gamma' = R\gamma + \alpha$, of the path space $\mathscr{H}$. One is naturally interested in other 'change-of-variable' transformations. For the present, we restrict our attention to the cylindrical functions. Let $f \in \mathscr{F}^t(\mathscr{H})$ be based in $P\mathscr{H}$, and consider an operator $B \in \mathscr{B}(\mathscr{H})$ with the following properties: B is regular, reduced by P and its part in $(I - P)\mathscr{H}$ coincides with the unit operator. With future applications in mind, it is convenient to write $B = I + K$. It is easy to see that the function

$$e_s^K f_{I+K} : (e_s^K f_{I+K})(\gamma) = \exp\left\{\frac{i}{s}(\gamma, K\gamma) + \frac{i}{2s}\|K\gamma\|^2\right\} f((I+K)\gamma) \quad (2.20a)$$

is again cylindrical and based in $P\mathscr{H}$. Suppose $e_s^K f_{I+K} \in \mathscr{F}(\mathscr{H})$, and in addition, let the cylindrical projections of f, $e_s^K f_{I+K}$ be integrable if s is real. Then both $I_s^{\mathrm{f}}(f)$ and $I_s^{\mathrm{f}}(e_s^K f_{I+K})$ may be expressed from Theorem 2.5; we obtain particularly

$$I_s^{\mathrm{f}}(e_s^K f_{I+K}) = (2\pi i s)^{-N/2} \int_{P\mathscr{H}} \exp\left\{\frac{i}{2s}\|B\gamma_P\|^2\right\} f(B\gamma_P)\, dm(\gamma_P),$$

so changing variables γ_P to $B\gamma_P$, we get

$$\int_{\mathscr{H}}^{\mathrm{f}} \exp\left\{\frac{i}{s}(\gamma, K\gamma) + \frac{i}{2s}\|K\gamma\|^2\right\} f((I+K)\gamma)\, \mathrm{D}\phi_s(\gamma)$$

$$= |\det(I+K)|^{-1} \int_{\mathscr{H}}^{\mathrm{f}} f(\gamma)\, \mathrm{D}\phi_s(\gamma), \quad (2.20b)$$

where the determinant is taken actually from the finite-dimensional part of $I + K$ only. We call this result the **Cameron–Martin formula** adopting the name from the analogous transformation relation valid for Wiener integrals. Of course, both the function f and the operator K used in the above derivation are rather special, and we would like to prove a more general result. This will be done in Section 4.

Despite their nice integral-like properties, one should not forget that the F_s-maps are not integrals, and therefore some important assertions from the standard theory of integration need not be valid for them. This particularly concerns the *dominated-convergence theorem*:

EXAMPLE 5.2.10. Choose a sequence $\{E_n\}_{n=1}$ of projections on $\mathscr{H}$ with $\dim E_n = n$, and define

$$f_n : f_n(\gamma) = N_n \exp\left\{-\frac{\epsilon + is_1}{2|s|^2}\|E_n\gamma\|^2\right\}, \quad (2.21a)$$

where $s = s_1 - is_2$ with $s_1 \neq 0$ and N_n is a normalization factor to be specified. If $\epsilon > 0$, it is not difficult to check that $f_n \in \mathscr{F}(\mathscr{H}) \cap L(E_n\mathscr{H}, m)$ for each n. At the same time, if the sequence $\{N_n\}$ is bounded, say $|N_n| \leqslant 1$ for all n, then the functions (3.21a) are majorized (independently of n) by the unit function e, which is 'integrable' in view of Proposition 2.2. This is particularly true for

$$N_n = [is(\epsilon + s_2)\, |s|^{-2}]^{n/2} \tag{2.21b}$$

with a sufficiently small positive ϵ. If the normalization factors are to be chosen less than 1, then (2.21a) gives

$$\lim_{n \to \infty} \|f_n\|_\infty = 0. \tag{2.22}$$

Had the dominated-convergence theorem held, the last relation would imply $I_n^{\mathrm{f}}(f_n)$ to converge to zero as $n \to \infty$. Using Theorem 2.5, however, one calculates easily that with the normalization factors (2.21b) we have $I_s^{\mathrm{f}}(f_n) = 1$ for each n.

Notice that in order to illustrate the invalidity of the dominated-convergence theorem, we had only to check $\lim_{n\to\infty} f_n(\gamma) = 0$ a.e. in $\mathscr{H}$. The relation (2.22) says that actually a stronger conclusion can be made, and it shows simultaneously that the functional $I_s^{\mathrm{f}}(\,.\,)$, though continuous in $\|\,.\,\|_0$, is *not* continuous with respect to the weaker norm $\|\,.\,\|_\infty$ on $\mathscr{F}(\mathscr{H})$. For the particular case of F-integrals, one can illustrate these facts in an easier way (cf. Problems 44 and 47).

5.3. Hilbert Spaces of Paths

Now we are going to find suitable path spaces which could be used as $\mathscr{H}$ in the above-exposed theory. As mentioned in the introduction, we are interested primarily in the quantum-mechanical systems whose configuration space is $\mathbb{R}^d$. A comparison of the abstract F_S-maps with the formal expression (1.3) then suggests that the norm in $\mathscr{H}$ should be given by

$$\|\gamma\|^2 = \int_0^t |\dot{\gamma}(\tau)|^2 \, d\tau \tag{3.1a}$$

provided that the time interval under consideration is $J^t \equiv [0, t]$. Although natural, this choice raises some mathematical questions. The paths which should be taken into account are presumably continuous, but this set is too large from the viewpoint of (3.1a). A continuous path $\gamma : J^t \to \mathbb{R}^d$ is determined completely by its derivative (up to an additive constant, but the end of the path is kept fixed, $\gamma(t) = x$) iff it is absolutely continuous. The reason is that $\dot{\gamma}$ does not feel the singularly continuous part which γ might eventually have, and which must be therefore factorized out of $\mathscr{H}$.

One may ponder on other possibilities of determining $\mathscr{H}$. One of them is to replace $\dot{\gamma}$ in (3.1a) by the weak (distributional) derivative of γ, because the latter determines γ fully. Even in this case, however, we cannot consider all continuous paths because of the well-known difficulty with the squaring of distributions.

Another possibility is based on use on Fourier series: we can treat the paths γ expressed as

$$\gamma(\tau) = \alpha_0(\tau - t) + \sum_{n=1}^{\infty} \frac{\alpha_n t}{2\pi n} \sin\left(\frac{2\pi n\tau}{t}\right) + \sum_{n=1}^{\infty} \frac{\beta_n t}{2\pi n}\left[1 - \cos\left(\frac{2\pi n\tau}{t}\right)\right], \tag{3.2a}$$

where $\alpha_0, \alpha_n, \beta_n, n = 1, 2, \ldots$, are some $\mathbb{R}^d$-valued coefficients which obey

$$\sum_n (|\alpha_n|^2 + |\beta_n|^2) < \infty. \tag{3.2b}$$

Our main aim here is to show that all the mentioned ways result, in effect, in the same Hilbert spaces of paths.

REMARK 5.3.1. Before proceeding further, it is useful to list some notations. The set of all continuous paths $\gamma : J^t \to \mathbb{R}^d$ which fulfil $\gamma(t) = x$ is denoted by $C_x[J^t; \mathbb{R}^d]$. By $AC_x[J^t; \mathbb{R}^d]$ we denote its subset containing all γ which are (componentwise) absolutely continuous with $\dot\gamma \in L^2(J^t; \mathbb{R}^d)$. In particular, $AC_0[J^t; \mathbb{R}^d]$ is a vector space which can be equipped with the inner product

$$(\gamma, \tilde\gamma) = \int_0^t \dot\gamma(\tau) \cdot \dot{\tilde\gamma}(\tau)\, d\tau \tag{3.1b}$$

referring to the norm (3.1a). By $\dot\gamma$ we mean the usual derivative of $\gamma : \dot\gamma(\tau) = \lim_{h\to 0} [\gamma(\tau + h) - \gamma(\tau)]h^{-1}$, while $\dot\gamma_{\mathrm{w}}$ is the weak derivative, $\int_0^t \dot\gamma_{\mathrm{w}}(\tau) \times \varphi(\tau)\, d\tau = -\int_0^t \gamma(\tau) \cdot \dot\varphi(\tau)\, d\tau$ for all φ from $C_0^\infty(J^t; \mathbb{R}^d)$, the set of all infinitely differentiable functions with a compact support contained in $(0, t)$. The symbol $\mathscr{F}[J^t; \mathbb{R}^d]$ denotes the set of all paths γ which are of the form (3.2). We define an inner product on $\mathscr{F}[J^t; \mathbb{R}^d]$ by

$$(\gamma, \tilde\gamma)_F = t\alpha_0 \cdot \tilde\alpha_0 + \tfrac{1}{2} t \sum_{n=1}^{\infty} (\alpha_n \cdot \tilde\alpha_n + \beta_n \cdot \tilde\beta_n). \tag{3.3}$$

In $\mathscr{F}[J^t; \mathbb{R}^d]$, other two derivatives can be defined. We set

$$s_N(\tau) = \alpha_0 + \sum_{n=1}^{N} \alpha_n \cos\left(\frac{2\pi n\tau}{t}\right) + \sum_{n=1}^{N} \beta_n \sin\left(\frac{2\pi n\tau}{t}\right), \tag{3.4}$$

then we may denote by $\dot\gamma_{\mathrm{m}}$ the L^2-norm limit of $\{s_N\}$, $\dot\gamma_{\mathrm{m}} = \lim_{N\to\infty} s_N$, and by $\dot\gamma_{\mathrm{p}}$ the pointwise limit, $\dot\gamma_{\mathrm{p}}(\tau) = \lim_{N\to\infty} s_N(\tau)$ whenever the limit exists.

Notice first that $AC_0[J^t; \mathbb{R}^d]$ with the inner product (3.1b) is obviously isomorphic to $L^2(J^t; \mathbb{R}^d)$ so it forms a real separable Hilbert space. The same is true for $\mathscr{F}[J^t; \mathbb{R}^d]$.

PROPOSITION 5.3.2. *$\mathscr{F}[J^t; \mathbb{R}^d]$ is a subset in $C_0[J^t; \mathbb{R}^d]$; it is a real separable Hilbert space when equipped with the inner product* (3.3).

Proof An arbitrary $\gamma \in \mathscr{F}[J^t; \mathbb{R}^d]$ is real-valued and fulfils $\gamma(t) = 0$. Further, the inequalities

$$\left| \sum_{n=1}^{\infty} \frac{\alpha_n t}{2\pi n} \sin\left(\frac{2\pi n\tau}{t}\right) \right|$$

$$\leqslant \sum_{n=1}^{\infty} \frac{|\alpha_n| t}{2\pi n} \leqslant \left(\sum_{n=1}^{\infty} |\alpha_n|^2 \right)^{1/2} \left(\sum_{n=1}^{\infty} \frac{t^2}{4\pi^2 n^2} \right)^{1/2}$$

hold together with the analogous estimate for the other series in (3.2a). In view of (3.2b), both of them are majorized by some convergent τ-independent series, and converge therefore uniformly in J^t. Continuity of the partial sums then implies the continuity of γ. Next we want to show that $\mathscr{F}[J^t; \mathbb{R}^d]$ is isomorphic to $l^2(\mathbb{Z}_+; \mathbb{R}^d)$, the Hilbert space of $\mathbb{R}^d$-valued square-summable sequences. To this end, we have only to establish injectivity of the mapping $(\alpha_0, \{\alpha_n\}, \{\beta_n\}) \mapsto \gamma$. Each $\gamma \in \mathscr{F}[J^t; \mathbb{R}^d]$ is continuous in the closed interval J^t so it is integrable there and has a unique set of Fourier coefficients,

$$\gamma(\tau) = A_0 + \sum_{n=1}^{\infty} A_n \cos\left(\frac{2\pi n\tau}{t}\right) + \sum_{n=1}^{\infty} B_n \sin\left(\frac{2\pi n\tau}{t}\right). \qquad (*)$$

Suppose $\gamma = 0$, then $\gamma(0) = 0$ gives $\alpha_0 = 0$. Comparing $(*)$ to (3.2a), we get

$$A_0 = \sum_{n=1}^{\infty} \frac{\beta_n t}{2\pi n}, \qquad A_n = -\frac{\beta_n t}{2\pi n}, \qquad B_n = \frac{\alpha_n t}{2\pi n},$$

but $A_0 = A_n = B_n = 0$ holds for $\gamma = 0$, so we obtain the desired result. ∎

The main result that we are going to prove here can be formulated as follows:

THEOREM 5.3.3. *The Hilbert spaces $AC_0[J^t; \mathbb{R}^d]$ and $\mathscr{F}[J^t; \mathbb{R}^d]$ are identical. For each $\gamma \in AC_0[J^t; \mathbb{R}^d]$, the functions $\dot{\gamma}$, $\dot{\gamma}_w$, $\dot{\gamma}_m$ and $\dot{\gamma}_p$ exist and equal each other a.e. in J^t.*

Notice that there is a natural orthogonal decomposition, namely

$$AC_0[J^t; \mathbb{R}^d] = \sum_{j=1}^{n}{}^{\oplus} AC_0[J^t; \mathbb{R}] \qquad (3.5)$$

corresponding to the expression of γ through its components in a fixed orthonormal basis of $\mathbb{R}^d$. The analogous relation holds for $\mathscr{F}[J^t; \mathbb{R}^d]$, thus the proof reduces to the case $d = 1$ only. First we shall prove some lemmas.

LEMMA 5.3.4. *Suppose that* $f \in L(J^t; \mathbb{R})$ *has the Fourier series*

$$\alpha_0 + \sum_{n=1}^{\infty} \alpha_n \cos\left(\frac{2\pi n\tau}{t}\right) + \sum_{n=1}^{\infty} \beta_n \sin\left(\frac{2\pi n\tau}{t}\right) \tag{3.6}$$

(which is not assumed to converge in any sense), then the function

$$F : F(\tau) = -\alpha_0\tau + \int_0^{\tau} f(\xi)\,\mathrm{d}\xi \tag{3.7a}$$

fulfils $F(0) = F(t) = 0$ *and its Fourier series*

$$\sum_{n=1}^{\infty} \frac{\alpha_n t}{2\pi n} \sin\left(\frac{2\pi n\tau}{t}\right) + \sum_{n=1}^{\infty} \frac{\beta_n t}{2\pi n}\left[1 - \cos\left(\frac{2\pi n\tau}{t}\right)\right] \tag{3.7b}$$

converges to $F(\tau)$ *uniformly in* J^t.

Proof. The assertion can be checked using Dirichlet–Jordan theorem (cf. (A.18)). Since $f \in L(J^t; \mathbb{R})$, the function F is absolutely continuous so it has a bounded variation in J^t. The equality $F(0) = 0$ is obvious, further $\alpha_0 \equiv t^{-1}\int_0^t f(\tau)\,\mathrm{d}\tau$ implies $F(t) = 0$. Continuity of F implies that its Fourier series converges to $F(\tau)$ for each $\tau \in J^t$; moreover, uniformly in J^t, because $F(0) = F(t)$ and the Dirichlet–Jordan theorem may be by the same right applied to a periodic extension of F, say, to the interval $[-t, 2t]$. It remains to check that (3.7b) is actually the Fourier series of F. The Fourier coefficients are obtained through integration by parts

$$A_n \equiv \frac{2}{t}\int_0^t F(\tau)\cos\left(\frac{2\pi n\tau}{t}\right)\mathrm{d}\tau$$

$$= -\frac{2}{t}\,\frac{t}{2\pi n}\int_0^t (-\alpha_0 + f(\tau))\sin\left(\frac{2\pi n\tau}{t}\right)\mathrm{d}\tau = -\frac{\beta_n t}{2\pi n}.$$

In the same way, we get $B_n = \alpha_n t/2\pi n$; and finally, $F(0) = 0$ implies $A_0 = \sum_{n=1}^{\infty} A_n$. ■

LEMMA 5.3.5. $AC_0[J^t; \mathbb{R}] \subset \mathscr{F}[J^t; \mathbb{R}]$.

Proof. For $\gamma \in AC_0[J^t; \mathbb{R}]$, the derivative $\dot{\gamma}$ belongs to $L^2(J^t; \mathbb{R}) \subset L(J^t; \mathbb{R})$. Consequently, $\dot{\gamma}$ is L^2-norm limit of some sequence (3.4) which obeys the condition (3.2b), and its Fourier series is of the form (3.6). The preceding lemma applied to $f = \dot{\gamma}$ states that $F(\tau) = -\alpha_0\tau + \gamma(\tau) - \gamma(0)$ equals (3.7b). Finally, $F(t) = \gamma(t) = 0$ gives $\gamma(0) = -\alpha_0 t$, so $\gamma \in \mathscr{F}[J^t; \mathbb{R}]$. ■

LEMMA 5.3.6. *The function* $\dot{\gamma}_{\mathrm{p}}$ *is defined a.e. in* J^t *and* $\dot{\gamma}_{\mathrm{p}}(\tau) = \dot{\gamma}_{\mathrm{m}}(\tau)$ *holds for almost all* $\tau \in J^t$ *provided that the condition* (3.2b) *is valid.*

Proof. The first assertion is highly non-trivial and follows from the theorem proved by Carleson (1966). We have therefore $\dot{\gamma}_{\mathrm{p}}(\tau) = \lim_{N\to\infty} s_N(\tau)$ for all $\tau \in J^t \backslash M_{\mathrm{p}}$, where the Lebesgue measure $m(M_{\mathrm{p}}) = 0$. On the other hand, if $\dot{\gamma}_{\mathrm{m}} =$

$\lim_{N\to\infty} s_N$, then there is a subsequence $\{s_{N_k}\}$ such that $\dot{\gamma}_{\mathrm{m}}(\tau) = \lim_{k\to\infty} s_{N_k}(\tau)$ with possible exception of an m-zero set M_{m} (see, e.g. Kolmogorov and Fomin (1976, §VII.2)). Consequently, $\dot{\gamma}_{\mathrm{p}}(\tau) = \dot{\gamma}_{\mathrm{m}}(\tau)$ holds for all $\tau \in J^t \backslash (M_{\mathrm{p}} \cup M_{\mathrm{m}})$. ∎

Proof of Theorem 3.3. As we have noticed, it is sufficient to consider the case $d = 1$ only. Let γ be an arbitrary element of $\mathscr{F}[J^t;\mathbb{R}]$ corresponding to some coefficients α_0, $\{\alpha_n\}$, $\{\beta_n\}$ which fulfil the condition (3.2b), then $\dot{\gamma}_{\mathrm{m}}$ exists and belongs to $L^2(J^t;\mathbb{R}) \subset L(J^t;\mathbb{R})$. Lemma 3.4 applied to the functions $f = \dot{\gamma}_{\mathrm{m}}$ and the corresponding F defined by (3.7a) asserts that $F(0) = F(t) = 0$ and $F(\tau)$ is expressed by (3.7b). Since $\gamma(\tau)$ is given by (3.2a), we get

$$\gamma(\tau) = \alpha_0(\tau - t) + F(\tau) = -\alpha_0 t + \int_0^{\tau} \dot{\gamma}_{\mathrm{m}}(\xi)\, \mathrm{d}\xi,$$

but $\dot{\gamma}_{\mathrm{m}} \in L(J^t;\mathbb{R})$ so γ is absolutely continuous. The last relation further shows that γ is differentiable a.e. in J^t and

$$\dot{\gamma}(\tau) = \dot{\gamma}_{\mathrm{m}}(\tau) \quad \text{for almost all } \tau \in J^t; \tag{$*$}$$

then $\dot{\gamma}$ belongs to $L^2(J^t;\mathbb{R})$. Since $\gamma(t) = 0$, we obtain $\mathscr{F}[J^t;\mathbb{R}] \subset AC_0[J^t;\mathbb{R}]$, so according to Lemma 3.5, the two sets coincide.

Let us denote by $\gamma_N(\tau)$ the Nth partial sum of the series (3.2a). We have shown in Proposition 3.2 that the sequence $\{\gamma_N(\tau)\}$ converges uniformly in J^t, so

$$\int_0^t \gamma(\tau)\varphi(\tau)\, \mathrm{d}\tau = \lim_{N\to\infty} \int_0^t \gamma_N(\tau)\varphi(\tau)\, \mathrm{d}\tau$$

holds for each $\varphi \in C_0^{\infty}(J^t;\mathbb{R})$. Since $\dot{\varphi}$ belongs to $C_0^{\infty}(J^t;\mathbb{R})$ too and $\dot{\gamma}_N = s_N$, we have

$$\int_0^t \gamma(\tau)\dot{\varphi}(\tau)\, \mathrm{d}\tau = \lim_{N\to\infty} \int_0^t \gamma_N(\tau)\dot{\varphi}(\tau)\, \mathrm{d}\tau = -\lim_{N\to\infty} \int_0^t s_N(\tau)\varphi(\tau)\, \mathrm{d}\tau.$$

However, s_N converges to $\dot{\gamma}_{\mathrm{m}}$ in the L^2-norm, so the last limit can be easily expressed: we obtain the equality

$$\int_0^t \gamma(\tau)\dot{\varphi}(\tau)\, \mathrm{d}\tau = -\int_0^t \dot{\gamma}_{\mathrm{m}}(\tau)\varphi(\tau)\, \mathrm{d}\tau$$

valid for all $\varphi \in C_0^{\infty}(J^t;\mathbb{R})$, i.e. $\dot{\gamma}_{\mathrm{w}} = \dot{\gamma}_{\mathrm{m}}$ in the L^2-sense. Combining this result with ($*$) and Lemma 3.6, we get

$$\dot{\gamma}(\tau) = \dot{\gamma}_{\mathrm{w}}(\tau) = \dot{\gamma}_{\mathrm{m}}(\tau) = \dot{\gamma}_{\mathrm{p}}(\tau) \quad \text{a.e. in } J^t. \tag{3.8}$$

It remains to check that $AC_0[J^t; \mathbb{R}]$ and $\mathscr{F}[J^t; \mathbb{R}]$ are identical as Hilbert spaces. The relations (3.1), (3.3) and (3.8), together with Parseval equality, yield

$$\|\gamma\|^2 = \int_0^t |\dot{\gamma}(\tau)|^2 \, d\tau = \|\dot{\gamma}_m\|^2_{L^2(J^t)}$$

$$= t\alpha_0 + \frac{t}{2} \sum_{n=1}^{\infty} (\alpha_n^2 + \beta_n^2) = \|\gamma\|_F^2,$$

and the proof is completed by the polarization identity. ∎

In the following we shall work mostly with $\mathscr{H} = AC_0[J^t; \mathbb{R}^d]$, then Theorem 3.3 provides us with a useful equivalent expression for elements of these path spaces. Another important property of $\mathscr{H}$ concerns the existence of the *reproducing kernel*. Let us introduce the function $G : J^t \times J^t \to \mathscr{B}(\mathbb{R}^d)$ by

$$G(\tau, \xi) = g(\tau, \xi) I_d, \tag{3.9a}$$

$$g(\tau, \xi) = t - \max(\tau, \xi), \tag{3.9b}$$

where I_d is the unit operator on $\mathbb{R}^d$. It is easy to see that $G(\,.\,,.\,)$ is the Green's function of $-(d^2/d\tau^2) \otimes I_d$ with suitable boundary conditions (Problem 45). This fact suggests that it could be used as a reproducing kernel if the inner product in $\mathscr{H}$ is given by (3.1).

PROPOSITION 5.3.7. *Let γ be an arbitrary element of $AC_0[J^t; \mathbb{R}^d]$, then the relation*

$$(\gamma, G(\,.\,,)\beta) = \beta \cdot \gamma(\tau) \tag{3.10}$$

holds for all $\tau \in J^t$ and $\beta \in \mathbb{R}^d$.

Proof. If $\beta \in \mathbb{R}^d$, then $G(\tau, .\,)\beta = \beta g(\tau, .\,)$ belongs obviously to $AC_0[J^t; \mathbb{R}^d]$. Hence we have

$$(\gamma, G(\tau, .\,)\beta) = \int_0^t \dot{\gamma}(\xi) \cdot \beta \, \frac{\partial g(\tau, \xi)}{\partial \xi} \, d\xi = -\int_\tau^t \dot{\gamma}(\xi) \cdot \beta \, d\xi$$

$$= \gamma(\tau) \cdot \beta,$$

because $\gamma(t) = 0$. ∎

Concluding this section, let us comment on the relations between $\mathscr{H} = AC_0[J^t; \mathbb{R}^d]$ and the path space $X = C_0[J^t; \mathbb{R}^d]$ used in the W-measure theory. The space $\mathscr{H}$ determines X in the sense that the latter is obtained by completion of $\mathscr{H}$ with respect to the norm $\|\,.\,\|_\infty$. This fact reflects in the mathematical concept of Wiener spaces, of which the pair $\mathscr{H}, X$ together with the natural embedding $i : \mathscr{H} \to X$ is the canonical example.

On the other hand, $\mathscr{H}$ is only a small subset in X from the viewpoint of the W-measure. In order to formulate this assertion more exactly, recall that the function $\gamma : J^t \to \mathbb{R}^d$ is said to be *Hölder continuous* of order α for some $\alpha \in (0, 1]$ if there is a constant M_γ such that

$$|\gamma(\tau) - \gamma(\xi)| \leqslant M_\gamma |\tau - \xi|^\alpha \tag{3.11}$$

holds for all $\tau, \xi \in J^t$. The set of all such functions in X is denoted by Ω_α. Now the well-known result claims that the W-measure is supported by the quiver-behaved functions: $w(\Omega_\alpha) = 1$ holds for $0 < \alpha < \frac{1}{2}$, while $w(\Omega_\alpha) = 0$ if $\alpha > \frac{1}{2}$. In particular, the set of all $\gamma \in \mathscr{H}$ with bounded derivatives is certainly dense in $\mathscr{H}$, but its Wiener measure is zero. On a more heuristic level, one can argue as follows. For a function $\gamma \in \mathscr{H}$, the derivative $\dot{\gamma}$ may have isolated singularities in J^t. Suppose that $\gamma(\tau)$ behaves around such a singular point τ_0 locally as $(\tau - \tau_0)^\alpha$, then the square-integrability of $\dot{\gamma}$ requires $\alpha > \frac{1}{2}$ so γ falls into a w-zero set in X.

These facts are particularly important because we would like to relate the F_{-ic}-maps with a positive c to the W-integrals. One might ask how it can be achieved when the functions involved in the F_s-maps are defined on such a small subset in X only. But there is no paradox here. The comparison would certainly be impossible when the whole $L(X, w)$ was considered. However, we deal actually with a rather small part of it; the corresponding functions are smooth enough to be determined by their values on $\mathscr{H}$. As a (not wholly fitting) illustration, consider the integral $\int_0^1 f(x)\,dx$ of an $\mathbb{R}$-valued function f. The set $\mathbb{Q}$ of all rational numbers in $[0, 1]$ is of zero measure, but once the function f is piecewise continuous, the knowledge of $f \restriction \mathbb{Q}$ is sufficient to determine value of the integral. Of course, there is a connection between these considerations and the remark made in Section 1: some smoothness properties of the 'integrated' function are at the same time needed in order to suppress the oscillations leading to infinite variation of the 'Feynman measure'.

5.4. Polygonal-Path Approximations

In this section, we are going to discuss some methods which follow closely the original Feynman's suggestion in the sense that they determine the sought path integral (1.3) through approximations constructed with the help of the polygonal paths. In particular, we shall be able in this way to extend for $\mathscr{H} = AC_0[J^t; \mathbb{R}^d]$ the definition of Feynman maps given in Section 2.

Since each method of this kind involves some time-slicing, we shall start the discussion from this point. By **partition** of J^t we mean any ordered finite sequence $\sigma = \{\tau_j : 0 = \tau_0 < \tau_1 < \cdots < \tau_n = t\}$. The family of all partitions for a given interval J^t will be denoted by $\mathscr{P}(J^t)$. Further, it is useful to introduce the symbols $\Delta_j = [\tau_j, \tau_{j+1}]$ and $\delta_j = \tau_{j+1} - \tau_j, j = 0, 1, \ldots, n-1$; occasionally we shall write $\delta_j(\sigma)$, $n(\sigma)$, etc., when specification of the partition used is needed. We also set

$$\delta(\sigma) = \max_{1 \leqslant j \leqslant n-1} \delta_j(\sigma). \tag{4.1}$$

A partition σ' is said to be the **refinement** of σ, $\sigma' \supset \sigma$, if each of the intervals $\Delta_k(\sigma')$ is contained in some $\Delta_j(\sigma)$. It is clear that the set $\mathscr{P}(J^t)$ is partially ordered by the relation $\supset$ but has no maximal elements. The symbols $\sigma \cap \sigma'$ and $\sigma \cup \sigma'$ mean the partitions obtained by the natural ordering of the intersection and union of σ, σ', respectively. A partition σ is said to **decompose to subpartitions**, $\sigma = \{\sigma^{(1)}, \dots, \sigma^{(r)}\}$, if $\sigma^{(k)} = \{\tau_j : j = j_{k-1}, j_{k-1} + 1, \dots, j_k\}$ for some numbers $0 = j_0 < j_1 < \cdots < j_r = n$ (endpoints of the neighbouring subpartitions coincide). In particular, if $\sigma \supset \sigma'$ and $\tau_{j_k} = \tau'_k$ for all $k = 0, 1, \dots, r = n(\sigma')$, then the decomposition is said to be **generated** by σ'. Decompositions $\sigma = \{\sigma^{(1)}, \dots, \sigma^{(r)}\}$ and $\tilde\sigma = \{\tilde\sigma^{(1)}, \dots, \tilde\sigma^{(r)}\}$ are **comparable** if $\tau_{j_k} = \tilde\tau_{i_k}$ holds for $k = 1, \dots, r-1$, i.e. if the subpartitions of σ and $\tilde\sigma$ refer to the same subintervals in J^t. Partitions $\sigma, \tilde\sigma \in \mathscr{P}(J^t)$ are said to be **commuting** if they have comparable decompositions such that at least one of the following inclusions, $\sigma^{(k)} \subset \tilde\sigma^{(k)}$ or $\sigma^{(k)} \supset \tilde\sigma^{(k)}$, holds for each $k = 1, \dots, r$; in particular, σ and $\tilde\sigma$ commute if one of them refines the other.

Finally, the set $\mathscr{P}(J^t)$ can be equipped with a sort of product: we may define $\sigma\sigma' = \{0, t\} \cup \{\tau_j : [\tau_{j-1}, \tau_{j+1}] \not\subset \Delta_k(\sigma'), k = 0, 1, \dots, n(\sigma') - 1\}$. In other words, $\sigma\sigma'$ is obtained by stripping σ of the points whose neighbours are both contained in the same subinterval of σ'. This operation is easily seen to be associative but not commutative. Now we can formulate the following assertion:

PROPOSITION 5.4.1. *Partitions $\sigma, \sigma' \in \mathscr{P}(J^t)$ commute iff $\sigma\sigma' = \sigma'\sigma$; in that case we have $\sigma\sigma' = \sigma'\sigma = \sigma \cap \sigma'$.*

Proof. Sufficient condition: Let $\sigma, \tilde\sigma$ commute and assume that $\sigma = \{\sigma^{(1)}, \dots, \sigma^{(r)}\}$ and $\tilde\sigma = \{\tilde\sigma^{(1)}, \dots, \tilde\sigma^{(r)}\}$ are the corresponding comparable decompositions. The endpoints $\tau_{j_{k-1}} = \tilde\tau_{i_{k-1}}$ and $\tau_{j_k} = \tilde\tau_{i_k}$ of the subpartitions $\sigma^{(k)}$, $\tilde\sigma^{(k)}$ belong clearly to $\sigma \cap \tilde\sigma$, and also to $\sigma\tilde\sigma$ and $\tilde\sigma\sigma$, because their neighbouring points lie in different intervals of the other partition. Let, for example, the inclusion $\sigma^{(k)} \supset \tilde\sigma^{(k)}$ hold; then the points of $\sigma^{(k)} \backslash \tilde\sigma^{(k)}$ are contained neither in $\sigma \cap \tilde\sigma$, nor in $\sigma\tilde\sigma$, nor of course in $\tilde\sigma\sigma$. On the other hand, the subpartition $\tilde\sigma^{(k)}$ is easily seen to belong to $\sigma \cap \tilde\sigma$, $\sigma\tilde\sigma$ and $\tilde\sigma\sigma$. Hence these sets coincide with the more rough subpartition in each particular interval, and therefore they equal each other.

Necessary condition: Let now $\sigma\sigma' = \sigma'\sigma$ and consider any two neighbouring points τ_{j_0} and τ_{j_0+1} of σ. If there is some $\tau'_k \in \sigma'$ such that $\tau_{j_0} < \tau'_k < \tau_{j_0+1}$, then the following possibilities arise: either $\tau_{j_0} \leqslant \tau'_{k-1} < \tau'_{k+1} \leqslant \tau_{j_0+1}$ or $\tau'_k \in \sigma'\sigma$, but the latter contradicts to the assumption, because $\tau'_k \notin \sigma\sigma'$. This procedure can be continued: either $\tau'_{k-1} = \tau_{j_0}$ and $\tau'_{k+1} = \tau_{j_0+1}$, or at least one of them belongs to $(\tau_{j_0}, \tau_{j_0+1})$ and the same argument applies. Since the partition σ' is finite, we are able to make the following conclusion: there are some k_1 and k_2, $k_1 < k < k_2$, such that $\tau'_{k_1} = \tau_{j_0}$ and $\tau'_{k_2} = \tau_{j_0+1}$. In the same way we find that for $\tau'_{k_0} < \tau_j < \tau'_{k_0+1}$ there are $j_1 < j < j_2$ such that $\tau_{j_1} = \tau'_{k_0}$ and $\tau_{j_2} = \tau'_{k_0+1}$.

Consider now the decompositions of σ and σ' generated by $\sigma \cap \sigma'$, which are obviously comparable. Let $\Delta = [\xi, \eta]$ be an arbitrary subinterval of the partition $\sigma \cap \sigma'$ with $\xi = \tau_{j_0} = \tau'_{k_0}$. In that case, the interior of Δ contains points of at most one of the partitions σ, σ'. Suppose that this is not true, then either $\xi = \tau_{j_0} < \tau'_{k_0+1} < \tau_{j_0+1} < \eta$ or $\xi = \tau'_{k_0} < \tau_{j_0+1} < \tau'_{k_0+1} < \eta$. In both these cases, the above conclusion implies the existence of some $\zeta \in (\xi, \eta)$ that belongs to $\sigma \cap \sigma'$, but ξ and η are neighbouring points of $\sigma \cap \sigma'$ due to the assumption. Consequently, one of the subpartitions referring to Δ is trivial, i.e. consisting of ξ, η only; the other is clearly its refinement so σ and σ' commute. ■

Let us now turn to the *polygonal paths*, i.e. those that are continuous and piecewise linear. We denote $\mathscr{P}_x[J^t; \mathbb{R}^d]$ the subset of all such paths in $C_x[J^t; \mathbb{R}^d]$; obviously $\mathscr{P}_x[J^t; \mathbb{R}^d] \subset AC_x[J^t; \mathbb{R}^d]$. The methods discussed below are based on approximations of the continuous paths by the polygonal ones. To begin with, we take an arbitrary $\sigma \in \mathscr{P}(J^t)$ and define the operator $P^c(\sigma) : C_x[J^t; \mathbb{R}^d] \to \mathscr{P}_x[J^t; \mathbb{R}^d]$ by

$$(P^c(\sigma)\gamma)(\tau) = \gamma(\tau_j) + [\gamma(\tau_{j+1}) - \gamma(\tau_j)]\, \delta_j^{-1}(\tau - \tau_j) \tag{4.2}$$

for $\tau \in \Delta_j$, $j = 0, 1, \ldots, n-1$. This operator assigns to each $\gamma \in C_x[J^t; \mathbb{R}^d]$ a polygonal path with apices at the points $\gamma(\tau_j)$, $j = 0, 1, \ldots, n$ (cf. Figure 4). It is clear that $P^c(\sigma)$ acts as a linear operator if $x = 0$. We are particularly interested in its restriction to the Hilbert space treated in the previous section,

$$P(\sigma) := P^c(\sigma) \restriction AC_0[J^t; \mathbb{R}^d]. \tag{4.3a}$$

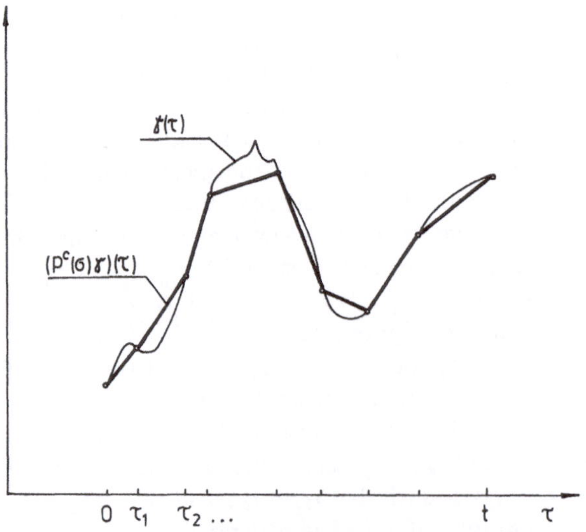

5.4. Approximation by the polygonal paths.

One would like to write the action of this operator in a more compact form than that provided by (4.2). This can be achieved with the help of the reproducing kernel introduced by (3.9). The relation

$$g(\tau_{j+1},\tau)-g(\tau_j,\tau)=\begin{cases}-\delta_j & \dots \quad \tau\leqslant\tau_j\\ \tau-\tau_{j+1} & \dots \quad \tau\in\Delta_j\\ 0 & \dots \quad \tau\geqslant\tau_j\end{cases} \tag{4.4}$$

implies easily the identity

$$(P(\sigma)\gamma)(\tau)=\sum_{j=0}^{n-1}[G(\tau_{j+1},\tau)-G(\tau_j,\tau)]\delta_j^{-1}[\gamma(\tau_{j+1})-\gamma(\tau_j)]\,. \tag{4.3b}$$

Of course, action of $P^c(\sigma)$ on the paths from $C_x[J^t;\mathbb{R}^d]$ is given by the same formula, but the main importance of (4.3b) is that it allows us to apply Proposition 3.7. Let us now list some properties of the operators $P(\sigma)$:

THEOREM 5.4.2. (a) *$P(\sigma)$ is an orthogonal projection on $\mathcal{H}=AC_0[J^t;\mathbb{R}^d]$ for each $\sigma\in\mathcal{P}(J^t)$, and* dim $P(\sigma)=dn(\sigma)$;

(b) *$P(\sigma)$ commutes with $P(\sigma')$ iff the partitions σ and σ' commute*;

(c) *$P(\sigma)\geqslant P(\sigma')$ iff $\sigma\supset\sigma'$.*

Proof. (a) Fix $\sigma\in\mathcal{P}(J^t)$. For an arbitrary $\gamma\in\mathcal{H}$, we have $(P(\sigma)\gamma)(\tau_j)=\gamma(\tau_j)\equiv\gamma^j$ so $P(\sigma)^2=P(\sigma)$. The relation (4.2b) gives

$$(\tilde\gamma,P(\sigma)\gamma)=\sum_{j=0}^{n-1}(\tilde\gamma,G(\tau_{j+1},.)(\gamma^{j+1}-\gamma^j))\delta_j^{-1}-$$

$$-\sum_{j=0}^{n-1}(\tilde\gamma,G(\tau_j,.)(\gamma^{j+1}-\gamma^j))\delta_j^{-1},$$

hence Proposition 3.7 yields

$$(\tilde\gamma,P(\sigma)\gamma)=\sum_{j=0}^{n-1}[\tilde\gamma(\tau_{j+1})-\tilde\gamma(\tau_j)]\cdot[\gamma(\tau_{j+1})-\gamma(\tau_j)]=(P(\sigma)\tilde\gamma,\gamma) \tag{4.5}$$

for all $\gamma,\tilde\gamma\in\mathcal{H}$. Consequently, the operator $P(\sigma)$ is symmetric, idempotent and defined everywhere in $\mathcal{H}$, i.e. an orthogonal projection. Moreover, let $\{e_k\}_{k=1}^d$ be an orthonormal basis in $\mathbb{R}^d$, then the vectors

$$\gamma_{jk}:\gamma_{jk}(\tau)=e_k[g(\tau_{j+1},\tau)-g(\tau_j,\tau)]\delta_j^{-1/2} \tag{4.6}$$

with $j=0,1,\dots,n-1$ and $k=1,\dots,d$ are orthonormal and span the subspace $P(\sigma)\mathcal{H}$ in view of (4.3b).

(b) The mapping $\sigma \mapsto P(\sigma)$ is obviously injective. If σ, σ' do not commute, then $\sigma\sigma' \neq \sigma'\sigma$ holds due to Proposition 4.1, and the mentioned injectivity together with the relation $\mathrm{Ran}\, P(\sigma)P(\sigma') = \mathrm{Ran}\, P(\sigma\sigma')$ show that $P(\sigma)$ and $P(\sigma')$ do not commute. Conversely, let σ, σ' commute. Using again (4.3b) in combination with the reproducing-kernel property, we get

$$(\tilde{\gamma}, P(\sigma')P(\sigma)\gamma)$$
$$= \sum_{k=0}^{n'-1} \sum_{j=0}^{n-1} \Gamma_{jk}[\tilde{\gamma}(\tau'_{k+1}) - \tilde{\gamma}(\tau'_k)] \cdot [\gamma(\tau_{j+1}) - \gamma(\tau_j)]\, \delta_j^{-1}(\delta'_k)^{-1} \tag{4.7a}$$

for arbitrary $\gamma, \tilde{\gamma} \in \mathscr{H}$, where

$$\Gamma_{jk} = g(\tau_{j+1}, \tau'_{k+1}) - g(\tau_{j+1}, \tau'_k) - g(\tau_j, \tau'_{k+1}) + g(\tau_j, \tau'_k),$$

and analogously

$$(\tilde{\gamma}, P(\sigma)P(\sigma')\gamma)$$
$$= \sum_{j=0}^{n-1} \sum_{k=0}^{n'-1} \Gamma'_{kj}[\tilde{\gamma}(\tau_{j+1}) - \tilde{\gamma}(\tau_j)] \cdot [\gamma(\tau'_{k+1}) - \gamma(\tau'_k)]\delta_j^{-1}(\delta'_k)^{-1} \tag{4.7b}$$

with

$$\Gamma'_{kj} = g(\tau'_{k+1}, \tau_{j+1}) - g(\tau'_{k+1}, \tau_j) - g(\tau'_k, \tau_{j+1}) + g(\tau'_k, \tau_j).$$

The interval J^t decomposes due to the assumption into subintervals $\Delta^{(s)}$, $s = 1, \dots, r$, such that in each of them the corresponding subpartitions fulfil either $\sigma^{(s)} \supset (\sigma')^{(s)}$ or $\sigma^{(s)} \subset (\sigma')^{(s)}$. In the first case, the relation (4.4) implies

$$\Gamma_{jk} = \Gamma'_{kj} = \begin{cases} \delta_j & \dots \quad \Delta_j \subset \Delta'_k \\ 0 & \dots \quad \text{otherwise} \end{cases} \tag{4.8a}$$

for each $\Delta'_k \subset \Delta^{(s)}$. Similarly, the inclusion $\sigma^{(s)} \subset (\sigma')^{(s)}$ gives

$$\Gamma_{jk} = \Gamma'_{kj} = \begin{cases} \delta'_k & \dots \quad \Delta'_k \subset \Delta_j \\ 0 & \dots \quad \text{otherwise} \end{cases} \tag{4.8b}$$

for each $\Delta_j \subset \Delta^{(s)}$. In particular, $\Gamma_{jk} = \Gamma'_{kj} = 0$ holds whenever the intervals Δ_j and Δ'_k are disjoint; then (4.7a) may be rewritten in the form

$$(\tilde{\gamma}, P(\sigma')P(\sigma)\gamma) = \sum_{s=1}^{r} (\tilde{\gamma}, P(\sigma')P(\sigma)\gamma)_s, \tag{4.9a}$$

where the sth term is in the case $\sigma^{(s)} \supset (\sigma')^{(s)}$ given by

$$\sum_{\Delta'_k \subset \Delta^{(s)}} \sum_{\Delta_j \subset \Delta'_k} \Gamma_{jk}[\tilde{\gamma}(\tau'_{k+1}) - \tilde{\gamma}(\tau'_k)] \cdot [\gamma(\tau_{j+1}) - \gamma(\tau_j)]\delta_j^{-1}(\delta'_k)^{-1}.$$

Using (4.7a), the last relation simplifies to the form

$$(\tilde{\gamma}, P(\sigma')P(\sigma)\gamma)_s$$

$$= \sum_{\Delta'_k \subset \Delta^{(s)}} [\tilde{\gamma}(\tau'_{k+1}) - \tilde{\gamma}(\tau'_k)] \cdot [\gamma(\tau'_{k+1}) - \gamma(\tau'_k)](\delta'_k)^{-1}. \tag{4.9b}$$

On the other hand, for $\sigma^{(s)} \subset (\sigma')^{(s)}$ we use (4.8b) to get

$$(\tilde{\gamma}, P(\sigma')P(\sigma)\gamma)_s$$

$$= \sum_{\Delta_j \subset \Delta^{(s)}} [\tilde{\gamma}(\tau_{j+1}) - \tilde{\gamma}(\tau_j)] \cdot [\gamma(\tau_{j+1}) - \gamma(\tau_j)]\delta_j^{-1}. \tag{4.9c}$$

Finally, $\Gamma_{jk} = \Gamma'_{kj}$ so $(\tilde{\gamma}, P(\sigma)P(\sigma')\gamma)$ is expressed again by the relations (4.9). Thus $(\tilde{\gamma}, P(\sigma)P(\sigma')\gamma) = (\tilde{\gamma}, P(\sigma')P(\sigma)\gamma)$ holds for all $\gamma, \tilde{\gamma} \in \mathscr{H}$, i.e. the projections $P(\sigma)$ and $P(\sigma')$ commute.

(c) If $\sigma \supset \sigma'$, then $P(\sigma)$ and $P(\sigma')$ commute due to (b). The relations (4.9) in combination with (4.5) now yield

$$(\tilde{\gamma}, P(\sigma')P(\sigma)\gamma)$$

$$= \sum_{\Delta'_k \subset J^t} [\tilde{\gamma}(\tau'_{k+1}) - \tilde{\gamma}(\tau'_k)] \cdot [\gamma(\tau'_{k+1}) - \gamma(\tau'_k)](\delta'_k)^{-1} = (\tilde{\gamma}, P(\sigma')\gamma)$$

for arbitrary $\gamma, \tilde{\gamma} \in \mathscr{H}$ so $P(\sigma)P(\sigma') = P(\sigma')P(\sigma) = P(\sigma')$, i.e. $P(\sigma) \geqslant P(\sigma')$. Another equivalent formulation of the last inequality reads Ran $P(\sigma) \supset$ Ran $P(\sigma')$. If $\sigma \not\supset \sigma'$, then there are some Δ_{j_0} and τ'_{k_0} such that $\tau_{j_0} < \tau'_{k_0} < \tau_{j_0+1}$. Each function $\gamma \in$ Ran $P(\sigma)$ is linear in Δ_{j_0}, but this is not true for Ran $P(\sigma')$ which contains paths having a 'corner' at $\tau = \tau'_{k_0}$; hence Ran $P(\sigma) \not\supset$ Ran $P(\sigma')$. ∎

It is intuitively clear that a path γ can be approximated by a family of polygonal paths $P(\sigma)\gamma$ referring to some successively refining partitions σ. Of course, we need a more rigorous statement. For our Hilbert spaces of paths, it may be formulated as follows:

THEOREM 5.4.3. *The family of projections $P(\sigma)$ on $\mathscr{H} = AC_0[J^t; \mathbb{R}^d]$ approximates the unit operator in the strong operator topology in the following sense: to any $\epsilon > 0$ and $\gamma \in \mathscr{H}$, there is a positive $\eta(\epsilon)$ such that $\|P(\sigma)\gamma - \gamma\| < \epsilon$ holds for all $\sigma \in \mathscr{P}(J^t)$ which fulfil $\delta(\sigma) < \eta(\epsilon)$, or symbolically*

$$\operatorname*{s-lim}_{\delta(\sigma)\to 0} P(\sigma) = I. \tag{4.10a}$$

Proof. We have to show that the relation

$$\lim_{\delta(\sigma)\to 0} \|P(\sigma)\gamma - \gamma\| = 0 \tag{4.10b}$$

is valid for each $\gamma \in \mathscr{H}$. Let us denote by $\mathscr{G}$ the set of all $\gamma \in \mathscr{H}$ for which (4.10a) holds; it is obviously linear. By the $\frac{1}{3}\epsilon$-trick, one can check its closedness:

we take an arbitrary Cauchy sequence $\{\gamma^{(k)}\}_{k=1}^{\infty} \subset \mathscr{G}$ which converges to some $\gamma \in \mathscr{H}$, and employ the inequality

$$\|P(\sigma)\gamma - \gamma\| \leqslant 2\,\|\gamma^{(k)} - \gamma\| + \|P(\sigma)\gamma^{(k)} - \gamma^{(k)}\|. \tag{4.11a}$$

To any $\epsilon > 0$, there are $k_0(\epsilon)$ and $\eta(k, \epsilon)$ such that the inequalities

$$\|\gamma^{(k)} - \gamma\| < \tfrac{1}{3}\epsilon, \qquad \|P(\sigma)\gamma^{(k)} - \gamma^{(k)}\| < \tfrac{1}{3}\epsilon \tag{4.11b}$$

hold for all $k > k_0(\epsilon)$ and $\sigma \in \mathscr{P}(J^t)$ with $\delta(\sigma) < \eta(k, \epsilon)$. Hence for an arbitrary partition σ with $\delta(\sigma) < \eta(k_0(\epsilon) + 1, \epsilon)$ the relations (4.11) give $\|P(\sigma)\gamma - \gamma\| < \epsilon$ so $\gamma \in \mathscr{G}$. In other words, $\mathscr{G}$ is a (closed) subspace in $\mathscr{H}$.

By Theorem 3.3, $\mathscr{H}$ possesses a trigonometric orthonormal basis which consists of the functions

$$\tilde{\gamma}_{jk} : \tilde{\gamma}_{jk}(\tau) = e_k v_j(\tau), \tag{4.12}$$

where $\{e_k\}_{k=1}^{d}$ is a fixed orthonormal basis in $\mathbb{R}^d$ and

$$v_1(\tau) = \tau - t,$$

$$v_{2N}(\tau) = \sin\left(\frac{2\pi N\tau}{t}\right),$$

$$v_{2N+1}(\tau) = 1 - \cos\left(\frac{2\pi N\tau}{t}\right)$$

with $N = 1, 2, \ldots$. We shall show that $\beta v_k(\,.\,) \in \mathscr{G}$ for $\beta \in \mathbb{R}^d$ and $k = 1, 2, \ldots$. It is elementary for $k = 1$. Assume further $k = 2N$ and $\tau \in \Delta_j$, then we have

$$(P(\sigma)\beta v_{2N})(\tau) = \beta \sin\left(\frac{2\pi N}{t}\tau_j\right) + \\ + \beta\left[\sin\left(\frac{2\pi N}{t}\tau_{j+1}\right) - \sin\left(\frac{2\pi N}{t}\tau_j\right)\right](\tau - \tau_j)\delta_j^{-1}$$

so a simple calculation gives

$$\begin{aligned} F(\tau) &\equiv -\frac{\mathrm{d}}{\mathrm{d}\tau}[P(\sigma)\beta v_{2N} - \beta v_{2N}](\tau) \\ &= -2\beta\cos\left(\frac{\pi N}{t}(2\tau_j + \delta_j)\right)\sin\left(\frac{\pi N}{t}\delta_j\right)\delta_j^{-1} + \frac{2\pi N}{t}\beta\cos\left(\frac{2\pi N\tau}{t}\right) \\ &= -\beta\cos\left(\frac{2\pi N}{t}\tau_j\right)\sin\left(\frac{2\pi N}{t}\delta_j\right)\delta_j^{-1} + \\ &\quad + 2\beta\sin\left(\frac{2\pi N}{t}\tau_j\right)\sin^2\left(\frac{\pi N}{t}\delta_j\right)\delta_j^{-1} + \frac{2\pi N}{t}\beta\cos\left(\frac{2\pi N\tau}{t}\right) \\ &= \frac{2\pi N}{t}\beta\left\{\left[\cos\left(\frac{2\pi N}{t}\tau\right) - \cos\left(\frac{2\pi N}{t}\tau_j\right)\right] + \left[1 - \left(\frac{2\pi N}{t}\delta_j\right)^{-1}\sin\left(\frac{2\pi N}{t}\delta_j\right)\right]\times\right. \\ &\quad \left.\times\cos\left(\frac{2\pi N}{t}\delta_j\right) + \left(\frac{\pi N}{t}\delta_j\right)^{-1}\sin^2\left(\frac{\pi N}{t}\delta_j\right)\sin\left(\frac{2\pi N}{t}\tau_j\right)\right\}. \end{aligned}$$

The last expression may be estimated by means of the inequalities

$$|\cos x - \cos y| \leqslant |x - y|, \qquad \left|\frac{\sin^2 x}{x}\right| \leqslant |x|,$$

$$\left|1 - \frac{\sin x}{x}\right| \leqslant \tfrac{1}{6}x^2;$$

we obtain in this way

$$|F(\tau)| \leqslant \frac{2\pi N}{t}\,|\beta|\left\{\frac{2\pi N}{t}\,(\tau - \tau_j) + \tfrac{1}{6}\left(\frac{2\pi N}{t}\,\delta_j\right)^2 + \frac{\pi N}{t}\,\delta_j\right\}. \tag{4.13a}$$

One might derive the desired result from (4.13a) directly, but it is more convenient to use a simpler and more rough estimate. The inequalities

$$\frac{2\pi N}{t}\,\delta_j \leqslant 2\pi N, \qquad \frac{\pi N}{3} + \tfrac{1}{2} < \frac{2\pi N}{3} < \pi N$$

give

$$|F(\tau)| < \left(\frac{2\pi N}{t}\right)^2 |\beta|\,[(\tau - \tau_j) + \pi N\delta_j] \tag{4.13b}$$

so

$$\int_{\Delta_j} |F(\tau)|^2\,d\tau \leqslant \left(\frac{2\pi N}{t}\right)^4 |\beta|^2\,(\pi^2 N^2 + \pi N + \tfrac{1}{3})\,\delta_j^3$$
$$< \left(\frac{2\pi N}{t}\right)^4 |\beta|^2\,(\pi N + 1)^2\,\delta_j^3.$$

In order to get the square of the sought norm, we have to sum these integrals over j. Since $\Sigma_{j=0}^{n-1}\delta_j = t$ and $\delta_j \leqslant \delta(\sigma)$ in view of (4.1), we obtain

$$\|P(\sigma)\beta v_{2N} - \beta v_{2N}\| < (2\pi N)^2\, t^{-3/2}\,|\beta|\,\delta(\sigma). \tag{4.14}$$

A similar argument shows that $\beta v_{2N+1} \in \mathscr{G}$ for $N = 1, 2, \ldots$. Hence we have proved the subspace $\mathscr{G}$ to contain an orthonormal basis of $\mathscr{H}$, i.e. the equality $\mathscr{G} = \mathscr{H}$. ■

REMARK 5.4.4. A sequence $\{\sigma_m\}_{m=1}^{\infty} \subset \mathscr{P}(J^t)$ is said to be **crumbling** if the lengths of all subintervals $\Delta_j(\sigma_m)$ tend to zero as $m \to \infty$, i.e.

$$\lim_{m\to\infty} \delta(\sigma_m) = 0. \tag{4.15a}$$

An important special case is represented by the sequences of **regular** (or equidistant) partitions, $\sigma_n = \{\tau_j = jt/n : j = 0, 1, \ldots, n\}$. The proved theorem particularly asserts that

$$\operatorname*{s-lim}_{m\to\infty} P(\sigma_m) = I \tag{4.15b}$$

holds for each crumbling sequence of partitions. On the other hand, it is clear that if the last limit exists for a non-crumbling $\{\sigma_m\}_{m=1}^{\infty}$, it cannot be equal to the unit operator. To this end, consider a suitable $\gamma \in \mathscr{H}$, say $\gamma(\tau) = (2t - \tau)^2 - t^2$, for which one calculates easily

$$\|P(\sigma)\gamma - \gamma\|^2 = \sum_{j=0}^{n-1} \int_{\Delta_j} (2\tau - \tau_j - \tau_{j+1})^2 \, d\tau \geqslant \tfrac{2}{3}\delta(\sigma)^3 .$$

Some of the conclusions derived above for $P(\sigma)$ remain valid when we pass to the operators $P^c(\sigma)$ on the Banach space $X = C_0[J^t; \mathbb{R}^d]$. For example, it is easy to see that $P^c(\sigma)$ is a projection on X whose range has dimension $dn(\sigma)$. The analogue of Theorem 4.3 reads:

PROPOSITION 5.4.5. *For each* $\gamma \in C_0[J^t; \mathbb{R}^d]$, *we have*

$$\lim_{\delta(\sigma)\to 0} \|P^c(\sigma)\gamma - \gamma\|_\infty = 0. \tag{4.16}$$

Proof. To an arbitrary $\epsilon > 0$, the well-known Weierstrass theorem implies the existence of some polynomials π_j such that $|\pi_j(\tau) - \gamma_j(\tau)| < \frac{1}{3}\epsilon d^{-1/2}$ for all $\tau \in J^t$ and $j = 1, \ldots, d$. The path $\pi^\epsilon : \pi^\epsilon(\tau) = (\pi_1(\tau), \ldots, \pi_d(\tau))$ then fulfils

$$\|\pi^\epsilon - \gamma\|_\infty < \tfrac{1}{3}\epsilon. \tag{4.17a}$$

Further, each π_j is approximated by the piecewise-linear function $P^c(\sigma)\pi_j$: if $\pi_j(\tau) = \sum_{k=0}^{m} \alpha_k t^k$, then a short calculation using (4.2) yields $|\pi_j(\tau) - (P^c(\sigma)\pi_j)(\tau)| \leqslant 2\delta(\sigma) \sum_{k=1}^{m} k\,|\alpha_k|\, t^{k-1}$, thus there is a positive $\eta(\epsilon)$ such that

$$\|\pi^\epsilon - P^c(\sigma)\pi^\epsilon\|_\infty < \tfrac{1}{3}\epsilon \tag{4.17b}$$

holds for each σ with $\delta(\sigma) < \eta(\epsilon)$. Finally, the relation

$$\|P^c(\sigma)\gamma\|_\infty \leqslant \|\gamma\|_\infty \tag{4.18}$$

can be checked easily for all $\gamma \in C_0[J^t; \mathbb{R}^d]$. In view of (4.17a), this relation implies $\|P^c(\sigma)\pi^\epsilon - P^c(\sigma)\gamma\|_\infty < \frac{1}{3}\epsilon$; and combining it with the inequalities (4.17), we get the desired result. ■

Now we are going to use these results to extend the definition of Feynman maps given in Section 2. In what follows, we denote $\mathscr{H}_x = AC_x[J^t; \mathbb{R}^d]$, usually omitting the subscript when $x = 0$. Consider a function $f : \mathscr{H}_x \to \mathbb{C}$. We associate with f the family of its *cylindrical projections*,

$$f_\sigma : f_\sigma(\gamma) = f(P^c(\sigma)\gamma) = f(x + P(\sigma)(\gamma - x)) \tag{4.19a}$$

corresponding to $\sigma \in \mathscr{P}(J^t)$; for simplicity we have denoted here by x the 'constant' path ending at x. We shall use the symbol f_σ simultaneously for the restriction of the cylindrical function (4.19a) to its basis $P^c(\sigma)\mathscr{H}_x \equiv \mathscr{H}_x^\sigma$. The expressions chosen to approximate the path integrals under consideration are the following

$$I_s(f;\sigma) = (2\pi i s)^{-nd/2} \int_{\mathscr{H}_x^\sigma} \exp\left\{\frac{i}{2s}\|\gamma_\sigma\|^2\right\} f_\sigma(\gamma_\sigma)\,\mathrm{d}m(\gamma_\sigma), \tag{4.20a}$$

where m is the Lebesgue measure on $\mathscr{H}_x^\sigma$, and

$$\|\gamma\| := \|\gamma - x\| \qquad \text{for} \quad \gamma \in \mathscr{H}_x \tag{4.20b}$$

(of course, one should remember that this is not a norm if $x \neq 0$). Unless stated otherwise, the integral in (4.20a) is understood in the usual Lebesgue sense, $\exp\{(i/2s)\|\,.\,\|^2\}f_\sigma(\,.\,) \in L(\mathscr{H}_x^\sigma,\ m)$. Alternatively, one may consider it to be an improper one: we set

$$I_s'(f;\sigma) = (2\pi i s)^{-nd/2} \lim_{\alpha\to\infty} \int_{C_\alpha} \exp\left\{\frac{i}{2s}\|\gamma_\sigma\|^2\right\} f_\sigma(\gamma_\sigma)\,\mathrm{d}m(\gamma_\sigma), \tag{4.21a}$$

where

$$C_\alpha = \{\gamma \in \mathscr{H}_x^\sigma : |(\gamma_\sigma(\tau_{j+1}) - \gamma_\sigma(\tau_j))_k| \leqslant \alpha\delta_j^{1/2},\ j = 0, 1, \ldots, n-1, k = 1, \ldots, d\} \tag{4.21b}$$

One must specify also the way in which the sought path integrals are approximated by the expressions (4.20a) or (4.21a). A natural choice is the limit occurring in Theorem 4.3, which shall be dubbed 'uniform': for a fixed $s \in \mathbb{C}_F$, one can set

$$I_s^{\mathrm{uc}}(f) = \lim_{\delta(\sigma)\to 0} I_s(f;\sigma) \tag{4.22a}$$

whenever the r.h.s. makes sense. Let us denote $\mathscr{F}_s^{\mathrm{uc}}(\mathscr{H}_x)$ the set of all functions for which the limit exists, then $I_s^{\mathrm{uc}} : \mathscr{F}_s^{\mathrm{uc}}(\mathscr{H}_x) \to \mathbb{C}$ defined by (4.20a) and (4.22a) is called **uniform cylindrical Feynman map** with the parameter s, in particular, **uniform cylindrical F-integral** if s is real and positive. Analogously, the relation

$$I_s^{\mathrm{ui}} : I_s^{\mathrm{ui}}(f) = \lim_{\delta(\sigma)\to 0} I_s'(f;\sigma) \tag{4.22b}$$

defines the **uniform improperly cylindrical $\mathbf{F}_s$-map** on the set $\mathscr{F}_s^{\mathrm{ui}}(\mathscr{H}_x)$ of functions for which the limit in (4.22b) makes sense. Another frequently used limiting procedure is the one based on the sequence $\{\sigma_m\}_{m=1}^{\infty}$ of regular partitions (cf. Remark 4.4 above): the **regular (improperly) cylindrical $\mathbf{F}_s$-map** for a given $s \in \mathbb{C}_F$ is defined by the relations

$$I_s^{\mathrm{rc}} : I_s^{\mathrm{rc}}(f) = \lim_{m\to\infty} I_s(f;\sigma_m), \tag{4.23a}$$

$$I_s^{\mathrm{ri}} : I_s^{\mathrm{ri}}(f) = \lim_{m\to\infty} I_s'(f;\sigma_m), \tag{4.23b}$$

respectively. There are many more modifications of these definitions, but we postpone commenting of them to Section 6.

REMARK 5.4.6. (a) The function spaces $\mathscr{F}_s^{\alpha}(\mathscr{H}_x)$ with α = uc, ui, rc, ri are obviously affine; in particular, linear for $x = 0$. The following elementary relations hold:

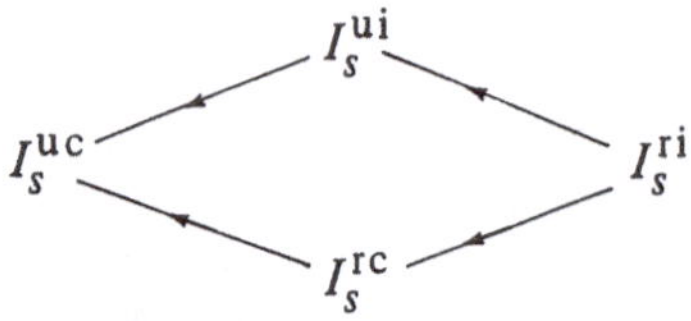

where the symbol $j \longleftarrow j'$ means that the mapping j' is extension of j.

(b) Translational invariance of the Lebesgue measure together with (4.20b) prove the validity of the relation (2.1). Let $f \in \mathscr{F}_s^{\mathrm{uc}}(\mathscr{H}_x)$, then the function $\gamma \mapsto f(\gamma + x)$ belongs to $\mathscr{F}_s^{\mathrm{uc}}(\mathscr{H}_0)$ and

$$\begin{aligned}\int_{\mathscr{H}_x}^{\mathrm{uc}} f(\gamma)\,\mathrm{D}\phi_s(\gamma) &= \lim_{\delta(\sigma)\to 0} (2\pi i s)^{-nd/2} \int_{\mathscr{H}_x^{\sigma}} \exp\left\{\frac{i}{2s}\|\gamma_\sigma\|^2\right\} f_\sigma(\gamma_\sigma)\,dm(\gamma_\sigma)\\ &= \lim_{\delta(\sigma)\to 0} (2\pi i s)^{-nd/2} \int_{\mathscr{H}_0^{\sigma}} \exp\left\{\frac{i}{2s}\|\gamma_\sigma\|^2\right\} f_\sigma(\gamma_\sigma + x)\,dm(\gamma_\sigma)\\ &= \int_{\mathscr{H}_0^{\sigma}}^{\mathrm{uc}} f(\gamma + x)\,\mathrm{D}\phi_s(\gamma);\end{aligned}$$

the same conclusion also holds for the other F-maps introduced above. We may therefore restrict our attention to the case $x = 0$ only.

The integrals (4.20a) and (4.21a) may be written in a more explicit form. To this end, recall that the elements of $\mathscr{H}^{\sigma} \equiv P(\sigma)\mathscr{H}$ express due to (4.3b) in terms of the orthonormal basis (4.6),

$$\gamma_\sigma(\tau) = \sum_{j=0}^{n-1} \sum_{k=1}^{d} (\gamma_\sigma(\tau_{j+1}) - \gamma_\sigma(\tau_j))_k\, \delta_j^{-1/2} \gamma_{jk}(\tau) \tag{4.24a}$$

so

$$I_s(f;\sigma) = (2\pi i s)^{-n/2} \int_{\mathbb{R}^{nd}} \exp\left\{\frac{i}{2s} \sum_{j=0}^{n-1} |\xi_j|^2\right\} f_\sigma\left(\sum_{j=0}^{n-1} \sum_{k=1}^{d} \xi_{jk}\gamma_{jk}\right) \times$$
$$\times\, d\xi_0 \ldots d\xi_{n-1}, \tag{4.24b}$$

where $\xi_j = (\xi_{j1}, \ldots, \xi_{jd})$. Moreover, there is an obvious linear bijection between the set of all polygonal paths $\gamma_\sigma \in \mathscr{H}^{\sigma}$ and the set of their apices $\gamma_\sigma(\tau_j)$, $j = 0, 1, \ldots, n-1$. Then we can set

$$f^{\sigma}(\gamma_\sigma(\tau_0), \ldots, \gamma_\sigma(\tau_{n-1})) := f_\sigma(\gamma_\sigma) \tag{4.19b}$$

so after the appropriate substitution, the integral (4.24b) becomes

$$I_s(f;\sigma) = \prod_{j=0}^{n-1} (2\pi i s\delta_j)^{-d/2} \int_{\mathbb{R}^{nd}} \exp\left\{\frac{i}{2s} \sum_{j=0}^{n-1} |\gamma_{j+1} - \gamma_j|^2 \,\delta_j^{-1}\right\} \times$$

$$\times f^\sigma(\gamma_0, \ldots, \gamma_{n-1})\,\mathrm{d}\gamma_0 \ldots \mathrm{d}\gamma_{n-1}, \tag{4.25}$$

where $\gamma_n = x$ is assumed to be zero.

REMARK 5.4.7. In this way, we have returned to the relation (1.4). Since the Feynman maps under consideration are defined by means of the cylindrical approximations (4.25), the definitions may be regarded as consistent with the heuristic requirement stated in Section 1. One is tempted to expect that for a cylindrical function $f: \mathscr{H} \to \mathbb{C}$ based in some $P(\tilde{\sigma})\mathscr{H}$, the relations

$$I_s^{\mathrm{uc}}(f) = I_s(f;\tilde{\sigma}), \qquad s \in \mathbb{C}_F^0 \tag{4.26a}$$

$$I_s^{\mathrm{ui}}(f) = I_s'(f;\tilde{\sigma}), \qquad s \in \mathbb{C}_F \tag{4.26b}$$

hold because of similar equalities obtained when the limit is taken over a suitably restricted set of partitions (Problem 46). But the situation is not so simple; in order to establish the relation $\lim_{\delta(\sigma)\to 0} I_s(f;\sigma) = I_s(f;\tilde{\sigma})$ without restrictions on σ, one needs to adopt an appropriate continuity assumption about f. In particular, Theorem 4.8 below, together with Theorem 2.5, imply that the relations (4.26) hold for $f \in \mathscr{F}^t(\mathscr{H})$.

Now we shall show that the uniform cylindrical F_s-maps extend the Fresnelian F_s-maps defined in Section 2. In view of Remark 4.6(a), the same is then true for the regular ones.

THEOREM 5.4.8. *For a fixed $s \in \mathbb{C}_F^0$, we have $\mathscr{F}_s^{\mathrm{uc}}(\mathscr{H}) \supset \mathscr{F}(\mathscr{H})$ and the relation*

$$I_s^{\mathrm{uc}}(f) = I_s^{\mathrm{f}}(f) \tag{4.27a}$$

holds for each $f \in \mathscr{F}(\mathscr{H})$. This conclusion fails if s is real, but $\mathscr{F}_s^{\mathrm{ui}}(\mathscr{H}) \supset \mathscr{F}(\mathscr{H})$ and

$$I_s^{\mathrm{ui}}(f) = I_s^{\mathrm{f}}(f) \tag{4.27b}$$

holds for an arbitrary $s \in \mathbb{C}_F$ and all $f \in \mathscr{F}(\mathscr{H})$.

Proof. Consider $f \in \mathscr{F}(\mathscr{H})$ and its cylindrical projections defined by (4.19a). Notice first that the relation

$$I_s^{\mathrm{f}}(f) = \lim_{\delta(\sigma)\to 0} I_s^{\mathrm{f}}(f_\sigma) \tag{4.28}$$

holds. Suppose it is not true; then one can easily construct a crumbling sequence $\{\sigma_m\} \subset \mathscr{P}(J^t)$ such that $I_s^{\mathrm{f}}(f_{\sigma_m})$ does not converge to $I_s^{\mathrm{f}}(f)$, which is impossible in view of (4.15b) and Proposition 2.7. If $s \in \mathbb{C}_F^{\mathrm{o}}$, then $I_s^{\mathrm{f}}(f_\sigma) = I_s(f;\sigma)$ due to Theorem 2.5, so the first assertion holds. In the case of real s, one must realize that the assumption of integrability was used in Theorem 2.5 only to establish the existence of the r.h.s. in (2.11) as a Lebesgue integral. The improper integral employed there is just the one we need, because according to (4.24), the same families of blowing-up cubes are used in (4.21) and in Lemma 2.6. Hence $I_s^{\mathrm{f}}(f_\sigma) = I_s'(f;\sigma)$ and (4.27b) follows. Finally, notice that I_s^{uc} does not extend I_s^{f} for real s: for instance, the unit function $e \in \mathscr{F}(\mathscr{H})$ is not contained in $\mathscr{F}_s^{\mathrm{uc}}(\mathscr{H})$. ■

Next we have to clarify how the F_s-maps with a purely imaginary parameter are related to the W-integral:

PROPOSITION 5.4.9. *For a fixed $c>0$ and $X_x = C_x[J^t;\mathbb{R}^d]$, let $f \in \mathscr{F}_{-ic}^{\mathrm{uc}}(\mathscr{H}_x)$ be the restriction to $\mathscr{H}_x$ of a w_c-measurable function $F: X_x \to \mathbb{C}$ which is $\|\,.\,\|_\infty$-continuous, with the possible exception of a w_c-zero subset of $X_x \setminus \mathscr{P}_x[J^t;\mathbb{R}^d]$. If there is a positive K such that $|f(\gamma)| \leqslant K$ holds for all $\gamma \in \mathscr{P}_x[J^t;\mathbb{R}^d]$, then*

$$\int_{\mathscr{H}_x}^{\mathrm{uc}} f(\gamma)\,\mathrm{D}\phi_{-ic}(\gamma) = \int_{X_x} F(\gamma)\,\mathrm{d}w_c(\gamma) \tag{4.29a}$$

and $|I_{-ic}^{\mathrm{uc}}(f)| \leqslant K$. The assertion is preserved when I_{-ic}^{uc} is replaced by I_{-ic}^{rc}. In particular, for $x = 0$ and $f \in \mathscr{F}(\mathscr{H})$ we have

$$\int_{X_0} F(\gamma)\,\mathrm{d}w_c(\gamma) = \int_{\mathscr{H}} \exp\left\{-\frac{c}{2}\|\gamma\|^2\right\} \mathrm{d}\mu_f(\gamma). \tag{4.29b}$$

Proof. In view of Remark 4.6(b) and the validity of (2.1) for the W-integral, we may restrict ourselves to the case $x = 0$. Due to Proposition 4.5, the continuity assumption implies

$$\lim_{\delta(\sigma)\to 0} F(P^c(\sigma)\gamma) = F(\gamma) \qquad w_c\text{-a.e. in } X_0; \tag{4.30a}$$

it further gives $|F(\gamma)| \leqslant K$ for w_c-almost all $\gamma \in X_0$, and therefore $F \in L(X_0, w_c)$. By the standard procedure, the W-integral of the cylindrical function $F(P^c(\sigma)\,.\,)$ is found to coincide with the r.h.s. of (4.25) for $s = -ic$, i.e.

$$\int_{X_0} F(P^c(\sigma)\gamma)\,\mathrm{d}w_c(\gamma) = I_{-ic}(f;\sigma). \tag{4.30b}$$

The r.h.s. of this equality tends to $I_{-ic}^{\mathrm{uc}}(f)$ as $\delta(\sigma) \to 0$ due to the assumption; the limit of the l.h.s. is obtained from (4.30a) and the dominated-convergence theorem. Normalization of w_c yields the bound on $I_{-ic}^{\mathrm{uc}}(f)$. The relations (4.30) may be used particularly for the sequence of regular partitions, so the argument works for the regular cylindrical F_{-ic}-maps as well. Finally, (4.29) follows from (4.27a) and the definition of $I_{-ic}^{\mathrm{f}}(f)$. ■

REMARK 5.4.10. The hypotheses of Proposition 4.9 cover most cases of physical interest but they are not the weakest possible. In particular, f might be not bounded but of a limited order of growth. Notice that the boundedness assumption was used only to justify application of the dominated-convergence theorem to the l.h.s. of (4.30b). Suppose, for example, that there is a non-decreasing function $\varphi : \mathbb{R}_+ \to \mathbb{R}_+$ such that

$$|f(\gamma)| \leqslant \varphi(\|\gamma\|_\infty)$$

holds for all $\gamma \in \mathscr{P}_0[J^t; \mathbb{R}^d]$. Monotonicity of φ together with (4.18) yield $|F(P^c(\sigma)\gamma)| \leqslant \varphi(\|\gamma\|_\infty)$, so the image-measure theorem implies

$$\int_{X_0} |F(P^c(\sigma)\gamma)| \, dw_c(\gamma) \leqslant \int_0^\infty \varphi(\alpha)\, \omega_d(\alpha, ct)\, d\alpha,$$

where

$$\omega_d(\alpha, ct) = \frac{d}{d\alpha}\, w_c(\gamma \in C_0[J^t; \mathbb{R}^d] \,:\, \|\gamma\|_\infty \leqslant \alpha) \tag{4.31}$$

(it is easy to see that this distribution depends on c and t only through their product). Hence the assertion of Proposition 4.9 can be strengthened with the knowledge of $\omega_d(\,.\,, ct)$; one admits those f which grow not faster than $\exp(a\,\|\,.\,\|_\infty^2)$ for a certain positive constant a (for more details, see the notes).

Let us now turn to the 'change-of-variable' transformations. As in Section 2, we shall start with the translations.

PROPOSITION 5.4.11. *Let $f \in \mathscr{F}_s^{\mathrm{ui}}(\mathscr{H}_x)$ for a fixed $s \in \mathbb{C}_F$. Let further $\alpha \in \mathscr{H}_0$, then the function f_α defined by* (2.18a) *belongs to $\mathscr{F}_s^{\mathrm{ui}}(\mathscr{H}_x)$ and*

$$\int_{\mathscr{H}_x}^{\mathrm{ui}} e^{i(\gamma,\alpha)/s} f(\gamma+\alpha)\, D\phi_s(\gamma) = \exp\left\{-\frac{i}{2s}\,\|\alpha\|^2\right\} \int_{\mathscr{H}_x}^{\mathrm{ui}} f(\gamma)\, D\phi_s(\gamma), \tag{4.32}$$

where we have set $(\gamma, \alpha) := (\gamma - x, \alpha)$ in analogy with (4.20b). *The same is true if I_s^{ui} is replaced by I_s^{uc}, or 'uniform' by 'regular'.*

Proof. We may again restrict ourselves to the case $x = 0$. The argument is similar to that of Proposition 2.8: we have

$$I_s^{\mathrm{ui}}(f_\alpha) = \lim_{\delta(\sigma)\to 0} I_s'(f_\alpha; \sigma)$$

$$= \lim_{\delta(\sigma)\to 0} \lim_{\beta\to 0} (2\pi i s)^{-nd/2} \int_{C_\beta} \exp\left\{\frac{i}{2s}\,\|\gamma_\sigma\|^2\right\} f_{\alpha\sigma}(\gamma_\sigma)\, dm(\gamma_\sigma);$$

using (4.19a) and Theorem 4.2(a), one can write

$$I_s^{\mathrm{ui}}(f_\alpha) = \lim_{\delta(\sigma)\to 0} \lim_{\beta\to 0} (2\pi i s)^{-nd/2} \int_{C_\beta} \exp\left\{ \frac{i}{2s} \|\gamma_\sigma\|^2 + \right.$$

$$\left. + \frac{i}{s} (\gamma_\sigma, P(\sigma)\alpha) \right\} f_\sigma(\gamma_\sigma + P(\sigma)_\alpha)\, dm(\gamma_\sigma).$$

Next we substitute $\tilde{\gamma}_\sigma = \gamma_\sigma + P(\sigma)_\alpha$ and employ the translational invariance of the Lebesgue measure. What is important is that in view of (4.21b) we integrate in $\tilde{\gamma}_\sigma$ over the *same* cube C_β:

$$I_s^{\mathrm{ui}}(f_\alpha) = \lim_{\delta(\sigma)\to 0} \lim_{\beta\to 0} (2\pi i s)^{-nd/2} \exp\left\{ -\frac{i}{2s} \|P(\sigma)\alpha\|^2 \right\} \times$$

$$\times \int_{C_\beta} \exp\left\{ \frac{i}{2s} \|\tilde{\gamma}_\sigma\|^2 \right\} f_\sigma(\tilde{\gamma}_\sigma)\, dm(\tilde{\gamma}_\sigma)$$

$$= \lim_{\delta(\sigma)\to 0} \exp\left\{ -\frac{i}{2s} \|P(\sigma)\alpha\|^2 \right\} \lim_{\delta(\sigma)\to 0} I_s'(f;\sigma).$$

According to the assumption, $I_s^{\mathrm{ui}}(f)$ exists; further, $\|P(\sigma)\alpha\|$ tends to $\|\alpha\|$ as $\delta(\sigma) \to 0$, so we arrive at the relation (4.32). An analogous argument applies to I_s^{uc}: one has only to realize that for each $\sigma \in \mathscr{P}(J^t)$ the integrals (4.20a) of f and f_α exist simultaneously and

$$I_s(f_\alpha;\sigma) = \exp\left\{ -\frac{i}{2s} \|P(\sigma)\alpha\|^2 \right\} I_s(f;\sigma).$$

It is also clear that the uniform limits may be replaced by the regular ones. ■

In combination with Theorem 4.8, the proved result particularly shows how the assumptions of Proposition 2.8 may be weakened: the function f need not be of $\mathscr{F}(\mathscr{H})$ when $s \in \mathbb{C}_F^0$, but it still belongs to $\mathscr{F}_s^{\mathrm{uc}}(\mathscr{H})$ and the formula (2.18b) with 'f' replaced by 'uc' holds.

By a similar argument, one can extend the assertion of Proposition 2.9 to the uniform (or regular) cylindrical F_s-maps (Problem 49). On the other hand, the same may not be true for the improperly cylindrical F_s-maps: a regular isometric operator R on $\mathscr{H}$ does not necessarily map the cube $C_\alpha \subset \mathscr{H}^\sigma$ on a cube in $R\mathscr{H}^\sigma$, and therefore the substitution $\gamma_\sigma \to R\gamma_\sigma$ leads in general to *another* improper integral.

Next we are going to generalize the Cameron–Martin formula (2.20b). Of course, when the 'infinite-dimensional' transformations are considered, one must choose a suitable class for which $\det(I + K)$ still makes sense.

THEOREM 5.4.12. *Let $I+K$ be invertible with K of the trace class. If $f \in \mathscr{F}_s^{\mathrm{uc}}(\mathscr{H})$ for a fixed $s \in \mathbb{C}_F$, then the function $e_s^K f_{I+K}$ defined by* (2.20a) *belongs again to $\mathscr{F}_s^{\mathrm{uc}}(\mathscr{H})$ and*

$$\int_{\mathscr{H}}^{\mathrm{uc}} \exp\left\{\frac{i}{s}(\gamma, K\gamma) + \frac{i}{2s}\|K\gamma\|^2\right\} f((I+K)\gamma)\, \mathrm{D}\phi_s(\gamma)$$

$$= |\det(I+K)|^{-1} \int_{\mathscr{H}}^{\mathrm{uc}} f(\gamma)\, \mathrm{D}\phi_s(\gamma). \tag{4.33}$$

Proof. We have

$$I_s^{\mathrm{uc}}(e_s^K f_{I+K}) = \lim_{\delta(\sigma)\to 0} (2\pi i s)^{-nd/2} \int_{\mathscr{H}^\sigma} \exp\left\{\frac{i}{2s}\|\gamma_\sigma\|^2\right\} \times$$

$$\times\, (e_s^K f_{I+K})(\gamma_\sigma)\, \mathrm{d}m(\gamma_\sigma)$$

$$= \lim_{\delta(\sigma)\to 0} (2\pi i s)^{-nd/2} \int_{\mathscr{H}^\sigma} \exp\left\{\frac{i}{2s}\|\gamma_\sigma + K\gamma_\sigma\|^2\right\} \times$$

$$\times\, f_\sigma(\gamma_\sigma + K\gamma_\sigma)\, \mathrm{d}m(\gamma_\sigma).$$

The existence of these integrals is established by the substitution $\gamma_\sigma \to \tilde{\gamma}_\sigma = (I+K)\gamma_\sigma$. In order to perform it, one must first find an orthonormal basis in $\widetilde{\mathscr{H}}^\sigma = \mathrm{Ran}[(I+K)P(\sigma)]$. We employ the orthonormal basis (4.6) in $\mathscr{H}^\sigma$. Since $I+K$ is injective by assumption, the vectors $\tilde{\gamma}_{jk} = (I+K)\gamma_{jk}$ are linearly independent so we obtain the required basis $\{\gamma'_{jk}\}$ by a suitable orthonormalization.

$$\gamma'_{jk} = \sum_{l=1}^{n} \sum_{m=1}^{d} C_{jk,lm}\tilde{\gamma}_{lm},$$

where C is a real regular $nd \times nd$ matrix. The orthonormality condition reads

$$(\gamma'_{jk}, \gamma'_{lm}) = \delta_{jl}\delta_{km} = \sum_{rs}\sum_{uv} C_{jk,rs}C_{lm,uv}((I+K)\gamma_{rs}, (I+K)\gamma_{uv}),$$

or equivalently $C(I+L)_\sigma C^{\mathrm{T}} = I$, where $L = K^*K + 2\,\mathrm{Re}\,K$ and $(I+L)_\sigma$ is the matrix representing the operator $P(\sigma)(I+L)P(\sigma)$ in the basis $\{\gamma_{jk}\}$. Since C is real, it follows easily

$$|\det C| = |\det(I+L)_\sigma|^{-1/2}.$$

Let us further check that the Jacobi matrix for the substitution under consideration is C^{T}. We write $\gamma_\sigma = \Sigma_{lm}\, \xi_{lm}\, \gamma_{lm}$, then the Fourier coefficients $\tilde{\xi}_{jk}$ of $\tilde{\gamma}_\sigma$ with respect to the basis $\{\gamma'_{jk}\}$ are given by

$$\tilde{\xi}_{jk} = (\gamma'_{jk}, \tilde{\gamma}_\sigma) = \sum_{lm} \xi_{lm}(\gamma'_{jk}, \tilde{\gamma}_{lm}) = \sum_{lm} \xi_{lm}(C^{-1})_{lm,jk}$$

so $\xi_{lm} = \Sigma_{jk}\, (C^{\mathrm{T}})_{lm,jk}\, \tilde{\xi}_{jk}$. Hence we have

$$I_s^{\mathrm{uc}}(e_s^K f_{I+K}) = \lim_{\delta(\sigma)\to 0} |\det(I + P(\sigma)LP(\sigma))|^{-1/2}\, (2\pi is)^{-nd/2} \times$$

$$\times \int_{\tilde{\mathcal{H}}^\sigma} \exp\left\{\frac{i}{2s}\, \|\tilde{\gamma}_\sigma\|^2\right\} f_\sigma(\tilde{\gamma}_\sigma)\, \mathrm{d}m(\tilde{\gamma}_\sigma).$$

The function f belongs to $\mathcal{F}_s^{\mathrm{uc}}(\mathcal{H})$ due to the assumption so it remains to be proved that the first factor has the limit required. To this end, notice that $P(\sigma)LP(\sigma)$ tends to L in the trace norm as $\delta(\sigma) \to 0$: since both L and $P(\sigma)$ are Hermitean, $\|P(\sigma)\| = 1$, we have

$$\|L - P(\sigma)LP(\sigma)\|_1 \leqslant 2\, \|(I - P(\sigma))L\|_1$$

and Theorem 4.3 together with Proposition 1.2.2 give the result. But $L \mapsto \det(I + L)$ is $\|\,.\,\|_1$-continuous and the factorization property holds for the determinants (see Reed and Simon (1978, theorem XIII.105, and Lemma 4 of the same section)) so

$$\lim_{\delta(\sigma)\to 0} |\det(I + P(\sigma)LP(\sigma))|^{-1/2} = |\det(I + L)|^{-1/2}$$

$$= |\det(I + K^*)\det(I + K)|^{-1/2}$$

$$= |\det(I + K)|^{-1};$$

the determinants here make sense because of the invertibility assumption. ■

REMARK 5.4.13. (a) It is evident that one may use the regular limit instead of the uniform one. Also, reformulation of the theorem for the path spaces $\mathcal{H}_x$ with $x \neq 0$ is straightforward.

(b) The situation is more complicated when the improperly cylindrical F_s-maps are considered. In that case, there is little hope to prove the analogue of (4.33) for the whole $\mathcal{F}_s^{\mathrm{ui}}(\mathcal{H})$. One would invite some reasonable weaker assertion, say, validity of the corresponding CM-formula for $f \in \mathcal{F}(\mathcal{H})$. According to Theorem 4.8, this is true for $s \in \mathbb{C}_F^0$. If s is real, one has to establish the relation

$$I'(e_s^K f_{I+K}; \sigma) = |\det(I + P(\sigma)LP(\sigma))|^{-1/2} \int_{\mathcal{H}} \exp\left\{-\frac{is}{2}\|\tilde{P}(\sigma)\gamma\|^2\right\} \mathrm{d}\mu_f(\gamma), \quad (4.34)$$

where $\tilde{P}(\sigma)$ is the orthogonal projection to $\mathrm{Ran}\,[(\mathrm{I} + K)P(\sigma)]$, which would allow us to perform the limit $\delta(\sigma) \to 0$. To this end, generalization of Lemma 2.6 is required, with the family of cubes C_α replaced by arbitrary blowing-up parallelepipeds centred at the origin. Such an assertion is presumably valid (cf. Example 6.1 below), but it seems to be not proven.

5.5. Product Formulae

The central problem concerning applications of Feynman path integrals in quantum physics is the expression of dynamics, i.e. an appropriately rigorous formulation of such formulae as (1.2). One natural way of doing this starts from the Trotter product formula or some of its modifications (cf. (A.7)). For simplicity, consider a Schrödinger operator $H_0 + V$ on $L^2(\mathbb{R}^d)$ which is assumed to be e.s.a. For a given $\psi \in L^2(\mathbb{R}^d)$, the Trotter formula gives

$$\mathrm{e}^{-i\bar{H}t}\,\psi = \lim_{n\to\infty} (\mathrm{e}^{-iH_0 t/n}\,\mathrm{e}^{-iVt/n})^n\,\psi. \tag{5.1a}$$

The l.h.s. may be written more explicitly. The free propagator $\exp(-iH_0 t)$ expresses as the following integral operator

$$(\mathrm{e}^{-iH_0 t}\,\psi)(x) = (2\pi it)^{-d/2}\int_{\mathbb{R}^d} \exp\left\{\frac{i}{2}\,\frac{|x-y|^2}{t}\right\}\psi(y)\,\mathrm{d}y \tag{5.2a}$$

(in fact, this relation holds for all t with $\operatorname{Im} t \leqslant 0$ – cf. Reed and Simon (1975, §IX.7)), where the r.h.s. for $\psi \in L^2(\mathbb{R}^d)$ is defined as the L^2-limit of some sequence $\{\psi_n\} \subset L^2 \cap L^1$; conventionally we set

$$\int_{\mathbb{R}^d} \mathrm{d}x = \lim_{m\to\infty} \int_{|x|\leqslant m} \mathrm{d}x. \tag{5.2b}$$

Then we may substitute from (5.2) to (5.1a), obtaining in this way

$$(\mathrm{e}^{-i\bar{H}t}\,\psi)(x) = \lim_{n\to\infty} (2\pi it/n)^{-nd/2}\int_{\mathbb{R}^d}\cdots\int_{\mathbb{R}^d} \exp\left\{\frac{in}{2t}\sum_{j=0}^{n-1} |\gamma_{j+1}-\gamma_j|^2 - \right.$$

$$\left. -\frac{it}{n}\sum_{j=0}^{n-1} v(\gamma_j)\right\}\psi(\gamma_0)\,\mathrm{d}\gamma_0\ldots\mathrm{d}\gamma_{n-1}, \tag{5.1b}$$

where $\gamma_n = x$ and the limit is again understood in the L^2-sense.

Comparing the r.h.s. of (5.1b) with (4.25) we observe its close similarity to the regular (improperly) cylindrical Feynman integral. The main difference concerns the replacement of the 'integrated' function (1.10a) by the approximative expression (1.10b). This may also be interpreted as follows: each path γ is replaced by the family of piecewise constant 'paths' $\gamma^{(\sigma)}$:

$$\gamma^{(\sigma)}(\tau) = \begin{cases} \gamma(\tau_j) & \ldots\quad \tau \in [\tau_j, \tau_{j+1}), \qquad j = 0, 1, \ldots, n-1 \\ \gamma(t) & \ldots\quad \tau = t \end{cases} \tag{5.3a}$$

referring to $\sigma \in \mathscr{P}(J^t)$. Comparison with the polygonal-path approximations is then very illustrative; of course, the kinetical part of the exponent entering the 'Feynman measure' is calculated as previously.

In this way, we are naturally led to the definition of the **product Feynman integral** for 'integrated' functions (1.10a) by the r.h.s. of (5.1b). At once various generalizations come to mind. One may introduce the mass parameter s, in general complex, to get the **product Feynman maps** whose values

$$\int_{\mathscr{H}_x}^{p} \exp\left\{-i\int_0^t v(\gamma(\tau))\,d\tau\right\} D\phi_s(\gamma)$$

are given by (5.1b) with t, v replaced by st and $s^{-1}v$, respectively. There is also no problem in introducing the **uniform product Feynman maps**: one has to use arbitrary partitions $\sigma \in \mathscr{P}(J^t)$ instead of the regular ones employed in (5.1b), and to consider the limit $\delta(\sigma) \to 0$.

The choice of domains for the maps introduced above is a more delicate problem. For instance, the above considerations show that the (regular) product F-maps may be defined if the Trotter formula applies, in particular, for functions of the type (1.10a) with some reasonable class of the potentials v, or eventually for their linear combinations. Of course, for a complex-valued function f on $\mathscr{H}_x$ (or another suitable space of continuous paths ending at x) one can always formally write

$$\begin{aligned}\int_{\mathscr{H}_x}^{up} f(\gamma)\, D\phi_s(\gamma) = \lim_{\delta(\sigma)\to 0} \prod_{j=0}^{n-1} (2\pi i s\delta_j)^{-d/2} \times \\ \times \int_{\mathbb{R}^{nd}} \exp\left\{\frac{i}{2s}\sum_{j=0}^{n-1} |\gamma_{j+1}-\gamma_j|^2\,\delta_j^{-1}\right\} \times \\ \times f^{(\sigma)}(\gamma_0,\ldots,\gamma_{n-1})\, d\gamma_0 \ldots d\gamma_{n-1},\end{aligned} \tag{5.4}$$

where

$$f^{(\sigma)}(\gamma(\tau_0),\ldots,\gamma(\tau_{n-1})) := f(\gamma^{(\sigma)}), \tag{5.3b}$$

but in order to give meaning to such a definition one must specify how the limits and eventual improper integrals should be understood, and to establish their existence for the chosen class of 'integrated' functions. Some comments on this problem will be given in the next section.

Motivated by the relations (5.1), we are now going to prove a theorem which shows that the product formulae may be applied to express the solutions of much more general evolution equations. In a sense, this result further develops the method described in (A.12); its particular form will be used in the following chapter. Let us start with some preliminaries.

LEMMA 5.5.1. *Let X, Y be Banach spaces. Consider a linear manifold $D \subset X$ which is the domain of some closed linear Y-valued operator, and the natural* *injection $i : D \to X$. Then D can be equipped with a (unique up to isomorphism) Banach-space structure such that $i \in \mathscr{B}(D, X)$* and *$C \in \mathscr{B}(D, Y)$ for each closable operator C from X to Y with the domain D.*

Proof. Let C be a closed operator of the described type. The Banach-space structure on D may be defined, for example, by means of the norm $\|x\|_C := \|x\|_X + \|Cx\|_Y$, then the check of $i \in \mathscr{B}(D, X)$ and $C \in \mathscr{B}(D, Y)$ is elementary. Let D' be the same set given by another Banach-space structure such that $i \in \mathscr{B}(D', X)$. Then the identical mapping is continuous from D' to D so each open set $\subset D$ is also open in D'; at the same time 'id' is surjective, so the open sets of D' remain open in D due to the open-mapping theorem. Consequently, 'id' is a linear homeomorphism of D and D'. Let C' be another operator with the domain D whose graph $\mathscr{G}(C')$ is closed in $X \oplus Y$. Since $i \in \mathscr{B}(D, X)$, the injection $j : D \oplus Y \to X \oplus Y$ is continuous; thus $\mathscr{G}(C')$ is also closed in $D \oplus Y$ and $C' \in \mathscr{B}(D, Y)$ follows from the closed-graph theorem. Finally, if C' is closable, then $\mathscr{G}(\bar{C}')$ is closed so the inverse image $j^{(-1)}(\mathscr{G}(\bar{C}')) = \mathscr{G}(C')$ is closed in $D \oplus Y$ and the same argument applies. ■

LEMMA 5.5.2. *Let X, Y be Banach spaces and consider a bounded two-parameter family $\{P(\tau, \delta)\} \subset \mathscr{B}(Y, X)$. Assume that* $\lim_{\delta \to 0} P(\tau, \delta)u = 0$ *holds for all u from some compact set $M \subset Y$ uniformly in τ, then the convergence is also uniform with respect to u in M.*

Proof. Suppose the converse is true, i.e. the convergence is not uniform in u. Then there is $\epsilon > 0$ such that to each natural number k, there are a positive $\delta_k < k^{-1}$ and $u_k \in M$ which fulfil $\sup_\tau \|P(\tau, \delta_k)u_k\| \geq \epsilon$. However, M is compact so $\{u_k\}$ has some limit point $u_0 \in M$; one may assume $u_k \to u_0$ without loss of generality. We have

$$\sup_\tau \|P(\tau, \delta_k)u_k\| \leq \sup_\tau \|P(\tau, \delta_k)\| \, \|u_k - u_0\| + \sup_\tau \|P(\tau, \delta_k)u_0\|$$

and $\|P(\tau, \delta)\|$ has a bound independent of τ, δ so $\liminf_{k \to \infty} \sup_\tau \|P(\tau, \delta_k)u_0\| \geq \epsilon$, but this contradicts the assumption. ■

Finally, notice that the Hilbert-space structure is not essential in the definition of the propagator given in Section 4.2; one may define *propagators on a Banach space X* in quite the same way. Now we are in position to formulate the announced assertion.

THEOREM 5.5.3. *Let the operators $A(t)$, $B(t)$ generate continuous contractive semigroups on a Banach space X for each $t \in J^b$. Assume further $C(t) = A(t) + B(t)$ to be a closed operator whose domain D is dense in X and independent of t. For each $u \in D$, let $A(\,.\,)u$ and $B(\,.\,)u$ be continuous on J^b. Assume that there is a contraction-valued propagator $V(\,.\,,.\,)$ on X which preserves D, i.e. $V(t, s)D \subset D$ for all $t, s \in J^b$. We denote $u(t) = V(t, 0)u$ for $u \in D$ and suppose that $C(\,.\,)u(\,.\,)$ is continuously differentiable on* $(0, b)$ *and there obeys the equation*

$$\frac{du(t)}{dt} + C(t)u(t) = 0. \tag{5.5}$$

Finally, let $u(\,.\,)$ be continuous also as the mapping from J^b to D equipped with the natural Banach-space structure determined by the operators $C(t)$ (cf. Lemma 5.1). *Then the relation*

$$V(t,0) = \operatorname*{s-lim}_{\delta(\sigma)\to 0} R(\tau_{n-1},\delta_{n-1})R(\tau_{n-2},\delta_{n-2})\ldots R(0,\delta_0) \tag{5.6a}$$

holds for each $t \in (0, b]$, where

$$R(\tau,\delta) := e^{-A(\tau)\delta}\, e^{-B(\tau)\delta} \tag{5.6b}$$

and τ_k, δ_k in (5.6a) *refer to the partition $\sigma \in \mathscr{P}(J^t)$.*

Proof. For a given $\sigma \in \mathscr{P}(J^t)$, we abbreviate $P_k = \exp(-A(\tau_k)\delta_k)$, $Q_k = \exp(-B(\tau_k)\delta_k)$ and $R_k = P_kQ_k = R(\tau_k, \delta_k)$, $k = 0, 1, \ldots, n-1$. Further, we set $S(\sigma) = R_{n-1}R_{n-2}\ldots R_0 - V(t,0)$ so the relation (5.6a) may be rewritten as

$$\lim_{\delta(\sigma)\to 0} S(\sigma)u = 0 \tag{5.6c}$$

for all $u \in X$. In fact, it is enough to check the relation (5.6c) for $u \in D$ only, because the family under consideration is uniformly bounded, $\|S(\sigma)\| \leqslant \|R_{n-1}\ldots R_0\| + \|V(t, 0)\| \leqslant 2$, and the required conclusion is obtained by the $\frac{1}{3}\varepsilon$-trick. Using the equality

$$S(\sigma) = \sum_{k=0}^{n-1} R_{n-1}\ldots R_{k+1}[R_k - V(\tau_{k+1},\tau_k)]\, V(\tau_k,0)$$

in combination with $\|R_k\| \leqslant 1$ and $A(\tau_k) + B(\tau_k) = C(\tau_k)$, we obtain

$$\|S(\sigma)u\| \leqslant \sum_{k=0}^{n-1} \delta_k \|[E_1(\tau_k,\delta_k) + E_2(\tau_k,\delta_k) - E_3(\tau_k,\delta_k)]\, V(\tau_k,0)u\|,$$

where

$$E_1(\tau,\delta) = [e^{-A(\tau)\delta} - I]\delta^{-1} + A(\tau),$$

$$E_2(\tau,\delta) = e^{-A(\tau)\delta}[e^{-B(\tau)\delta} - I]\delta^{-1} + B(\tau),$$

$$E_3(\tau,\delta) = [V(\tau+\delta,\tau) - I]\delta^{-1} + C(\tau).$$

Further, we employ the fact that $\Sigma_{k=0}^{n-1}\delta_k = t$, then the last estimate gives

$$\|S(\sigma)u\| \leqslant t \sum_{j=1}^{3} \sup\{\|E_j(\tau,\delta)V(\tau,0)u\| : \tau \in J^t, 0 < \delta < \delta(\sigma)\}.$$

Hence it is sufficient to show that

$$\lim_{\delta\to 0} E_j(\tau,\delta)V(\tau,0)u = 0, \quad j = 1, 2, 3, \tag{5.7}$$

holds for each $u \in D$ uniformly in $\tau \in J^t$.

According to the assumption, $\{e^{-A(\tau)\delta}\}$ is a continuous semigroup for each $\tau \in J^t$ and the domain of $A(\tau)$ contains D, so

$$\lim_{\delta \to 0} \, [e^{-A(\tau)\delta} - I] A(\tau) v = 0 \tag{5.8a}$$

holds for all $v \in D$. We shall prove that the convergence in (5.8a) is uniform in $\tau \in J^t$. Consider first the identity

$$e^{-A(\tau)\delta} \, v = v - \int_0^t e^{-A(\tau)\eta} \, A(\tau) v \, d\eta$$

valid for each $v \in D$. The function $\|A(\,.\,)v\|$ is by assumption continuous in $[0, b]$, and therefore bounded by some constant K so $\|e^{-A(\tau)\delta} \, v - v\| \leqslant K\delta$, i.e. $\lim_{\delta \to 0} e^{-A(\tau)\delta} \, v = v$ uniformly in τ. The conclusion extends easily to all $v \in X$, because D is dense and the operators involved are uniformly bounded. Further, the set $\{e^{-A(\tau)\delta} - I : \tau \in J^t, \delta \geqslant 0\}$ is bounded in $\mathscr{B}(X)$ and $\{A(\tau)v : \tau \in J^t\}$ is compact (as the continuous image of the compact J^t) so the uniform convergence in (5.8a) follows from Lemma 5.2. Then the following estimate is possible,

$$\|E_1(\tau, \delta)v\| = \left\| \frac{1}{\delta} \int_0^\delta \, [e^{-A(\tau)\eta} - I] \, A(\tau) v \, d\eta \right\|$$

$$\leqslant \sup_{0 \leqslant \eta \leqslant \delta} \| [e^{-A(\tau)\eta} - I] \, A(\tau) v \|$$

for each $v \in D$, so (5.8a) yields

$$\lim_{\delta \to 0} \, E_1(\tau, \delta) v = 0 \quad \text{uniformly in } \tau \in J^t. \tag{5.9a}$$

The second term may be treated in a similar way: first we must check that

$$\lim_{\delta \to 0} \, [e^{-A(\tau)\delta} \, e^{-B(\tau)\eta(\delta)} - I] \, B(\tau) v = 0 \tag{5.8b}$$

holds for all $v \in D$ uniformly in τ, where $\eta(\delta)$ is an arbitrary number from $[0, \delta]$. The same argument as above yields $\lim_{\delta \to 0} e^{-A(\tau)\delta} \, v = v$ and $\lim_{\delta \to 0} e^{-B(\tau)\eta(\delta)} \, v = v$ for $v \in X$ uniformly in τ; further, $\{e^{-A(\tau)\delta}\}$ is uniformly bounded, so

$$\lim_{\delta \to 0} \, e^{-A(\tau)\delta} \, e^{-B(\tau)\eta(\delta)} \, v = v$$

holds for $v \in X$ uniformly in τ. The set $\{B(\tau)v : \tau \in J^t\}$ is compact and $\{e^{-A(\tau)\delta} \, e^{-B(\tau)\eta(\delta)} - I : \tau \in J^t, \delta \geqslant 0\} \subset \mathscr{B}(X)$ is bounded, thus Lemma 5.2 applies again giving (5.8b). Combining this with the estimate

$$\|E_2(\tau, \delta)v\| = \left\| \frac{1}{\delta} \int_0^\delta \, [e^{-A(\tau)\delta} \, e^{-B(\tau)\eta} - I] \, B(\tau) v \, d\eta \right\|$$

$$\leqslant \sup_{0 \leqslant \eta \leqslant \delta} \| [e^{-A(\tau)\delta} \, e^{-B(\tau)\eta} - I] \, B(\tau) v \|,$$

valid for an arbitrary $v \in D$, we get

$$\lim_{\delta \to 0} E_2(\tau, \delta)v = 0 \quad \text{uniformly in } \tau \in J^t. \tag{5.9b}$$

Now we fix $u \in D$ and denote $M_u = \cup_{\tau \in J^t} V(\tau, 0)u$. Clearly $M_u \subset D$, further M_u is compact in D, because J^t is compact and $u(\,.\,)$ is assumed to be continuous in D. The relation (5.7) will be established for $j = 1, 2$, if we check that the convergence in (5.9) is also uniform in $v \in M_u$. Lemma 5.2 is not directly applicable to this case, though the families $\{E_j(\tau, \delta) : \tau \in J^t, \delta > 0\} \subset \mathscr{B}(D, X)$, $j = 1, 2$, are bounded due to the uniform boundedness principle; the reason is that the convergence in (5.9) refers to the norm of X. Fortunately, the argument modifies easily to the present situation. Suppose the assertion to be false, then there are $\epsilon > 0$ and $\delta_k < k^{-1}$, $u_k \in M_u$ to each natural number k such that $\sup_\tau \|E_j(\tau, \delta_k)u_k\| \geqslant \epsilon$. Let $\|\,.\,\|_D$ be a norm generating the Banach-space structure of D; one can choose it so that $\|x\|_D \geqslant \|x\|$ holds for all $x \in D$. Since M_u is compact in D, the family $\{u_k\}$ can be chosen in such a way that $\|u_k - u_0\|_D \to 0$ as $k \to \infty$ for some $u_0 \in M_u$. Then $\{u_k\}$ converges to u_0 in the $\|\,.\,\|$-norm too and the needed contradiction is obtained as in the proof of Lemma 5.2.

It remains to cope with the third term. Since $u(\,.\,)$ obeys equation (5.5), we have

$$\|E_3(\tau, \delta)V(\tau, 0)u\| = \left\| \frac{1}{\delta} \int_0^\delta [C(\tau + \eta)V(\tau + \eta, 0) - C(\tau)V(\tau, 0)]u \, d\eta \right\|$$

$$\leqslant \sup_{0 \leqslant \eta \leqslant \delta} \|C(\tau + \eta)V(\tau + \eta, 0)u - C(\tau)V(\tau, 0)u\|,$$

but $C(\,.\,)u(\,.\,)$ is continuous due to the assumption, so (5.7) also holds for $j = 3$. ∎

In particular, we have obtained in this way the following 'uniform' version of the Trotter product formula in an important special case (cf. (A.7)).

COROLLARY 5.5.4. *Let the operators A, B generate continuous contractive semigroups on a Banach space X. If the sum C = A + B generates a continuous contractive semigroup too, then*

$$e^{-Ct} = \operatorname*{s-lim}_{\delta(\sigma) \to 0} \prod_{k=0}^{n-1} e^{-A\delta_k} e^{-B\delta_k} \tag{5.10}$$

holds for each $t > 0$, *where* δ_k *and* n *in* (5.10) *refer to the partition* $\sigma \in \mathscr{P}(J^t)$.

Proof. One has only to use some simple properties of continuous semigroups on a Banach space (see, e.g. Davies (1980a, §1.1)) to check that the hypotheses of the theorem are fulfilled. ∎

The application of these results to the expression of propagators by means of the uniform product F-integrals is left to Section 6.1.

5.6. More about Other F-Integral Theories

The theory exposed in this chapter does not cover all possible approaches to Feynman path integrals, even if we stay within the limits set in the introduction. Moreover, each of the described methods admits many modifications. In this section we are going to give some comments on this wide variety of different F-integrals.

Let us start with the *sequential methods*, i.e. those in which the path integral (1.3) is defined as a limit over a family of 'finite-dimensional' integrals corresponding to a successively refined time-slicing. There are two main points of view from which these methods may be classified; namely:

(i) the choice of the approximating expressions $I_s(f;\sigma)$ for the partitions σ from $\mathscr{P}(J^t)$ or a suitable subset of it;
(ii) the choice of the limiting prescription.

The first criterion offers various possibilities, in particular, those listed below (the attached letter may be used to specify the respective F-maps).

(f) *If the cylindrical projections* (4.19) *are of the form* $f_\sigma(\gamma) = g_\sigma(\gamma - x)$ *with* $g_\sigma \in \mathscr{F}(\mathscr{H})$, *then we may set* $I_s(f;\sigma) := I_s^{\mathrm{f}}(g_\sigma)$.
(c) $I_s(f;\sigma)$ *are given by the integrals* (4.20a) *if the latter exist.*
(i) $I_s(f;\sigma)$ *are given by the improper integrals* (4.21a) *if the latter exist.*
(o) *Alternately, one may define* $I_s(f;\sigma)$ *by means of the 'oscillatory integrals'. For arbitrary* $\epsilon > 0$ *and* $\varphi \in \mathscr{S}(\mathbb{R}^{nd})$ *such that* $\varphi(0) = 1$, *we denote*

$$I_s^\epsilon(f;\sigma,\varphi) = (2\pi i s)^{-nd/2} \int_{\mathscr{H}_x^\sigma} \exp\left\{\frac{i}{2s}\|\gamma_\sigma - x\|^2\right\} \times$$

$$\times f_\sigma(\gamma_\sigma)\varphi(\epsilon(\gamma_\sigma - x))\,\mathrm{d}m(\gamma_\sigma) \tag{6.1a}$$

and set

$$I_s(f;\sigma) := \lim_{\epsilon \to 0+} I_s^\epsilon(f;\sigma,\varphi) \tag{6.1b}$$

if the limit exists and its value does not depend on φ.
(p) $I_s(f;\sigma)$ *are given by the expressions appearing on the r.h.s. of* (5.4), *the corresponding integrals being understood as in* (5.2b). *Of course, one can also restrict attention to the case when they exist in the proper Lebesgue sense, or eventually to employ some other improper integral.*

Except the last one, the above choices of $I_s(f;\sigma)$ rely in different ways on the approximation of γ by the family of polygonal paths γ_σ. Each of the choices has advantages and disadvantages of its own. The first provides the most natural basis for an extension of the Fresnelian F-maps, but it is necessary, at least in principle, to find the inverse Fourier transform to each f_σ. This makes the corresponding method clumsy, especially from a computational point of view. In the next three cases, $I_s(f;\,\sigma)$ is calculated directly. The 'proper' approximation (c), which one would certainly prefer, unfortunately has a limited range of applicability in the case of real s that is most important in practice. For any real-valued potential v, the modulus of the 'integrated' function (1.10a) is independent of the values $\gamma(\tau)$ with $\tau \in (0, t]$, and therefore the respective integral (4.20a) cannot exist. On the other hand, the use of improper integrals forces us to keep strictly the defining prescription, because two quite reasonable limiting procedures can eventually yield completely different values of the improper integral.

EXAMPLE 5.6.1. Let us consider again integrals of the type (2.12a) whose behaviour in the limit $\alpha \to \infty$ we have used to establish existence of $I_s^{\mathrm{ui}}(f)$ for $f \in \dot{\mathcal{F}}(\mathcal{H})$. We now replace the cube C_α by a convex body B_α. The latter can be characterized by a function $r_\alpha : S^{N-1} \to \mathbb{R}_+$ on the unit sphere in $\mathbb{R}^N$; the value $r_\alpha(\Theta)$ means the radius of the surface point specified by the angular coordinates Θ. For simplicity, we set $y = 0$ and restrict ourselves to even $N = 2k$ only, then the integrals under consideration express as

$$(2\pi i s)^{-N/2} \int_{B_\alpha} \exp\left\{ \frac{i}{2s}\, |x|^2 \right\} \mathrm{d}x$$

$$= \tfrac{1}{2}(2\pi i s)^{-k} \int_{S^{2k-1}} \mathrm{d}\Omega \int_0^{r_\alpha(\Theta)^2} e^{iu/2s}\, u^{k-1}\, \mathrm{d}u$$

$$= 1 + (2\pi i s)^{-k} \int_{S^{2k-1}} \exp\left\{ \frac{i}{2s}\, r_\alpha(\Theta)^2 \right\} P_k(r_\alpha(\Theta)^2)\, \mathrm{d}\Omega,$$

where P_k is a polynomial of order $k - 1$. If, for example, the functions $r_\alpha(\,.\,)$ are constant on some non-zero set of angular coordinates independent of α, then the respective part of the last integral is oscillating and has no limit when B_α blows up and s is real. As the simplest illustration, choose $N = 2$ and B_α as the circle of radius α, i.e. $r_\alpha(\varphi) = \alpha$ for all $\varphi \in [0, 2\pi)$. Then

$$(2\pi i s)^{-1} \int_{B_\alpha} \exp\left\{ \frac{i}{2s}\, |x|^2 \right\} \mathrm{d}x = 1 - e^{i\alpha^2/2s},$$

so the corresponding improper integral over $\mathbb{R}^2$ either does not exist or it eventually assumes some value a fulfilling $|a - 1| = 1$ if the limit is taken over a suitably chosen sequence of blowing-up circles.

The use of oscillatory integrals for the rigorous treatment of F-integrals is of a very recent date. It should be noticed that this approach corresponds to the convention often used by the physicists, namely $\exp(i\infty) = 0$. The oscillatory integrals make sense in some cases when the corresponding (proper or improper) integrals do not exist, in particular, for unbounded functions (Problem 50).

In the last named choice of $I_s(f;\,\sigma)$, the polygonal paths are replaced by the piecewise-constant 'paths' $\gamma^{(\sigma)}$ defined by (5.3a). In particular, for the 'integrated' functions (1.10a) it means their replacement by (1.10b) in the approximating expressions. This makes it possible to use the powerful product–formulae machinery, but on the other hand, the use of the particular Riemannian approximation to $\int_0^t v(\gamma(\tau))\,d\tau$ burdens this method by an additional arbitrariness. Besides those mentioned, one could construct other approximating expressions starting from the observation that a path γ can be approximated not only by the families γ_σ or $\gamma^{(\sigma)}$, but also by the functions of another class, for example, by the piecewise quadratic ones. However, the more complicated approximation we choose, the more difficult will be the respective integrals $I_s(f;\sigma)$ to compute; this fact prefers the choices listed above.

Let us now turn to a discussion of possible limits of $I_s(f;\sigma)$ with respect to the partitions σ. Again various possibilities arise:

(r) *The simplest choice is to take the sequence* $\{\sigma_n\}_{n=1}^{\infty}$ *of regular partitions and use the limit* $n\to\infty$. *Alternatively, a suitable subsequence can be used, say* $\{\sigma_{2^n}\}_{n=0}^{\infty}$.

($\mathscr{C}$) *One may consider the set* $\mathscr{C}(J^t)$ *of all crumbling sequences in* $\mathscr{P}(J^t)$, *and define the F-integral of f as* $\lim_{m\to\infty} I_s(f;\sigma_m)$ *if the limit exists for each* $\{\sigma_m\}_{m=1}^{\infty}\subset\mathscr{C}(J^t)$ *and its value does not depend on the particular sequence used.*

($\mathscr{R}$) *One may replace* $\mathscr{C}(J^t)$ *in the preceding prescription by* $\mathscr{R}(J^t)$, *the set of all refining sequences* $\{\sigma_m\}_{m=1}^{\infty}\subset\mathscr{C}(J^t)$, $\sigma_m\subset\sigma_{m+1}$ *for all m, or by some subset* $\mathscr{R}'\subset\mathscr{R}(J^t)$.

($\mathscr{P}$) *The set* $\mathscr{P}(J^t)$ *is partially ordered by the relation* $\supset$, *and to each* σ, $\sigma'\in\mathscr{P}(J^t)$ *there is some* $\sigma''\in\mathscr{P}(J^t)$, *say* $\sigma''=\sigma\cup\sigma'$, *such that* $\sigma\subset\sigma''$ *and* $\sigma'\subset\sigma''$. *In other words,* $\mathscr{P}(J^t)$ *equipped with the ordering* $\supset$ *is a directed set. Then one can define the F-integral as the limit along it,* $I_s(f):=\lim_{\mathscr{P}} I_s(f;\sigma)$, *or more explicitly stated, there exists some* $\sigma_\epsilon\in\mathscr{P}(J^t)$ *to each* $\epsilon>0$ *such that* $|I_s(f)-I_s(f;\sigma)|<\epsilon$ *holds for all* $\sigma\supset\sigma_\epsilon$.

(u) *The uniform approximation: the F-integral under consideration is defined as* $\lim_{\delta(\sigma)\to 0} I_s(f;\sigma)$.

PROPOSITION 5.6.2. *The above-listed limiting prescriptions are related as follows:*

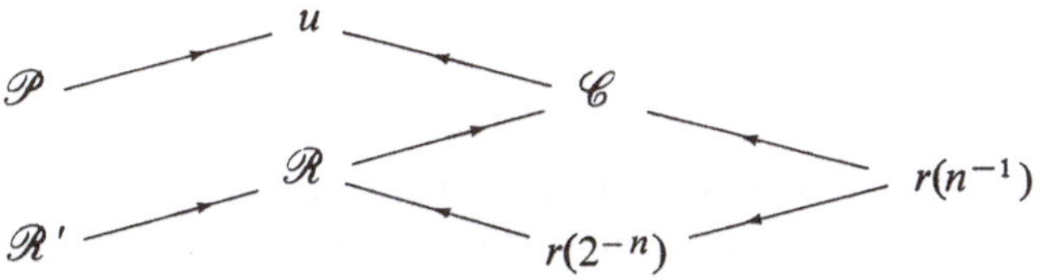

where the arrows have the same meaning as in Remark 4.6(a).

Proof. Suppose $I_s^{\mathrm{u}}(f)\equiv\lim_{\delta(\sigma)\to 0} I_s(f;\sigma)$ exists, then there is $\eta(\epsilon)>0$ to each $\epsilon>0$ such that $|I_s^{\mathrm{u}}(f)-I_s(f;\sigma)|<\epsilon$ if $\delta(\sigma)<\eta(\epsilon)$. One can choose σ_ϵ

so that $\delta(\sigma_\epsilon) < \eta(\epsilon)$, then $\delta(\sigma) < \eta(\epsilon)$ holds for each $\sigma \supset \sigma_\epsilon$, and therefore $\lim_{\mathscr{P}} I_s(f;\sigma) = I_s^{\mathrm{u}}(f)$. Consider further a sequence $\{\sigma_m\}_{m=1}^\infty \subset \mathscr{C}(J^t)$. Since it is crumbling, there is $m_0(\epsilon)$ such that $\delta(\sigma_m) < \eta(\epsilon)$ for all $m > m_0(\epsilon)$. Then $|I_s^{\mathrm{u}}(f) - I_s(f;\sigma_m)| < \epsilon$ holds for $m > m_0(\epsilon)$ so $\lim_{m\to\infty} I_s(f;\sigma_m) = I_s^{\mathrm{u}}(f)$ independently of the sequence used. The remaining relations are elementary. ■

It should be noted that the existence of the limit according to ($\mathscr{P}$) does not necessarily imply the same for ($\mathscr{R}$). The reason is that the index set of the net $\{I_s(f;\sigma) : \sigma \in \mathscr{P}(J^t)\}$ is not fully ordered so the convergence of a subnet (in particular, a subsequence) may not follow from convergence of the whole net.

This brief survey of the limiting prescriptions shows that there are two among them which are presumably the most important. The regular approximation is distinguished by its simplicity, which makes it suitable for computational purposes. On the other hand, Proposition 6.2 illustrates the role played by the uniform approximation among the considered limits. One can say that this limiting prescription suits best to the Riemann-integral ideology inherent to the sequential methods.

Let us next mention one other approach based on the sequential methods. We have observed in Section 4 that the domains of various F_s-maps are considerably wider when $s \in \mathbb{C}_F$ is non-real. For instance, the functions (1.10a) with a real-valued potential v do not belong to $\mathscr{F}_s^{\mathrm{uc}}(\mathscr{H}_x)$ for a real s, but they often do so if $\operatorname{Im} s < 0$. This fact suggests that one can try to treat the case of real s separately; the corresponding F-integrals are then given by a non-tangential limit in s. For example, we can define the **limiting uniform cylindrical F-integral** for a positive s by

$$I_s^{\mathrm{luc}}(f) := \lim_{\epsilon\to 0+} I_{s-i\epsilon}^{\mathrm{uc}}(f) \tag{6.2}$$

for those $f : \mathscr{H}_x \to \mathbb{C}$ for which the limit exists. In other words, methods of this type employ the trick mentioned in connection with Theorem 1.1, but the inconsistent measure-theoretic definition of $I_{s-i\epsilon}(f)$ is replaced by a sequential one which makes sense.

PROPOSITION 5.6.3. *The F-integral defined by* (6.2) *obeys the relation* (2.1). *If* $x = 0$, *then its domain fulfils* $\mathscr{F}_s^{\mathrm{luc}}(\mathscr{H}) \supset \mathscr{F}(\mathscr{H})$ *and the equality*

$$I_s^{\mathrm{luc}}(f) = I_s^{\mathrm{f}}(f) \tag{6.3}$$

holds for each $f \in \mathscr{F}(\mathscr{H})$.

Proof. The first assertion follows from Remark 4.6(b). For a fixed $f \in \mathscr{F}(\mathscr{H})$ and an arbitrary $s \in \mathbb{C}_F^0$, the uniform cylindrical F-integral of f equals $I_s^{\mathrm{f}}(f)$ in view of (4.27a), so the result follows from Proposition 2.3. ■

Let us now turn to the non-sequential F-integral theories. We shall be concerned mostly with the path space $\mathscr{H}$, the extension to the case $x \neq 0$ being straightforward. In order to describe the Itô's method, recall first that the *Gaussian measure* μ on a (real, separable) Hilbert space $\mathcal{H}$ is a Borel measure such that for each $\gamma \in \mathcal{H}$, the function $(\gamma, .)$ is normally distributed,

$$\mu(\gamma' \in \mathcal{H} : (\gamma, \gamma') \leqslant a) = (2\pi\sigma_\gamma)^{-1/2} \int_{-\infty}^{a} \exp\{-(t - m_\gamma)^2/2\sigma_\gamma\}\, dt. \quad (6.4a)$$

According to Prokhorov theorem, each Gaussian measure is uniquely specified by its *mean* $\beta \in \mathcal{H}$ and *correlation operator* S on $\mathcal{H}$, which is positive symmetric and of the trace class; we have

$$m_\gamma = (\beta, \gamma), \qquad \sigma_\gamma = (\gamma, S\gamma) \quad (6.4b)$$

for all $\gamma \in \mathcal{H}$. If the space $\mathcal{H}$ is finite-dimensional, we have

$$\int_{\mathcal{H}} f(\gamma)\, d\mu_{S,\beta}(\gamma) = (\det(2\pi S))^{-1/2} \int_{\mathcal{H}} \exp\{-\tfrac{1}{2}(\gamma - \beta, S^{-1}(\gamma - \beta))\} f(\gamma)\, d\gamma \quad (6.4c)$$

but this expression becomes formal when passing to an infinite-dimensional $\mathcal{H}$. The idea is to approximate the F-integral of $f : \mathcal{H} \to \mathbb{C}$ by the family of Gaussian integrals

$$I_s(f; S, \beta) = \left[\det\left(I - \frac{i}{s} S\right)\right]^{1/2} \int_{\mathcal{H}} \exp\left\{\frac{i}{2s} \|\gamma\|^2\right\} f(\gamma)\, d\mu_{S,\beta}(\gamma) \quad (6.5a)$$

whose normalization is taken from the finite-dimensional case. There are various ways in which the added exponential factor can be removed by making the correlation operator S large. One of them is to realize that the set $\mathcal{S}$ of all positive symmetric trace-class operators is partially ordered and directed, because $S_j \leqslant S_1 + S_2$ for $j = 1, 2$. Hence it is possible to define the **Gaussian-regularized F-integral** of f by

$$I_s^{\mathrm{g}}(f) := \lim_{\mathcal{S}} I_s(f; S, \beta) \quad (6.5b)$$

if the limit exists and is independent of β. We have the following assertion:

THEOREM 5.6.4 (Itô). *Let $f \in \mathcal{F}(\mathcal{H})$ on $\mathcal{H} = AC_0[J^t; \mathbb{R}]$ and $s > 0$, then $I_s^{\mathrm{g}}(f)$ exists and equals $I_s^{\mathrm{f}}(f)$.*

Furthermore $\int_{\mathcal{H}}^{\mathrm{g}} f(\gamma)\, D\phi_s(\gamma)$ admits certain change-of-variable transformations, which eventually may lead us out of $\mathcal{F}(\mathcal{H})$. It is invariant under Euclidean motions of $\mathcal{H}$ and the Cameron–Martin formula holds for a class of 'nearly isometric' transformations. Reference to the proofs together with some further comments are given in the notes.

Our next remark concerns *generalizations of the Fresnelian F-maps*. A straightforward way is to extend $\mathcal{F}(\mathcal{H})$ by adding to it Fourier transforms of a suitable class of distributions on $\mathcal{H}$, and to modify correspondingly the definition (2.7). The main obstacle for such a method is confined in the fact that the theory of distributions on infinite-dimensional spaces is far less developed than its finite-dimensional counterpart. Another possibility is to modify the Parseval-type identity on which the construction relies. For example, let B be a regular positive operator on $\mathbb{R}^n$, then the equality

$$(2\pi i s)^{-n/2}\, |\det B|^{1/2} \int_{\mathbb{R}^n} \exp\left\{\frac{i}{2s}\, x \cdot Bx\right\} f(x)\, dx$$

$$= \int_{\mathbb{R}^n} \exp\left\{-\frac{is}{2} y \cdot B^{-1} y\right\} \tilde{f}(y)\, dy \tag{6.6}$$

holds for each $f \in \mathscr{S}(\mathbb{R}^n)$, where $\tilde{f}$ is given by (1.11b); it reduces to (1.11a) if $B = I$ and $s = 1$. The relation (6.6) suggests how one can define F-integrals with $\|\gamma\|^2$ in the exponential term replaced by $(\gamma, B\gamma)$ for a regular $B \in \mathscr{B}(\mathscr{H})$, or more generally, by the values $\Delta(\gamma)$ of some closed non-singular quadratic form Δ on $\mathscr{H}$. This method was initiated again by Albeverio and Høegh-Krohn. It is suitable especially for certain functions of the type (1.10a), for example, the one corresponding to an anharmonic oscillator, because it allows us to include a part of the potential term into the 'measure'.

On the other hand, there are some problems connected with this method. The normalization factor of the l.h.s. in (6.6) contains det B, and therefore the generalized Fresnelian F-integrals cannot be generally expected to obey the heuristic criterion expressed by the relation (1.4). This means, particularly, that they do not represent an extension of $I_s^{\mathrm{f}}(\,.\,)$ unless the form is generated by a multiple of the unit operator. In fact, it is one more way in which non-existence of the Feynman measure is manifested. Had the latter existed, the two F-integrals would exist simultaneously and fulfil

$$\int_{\mathscr{H}} f(\gamma)\, e^{(i/2s)\Delta(\gamma)}\, D\gamma = \int_{\mathscr{H}} f(\gamma)\, e^{(i/2s)\Delta_1(\gamma)}\, D\phi_s(\gamma),$$

where $\Delta_1(\gamma) := \Delta(\gamma) - \|\gamma\|^2$ (cf. (A.5)).

Nevertheless, there are cases when one can write simple relations between some of the F-maps treated above (which extend I_s^{f}) and the generalized Fresnelian F-maps relative to a form Δ. The formula (6.6) suggests that it might be so if Δ is generated by an operator B whose determinant makes sense, say, $B = I + L$ with L symmetric and of the trace class: the two F-maps should differ mutually by the multiplicative factor $|\det(I + L)|^{1/2}$. An additional problem arises when the operator $I + L$ is regular but not positive; then one must choose properly the branch of square root involved. The following formula is expected to be valid for $f \in \mathscr{F}(\mathscr{H})$,

$$\int_{\mathscr{H}} \exp\left\{\frac{i}{2s}(\gamma, L\gamma)\right\} f(\gamma)\, D\phi_s(\gamma) = |\det(I+L)|^{-1/2} \exp\left\{-\frac{\pi i}{2}\, \mathrm{ind}(I+L)\right\} \times$$

$$\times \int_{\mathscr{H}} \exp\left\{-\frac{is}{2}(\gamma, (I+L)^{-1}\gamma)\right\} d\mu_f(\gamma) \tag{6.7}$$

with a suitably chosen definition of the l.h.s. Here ind($I + L$) denotes the number of negative eigenvalues of $I + L$ counted according to their multiplicity; it is finite if only L is compact. There are two features of the formula (6.7) which should be stressed. The first concerns its close relation to the Cameron–Martin formula discussed above, which emerges formally if we insert there a positive $I + L = (I + K)^*(I + K)$ and set $g((I + K)\gamma) = f(\gamma)$. The other interesting property is the presence of the exponential phase-changing factor on the r.h.s. It is just this factor which is responsible for the occurrence of Morse or Maslov indices; an illustration will be given in Section 6.2.

This survey would be certainly incomplete without mentioning the methods based on *analytical continuation*. Since they start from the knowledge of some Wiener integrals, the appropriate path space will be now $X = C_0[J^t; \mathbb{R}^d]$ (the extension to the case $x \neq 0$ is again simple). In order to formulate the definition, a few notions are needed. A set $M \subset X$ is said to be *scale-invariant measurable* if αM is measurable for each $\alpha > 0$, and in the same way one introduces the *scale-invariant zero set* with respect to a given measure. A property which holds with exception of such a set is said to hold *scale-invariant a.e.* Notice that the concept of invariance with respect to changes of the scale is rather alien to the W-integral theory.

Consider now a function $F : X \to \mathbb{C}$ which is scale-invariant a.e. defined and w-measurable, and suppose the W-integral

$$J(F, \lambda) = \int_X F(\lambda^{-1/2}\gamma)\, dw(\gamma) \tag{6.8a}$$

exists for all $\lambda > 0$. Further, we assume that there is a function analytic in $\{\lambda : \mathrm{Re}\,\lambda > 0\}$ whose values coincide with (6.8a) on $(0, \infty)$; for simplicity we denote it again as $J(F, .)$. For a fixed $s > 0$, we can then define the **analytic F-integral** of F by

$$\int_X^{\mathrm{a}} F(\gamma)\, D\phi_s(\gamma) := \lim_{\substack{\lambda \to -i/s \\ \mathrm{Re}\,\lambda > 0}} J(F, \lambda) \tag{6.8b}$$

if the limit makes sense. Of course, one is primarily interested in a comparison of (6.8) to the other above-discussed definitions of the F-integral. Moreover, the present definition should be related to the one described on the heuristic level in Section 1. Their correspondence is not quite obvious, because the parameter used for the analytical continuation enters there into Wiener measure as its dispersion, while here it is connected with the scale-changing transformations of the configuration space.

PROPOSITION 5.6.5. *Let $F : X \to \mathbb{C}$ be $\| . \|_\infty$-continuous. Let further the restriction $f = F \upharpoonright \mathcal{H}$ on $\mathcal{H} = AC_0[J^t; \mathbb{R}^d]$ belong to $\mathcal{F}(\mathcal{H})$, then the analytic F-integral* (6.8b) *exists for each $s > 0$ and*

$$\int_X^{\mathrm{a}} F(\gamma)\, D\phi_s(\gamma) = \int_{\mathcal{H}}^{\mathrm{f}} f(\gamma)\, D\phi_s(\gamma). \tag{6.9}$$

Proof. The continuity assumption implies F to be defined everywhere and scale-invariant w-measurable. The function f is bounded in view of Proposition 2.2, so Proposition 4.9 may be applied. We set $c = \lambda^{-1}$ for a fixed $\lambda > 0$, then (4.29a) yields

$$\int_X F(\gamma)\, dw_c(\gamma) = \lim_{\delta(\sigma)\to 0} (2\pi c)^{-nd/2} \int_{\mathscr{H}_X^\sigma} \exp\left\{-\frac{1}{2c}\|\gamma_\sigma\|^2\right\} f_\sigma(\gamma_\sigma)\, dm(\gamma_\sigma).$$

Substituting $\tilde{\gamma}_\sigma = c^{-1/2}\gamma_\sigma$ into the last integral, we obtain

$$\int_X F(\gamma)\, dw_c(\gamma) = \int_X F(c^{1/2}\gamma)\, dw(\gamma), \tag{6.10}$$

where $w = w_1$ as usually. The r.h.s. can be expressed from (4.29b) so we have

$$\int_X F(\lambda^{-1/2}\gamma)\, dw(\gamma) = I^{\mathrm{f}}_{-i/\lambda}(f).$$

Since $\lambda \mapsto -i\lambda^{-1}$ is analytic with exception of the point $\lambda = 0$, the result follows from Proposition 2.3. ∎

The relation (6.10) used in the proof shows that the two above-mentioned 'analytic' definitions are essentially identical, up to the choice of the required region of analycity and other technicalities. The assumptions might be eventually weakened by considering the functions which are $\|\,.\,\|_\infty$-continuous with exception of a (scale-invariant) w-zero set. Even such a modification, however, leaves some important questions open. One would like to know, under which assumptions about F the restriction $f = F \upharpoonright \mathscr{H}$ belongs to $\mathscr{F}(\mathscr{H})$, and conversely, whether each $f \in \mathscr{F}(\mathscr{H})$ can be extended to some F on X for which the analytic F-integral (6.8b) exists. These questions are not trivial; notice, for example, that the continuity of $f \in \mathscr{F}(\mathscr{H})$ with respect to the Hilbert-space norm of $\mathscr{H}$ does not imply its $\|\,.\,\|_\infty$-continuity on $\mathscr{H}$ (cf. Problem 51).

A complete solution to the described problem has been given recently by G. Johnson, whose results can be summed briefly as follows:

(i) There is a (fully specified) Banach algebra $S(X)$ on which the analytic F-integral (6.8) is defined. The elements of $S(X)$ are equivalence classes of the functions $F : X \to \mathbb{C}$ which coincide scale-invariant a.e. in X.

(ii) The Banach algebras $S(X)$ and $\mathscr{F}(\mathscr{H})$ are isomorphic by the mapping which assigns to each equivalence class $\tilde{F} \in S(X)$ the restriction $f = F \upharpoonright \mathscr{H}$ of an arbitrary representative F of $\tilde{F}$. Moreover, for every such pair f, F the relation (6.9) is valid.

Each of the two settings of the F-integral theory has its advantages. In this chapter, we have dealt mostly with the Hilbert-space one which makes it possible to avoid many subtle measure-theoretical problems. On the other hand, the Banach-space setting can rely on numerous results which are available for Wiener integrals.

It is therefore clear that equivalence results of the above type are likely to give a new impetus to the theory.

Concluding this section, let us mention one more generalization of the concepts discussed here. The primary physical aim of introducing F-integrals was to express certain operators, especially e^{-iHt} for Schrödinger operators H on $L^2(\mathbb{R}^d)$. This is usually achieved by writing them as integral operators whose kernels admit a path-integral representation. However, it might be sometimes preferable to interpret this procedure as the construction of a path-integral expression to the operators themselves. A simple example is provided by the considerations of the preceding section: the formulae like (1.1) represent primarily the relations between operators. In this way, one is led to the concept of *operator-valued F-integrals*.

In order to illustrate this, consider a function $f : \mathscr{H} \to \mathbb{C}$ on $\mathscr{H} = AC_0[J^t; \mathbb{R}^d]$. To each $\sigma \in \mathscr{P}(J^t)$ and $s \in \mathbb{C}_F$, one can define the operator $I_s[f; \sigma]$ on $L^2(\mathbb{R}^d)$ by

$$(I_s[f;\sigma]\psi)(x) = \prod_{j=0}^{n-1} (2\pi i s\delta_j)^{-d/2} \int_{\mathbb{R}^{nd}} \exp\left\{ \frac{i}{2s} \sum_{j=0}^{n-1} |\gamma_{j+1} - \gamma_j|^2 \, \delta_j^{-1} \right\} \times$$

$$\times f^{(\sigma)}(\gamma_0 + x, \ldots, \gamma_{n-1} + x)\psi(\gamma_0 + x)\, d\gamma_0 \ldots d\gamma_{n-1}, \tag{6.11a}$$

where $\gamma_n = 0$ and $f^{(\sigma)}$ is given by (5.3). The r.h.s. need not necessarily exist for all $x \in \mathbb{R}^d$, we require only $I_s[f;\sigma]\psi \in L^2(\mathbb{R}^d)$. The operator-valued uniform product F-integral of f is then defined by

$$I_s^{\mathrm{up}}[f] := \operatorname*{s-lim}_{\delta(\sigma)\to 0} I_s[f;\sigma]. \tag{6.11b}$$

Similarly, the operator-valued analytic F-integral of a function $F : X \to \mathbb{C}$ on $X = C_0[J^t; \mathbb{R}^d]$ is defined as the continuation of the operator-valued function $J[F, .]$ given by

$$(J[F,\lambda]\psi)(x) = \int_X F(\lambda^{-1/2}\gamma + x)\psi(\lambda^{-1/2}\gamma(0) + x)\, dw(\gamma) \tag{6.12}$$

from $(0, \infty)$ to the point $\lambda = -is^{-1}$. Some results concerning the operator-valued F-integrals are mentioned in the notes.

Notes to Chapter 5

SECTION 5.1. The beginning of the path-integral concept under consideration is connected with Feynman's paper (1948) based on his unpublished thesis written six years earlier. The inspiration came from the work of Dirac (1933; see also 1945 and 1959, §32), which in turn was motivated by the classical Hamilton–Jacobi equation. Feynman applied his idea to quantum mechanics and quantum electrodynamics (1949–1951) and later to statistical physics. His results are reviewed in Feynman and Hibbs (1965); this book may be regarded as a good introduction to the physical philosophy of the method. Physical applications of the (non-rigorous) F-integral techniques are very numerous. We limit ourselves to quoting a few reviews, proceedings and monographs from which detailed information and further references can be derived: Gel'fand and Yaglom (1956), Blokhintsev and Barbashov (1972), Bogoliubov and Shirkov (1976, chap. VIII), Popov (1976), Vasiliev (1976), Papadopoulos and Devreese (1978), Slavnov and Fadeev (1978), DeWitt-Morette *et al.* (1979), Marinov (1980), Schulman (1981), Prokhorov (1982) and Mensky (1983).

• If we are interested in the kernel of the evolution operator, then the path integral is taken over the set of paths with both ends fixed. Feynman's heuristic definition of the path integral uses the regular partitions only, $\tau_j = jt/n$ with $j = 0, 1, \dots, n$, together with the limit $n \to \infty$; but there is no *a priori* reason why other sequences of partitions should be excluded, if only the maximal subinterval length tends to zero. This question will be discussed in detail later in this chapter. Among rigorous approaches to the construction of path integrals (1.3), the one closely following Feynman's original idea was initiated by Cameron (1960); his effort was later continued by Truman (1977, 1979). Notice that in the physical literature the 'integrated' functions are mostly named functionals.

• In spite of the strong formal similarity between the F- and W-integrals, one should notice possible differences in interpretation. For example, in the quantum-theoretical applications of the Feynman–Kac formula we are interested in the probability *amplitudes* similarly as in (1.2); on the other hand, in its applications to the problems of classical physics (Brownian motion, Feynman–Kac formula for the heat equation), the Wiener measure is understood as a probability measure. A compact exposition of the Wiener-measure theory is due to Kuo (1975); a more detailed presentation with numerous applications in quantum physics and extensive bibliography can be found in Simon (1979a), Glimm and Jaffe (1981). See also older reviews by Kovalchik (1963) and Shilov (1963). Let us remark that the W-measure is often defined on the $\mathbb{R}$-valued paths only, but it is easy to construct the 'd-dimensional' W-measure as the product measure of the 'one-dimensional' ones.

• There is a vast amount of literature devoted to the mathematical aspects of Feynman path integrals. The list of references given here is representative by intention, but we do not pretend it to be exhaustive. The bibliography can be completed from Albeverio and Høegh-Krohn (1976), DeWitt-Morette *et al.* (1979), Albeverio *et al.* (1979) and from the papers quoted below.

• The proposal to define the F-integrals as a limit when the parameter s approaches the real axis from below was given by Gel'fand and Yaglom (1956). Theorem 1.1, which illustrates its failure, is due to Cameron (1960); see also Daletskii (1961, 1962) and Truman (1978a). It is sufficient for the proof of choose $\epsilon = 0$; however, it is interesting that the argument works even if the 'cylindrical projections' of the considered functions are required to be integrable.

• The definition of F-integrals by analytical continuation in mass was proposed first by Cameron (1962–1963), and later independently by Nelson (1964). It was further studied, e.g. by Cameron and Storvick (1966, 1968), Johnson (1982), Johnson and Skoug (1973–1981); see also the more general problem considered by Beekman (1965). Of course, the complex mass presumably has no direct physical meaning (Kyselka, 1981). The definition of F-integrals for the functions of the type (1.10a) by analytic continuation in time appeared in the papers by Babbitt (1963, 1965) and Feldman (1963). The idea can be traced to the work of Kac (1949), where the Feynman–Kac formula was proved; see Reed and Simon (1975, §X.11) or Simon (1979a, §II.6). The method starting from the imaginary-time description of dynamics, usually dubbed the Euclidean approach, provided a powerful tool especially for the constructive quantum field theory, which made it possible to prove the existence of non-trivial realizations for Wightman axioms – see, e.g. Simon (1974), Glimm and Jaffe (1981). Also, there are numerous applications in quantum mechanics and statistics for which we refer to Simon's book (1979a).

• The product-formulae definitions of F-integrals were initiated by Nelson (1964) who employed a particular version of the Trotter formula (cf. (A.7)). In this way, he was able to prove the basic dynamical formula (1.2) for the potentials of the class $L^p + L^\infty$, where $p \geqslant 2$

and $p > \frac{1}{2}d$. This result was later generalized for time-dependent potentials (Faris, 1967b) and for the harmonic oscillator potential (Combe *et al.*, 1978). A more general product formula was used by Ichinose (1980) for the case when v was bounded below and the corresponding Hamiltonian existed in the sense of quadratic forms. In this case, however, the F-integral used was the 'phase space' one.

• Itô's method was first formulated in Itô (1961), where he proved the dynamical formula for an implicitly specified subset in the set of all bounded continuous potentials in $\mathbb{R}$. In his next paper (1967) he used another limiting procedure (the one mentioned here) to get the result for the following three cases: the potentials v which belong to $\mathcal{F}(\mathbb{R})$, the set of Fourier transforms of bounded measures on $\mathbb{R}$, and for the linear and quadratic potentials. Itô was also the first who systematically used Hilbert-space metrics in the path space. Recently Tarski (1979) proposed a method which unifies, up to a certain degree, Itô's approach with the 'sequential' ones.

• Next we have to mention the F-integral theory of DeWitt-Morette (1972, 1974). She replaced the non-existent 'Feynman measure' by a 'prodistribution' – i.e. a generalized projective measure on the path space X, which was assumed to be a locally convex Hausdorff space. The prodistribution relevant to F-integrals is determined by its Fourier transform, which is (up to a normalization factor) equal to $\exp\{-(i/2)W(x', x')\}$, where W is a bilinear form on the dual X' to X. The method was used to compute various path integrals of physical interest, in particular, with a non-flat configuration-space metrics, and to the semiclassical approximation (DeWitt-Morette, 1976, 1979). We do not pretend to expose this theory here, and refer the reader to the recent monograph (DeWitt-Morette *et al.*, 1979), see also Nelson and Sheeks (1982). For a related definition to the F-integral in terms of forms, see Kree (1979).

• The method of Albeverio and Høegh-Krohn (1976, 1979) based on an analogue to the formula (1.11) will be discussed in the next section. In this way, the dynamical formula can be proved for the potentials from $\mathcal{F}(\mathbb{R}^d)$. Attempts to extend the applicability of the method, particularly to some quantum field-theoretical models, inspired various generalizations (Albeverio and Høegh-Krohn, 1976, 1979; Truman, 1976–1979; Uglanov, 1978; Berg and Tarski, 1981; Parthasarathy and Sinha, 1982), which will be also commented on later. Recently Johnson (1982) gave an equivalence relation between these F-integrals and the 'analytical' ones.

• The 'phase space' path integrals were also introduced by Feynman (1951, appendix), and studied further by, e.g., Garrod (1966), Berezin (1971, 1980), Mizrahi (1975, 1980) or Tarski (1981). The Hamiltonian F-integrals can be also defined with the help of Fourier transformation (Maslov and Chebotarev, 1976); in this method the exponent of the potential term yields a generalized measure which is used to 'integrate' the exponent of the kinetical part of the action; see also Combe *et al.* (1981). For applications of the phase-space F-integrals to quantization see, e.g. Gutzwiller (1967–1971, 1978), Fadeev (1969), Berezin (1971, 1980), Mizrahi (1979, 1981a), and also Smolianov (1982).

• As we have mentioned, the idea of obtaining the classical limit or, more generally, the semiclassical approximation to quantum dynamics from its path-integral expression is intuitively very natural. Of course, realizing this program mathematically, one must deal with certain technical subtleties: see, e.g., DeWitt-Morette (1976), Albeverio and Høegh-Krohn (1977), Truman (1976, 1977b, 1979) or Mizrahi (1977, 1981b). There is a close connection between these considerations and the standard semiclassical methods of quantum mechanics (Berry and Mount, 1972).

• For applications of rigorous functional-integration methods in quantum field theory and statistical physics see, e.g., Simon (1974). Albeverio and Hoegh-Krohn (1976, 1979) or Glimm and Jaffe (1981), where many other references are given. The Treatment of F-integrals in a curved-space background can be found, e.g., in DeWitt-Morette *et al.* (1979) or Elworthy and Truman (1981). For some constraint problems connected with the F-integrals see Clark *et al.* (1980), Bernido and Inomata (1981) or Goodman (1981); in fact the considerations concerning the 'Feynman paths' presented in Section 6.3 also fall into this category.

• Recently some methods have been devised in which functional integration is simulated on a computer. Their main aims are the lattice gauge field theories, but they have been applied also to quantum-mechanical problems (Scher *et al.*, 1980; Creutz and Freedman, 1981). There is another 'finite' framework for functional integration, namely the variant of quantum mechanics which is obtained when one replaces the continuous group of space translations appearing in Weyl form of the canonical commutation relations by a finite Abelian group (see, e.g. Santhanam and Sinha (1978), Šťovíček and Tolar (1979), Gudder and Naroditzky (1981)). The phase-space F-integrals were introduced in this context by Šťovíček (1980).

SECTION 5.2. The theory of F-integrals defined by a Parseval-type equality was formulated by Albeverio and Høegh-Krohn (1976, chap. 2; 1979). Their work was influenced especially by Itô's treatment of F-integrals on Hilbertian path spaces, and by DeWitt-Morette (1972, 1974) who emphasized the role which the Fourier transformation can play. The concept of Feynman maps is due to Truman (1978a); it is essentially identical with that of the complex Wiener integral introduced by Cameron (1960). Their definitions are sequential and we shall discuss them in Section 4. The extension of the abstract F-integral of Albeverio and Høegh-Krohn to the complex-mass case is straightforward; our exposition follows Exner and Kolerov (1981b), where it was first given. Notice that the continuous extension of the F-integral definition to the point $s = 0$ is of interest when one considers the classical limit in the path-integral expression of quantum dynamics, because the parameter is in fact $s = \hbar/m$ – cf. Truman (1979) and the other references to this problem given in the notes to the preceding section. The construction used in Example 2.10 is clearly due to Theorem 1.1; for the particular case of real s, an alternative possibility is given in Problem 44. A sort of bounded-convergence theorem for functions of the type (1.10a) and the analytic F-integral has been proved recently by Johnson (1983).

SECTION 5.3. Once we have decided to equip the set of paths (for a quantum-mechanical system with $\mathbb{R}^d$ as configuration space) by a Hilbert-space structure, then $AC_0[J^t; \mathbb{R}^d]$ is the natural candidate (Itô, 1961, 1967). The fact that the singularly continuous paths must be factorized out led Truman (1976) to propose the replacement of $\dot{\gamma}$ by $\dot{\gamma}_W$. It is not difficult to prove that the weak derivative determines the path γ (with fixed end) uniquely (Truman, 1977a). The idea of expressing paths by their Fourier series is due to Feynman (see Feynman and Hibbs (1965, §3.11)); its first rigorous treatment can be found in Truman (1976). This author also proved (1977a) the equality $\dot{\gamma}_W = \dot{\gamma}_m$, and later found (1978b) that the paths $\gamma \in \mathscr{F}[J^t; \mathbb{R}^d]$ are absolutely continuous. Theorem 3.3 together with the presented proof is taken from Exner and Kolerov (1981a). It is worth mentioning that the result of Carleson (1966) used in Lemma 3.6 represents a solution to the problem formulated by Luzin in 1915 (see Kolmogorov and Fomin (1976, §VIII.1); a detailed account on the pointwise convergence of a Fourier series can be found in Mozzochi (1971)). The reproducing-kernel property was noticed and used by Albeverio and Høegh-Krohn (1976) and by Truman (1976–1979).

• A concise exposition of the theory of Wiener spaces can be found in Kuo (1975). The fact that the W-measure is supported by the paths which are continuous but not smooth was noticed in 1933 by Paley, Wiener and Zygmund – cf. Simon (1979a, §II.5) and Kuo (1975, theorem I.3.2). The characteristic 'quiver' causes that a typical Brownian path should be regarded in certain sense as an object whose dimension is two rather than one (Simon, 1979a, §VII.22); there are heuristic arguments showing that the 'paths' of quantum-mechanical systems might have an analogous property (e.g. Abbot and Wise, (1981), Campesino-Romeo *et al.* (1982)).

SECTION 5.4. The polygonal-path methods, which are probably the most popular tool used by physicists for treating F-integrals, can be traced to the very beginning of the concept of the Feynman path integral. There are not so many papers, however, in which these methods are treated rigorously: we should mention especially the one by Cameron (1960) and the pioneering work of Truman (1976–1979) on polygonal-path approximations on the Hilbert spaces of paths. In particular, this author proved (1976) Theorem 4.3 for the sequence of regular partitions and $d = 1$; the present generalization together with the corresponding analysis of the sets $\mathscr{P}(\mathscr{I}^t)$ and $\mathscr{P}_x[\mathscr{I}^t; \mathbb{R}^d]$ is taken from Exner and Kolerov (1982a). Proposition 4.5 is essentially due to Cameron (1960).

• The term 'uniform' is probably not ideal, but it reflects well the fact that the limiting procedure under consideration does not rely on some particular time-slicings. The definitions of the cylindrical F_S-maps given here are just a few from a wide variety of possible modifications; this will be discussed in Section 6. In particular, I_S^{ri} is the mapping introduced and studied by Truman (1978a). On the other hand, Cameron's (1960) concept of complex sequential Wiener integral corresponds rather to I_S^{uc}. The two last-mentioned maps are not strictly identical being not defined on the same path spaces, but it makes little difference in view of the continuity restrictions, as we have mentioned in the conclusions of the previous section. The idea of Theorem 4.8 belongs to Truman (1976).

• In order to extend the validity of Proposition 4.9 to the functions with a limited order of growth, one may use the Fernique theorem (Kuo, 1975, theorem III.3.1) according to which there is a positive a such that $\exp(a \| \,.\, \|_\infty^2) \in L(X_0, w_c)$. The explicit form of the distribution (4.31) can be found through the kernel of $\exp(-ctH_\alpha)$, where H_α is a multiple of the Dirichlet Laplacian relative to the ball of radius α in $\mathbb{R}^d$ (cf. Simon (1979a, Lemma 7.10)). This application of the Feynman–Kac formula is illustrated in Problem 48 for $d = 1$. In that case, the result has long been known (see Erdös and Kac (1946)), but it is expressed through the sum of a series so the behaviour of $\omega_1(\,.\,, ct)$ is not quite obvious. A closer inspection suggests that the value of the constant a here might be $(2ct)^{-1}$; this fact was already used by Cameron (1960) to prove an assertion analogous to Proposition 4.9 for the sequential Wiener integral. Notice also that there are examples of functions for which the standard and sequential W-integrals do not coincide (Maikov, 1958).

• Proposition 4.11 generalizes the result obtained by Truman (1978a) for I_S^{ri} and $f \in \mathscr{F}(\mathscr{H})$, as well as Proposition 2.8. All these assertions are inspired by the transformation properties of the W-integral under translations (which are, however, more complicated in view of the larger path spaces involved – cf. Kuo (1975, chap. 2)). Also Theorem 4.12 is due to the W-integral transformation formula obtained by Cameron and Martin (1945). Our proof has been borrowed from Truman (1978a) who attempted the analogous assertion for I_S^{ri} and $f \in \mathscr{F}(\mathscr{H})$ (cf. Remark 4.13(b)). Basic facts about infinite-dimensional determinants can be found, e.g. in Dunford and Schwartz (1973, §XI.9) or Reed and Simon (1978, §XIII.17). If K is not of the trace class, (4.33) may be still valid with a suitable substitute for $\det(I + K)$ (Truman, 1979). The CM-formula for an alternative definition of F-maps with application to Maslov indices may be found in Elworthy and Truman (1982).

SECTION 5.5. The product F-maps introduced here may also be labelled as 'regular product' if one wants to emphasize the character of the limit involved. There is a close connection between the formalism under consideration and the classical Volterra product integrals – cf. DeWitt-Morette *et al.* (1979, chap. 2). As mentioned above, the idea of defining F-integrals by means of the product formulae is due to Nelson (1964); for further references see notes to Section 1. The main theorem of this section was proved by Faris (1967b) with the limit taken over the sequence of regular partitions; the 'uniform' generalization is due to Exner and Kolerov (1982b).

SECTION 5.6. Our classification of sequential methods follows essentially Exner and Kolerov (1982a). Among various choices of the approximating expressions, the proper integrals are used mostly in combination with the limit in s (Cameron, 1960). The improper integrals were studied and used in the series of papers by Truman (1976–1979). The oscillatory integrals have been used recently for the present purpose by Elworthy and Truman (1982). The approximating expressions (p) usually appear in connection with the product formulae, i.e. for the 'integrands' of the type (1.10a) (Babbitt, 1963; Nelson, 1964; Reed and Simon, 1975, §X.11; Combe *et al.*, 1978; Tarski, 1979; Exner and Kolerov, 1982a, b); the more general formulation can be found in Cameron and Storvick (1968) or Johnson and Skoug (1973). Notice that the replacement of the action by a Riemannian approximation to it may not be a harmless operation: e.g. in the phase-space formulation, different choices of this approximation are known to yield different quantizations (Berezin, 1971). Though such problems do not arise here, there is another disturbing question; namely, how fast is the relative convergence between the methods which determine F-integrals using the polygonal paths and the 'paths' (5.3a). The only serious inquiry to this problem was undertaken by Cameron (1968), who showed that for the functions (1.10) with a (possibly time-dependent but) sufficiently regular potential and the regular limit, the difference is of order $O(n^{-1})$. Example 6.1 with $N = 2$ can be found in DeWitt-Morette *et al.* (1979). For the convention $e^{i\infty} = 0$ see, e.g., Slavnov and Faddeev (1978, §II.3). More general approximations of paths have been studied in the phase-space setting by Fiziev (1983).

• As for the limiting prescriptions, the regular one has already been used by Feynman in his heuristic definition (Feynman and Hibbs, 1965, §2.4); it appears in nearly all physical papers dealing with F-integrals. Among the rigorous treatments, it can be found, e.g., in Gel'fand and Yaglom (1956), Babbitt (1963), Nelson (1964), Truman (1976–1979) or Combe *et al.* (1978). The $\mathscr{C}$-approximation has been used recently by Elworthy and Truman (1982). The $\mathscr{R}$-approximation represents a specification of the prescription proposed by Tarski (1979). He employs increasing sequences of projections tending to I which is just the case in view of Theorems 4.2 and 4.3. His reference families correspond, of course, to subsets $\mathscr{R}' \subset \mathscr{R}(J^t)$. To our knowledge, the $\mathscr{P}$-approximation has not been used in the literature, but it represents one of the natural choices. The uniform approximation appears, e.g., in Cameron (1960), Cameron and Storvick (1968), Johnson and Skoug (1973) or Exner and Kolerov (1982a, b).

• As mentioned above, the original idea of defining the limiting F-integral (Gel'fand and Yaglom, 1956) failed on Theorem 1.1. A meaningful definition in which the complex F-integrals are determined sequentially was given by Cameron (1960), and studied further, e.g., in Cameron and Storvick (1968) or Johnson and Skoug (1973). Notice that the limiting approach, too, offers various modifications: instead of (6.2), one might consider the limit when s reaches a given point of $(0, \infty)$ in any non-tangential direction from the lower halfplane, etc. It should be stressed, however, that the conclusion of Proposition 6.2 is not altered by such modifications.

• The basic facts about Gaussian measures on Hilbert spaces may be found in Kuo (1975, §I.2). The correlation operator is called covariance if the mean is zero; however, some authors do not distinguish these notions at all. If the correlation operator is not of the trace class, then the corresponding measure is additive but generally not σ-additive; in order to ensure σ-additivity, the underlying Hilbert space has to be extended. In particular, this is the case of Wiener measures (Kuo, 1975, §I.4). Our definition of Gaussian-regularized F-integrals is taken from Itô (1967). In his earlier paper (1961), this author assumed $\beta = 0$ and used the sequence of correlation operators $\{nS\}_{n=1}^{\infty}$ for a fixed S of the trace class. The relation between these limits is similar to that between the $\mathscr{P}$- and $\mathscr{R}$-approximations considered above; one would invite to have a limiting prescription 'stronger' than both of them, as the uniform approximation was in Proposition 6.2. Theorem 6.4 was proved in Itô (1967) for $d = 1$, but the argument is likely to work for $d > 1$ too. In this paper also, the mentioned change-of-variable transformations were derived.

• The extension of Fresnelian F-maps to Fourier transforms of a class of distributions (derivatives of bounded measures) was given by Berg and Tarski (1981). The Fresnelian F-integrals relative to a non-singular quadratic form were introduced by Albeverio and Høegh-Krohn (1976, chap. 4; 1979); for a related construction see Uglanov (1978). Proof of the formula (6.7), with the l.h.s. understood as $I_s^{\mathscr{C}o}$, has been announced recently by Elworthy and Truman (1982).

• The references to the F-integral theories based on analytical continuation in mass are given in the notes to Section 1. For a detailed treatment of scale-invariant measurability see Johnson and Skoug (1979). The presented definition of analytic F-integral again admits of modifications, especially concerning the required region of analycity. The Banach algebra $S(X)$ was introduced by Cameron and Storvick (1980); the quoted equivalence result is due to Johnson (1982).

• The operator-valued F-integrals, both analytical and sequential, were introduced by Cameron and Storvick (1968) and studied further by, e.g., Johnson and Skoug in a series of papers (1973–1979). In the sequential case, the original definition employed a weak rather than strong limit $\delta(\sigma) \to 0$ together with the limit in s. The presented definitions (6.11) and (6.12) are formulated for operators on $L^2(\mathbb{R}^d)$. This is certainly the most interesting case from the viewpoint of applications, however, in a similar way one can study also F-integrals whose values are operators from $L(\mathbb{R})$ to $L^\infty(\mathbb{R})$ (Cameron and Storvick, 1973; Johnson and Skoug, 1975) or from $L^p(\mathbb{R}^d)$ to $L^q(\mathbb{R}^d)$, $1 < p \leqslant 2, p^{-1} + q^{-1} = 1$ (Johnson and Skoug, 1976). Notice also that the 'integrands' in all these cases are complex-valued functions. In distinction, the Fresnelian operator-valued F-integral introduced recently by Parthasarathy and Sinha (1982) concerns a class of operator-valued functions on the path space $\mathscr{H}$.

Chapter 6

Application to Schrödinger Pseudo-Hamiltonians

"In part, the point of functional integration is a less cumbersome notation, but there is a larger point: like any other successful language, its existence tends to lead us to a different and very special way of thinking."

B. SIMON

In the final chapter, we are going to use the path-integral methods discussed above to treat the dissipative systems whose time evolution is governed by Schrödinger pseudo-Hamiltonians. First we prove the formula that expresses the propagator of such a system by means of an appropriate F-integral for various classes of absorptive complex potentials, including the time-dependent ones. As an illustration, we present in Section 2 a detailed discussion of a multidimensional damped harmonic oscillator described by a complex quadratic potential; its propagator is obtained explicitly by evaluating the respective product F-integral. Finally, in Section 3 we return to the argument which is employed usually to motivate the path-integral expression of time-evolution operators. We show that it is also applicable to dissipative systems, though they may have no classical counterpart.

6.1. Feynman–Cameron–Itô Formula

Feynman path integrals are physically interesting primarily as a tool for expressing the time evolution. This fact was closely tied with the origin of the concept, and also later most efforts have been concerned with the justification and/or application of such formulae as (5.2.1). The ways in which the path-integral methods are used in quantum theory are numerous. In accordance with the restrictions set in Section 5.1, we are going to discuss here only the systems which are effectively quantum-mechanical, primarily those where the time evolution is governed by a Schrödinger operator H of the type (4.3.1) on $L^2(\mathbb{R}^d)$.

Feynman's heuristic method of expressing the solution to the Schrödinger equation has stimulated many studies aimed to set this result on a more rigorous basis, i.e. to prove the relation

$$\left(\exp\left(-\frac{i}{\hbar}Ht\right)\psi\right)(x) = \int_{\mathscr{H}_x}^{\alpha} \exp\left\{-\frac{i}{\hbar}\int_0^t u(\gamma(\tau))\,d\tau\right\}\psi(\gamma(0))\,D\phi_{\hbar/m}(\gamma), \quad (1.1a)$$

where α specifies some well-defined Feynman integral over $\mathscr{H}_x = AC_x[J^t; \mathbb{R}^d]$ or another space of paths ending at x, and the operator H is a suitable extension of $-(\hbar^2/2m)\Delta + U$. The first principal result was obtained in 1951 by Kac who proved the formula expressing solution to the diffusion equation by means of the Wiener integral, i.e. the imaginary-time version of (1.1a). Only a decade later, rigorous real-time results appeared. The first of these are due to R. Cameron and K. Itô who have been followed by an array of authors; this is why we shall use the term **Feynman–Cameron–Itô formula** (or FCI-formula) to denote rigorous (real-time) relations of the type (1.1a), or more general.

Of course, one may ask whether the FCI-formula can also hold for some operators H which are not self-adjoint; in particular, which are properly dissipative. The problems that might arise in connection with this question are of a physical rather than mathematical nature. In the self-adjoint case, we have a natural heuristic interpretation of the r.h.s. in (1.1a) – namely, that it represents the sum over all paths γ of the probability amplitudes $e^{iS(\gamma)}$, where $S(\gamma)$ is the action referring to the potential u. This interpretation no longer suits a properly dissipative H due to the absence of a classical counterpart. Treating the classical dissipative systems, one usually works with velocity-dependent forces (the Rayleigh dissipation function, etc.) rather than with a complex-valued action. However, there is no real problem here. The lack of a heuristic interpretation for the Feynman integral under consideration reflects merely the fact mentioned in Section 1.2 that the quantum-theoretical description of an open system is seldom obtained through quantization of some classical dissipative system; more details will be given in Section 3.

With this fact in mind, we are going to attempt the FCI-formula for Schrödinger pseudo-Hamiltonians with potentials of some reasonably large class. In view of (5.2.1), one can write it with the Feynman integral taken over $\mathscr{H} = AC_0[J^t; \mathbb{R}^d]$; setting $m = \hbar = 1$ as usual, we get

$$(e^{-iHt}\psi)(x) = \int_{\mathscr{H}}^{\alpha} \exp\left\{-i\int_0^t u(\gamma(\tau)+x)\,d\tau\right\}\psi(\gamma(0)+x)\,D\phi(\gamma), \qquad (1.1b)$$

where $D\phi(\gamma) \equiv D\phi_1(\gamma)$. Let us consider first a family of sufficiently nice potentials.

THEOREM 6.1.1. *Let $H = H_0 + U$ be a dissipative Schrödinger operator on $L^2(\mathbb{R}^d)$ referring to a potential $u \in \mathscr{F}(\mathbb{R}^d)$. For an arbitrary $\psi \in \mathscr{F}(\mathbb{R}^d) \cap L^2(\mathbb{R}^d)$, the function*

$$f_{x,t}^{u,\psi} : f_{x,t}^{u,\psi}(\gamma) = \exp\left\{-i\int_0^t u(\gamma(\tau)+x)\,d\tau\right\}\psi(\gamma(0)+x) \qquad (1.2)$$

belongs to $\mathscr{F}(\mathscr{H})$, and the formula (1.1b) *holds with the Fresnelian F-integral, $\alpha = f$, on its r.h.s.*

REMARK 6.1.2. (a) The set $\mathscr{F}(\mathbb{R}^d)$ consists of Fourier transforms of (finite, complex) Borel measure on $\mathbb{R}^d$. Since $\mathscr{S}(\mathbb{R}^d)$, e.g., is contained in it, the set

$\mathscr{F}(\mathbb{R}^d)$ is dense in $L^2(\mathbb{R}^d)$. Hence the semigroup $\{e^{-iHt} : t \geqslant 0\}$ is determined uniquely by the FCI-formula.

(b) Combining the above assertion with the result of the previous chapter, one can establish the FCI-formula for various other F-integrals. In particular, Theorems 5.4.8 and 5.6.4, and Proposition 5.6.3 show that it holds with $\alpha =$ ui, g, luc, respectively. Moreover, the functions $u, \psi \in \mathscr{F}(\mathbb{R}^d)$ are continuous and bounded so that function (1.2) is $\|\,.\,\|_\infty$-continuous on $C_0[J^t; \mathbb{R}^d]$ and the relation (1.1b) also holds with the analytic F-integral in view of Proposition 5.6.5. On the other hand, a more general result for $\alpha =$ up will be established later.

Proof. Consider first a function $g \in \mathscr{F}(\mathbb{R}^d)$ which refers to a (finite, complex) Borel measure ν_g on $\mathbb{R}^d$. For fixed x, t, one has

$$g(\gamma(\tau) + x) = \int_{\mathbb{R}^d} e^{i\gamma(\tau)\cdot y}\, e^{ix\cdot y}\, d\nu_g(y).$$

We introduce the measure $\nu_{g,x}$ by $\nu_{g,x}(M) = \int_M e^{ix\cdot y}\, d\nu_g(y)$; according to (A.5), it is again Borel and of the same total variation as ν_g. The first exponent can be expressed by means of the reproducing-kernel property (5.3.10):

$$g(\gamma(\tau) + x) = \int_{\mathbb{R}^d} e^{i(\gamma, G(\,.\,,\tau)y)}\, d\nu_{g,x}(y).$$

This integral can be further rewritten using the image-measure theorem. To this end, we introduce

$$\varphi_\tau : \mathbb{R}^d \to \mathscr{H}, \qquad \varphi_\tau(y) = G(\,.\,,\tau)y,$$
$$\mu_{g,\tau,x} : \mu_{g,\tau,x}(M) = \nu_{g,x}(\varphi_\tau^{(-1)}(M)). \tag{1.3a}$$

In view of (A.4), $\mu_{g,\tau,x}$ is a finite complex σ-additive measure on $\mathscr{H}$; it remains to check that it is Borel. It holds that $\|\varphi_\tau(y) - \varphi_\tau(y')\| = |t - \tau|\, |y - y'|$, so the mapping φ_τ is continuous, then $\varphi_\tau^{(-1)}(M)$ is open in $\mathbb{R}^d$ if $M \subset \mathscr{H}$ is open. Hence $\mu_{g,\tau,x}$ makes sense as a Borel measure, because the σ-algebra on which it is defined contains all the open sets in $\mathscr{H}$. The image-measure theorem then gives

$$g(\gamma(\tau) + x) = \int_{\mathscr{H}} e^{i(\gamma,\gamma')}\, d\mu_{g,\tau,x}(\gamma'). \tag{1.3b}$$

In particular, the function $\gamma \mapsto \psi(\gamma(0) + x)$ belongs to $\mathscr{F}(\mathscr{H})$, the corresponding measure being $\mu_{\psi,0,x} \in \mathscr{M}(\mathscr{H})$.

The function $\gamma \mapsto \int_0^t u(\gamma(\tau) + x)\, d\tau$ may be handled in a similar manner. First we remove $e^{ix\cdot y}$ by passing from ν_u to $\nu_{u,x}$. The function $(\tau, y) \mapsto e^{i\gamma(\tau)\cdot y}$ is continuous and bounded so it belongs to $L(J^t \times \mathbb{R}^d, m \otimes \nu_{u,x})$, where m denotes the Lebesgue measure, and the Fubini theorem may therefore be applied

$$\int_0^t u(\gamma(\tau) + x)\, d\tau = \int_{J^t \times \mathbb{R}^d} e^{i\gamma(\tau)\cdot y}\, d(m \otimes \nu_{u,x})(\tau, y).$$

Further, one has to employ the reproducing-kernel property and the image-measure theorem to get

$$\int_0^t u(\gamma(\tau)+x)\,\mathrm{d}\tau = \int_{\mathscr{H}} \mathrm{e}^{i(\gamma,\gamma')}\,\mathrm{d}\tilde{\mu}_{u,x}(\gamma'), \tag{1.4a}$$

$$\tilde{\mu}_{u,x} : \tilde{\mu}_{u,x}(M) = (m \otimes \nu_{u,x})\,(\varphi^{(-1)}(M)), \tag{1.4b}$$

where the mapping $\varphi : J^t \times \mathbb{R}^d \to \mathscr{H}$ is now given by $\varphi(\tau, y) = G(\,.\,,\tau)y$. It is not difficult to check the relation

$$\|\varphi(\tau,y)-\varphi(\tau',y')\| \leqslant (|y|+|y'|)\,|\tau'-\tau| + (t-\max(\tau,\tau'))\,|y-y'|,$$

which shows that φ is continuous. Consequently, (1.4b) is a finite Borel measure on $\mathscr{H}$ so the function under consideration belongs to $\mathscr{F}(\mathscr{H})$. In view of Proposition 5.2.2, the same is true for its composition with the exponential function, and one can conclude that $f^{u,\psi}_{x,t} \in \mathscr{F}(\mathscr{H})$.

The Feynman integral $I^{\mathrm{f}}(f^{u,\psi}_{x,t})$ may be calculated by expanding the exponent into a power series. The corresponding series of F-integrals converges because of the inequality

$$|I_s^{\mathrm{f}}(f^n g)| \leqslant \|f\|_0^n\,\|g\|_0$$

valid for all $f, g \in \mathscr{F}(\mathscr{H})$, which follows from $\|I_s^{\mathrm{f}}\| = 1$ and the fact that $\mathscr{F}(\mathscr{H})$ is a B-algebra. Hence we have

$$I^{\mathrm{f}}(f^{u,\psi}_{x,t}) = \sum_{n=0}^{\infty} (-i)^n\, I_n(u,\psi;x,t), \tag{1.5a}$$

$$I_n(u,\psi;x,t) = \frac{1}{n!}\int^{\mathrm{f}}_{\mathscr{H}} \left(\int_0^t u(\gamma(\tau)+x)\,\mathrm{d}\tau\right)^n \psi(\gamma(0)+x)\,\mathrm{D}\phi(\gamma). \tag{1.5b}$$

In order to calculate the F-integrals (1.5b), we use the measures (1.3a), (1.4b) together with the formula (5.2.4) which extends naturally to convolutions of more than two measures:

$$I_n(u,\psi;x,t) = \frac{1}{n!}\int_{\mathscr{H}} \exp\left\{-\frac{i}{2}\|\gamma\|^2\right\}\,\mathrm{d}(\tilde{\mu}_{u,x} * \cdots * \tilde{\mu}_{u,x} * \mu_{\psi,0,x})(\gamma)$$

$$= \frac{1}{n!}\int_{\mathscr{H}} \mathrm{d}\tilde{\mu}_{u,x}(\gamma_1)\ldots\int_{\mathscr{H}} \mathrm{d}\tilde{\mu}_{u,x}(\gamma_n)\,\times$$

$$\times \int_{\mathscr{H}} \mathrm{d}\mu_{\psi,0,x}(\gamma_{n+1})\exp\left\{-\tfrac{i}{2}\left\|\sum_{j=0}^{n+1}\gamma_j\right\|^2\right\}$$

Now we use the image-measure theorem and the Fubini theorem to get

$$I_n(u,\psi;x,t)=\frac{1}{n!}\int_0^t d\tau_1\ldots\int_0^t d\tau_n\int_{\mathbb{R}^d}d\nu_u(y_1)\ldots$$

$$\ldots\int_{\mathbb{R}^d}d\nu_u(y_n)\int_{\mathbb{R}^d}d\nu_\psi(y_{n+1})\times$$

$$\times\exp\left\{-\frac{i}{2}\left\|\sum_{j=1}^{n+1}G(\,.\,,\tau_j)y_j\right\|^2+i\sum_{j=1}^{n+1}x\cdot y_j\right\},\tag{1.6a}$$

where we have denoted $\tau_{n+1}=0$. If the τ_j's are ordered, $t=\tau_0\geqslant\tau_1\geqslant\cdots\geqslant\tau_n\geqslant\tau_{n+1}=0$, the first term in the exponent is easily seen to be

$$\left\|\sum_{j=1}^{n+1}G(\,.\,,\tau_j)y_j\right\|^2=\sum_{k=1}^{n+1}(\tau_{k-1}-\tau_k)\left|\sum_{j=k}^{n+1}y_j\right|^2.\tag{1.7a}$$

By induction in n, one is able to rewrite this expression in a more compact form, namely

$$\left\|\sum_{j=1}^{n+1}G(\,.\,,\tau_j)y_j\right\|^2=\sum_{j,k=1}^{n+1}[t-\max(\tau_j,\tau_k)]\,y_j\cdot y_k.\tag{1.7b}$$

The integrand in (1.6a) is symmetric with respect to interchanges of τ_j's, so one may use (1.7a) to get the expression

$$I_n(u,\psi;x,t)=\int_0^t d\tau_1\int_0^{\tau_1}d\tau_2\ldots\int_0^{\tau_{n-1}}d\tau_n\int_{\mathbb{R}^d}d\nu_u(y_1)\ldots$$

$$\ldots\int_{\mathbb{R}^d}d\nu_u(y_n)\int_{\mathbb{R}^d}d\nu_\psi(y_{n+1})\times$$

$$\times\exp\left\{-\frac{i}{2}\sum_{k=1}^{n+1}(\tau_{k-1}-\tau_k)\left|\sum_{j=k}^{n+1}y_j\right|^2+i\sum_{j=1}^{n+1}x\cdot y_j\right\}.\tag{1.6b}$$

Our aim is now to compare the obtained expression for $I^f(f^{u,\psi}_{x,t})$ with $(e^{-iHt}\psi)(x)$. Since the existence of e^{-iHt} is ensured by Corollary 4.2.3, we may express it by means of the Dyson expansion in the interaction picture (cf. (4.2.8), (4.2.12)):

$$(e^{-iHt}\psi)(x)=(e^{-iH_0t}\psi)(x)+\sum_{n=1}^{\infty}(-i)^n(\tilde V_n(t)\psi)(x),\tag{1.8a}$$

$$\tilde V_n(t)\psi=\int_0^t d\tau_1\int_0^{\tau_1}d\tau_2\ldots\int_0^{\tau_{n-1}}d\tau_n\,e^{-iH_0(t-\tau_1)}Ue^{-iH_0(\tau_1-\tau_2)}U\ldots$$

$$\ldots Ue^{-iH_0\tau_n}\psi.\tag{1.8b}$$

We need to express explicitly the action of $\exp(-iH_0\tau)$ on ψ. To this end, we use the relation (4.3.3) which gives

$$(F_d\, \mathrm{e}^{-iH_0\tau}\, \psi)\,(x) = \exp\left(-\frac{i}{2}\,|x|^2\tau\right)\,(F_d\psi)\,(x)$$

so

$$(\mathrm{e}^{-iH_0\tau}\,\psi)\,(x) = (2\pi)^{-d/2}\int_{\mathrm{IR}^d} \mathrm{e}^{ix\cdot y}\,\mathrm{e}^{-i|y|^2\tau/2}\,(F_d\psi)\,(y)\,\mathrm{d}y,$$

where the integral might be eventually understood as an improper one. Let f_ψ be the distribution generated by the measure ν_ψ. Since the function $\psi : \psi(x) = \int_{\mathrm{IR}^d} \mathrm{e}^{ix\cdot y} f_\psi(y)\,\mathrm{d}y$ belongs to $L^2(\mathrm{IR}^d)$ due to the assumption, so does its Fourier transform $(2\pi)^{d/2} f_\psi(\,.\,)$ and

$$(\mathrm{e}^{-iH_0\tau}\,\psi)\,(x) = \int_{\mathrm{IR}^d} \mathrm{e}^{ix\cdot y - i|y|^2\tau/2}\,\mathrm{d}\nu_\psi(y). \tag{1.9}$$

Now it is easy to express $U\mathrm{e}^{-iH_0\tau}\,\psi$, because U acts as a multiplication operator. Using the Fubini theorem, we can write the result as

$$(U\mathrm{e}^{-iH_0\tau_n}\,\psi)\,(x) = \int_{\mathrm{IR}^{2d}} \exp\left\{ix\cdot(y_{n+1}+y_n) - \right.$$
$$\left. -\frac{i}{2}\,|y_{n+1}|^2\,\tau_n\right\}\,\mathrm{d}(\nu_u\otimes\nu_\psi)\,(y_n, y_{n+1}).$$

By the image-measure theorem and (A.5), one can write the r.h.s. as a Fourier transform of a finite Borel measure on IR^d; then it is possible to apply $\exp(iH_0(\tau_{n-1} - \tau_n))$ using the formula (1.9). Continuing this procedure, we arrive finally at the relations

$$(\mathrm{e}^{-iH_0t}\,\psi)\,(x) = I_0(u, \psi; x, t),$$

$$(\tilde{V}_n(t)\,\psi)\,(x) = I_n(u, \psi; x, t), \qquad n = 1, 2, \ldots,$$

which show the expansions (1.5) and (1.8) to coincide term by term. ∎

The situation becomes naturally more complicated when we admit potentials which are discontinuous and/or unbounded. The above proof does not work in such a case; either the method of evaluating the F-integral or the Dyson expansion is not applicable. A substantially wider class of potentials can be handled using the product F-integrals. Let us list some cases in which this method may be used.

THEOREM 6.1.3. *Let H be a dissipative Schrödinger operator on $L^2(\mathbb{R}^d)$ corresponding to a potential u, and consider the following hypotheses:*

(a) $H_0 + U$ *is essentially J-self-adjoint;*
(b) *the potential u belongs to $L^p(\mathbb{R}^d) + L^\infty(\mathbb{R}^d)$, where $p \geqslant \max\{2, \frac{1}{2}d\}$, with the exception of the case $p = \frac{1}{2}d = 2$;*
(c) $d = 3n$, *and u is a sum of two-particle potentials belonging to $L^2(\mathbb{R}^3) + L^\infty(\mathbb{R}^3)$;*
(d) $u \in L^2_{\mathrm{loc}}(\mathbb{R}^d)$ *and* Re $u(x)$ *is bounded below by some constant a.e. in $\mathbb{R}^d$.*

If any of the conditions (a)–(d) *is valid, the relation* (1.1b) *holds with the* (*regular*) *product F-integral on its r.h.s. Moreover, the same is true with the uniform product F-integral,* α = up, *if* (b) *or* (c) *holds.*

Proof. In view of the definition of product F-integrals, one has only to justify use of the Trotter formula, i.e. to check that H is essentially maximal dissipative. This follows from Theorem 4.3.2 in case (a), from Theorem 4.3.3 and Problem 34 in (b), from Theorem 4.3.4 in case (c), and finally, from Theorem 4.3.5 when the condition (d) holds. In cases (b) and (c), moreover, H is maximal dissipative, and therefore the 'uniform' Trotter formula (Corollary 5.5.4) applies. ∎

The product F-integrals also suit some cases in which the integrand depends explicitly on the time parameter. Such situations occur, e.g., when the evolution of the system is governed by a time-dependent (pseudo-) Hamiltonian $H(t) = H_0 + U(t)$ on $L^2(\mathbb{R}^d)$ referring to a potential $u : \mathbb{R}^d \times J^b \to \mathbb{C}$ acting as $(U(t)\psi)(x) = u(x, t)\psi(x)$. The definition of the (uniform) product F-integral given in Section 5.5 generalizes naturally as follows

$$\int_{\mathcal{H}}^{\mathrm{up}} \exp\left\{-i\int_0^t u(\gamma(\tau)+x, \tau)\,\mathrm{d}\tau\right\} \psi(\gamma(0)+x)\,\mathrm{D}\phi_s(\gamma) :=$$

$$:= \lim_{\delta(\sigma)\to 0} \prod_{j=0}^{n-1} (2\pi i s\delta_j)^{-d/2} \int_{\mathbb{R}^{nd}} \exp\left\{\frac{i}{2s}\sum_{j=0}^{n-1} |\gamma_{j+1}-\gamma_j|^2\,\delta_j^{-1} - \right.$$

$$\left. - i\sum_{j=0}^{n-1} u(\gamma_j + x, \tau_j)\,\delta_j\right\} \psi(\gamma_0 + x)\,\mathrm{d}\gamma_0\,\mathrm{d}\gamma_1 \ldots \mathrm{d}\gamma_{n-1}, \tag{1.10}$$

where $\gamma_n = 0$. A similar generalization can be formulated easily to the operator-valued F-integral (5.6.11).

Our aim is now to use the F-integral (1.10) to prove the FCI-formula for a class of time-dependent potentials. It can be accomplished, e.g., if the potentials under consideration are modestly singular, bounded at infinity and the time dependence is smooth enough.

THEOREM 6.1.4. *Suppose there is a positive b such that for each $t \in J^b$, $H(t) = H_0 + U(t)$ is a dissipative Schrödinger operator on $L^2(\mathbb{R}^d)$ referring to a potential $U(t) = U_1(t) + U_2(t)$. Let $u_1(\,.\,,t) \in L^p(\mathbb{R}^d)$ and $u_2(\,.\,,t) \in L^\infty(\mathbb{R}^d)$ for all $t \in J^b$, where $p \geqslant 2$ and $p > \frac{1}{2}d$. Moreover, assume the functions $t \mapsto u_j(\,.\,,t)$, $j = 1, 2$, to be continuously differentiable in $L^p(\mathbb{R}^d)$ and $L^\infty(\mathbb{R}^d)$, respectively. Then for any $t \in [0, b)$, the F-integral* (1.10) *exists and defines a contraction-valued propagator on $L^2(\mathbb{R}^d)$,*

$$(V(t,0)\psi)(x) = \int_{\mathscr{H}}^{\mathrm{up}} \exp\left\{-i \int_0^t u(\gamma(\tau) + x, \tau)\, \mathrm{d}\tau\right\} \psi(\gamma(0) + x)\, \mathrm{D}\phi(\gamma) \quad (1.11a)$$

by which the solution to the corresponding Schrödinger equation is expressed,

$$i \frac{\mathrm{d}}{\mathrm{d}t} \psi_t = H(t)\psi_t, \quad (1.11b)$$

where $\psi_t = V(t,0)\psi$, for any initial datum $\psi_0 = \psi$ belonging to $D(H_0)$.

The proof relies essentially on Theorem 5.5.3. Let us specify its hypotheses to the present case, where $X = L^2(\mathbb{R}^d)$ and $A = -(i/2)\Delta = iH_0$ is independent of t. Clearly both A and $B(t) = iU(t)$ generate continuous contractive semigroups on $L^2(\mathbb{R}^d)$, further $H(t) = H_0 + U(t)$ is maximal dissipative for each $t \in J^b$ in view of Theorem 4.3.3 and its domain $D(H(t))$ equals $D(H_0) \equiv D$ independently of t. Hence we require fulfilment of the following assumptions:

(a) *For each $\psi \in D$, the function $U(\,.\,)\psi$ is continuous in J^b.*
(b) *There is a contraction-valued propagator V on $L^2(\mathbb{R}^d)$ which preserves D. For each $\psi \in D$, the corresponding function $t \mapsto H(t)\psi_t$ is continuous on J^b; further, $t \mapsto \psi_t$ is continuously differentiable on $(0, b)$ and fulfils there equation* (1.11b).
(c) *Finally, $t \mapsto \psi_t$ is continuous for each $t \in J^b$ with respect to the natural Banach-space topology of D* (cf. Lemma 5.5.1).

In order to check (a), we employ the inequalities (4.3.9). Since it is useful to treat the cases $d \leqslant 3$ and $d \geqslant 4$ simultaneously, we shall write them always in the form (4.3.9b); if $p = 2$, one has to drop the subscript, $\|u\|_2 \equiv \|u\|$, and set $q = \infty$. Hence we have

$$\|U(t)\psi - U(s)\psi\| \leqslant \|u_1(\,.\,,t) - u_1(\,.\,,s)\|_p \,\|\psi\|_q + \\ + \|u_2(\,.\,,t) - u_2(\,.\,,s)\|_\infty \,\|\psi\|, \quad (1.12)$$

where the restrictions imposed on p imply $q > 2$ for $d \leqslant 4$, and $2 < q < 2d/(d-4)$ for $d \geqslant 5$. However, each $\psi \in D(H_0)$ belongs to $L^q(\mathbb{R}^d)$ for these values of q (Reed and Simon, 1975, theorem IX.28), so (1.12) establishes continuity of $U(\,.\,)\psi$ for all $\psi \in D$.

Next we show that (c) follows from (a) and (b). For convenience, we denote $H_\lambda(t) = H(t) - i\lambda$. If $\mathrm{Re}\,\lambda > 0$, the inverse operators $H_\lambda(t)^{-1}$ are defined on the whole $L^2(\mathbb{R}^d)$, because $\lambda \in \rho(H(t))$ for each $t \in J^b$ due to Theorem 4.2.1. We have $H_\lambda(t) \in \mathscr{B}(D, L^2(\mathbb{R}^d))$, where D is understood to be equipped with the natural B-space structure, so the inverse-mapping theorem yields $H_\lambda(t)^{-1} \in \mathscr{B}(L^2(\mathbb{R}^d), D)$. Further, we use the identity $\psi_t = H_\lambda(t)^{-1} H_\lambda(t)\psi_t$, which gives

$$\begin{aligned}\|\psi_s - \psi_t\|_D &\leqslant \|(H_\lambda(s)^{-1} - H_\lambda(t)^{-1})H_\lambda(t)\psi_t\|_D + \\ &\quad + \|H_\lambda(s)^{-1}(H_\lambda(s)\psi_s - H_\lambda(t)\psi_t)\|_D \\ &\leqslant \|H_\lambda(s)^{-1}(H_\lambda(s) - H_\lambda(t))\psi_t\|_D + \\ &\quad + N_\lambda(s)\,\|H_\lambda(s)\psi_s - H_\lambda(t)\psi_t\| \\ &\leqslant N_\lambda(s)\,\{\|(H_\lambda(s) - H_\lambda(t))\psi_t\| + \|H_\lambda(s)\psi_s - H_\lambda(t)\psi_t\|\},\end{aligned}$$

where $\|\,.\,\|_D$ is some norm generating the B-space structure of D and $N_\lambda(s)$ is the norm of the operator $H_\lambda(s)^{-1} \in \mathscr{B}(L^2(\mathbb{R}^d), D)$. Under the assumption (b), the functions $s \mapsto \psi_s$ and $s \mapsto H(s)\psi_s$ are continuous so the same is true for $s \mapsto H_\lambda(s)\psi_s$. Further, $H_\lambda(\,.\,)\psi$ with $\psi \in D$ is continuous according to (a), and therefore one needs only to check the norm $N_\lambda(s)$ to be locally bounded with respect to s for a fixed λ.

Using the argument from the proof of Theorem 4.3.3, one obtains to each $a > 0$ and $s \in J^b$ the inequality

$$\|U(s)\psi\| \leqslant a\,\|H_0\psi\| + b\,\|\psi\|$$

valid for all $\psi \in D$. The number b depends, of course, on a. On the other hand, the relation (4.3.10) shows that b depends on s only through $\|u_1(\,.\,,s)\|_p$; since the latter is continuous with respect to s in J^b, one can choose b independent of s. In the same way as in the proof of Theorem 4.2.8, we get the inequality

$$\|U(s)\,(H_0 - i\lambda)^{-1}\psi\| \leqslant \left(a + \frac{b}{\lambda}\right)\|\psi\|$$

valid for all $\psi \in D$ and $\lambda > 0$. Hence if a, λ^{-1} are chosen small enough, we have, say, $\|U(s)\,(H_0 - i\lambda)^{-1}\| < \frac{1}{2}$ independently of s. By the same argument as above, $(H_0 - i\lambda)^{-1} \in \mathscr{B}(L^2(\mathbb{R}^d), D)$, then the identity

$$H_\lambda(s)^{-1} = (H_0 - i\lambda)^{-1}\,[I - U(s)\,(H_0 - i\lambda)^{-1}]^{-1}$$

shows that $N_\lambda(s)$ for the particular λ has a bound independent of s.

Thus it remains to check the assumption (b). To this end, we are going to employ the theorem quoted in (A.12) applying it to the operators $A(t) = iH_1(t)$, which generate continuous contractive semigroups and fulfil $0 \in \rho(iH_1(t))$ for each $t \in J^b$, rather than to $H(t)$ themselves. Once we establish the existence of a solution $\psi_t = \tilde{V}(0, t)\psi$ to the modified equation, the propagator referring to (6.11b) is given by $V(t, s) = e^{t-s}\,\tilde{V}(t, s)$; its contractivity verifies directly.

The operators $H_1(t)$ have $D = D(H_0)$ as their common domain. The function

$$C : C(t, s) = H_1(t)H_1(s)^{-1} - I = (U(t) - U(s))H_1(s)^{-1} \tag{1.13}$$

assumes values from $\mathscr{B}(L^2(\mathbb{R}^d))$, because all the operators $H_\lambda(t)$ have been shown to be bicontinuous bijections from D to $L^2(\mathbb{R}^d)$. In order to check that C has the required properties, let us formulate the assumption about differentiability of $U(.)$ more explicitly: there are functions $g_j : \mathbb{R}^d \times J^b \to \mathbb{C}, j = 1, 2$, such that $g_1(.\,,t) \in L^p(\mathbb{R}^d)$ and $g_2(.\,,t) \in L^\infty(\mathbb{R}^d)$ for each $t \in J^b$; further, the functions $t \mapsto g_j(.\,,t), j = 1, 2$, are continuous in $L^p(\mathbb{R}^d)$ and $L^\infty(\mathbb{R}^d)$, respectively; and finally, the relations

$$\lim_{s \to t} \left\| \frac{u_j(.\,,s) - u_j(.\,,t)}{s - t} - g_j(.\,,t) \right\|_{p_j} = 0, \qquad j = 1, 2, \tag{1.14}$$

hold for all $t \in (0, b)$, where $p_1 = p$ and $p_2 = \infty$. We shall need also the information about boundedness and continuity of $H_1(.)^{-1}\psi$ in $L^q(\mathbb{R}^d)$. For arbitrary $\psi \in L^2(\mathbb{R}^d)$ and $s \in J^b$, we have $H_1(s)^{-1}\psi \in D(H_0) \subset L^2(\mathbb{R}^d) \cap L^q(\mathbb{R}^d)$ and the relations (4.3.8), (4.3.9) together with (4.2.2a) give

$$\begin{aligned}
\|H_1(s)^{-1}\psi\|_q &\leqslant a\,\|H_0H_1(s)^{-1}\psi\| + b\,\|H_1(s)^{-1}\psi\| \\
&\leqslant a\,\|\psi\| + a\,\|U(s)H_1(s)^{-1}\psi\| + (a+b)\,\|H_1(s)^{-1}\psi\| \\
&\leqslant (2a+b)\,\|\psi\| + a\,\|u_1(.\,,s)\|_p\,\|H_1(s)^{-1}\psi\|_q + \\
&\quad + a\,\|u_2(.\,,s)\|_\infty\,\|H_1(s)^{-1}\psi\|
\end{aligned}$$

so

$$\|H_1(s)^{-1}\psi\|_q \leqslant \frac{a(2 + \|u_2(.\,,s)\|_\infty) + b}{1 - a\,\|u_1(.\,,s)\|_p}\,\|\psi\|. \tag{1.15a}$$

In view of the continuity assumptions and the fact that $a > 0$ may be chosen arbitrarily small, the l.h.s. has a bound independent of s,

$$\|H_1(s)^{-1}\psi\|_q \leqslant C_1\,\|\psi\|. \tag{1.15b}$$

Using, further, the identity

$$H_1(s)^{-1} - H_1(t)^{-1} = H_1(s)^{-1}(H(t) - H(s))H_1(t)^{-1}$$

together with the assumption (a), one can establish the continuity of $H_1(.)^{-1}\psi$ in $L^q(\mathbb{R}^d)$.

Let us check first the boundedness of $(t - s)^{-1}C(t, s)\psi$. The relations (1.13) and (4.3.9) together with (1.15b) give

$$\left\| \frac{C(t, s)}{t - s} \psi \right\| \leqslant \left\| \frac{u_1(\,.\,, t) - u_1(\,.\,, s)}{t - s} \right\|_p \|H_1(s)^{-1}\psi\|_q +$$

$$+ \left\| \frac{u_2(\,.\,, t) - u_2(\,.\,, s)}{t - s} \right\|_\infty \|H_1(s)^{-1}\psi\|$$

$$\leqslant \left\{ C_1 \left\| \frac{u_1(\,.\,, t) - u_1(\,.\,, s)}{t - s} \right\|_p + \left\| \frac{u_2(\,.\,, t) - u_2(\,.\,, s)}{t - s} \right\|_\infty \right\} \|\psi\|.$$

Both the norms in the curly bracket have bounds independent of s and t. In order to see this, one has to realize that they are jointly continuous in s, t with the exception of the diagonal points $s = t$ to which they can be continuously extended in view of (1.14), and that the square $J^b \times J^b$ is compact. Hence there is a positive C_2 such that

$$\left\| \frac{C(t, s)}{t - s} \psi \right\| \leqslant C_2 \|\psi\| \tag{1.16}$$

holds for all $\psi \in L^2(\mathbb{R}^d)$ and $s \neq t$ uniformly in s and t. By similar arguments, $(t - s)^{-1}C(t, s)\psi$ can be shown to be uniformly continuous in both t and s. Next we shall prove the relation

$$C(t) = \operatorname*{s-lim}_{s \to t} \frac{C(t, s)}{t - s} = G(t)H_1(t)^{-1}, \tag{1.17}$$

where $G(t)$ is the operator of multiplication by $g_1(\,.\,, t) + g_2(\,.\,, t)$. Using the same estimates as above, we get

$$\left\| \frac{C(t, s)}{t - s} \psi - G(t)H_1(t)^{-1}\psi \right\| \leqslant \left\| \left(\frac{U(t) - U(s)}{t - s} - G(t) \right) H_1(s)^{-1}\psi \right\| +$$

$$+ \|G(t)\,(H_1(s)^{-1} - H_1(t)^{-1})\psi\|$$

$$\leqslant \left\{ C_1 \left\| \frac{u_1(\,.\,, t) - u_1(\,.\,, s)}{t - s} - g_1(\,.\,, t) \right\|_p + \right.$$

$$\left. + \left\| \frac{u_2(\,.\,, t) - u_2(\,.\,, s)}{t - s} - g_2(\,.\,, t) \right\|_\infty \right\} \times$$

$$\times \|\psi\| + \|g_1(\,.\,, t)\|_p \, \|H_1(s)^{-1}\psi - H_1(t)^{-1}\psi\|_q +$$

$$+ \|g_2(\,.\,, t)\|_\infty \, \|H_1(s)^{-1}\psi - H_1(t)^{-1}\psi\|$$

so (1.17) holds and the limit is uniform in t. Since the continuity of the functions $t \mapsto g_j(\,.\,, t)$, $j = 1, 2$, from the compact J^b to $L^p(\mathbb{R}^d)$ and $L^\infty(\mathbb{R}^d)$, respectively, implies their boundedness and also the uniform continuity (Jarník, 1956, theorem 169), one can easily verify $\mathcal{C}(\,.\,)$ to be strongly continuous and bounded. This completes the proof. ∎

The above theorems certainly do not exhaust the list of cases in which the FCI-formula can be proved; other potentials and/or other Feynman integrals may be considered. Most of the results contained in the literature are formulated for real-valued potentials, but usually they can be adapted to the pseudo-Hamiltonian case only if existence of the corresponding contraction-valued propagator is ensured. One should mention, however, that some of the known existence results need not fit directly to the present context: often the existence of a solution to the Schrödinger equation is established in the sense of partial differential equations only, or eventually as a solution to the formally equivalent equation (4.2.13), etc.

Another extension of the above results concerns applications in the scattering theory. Once we have a path-integral expression for the finite-time evolution, it is natural to ask about the behaviour of the system in the large-time limit or, more concretely, to get path-integral expressions for the wave-operators and the S-matrix. References to these problems are given in the notes.

Let us finally note that the method based on FCI-formula for dissipative Schrödinger operators is not the only way in which the F-integrals are used to treat open quantum systems. On the heuristic level, the following alternatives might be mentioned:

(i) The decay probability and related quantities for unstable nuclei are often calculated from the quasiclassical expression for barrier penetration, which is, in essence, nothing else but an approximation to the F-integral solution of the corresponding Schrödinger equation.

(ii) The method based on the 'influence functional' is complementary in a sense to the approach discussed here. The main difference is confined in structure of the full state Hilbert space $\mathcal{H}$ referring to the minimal isolated system containing the open system under consideration. In this method, it is assumed to be $\mathcal{H} = \mathcal{H}_{\mathrm{open}} \otimes \mathcal{H}_{\mathrm{rest}}$ (cf. Section 1.2); the 'measure' on the paths, which take values from the configuration space related to $\mathcal{H}_{\mathrm{open}}$, is obtained by 'integrating' the exponent of the full action over the remaining 'path coordinates'.

6.2. The Damped Harmonic Oscillator

The harmonic oscillator is certainly one of the basic examples for various parts of quantum theory. In order to illustrate how the FCI-formula expresses the time evolution of dissipative systems, we are going to treat in this section the multidimensional damped harmonic oscillator described by the Schrödinger pseudo-Hamiltonian H with a complex-valued quadratic potential. We shall find the propagator e^{-iHt} by evaluation of the respective F-integral. Then we shall discuss the properties of this solution, for simplicity limiting ourselves to the one-dimensional case.

The pseudo-Hamiltonian $H = H_0 + U$ is assumed to be a generalized Schrödinger operator on $L^2(\mathbb{R}^d)$ referring to the potential

$$u : u(x) = v(x) - iw(x), \qquad v(x) = x \cdot Ax, \quad w(x) = x \cdot Bx, \tag{2.1}$$

where A and B are *positive symmetric* operators on $\mathbb{R}^d$; for brevity, we shall speak about them as $d \times d$ matrices. The positivity of B ensures dissipativity of H, and the positivity of A implies, by Theorem 4.3.5, that H is e.m.d. and $C_0^\infty(\mathbb{R}^d)$ is a core for it.

Below we shall give an alternative proof, which yields (under one additional hypothesis) a stronger result; namely, that the operator H itself is maximal dissipative. It is useful to introduce first some notation: we abbreviate $Q^2 = \Sigma_{j=1}^d Q_j^2$, where $(Q_j\psi)(x) = x_j\psi(x)$; further, P_j are components of the momentum, $P_j = F_d^{-1} Q_j F_d$, and $P^2 = \Sigma_{j=1}^d P_j^2 = 2H_0$ (cf. (4.3.3)). The canonical commutation relation may be written in the form

$$((P_jQ_k - Q_kP_j)\psi)(x) = -i\,\delta_{jk}\,\psi(x) \tag{2.2}$$

if the function ψ is smooth enough, in particular, for each $\psi \in \mathscr{S}(\mathbb{R}^d)$. We also denote $H_1 = H_0 + V$ and use the symbol H_1^c for the restriction of H_1 to a suitable core, say, $\mathscr{S}(\mathbb{R}^d)$. The following two auxiliary assertions are required:

LEMMA 6.2.1. *H_1 is self-adjoint.*

Proof. It is easy to see that H_1^c is e.s.a. In particular, for a strictly positive A it follows from the existence of a complete set of eigenvectors $\subset \mathscr{S}(\mathbb{R}^d)$. Both the operators P^2 and V are self-adjoint and therefore closed so $H_1 \subset \overline{H_1^c}$. In order to prove the opposite inclusion, we employ the relations (2.2) to show that there is a positive b such that

$$\tfrac{1}{4}\,\|P^2\psi\|^2 + \|V\psi\|^2 \leqslant \|H_1\psi\|^2 + b\,\|\psi\|^2 \tag{2.3}$$

holds for each $\psi \in \mathscr{S}(\mathbb{R}^d)$. We choose a basis in $\mathbb{R}^d$ so that A is diagonal and denote by α_j, $j = 1, \ldots, d$, its eigenvalues, then

$$(\psi, (P^2V + VP^2)\psi) \geqslant \sum_{j=1}^{d} \alpha_j(\psi, (P_j^2Q_j^2 + Q_j^2P_j^2)\psi),$$

because $(\psi, P_j^2Q_k^2\psi) \geqslant 0$ for $j \neq k$. Applying the relations (2.2) to the r.h.s., we get

$$(\psi, (P^2V + VP^2)\psi) \geqslant \tfrac{1}{2} \sum_{j=1}^{d} \alpha_j\,\|(P_jQ_j + Q_jP_j)\psi\|^2 - \tfrac{3}{2}\,\|\psi\|^2\,\mathrm{Tr}\,A$$

so the inequality (2.3) holds if $b \geqslant \frac{3}{4}\,\mathrm{Tr}\,A$. Consider now a sequence $\{\psi_n\} \subset \mathscr{S}(\mathbb{R}^d)$ which converges to a fixed $\psi \in D(\overline{H_1^c})$. Then $\{H_1\psi_n\}$ also converges, i.e. $\|H_1\psi_n - H_1\psi_m\| \to 0$ with $n, m \to \infty$. The inequality (2.3) then shows that also $\{P^2\psi_n\}$ and $\{V\psi_n\}$ converge; however, both P^2 and V are closed so $\psi \in D(P^2) \cap D(V) = D(H_1)$. ■

LEMMA 6.2.2. *Suppose that A is strictly positive and one of the following conditions is valid:*

(a) $a = 0$ *and* $b^2\|B\|^2 \leqslant \frac{1}{2}\alpha^2$, *where* $\alpha = \min_j \alpha_j$ *is the lower bound of* A;
(b) $a > 0$ *and* $b^2 \leqslant \frac{1}{2}a^2$.

Then there is a positive c such that the inequality

$$\|bW\psi\|^2 \leqslant \tfrac{1}{2}\,\|(H_1 - iaW)\psi\|^2 + c\,\|\psi\|^2 \tag{2.4}$$

holds for all $\psi \in \mathscr{S}(\mathbb{R}^d)$.

Proof. We have to find a constant c such that

$$I \equiv (\psi, [\tfrac{1}{2}(H_1 + iaW)(H_1 - iaW) - b^2W^2 + c]\psi)$$

is non-negative independently of $\psi \in \mathscr{S}(\mathbb{R}^d)$. We choose again a basis in $\mathbb{R}^d$ so that A is diagonal and denote by B_{jk} the corresponding matrix elements of B. The positive term $\frac{1}{4}(\psi, P^4\psi)$ may be omitted; further, we rewrite the terms $\frac{1}{4}i(WP^2 - P^2W)$ and $\frac{1}{4}(VP^2 + P^2V)$ using the canonical commutation relations (2.2) obtaining in this way

$$I \geqslant \left(\psi, \left[\frac{1}{8}\sum_{j,k=1}^{d} \alpha_j(P_kQ_j + Q_jP_k)^2 - \frac{3}{8}\,\mathrm{Tr}\,A + \frac{1}{2}\left(\sum_{j=1}^{d}\alpha_jQ_j\right)^2 + \right.\right.$$

$$\left.\left. + (\tfrac{1}{2}a^2 - b^2)\left(\sum_{j,k=1}^{d} B_{jk}Q_jQ_k\right)^2 - \frac{a}{4}\sum_{j,k=1}^{d} B_{jk}(P_kQ_j + Q_jP_k) + c\right]\psi\right).$$

Suppose first the condition (a) is valid, then the last inequality yields

$$I \geqslant (\psi, [c - \tfrac{3}{8}\,\mathrm{Tr}\,A + (\tfrac{1}{2}\alpha^2 - b^2\,\|B\|^2)Q^4]\psi) \geqslant (c - \tfrac{3}{8}\,\mathrm{Tr}\,A)\,\|\psi\|^2$$

so (2.4) holds if $c \geqslant \frac{3}{8}\,\mathrm{Tr}\,A$. On the other hand, if $a^2 \geqslant 2b^2$, we have

$$I \geqslant \left(\psi, \left[\sum_{j,k=1}^{d} (8\alpha_j)^{-1}[\alpha_j(P_kQ_j + Q_jP_k) - aB_{jk}]^2 - \right.\right.$$

$$\left.\left. - \frac{1}{8}a^2\sum_{j,k=1}^{d} \alpha_j^{-2}B_{jk}^2 + c - \frac{3}{8}\,\mathrm{Tr}\,A\right]\psi\right)$$

so (2.4) holds if $c \geqslant \frac{3}{8}\,\mathrm{Tr}\,A + \frac{1}{8}a^2\sum_{j,k=1}^{d}\alpha_j^{-2}B_{jk}^2$. ■

Now we can prove the announced result.

THEOREM 6.2.3. *The generalized Schrödinger operator* $H = H_0 + U$ *referring to the potential* (2.1) *is maximal dissipative, if at least one of the matrices* A, B *is strictly positive.*

Proof. Assume first that A is strictly positive, $\alpha > 0$, then the preceding lemma makes it possible to use Theorem 4.2.8. The operator W is positive, and therefore accretive; further, the strict positivity of A implies $D(V) = D(Q^2) \subset D(W)$ so $D(W) \supset D(H_1)$. Finally, H_1 is maximal dissipative according to Lemma 2.1 and $\mathscr{S}(\mathbb{R}^d)$ is a core for it. Hence if $\alpha^2 \geqslant \frac{1}{2} \|B\|^2$, the operator $H = H_1 - iW$ is maximal dissipative and $\mathscr{S}(\mathbb{R}^d)$ is its core.

If $\alpha^2 < 2\|B\|^2$, one has to apply Theorem 4.2.8 iteratively. We choose a positive γ and a natural number n such that $2\gamma^2 \|B\|^2 \leqslant \alpha^2$ and

$$\gamma(1 + 2^{-1/2})^{n-1} = 1. \tag{$*$}$$

The same argument as above shows that the operator $H_1 - i\gamma W$ is maximal dissipative, and that it is e.m.d. on $\mathscr{S}(\mathbb{R}^d)$. Assume now the assertion to be valid for the operators $H_{1j} = H_1 - i\gamma(1 + 2^{-1/2})^{j-1} W$, $j = 1, \ldots, k$. Since the assumption (b) of Lemma 2.2 is fulfilled for $a = 2^{1/2} b = \gamma(1 + 2^{-1/2})^{k-1}$, Theorem 4.2.8 may be applied again, and the assertion also holds for

$$H_{1k} - i\gamma 2^{-1/2}(1 + 2^{-1/2})^{k-1} = H_{1,k+1}.$$

In this way, we arrive to the operator H_{1n} which equals H in view of $(*)$.

Consider, further, the situation when A need not be strictly positive but B is strictly positive with the lower bound $\beta > 0$. In that case, we write $H = H_\epsilon - \epsilon Q^2$ for some $\epsilon > 0$, where the operator $H_\epsilon = H_0 + V + \epsilon Q^2 - iW$ with $D(H_\epsilon) = D(P^2) \cap \cap D(Q^2)$ is maximal dissipative according to the first part of the proof. Hence one has to check that ϵQ^2 is H_ϵ-bounded with the relative bound, say, equal to $\frac{1}{2}$. Using the essentially same argument as in Lemma 2.2, we see that this is true if

$$(\psi, [c - \tfrac{3}{8}\operatorname{Tr} A + \tfrac{1}{2}(\beta^2 - \epsilon^2)Q^4]\psi) \geqslant 0$$

holds for some $c > 0$ and all $\psi \in \mathscr{S}(\mathbb{R}^d)$. Choosing ϵ small enough, we get the desired result. ■

As an immediate consequence, we have:

COROLLARY 6.2.4. *The semigroup propagator corresponding to the Schrödinger pseudo-Hamiltonian* H *with the potential* (2.1) *is expressed by FCI-formula* (1.1) *with* $\alpha = p$. *If at least one of the matrices* A, B *is strictly positive, then the formula also holds with* $\alpha =$ up.

Our aim is now to find an explicit expression for the propagator by evaluating the respective product F-integral. Since we are interested primarily in the effect of damping, we shall assume that B is *strictly positive*. This assumption is presumably not necessary for the calculations performed below (the method is known, e.g., to work in the case $B = 0$; see the notes), but it helps to get rid of the complications connected with the presence of improper integrals. We have also to adopt a convention concerning the square roots which will appear frequently in the following considerations. We shall use the branch that is positive on positive reals and has a cut along the negative real axis.

With these preliminaries, we can formulate the sought result in the following way:

THEOREM 6.2.5. *Let A be positive, B strictly positive and denote $\Omega = (2C)^{1/2}$, where $C = A - iB$. Then for each $t > 0$, the propagator $V_t = e^{-iHt}$ corresponding to the Schrödinger pseudo-Hamiltonian H with the potential* (2.1) *acts on an arbitrary $\varphi \in L^2(\mathbb{R}^d)$ as*

$$(V_t\varphi)(x) = \int_{\mathbb{R}^d} G_t(x, y)\varphi(y)\, dy, \tag{2.5a}$$

$$G_t(x, y) = (2\pi i)^{-d/2}(\det(\Omega^{-1} \sin \Omega t))^{-1} \exp\left\{\frac{i}{2}\,[x \cdot (\Omega \operatorname{ctg} \Omega t)x + \right.$$
$$\left. + y \cdot (\Omega \operatorname{ctg} \Omega t)y\,] - iy \cdot (\Omega \operatorname{cosec} \Omega t)x\right\}. \tag{2.5b}$$

The proof needs some lemmas. First we derive a useful formula for the integral

$$I_N(M, \eta) = \int_{\mathbb{R}^N} \exp\left\{\frac{i}{2}\,\xi \cdot M\xi + i\xi \cdot \eta\right\} d\xi, \tag{2.6}$$

where M is a symmetric $N \times N$ matrix whose imaginary part is supposed to be strictly positive, $\operatorname{Im} \xi \cdot M\xi > 0$ for each non-zero $\xi \in \mathbb{R}^N$, and η belongs to $\mathbb{C}^N$.

LEMMA 6.2.6. *Under the stated assumptions, the integral* (2.6) *equals*

$$I_N(M, \eta) = (2\pi i)^{N/2}(\det M)^{-1/2} \exp\left\{-\frac{i}{2}\,\eta \cdot M^{-1}\eta\right\}. \tag{2.7}$$

The proof is based on analytic continuation: we denote $M_\lambda = \lambda M_1 + iM_2$ and $\eta_\lambda = \lambda\eta_1 + i\eta_2$, where $M_1 = \operatorname{Re} M = \frac{1}{2}(M + M^{\mathrm{T}})$, etc. Due to the assumption, $\lambda M_1 + iM_2$ is strictly positive for all real λ with small enough modulus. Then there is a positive δ such that $I_N(M_\lambda, \eta_\lambda)$ makes sense in the strip $S_\delta = \{\lambda : |\operatorname{Im} \lambda| < \delta\}$, and the function $\lambda \mapsto I_N(M_\lambda, \eta_\lambda)$ is easily seen to be analytic there. If $\lambda = i\epsilon$ with $|\epsilon| < \delta$, one can choose a basis in $\mathbb{R}^N$ in which M_λ is diagonal and use the Fubini theorem to get

$$I_N(M_{i\epsilon}, \eta_{i\epsilon}) = \prod_{j=1}^{N} \int_{\mathbb{R}} \exp\left\{\frac{i}{2}\, m_j\xi_j^2 + i\xi_j z_j\right\} d\xi_j$$
$$= \prod_{j=1}^{N} (2\pi i m_j^{-1})^{1/2} \exp\left\{-\frac{i}{2}\, m_j^{-1} z_j^2\right\},$$

where m_j and z_j are the eigenvalues of $M_{i\epsilon}$ and components of $\eta_{i\epsilon}$, respectively. Hence the analycity implies that the required relation holds if $\lambda \in S_\delta$, in particular, for $M_1 = M$ and $\eta_1 = \eta$. ■

Next we have to check that the r.h.s. of (2.5b) makes sense. Actually, we prove a little more.

LEMMA 6.2.7. *Let A be positive, B strictly positive, $t > 0$, then the matrix Ω is regular*, $\det(\Omega^{-1} \sin \Omega t)$ *and* $\det(\cos \Omega t)$ *are non-zero, and the real quadratic forms $x \mapsto x \cdot Mx$ with $M = -\Omega^{-1}$* tg $\Omega t, -\Omega$ tg Ωt, ctg Ωt *are strictly positive.*

Proof. Suppose first $d = 1$. We have $3\pi/2 \leqslant \arg C < 2\pi$ due to the assumption, so $0 < \nu \leqslant \omega$ holds for $\Omega = \omega - i\nu$. Then

$$-\mathrm{Im}\, \Omega^{-1} \,\mathrm{tg}\, \Omega t = C(t)\, [\omega \,\mathrm{th}\, \nu t \cos^{-2} \omega t - \nu \,\mathrm{tg}\, \omega t \,\mathrm{ch}^{-2}\, \nu t]\,,$$

where $C(t)^{-1} = |\Omega|^2\, |1 + i \,\mathrm{tg}\, \omega t \,\mathrm{th}\, \nu t|^2 > 0$. The inequalities $\alpha^{-1} \sin \alpha < 1 < \beta^{-1} \,\mathrm{sh}\, \beta$, valid for all non-zero real α, β, then imply

$$-\mathrm{Im}\, \Omega^{-1} \,\mathrm{tg}\, \Omega t = C(t)\, [2t \cos^2 \omega t \,\mathrm{ch}^2\, \nu t]^{-1}\, [\omega t \,\mathrm{sh}(2\nu t) -$$

$$- \nu t \sin(2\omega t)] > 0. \tag{$*$}$$

Positivity of $\mathrm{Im}\, \Omega$ ctg Ωt and $-\mathrm{Im}\, \Omega$ tg Ωt can be checked in the same way.

Let, further, $d > 1$. The regularity of Ω is obvious: $|\Omega x|^2 = x \cdot \Omega^{\mathrm{T}} \Omega x = 2x \cdot Cx \neq 0$ for each non-zero $x \in \mathbb{R}^d$, because Ω is symmetric (as a function of symmetric C) and B is strictly positive. A real quadratic form is strictly positive iff all eigenvalues of its matrix are positive (Gantmakher, 1966, §§IX.13, X.4). They are equal to $-\mathrm{Im}\, \omega_j^{-1}$ tg $\omega_j t$ in the first case (*ibid.*, §V.1), where ω_j are the eigenvalues of Ω. Further, each eigenvalue $\gamma_j = \frac{1}{2} \omega_j^2$ of C fulfils $\mathrm{Im}\, \gamma_j < 0, j = 1, \ldots, d$; otherwise k and a non-zero vector x_k would exist such that $x_k \cdot Bx_k = -\mathrm{Im}\, x_k \cdot Cx_k = -|x_k|^2 \,\mathrm{Im}\, \gamma_k \leqslant 0$, but this contradicts our assumptions. Hence ($*$) yields $-\mathrm{Im}\, \omega_j^{-1}$ tg $\omega_j t \leqslant 0$ for $j = 1, \ldots, d$; the same argument applies to the other two forms.

As for the determinants, it is sufficient to check that all eigenvalues of both the matrices are non-zero: they equal $\omega_j^{-1} \sin \omega_j t$ and $\cos \omega_j t$, $j = 1, \ldots, d$, respectively. Further, $\mathrm{Im}\, \gamma_j < 0$ implies $\mathrm{Im}\, \omega_j \neq 0$, but sin and cos have no zeros outside the real axis. ■

Now we can establish some of the properties required for the operators V_t.

PROPOSITION 6.2.8. *Let A, B be as in Theorem* 2.5; *let, further, V_t be given by* (2.5) *and $V_0 = I$. Then $\{V_t : t \geqslant 0\}$ is a semigroup of bounded operators on* $L^2(\mathbb{R}^d)$.

Proof. The preceding lemma implies the existence of positive numbers a, b (depending on t) such that

$$|G_t(x, y)| \leqslant a \exp\{-b(x^2 + y^2)\}. \tag{2.8}$$

This inequality shows that V_t belongs to $\mathscr{J}_2$ for each $t > 0$, with the Hilbert–Schmidt norm

$$\| V_t \|_2^2 \leqslant \int_{\mathbb{R}^{2d}} a^2 \exp\{-2b(x^2 + y^2)\} \, dx \, dy = a^2 \left(\frac{\pi}{2b}\right)^d,$$

and consequently, $\| V_t \| \leqslant \| V_t \|_2 \leqslant a(\pi/2b)^{d/2}$. As for the semigroup property, in view of $V_0 = I$ and the inequality (2.8), it is sufficient to verify

$$G_{t+s}(x, y) = \int_{\mathbb{R}^d} G_s(x, y) G_t(y, z) \, dy \tag{2.9}$$

for all $t, s > 0$. The r.h.s. of this relation can be written with the help of (2.6) as

$$(2\pi i)^{-d} \, [\det(\Omega^{-1} \sin \Omega t) \det(\Omega^{-1} \sin \Omega s)]^{-1/2} \times$$

$$\times \exp\left\{\frac{i}{2} \, [x \cdot (\Omega \operatorname{ctg} \Omega s) x + z \cdot (\Omega \operatorname{ctg} \Omega t) z]\right\} \times$$

$$\times \, I_d(\Omega(\operatorname{ctg} \Omega t + \operatorname{ctg} \Omega s), -\Omega[(\operatorname{cosec} \Omega s) x + (\operatorname{cosec} \Omega t) z]).$$

Now one has to apply Lemma 2.6 to this integral, and to use the identity $\det M_1 M_2 = \det M_1 \det M_2$, symmetry of the matrices involved and the matrix-functional-calculus rules (Gantmakher, 1966, §V.5) to get the relation (2.9). ■

We shall also need the following equivalent expression for V_t:

PROPOSITION 6.2.9. *Let A, B be as in Theorem* 2.5, *and let V_t be defined by* (2.5), *then the relations*

$$(V_t \varphi)(x) = \int_{\mathbb{R}^d} F_t(x, y) \, (F_d \varphi)(y) \, dy, \tag{2.10a}$$

$$F_t(x, y) = (2\pi)^{-d/2} [\det(\cos \Omega t)]^{-1/2} \exp\left\{-\frac{i}{2} \, [x \cdot (\Omega \operatorname{tg} \Omega t) x + \right.$$

$$\left. + \, y \cdot (\Omega^{-1} \operatorname{tg} \Omega t) y] + i y \cdot (\sec \Omega t) x \right\} \tag{2.10b}$$

hold for all $t > 0$ and $\varphi \in L^2(\mathbb{R}^d)$, where F_d is Fourier–Plancherel operator.

Proof. The r.h.s. of (2.10a) makes sense in view of Lemma 2.7. Let first $\varphi \in \mathscr{F}(\mathbb{R}^d) \cap L^2(\mathbb{R}^d)$, $\varphi(x) = \int_{\mathbb{R}^d} e^{ix \cdot y} \, d\nu(y)$ with some $\nu \in \mathscr{M}(\mathbb{R}^d)$, then (2.10a) may be rewritten as

$$(V_t \varphi)(x) = (2\pi)^{d/2} \int_{\mathbb{R}^d} F_t(x, y) \, d\nu(y). \tag{2.11}$$

In order to prove this assertion, we use (2.8) together with the boundedness of φ, $|\varphi(x)| \leqslant |\nu|\,(\mathbb{R}^d)$. Then the Fubini theorem applied to (2.5) gives (2.11), with

$$F_t(x, y) = (2\pi)^{-d/2} \int_{\mathbb{R}^d} G_t(x, z)\, e^{iy \cdot z}\, dz$$

$$= (4\pi^2 i)^{-d/2} \left[\det(\Omega^{-1} \sin \Omega t)\right]^{-1/2} \exp\left\{\frac{i}{2} x \cdot (\Omega \operatorname{ctg} \Omega t) x\right\} \times$$

$$\times I_d(\Omega \operatorname{ctg} \Omega t, y - (\Omega \operatorname{cosec} \Omega t) x).$$

Using now Lemma 2.6, symmetry of the matrices involved and the matrix-functional-calculus rules, we get (2.10b). Consider further an arbitrary $\varphi \in L^2(\mathbb{R}^d)$. We construct the following sequence

$$\varphi_n : \varphi_n(x) = (2\pi)^{-d/2} \int_{\mathbb{R}^d} e^{ix \cdot y}\, \hat{\varphi}_n(y)\, dy,$$

$$\hat{\varphi}_n(y) = \chi_{[-n,n]}(y) \cdot \max\{n, (F_d\varphi)(y)\}.$$

It holds obviously that $F_d\varphi_n = \hat{\varphi}_n \in L(\mathbb{R}^d)$, so the assertion is valid for each φ_n. The sequence $\{\hat{\varphi}_n\}$ converges pointwise to $F_d\varphi \in L^2(\mathbb{R}^d)$; further, $|\hat{\varphi}_n(y)| \leqslant |(F_d\varphi)(y)|$ holds everywhere in $L^2(\mathbb{R}^d)$ and $F_t(x, .) \in L^2(\mathbb{R}^d)$. Then the dominated-convergence theorem yields

$$\lim_{n \to \infty} (V_t\varphi_n)(x) = \int_{\mathbb{R}^d} F_t(x, y)\, (F_d\varphi)(y)\, dy. \qquad (*)$$

One verifies easily that $\hat{\varphi}_n$ converges to $F_d\varphi$ also in the L^2-norm. Since Fd is unitary and V_t is bounded due to Proposition 2.8, we obtain $V_t\varphi_n \to V_t\varphi$. In that case, however, one can pick a subsequence $\{V_t\varphi_{n_k}\}$ which converges to $V_t\varphi$ pointwise a.e. in $\mathbb{R}^d$, and the assertion follows from (*). ■

The last auxiliary result concerns the block-structured matrices:

LEMMA 6.2.10. *Let* $m = (m_{ij})$ *be a* $n \times n$ *matrix and assume that* $M = (M_{ij})$ *is a* $nd \times nd$ *matrix which consists of* $d \times d$ *blocks* M_{ij}, $i, j = 1, \ldots, n$. *Let us denote* $d(m) = \mathscr{D}(m_{11}, m_{12}, \ldots, m_{nn}) := \det m$. *If all the blocks of M commute,* $[M_{ij}, M_{kl}] = 0$ *for* $i, j, k, l = 1, \ldots, n$, *then*

$$\det M = \det(d(M)), \qquad (2.12a)$$

where $d(M)$ *is the 'block determinant' of M, i.e. the* $d \times d$ *matrix* $\mathscr{D}(M_{11}, M_{12}, \ldots, M_{nn})$. *Moreover, if M is regular, then*

$$(M^{-1})_{ij} = (-1)^{i+j} (d(M))^{-1} M \begin{bmatrix} 1, \ldots, j-1, j+1, \ldots, n \\ 1, \ldots, i-1, i+1, \ldots, n \end{bmatrix}, \qquad (2.12b)$$

where $M[\ldots]$ *is the respective 'block minor' of* M, *i.e.* $d(M^{(ij)})$ *with* $M^{(ij)}$ *obtained from* M *by dropping the* jth *block row and* ith *block column.*

Proof. In the simplest non-trivial case, $n = 2$, the assertion can be verified directly; in particular, (2.12b) follows from the Frobenius formula (Gantmakher, 1966, §II.5). The argument can be generalized to an arbitrary n by means of the block variant of the Gausss algorithm. We have

$$\det M = \det \tilde{M}, \quad \tilde{M} = M_{11} M_{22}^{(1)} M_{33}^{(2)} \ldots M_{nn}^{(n-1)},$$

where $M_{ij}^{(k)} = M_{ij}^{(k-1)} - M_{ik}^{(k-1)} (M_{kk}^{(k-1)})^{-1} M_{kj}^{(k-1)}$ for all $i, j = k + 1, \ldots, n$ and $k = 1, 2, \ldots, n - 1$. Since all the blocks commute mutually, we see that $\tilde{M}$ is the same polynomial function of the variables M_{ij} as $d(m)$ of m_{ij}. Hence we obtain $\tilde{M} = d(M)$, i.e. the relation (2.12a). Notice that it is true even if some $M_{kk}^{(k-1)}$ are singular: in such a case, one has to replace M by $M + \epsilon I$ and use the continuity of the determinant with respect to its entries. Furthermore, (2.12b) is equivalent to the relation

$$\sum_{j=1}^{n} (-1)^{i+j} M \begin{bmatrix} 1, \ldots, j-1, j+1, \ldots, n \\ 1, \ldots, i-1, i+1, \ldots, n \end{bmatrix} M_{jk} = \delta_{ik} d(M),$$

which follows similarly from the analogous equality for the matrix m. ■

Proof of Theorem 2.5. After the above preliminaries, we can prove the relations (2.5) by evaluating the appropriate product F-integral. Specification of the relations (5.5.1) to the present case gives

$$\mathrm{e}^{-iHt} = \operatorname*{s-lim}_{n \to \infty} S_n^t, \tag{2.13a}$$

where the operators S_n^t act on an arbitrary $\varphi \in L^2(\mathbb{R}^d)$ as

$$(S_n^t \varphi)(x) = (2\pi i \delta)^{-nd/2} \int_{\mathbb{R}^{nd}} \exp\left\{ \frac{i}{2\delta} \sum_{k=0}^{n-1} |\gamma_{k+1} - \gamma_k|^2 - \right.$$

$$\left. - i\delta \sum_{k=0}^{n-1} \gamma_k \cdot C\gamma_k \right\} \varphi(\gamma_0)\, \mathrm{d}\gamma_0 \ldots \mathrm{d}\gamma_{n-1} \tag{2.13b}$$

with $\gamma_n = x$ and $\delta = t/n$. Modulus of the integrand is majorized by

$$|\varphi(\gamma_0)| \exp\left\{ -\delta \sum_{k=0}^{n-1} \gamma_k \cdot B\gamma_k \right\},$$

thus the r.h.s. of (2.13b) makes sense and the integrations may be interchanged arbitrarily.

Suppose first $\varphi \in \mathscr{F}(\mathbb{R}^d)$, $\varphi(x) = \int_{\mathbb{R}^d} e^{ix \cdot y}\, d\nu(y)$. In that case, we can substitute $\gamma_k = \xi_k \delta^{1/2}$ and rearrange the integral to the form

$$(S_n^t \varphi)(x) = (2\pi i)^{-nd/2} \int_{\mathbb{R}^d} d\nu(y) \exp\left\{\frac{i}{2}\,\delta x^2\right\} I_{nd}(M_n, \eta), \tag{2.14a}$$

where $\eta = (y\delta^{1/2}, 0, \dots, 0, -x\delta^{-1/2})$ and $M_n = M_n(\delta)$ is the following $nd \times nd$ matrix

$$M_n = \begin{pmatrix} I - 2\delta^2 C & -I & 0 & 0 & \dots & 0 \\ -I & 2I - 2\delta^2 C & -I & 0 & \dots & 0 \\ 0 & -I & 2I - 2\delta^2 C & -I & \dots & 0 \\ \cdot & & & & & \\ \cdot & & & & & \\ \cdot & & & & & \\ 0 & 0 & 0 & 0 & \dots & 2I - 2\delta^2 C \end{pmatrix} \tag{2.14b}$$

that obviously fulfils the assumptions of Lemma 2.6. In order to make use of the latter, we have to calculate $\det M_n$ and M_n^{-1}, or at least the corner blocks of this matrix. This is accomplished easily with the help of Lemma 2.10; we obtain

$$(S_n^t \varphi)(x) = (\det[d(M_n)])^{-1/2} \int_{\mathbb{R}^d} d\nu(y) \exp\left\{-\frac{i}{2\delta}\, x \cdot d(M_n)^{-1}\,[d(M_{n-1}) - \right.$$

$$\left. - d(M_n)]\,x - \frac{i\delta}{2}\, y \cdot d(M_n)^{-1}\, d(K_{n-1})y + iy \cdot d(M_n)^{-1} x\right\}, \tag{2.15}$$

where $K_{n-1} = K_{n-1}(\delta)$ is the abbreviation for the lower-right $(n-1)d \times (n-1)d$ submatrix of M_n. The 'block determinants' under consideration obey the following relations

$$d(M_n) = (I - \delta^2\Omega^2)d(K_{n-1}) - d(K_{n-2}),$$

$$d(K_{n-1}) = (2I - \delta^2\Omega^2)d(K_{n-2}) - d(K_{n-3}).$$

One can check directly that the second relation is solved by

$$d(K_{n-1}) = \sum_{j=0}^{\infty} (-1)^j \binom{n+j}{2j+1} (\delta\Omega)^{2j}; \tag{2.16a}$$

substituting from here to the first one, we get

$$d(M_n) = \sum_{j=0}^{\infty} (-1)^j \binom{n+j}{2j} (\delta\Omega)^{2j}. \tag{2.16b}$$

Let us turn now to the limits; consider first $\delta d(K_{n-1}(\delta))$ with $\delta = t/n$. This sum obviously converges, because it is finite; however, one must verify that it converges uniformly with respect to n. The relation (2.16a) gives

$$\frac{t}{n}\, d\left(K_{n-1}\left(\frac{t}{n}\right)\right) = \Omega^{-1} \sum_{j=0}^{\infty} \frac{(-1)^j}{(2j+1)!}\, c_{nj}(\Omega t)^{2j+1},$$

where $c_{nj} = \Pi_{k=1}^{j} (1 - k^2/n^2)$; thus it holds $0 \leqslant c_{nj} \leqslant 1$, and the convergence is uniform. The limit then equals

$$\lim_{n\to\infty} \frac{t}{n}\, d\left(K_{n-1}\left(\frac{t}{n}\right)\right) = \Omega^{-1} \sin \Omega t. \tag{2.17a}$$

In the same way, the relation (2.16b) implies

$$\lim_{n\to\infty} d\left(M_n\left(\frac{t}{n}\right)\right) = \cos \Omega t, \tag{2.17b}$$

$$\lim_{n\to\infty} \frac{n}{t}\left[d\left(M_{n-1}\left(\frac{t}{n}\right)\right) - d\left(M_n\left(\frac{t}{n}\right)\right)\right] = \Omega \sin \Omega t. \tag{2.17c}$$

Combining the relations (2.17) with (2.11), (2.15) and the matrix-functional-calculus rules, we get

$$\lim_{n\to\infty} (S_n^t\varphi)(x) = (V_t\varphi)(x)$$

for each $\varphi \in \mathscr{F}(\mathbb{R}^d) \cap L^2(\mathbb{R}^d)$, where the r.h.s. is defined by (2.5). On the hand, (2.13a) implies the existence of a subsequence $\{S_{nk}^t\varphi\}$ that converges to $e^{-iHt}\varphi$ pointwise a.e. in $\mathbb{R}^d$. Consequently, we have

$$V_t\varphi = e^{-iHt}\varphi \tag{2.18}$$

for each $\varphi \in \mathscr{F}(\mathbb{R}^d) \cap L^2(\mathbb{R}^d)$, where the r.h.s. is defined by (2.5). On the other hand, (2.13a) implies the existence of a subsequence $\{S_{nk}^t\varphi\}$ that converges to $e^{-iHt}\varphi$ pointwise a.e. in $\mathbb{R}^d$. Consequently, we have
an arbitrary $\varphi \in L^2(\mathbb{R}^d)$, and the proof is complete. ∎

REMARK 6.2.11. (a) In order to prove Theorem 2.5 in a straightforward way, one should check first that the semigroup $\{V_t : t \geqslant 0\}$ from Proposition 2.8 is strongly continuous or, equivalently,

$$\lim_{t\to 0+} (\psi, V_t\varphi) = (\psi, \varphi) \tag{2.19}$$

for all $\psi, \varphi \in L^2(\mathbb{R}^d)$ (cf. (A.2)). Further, the generator of this semigroup must be calculated and shown to coincide with H. In view of Proposition 2.9, the relation (2.19) is valid for $\psi, \varphi \in L^2(\mathbb{R}^d) \cap L(\mathbb{R}^d)$. Further, using the matrix-functional-calculus rules together with the identity

$$\frac{\mathrm{d}}{\mathrm{d}t} \det[g(\Omega t)] = \det[g(\Omega t)] \operatorname{Tr}[\Omega g'(\Omega t)(g(\Omega t))^{-1}],$$

one can verify that the function $\psi : \psi(x, t) = (V_t\varphi)(x)$ solves in $\mathbb{R}^d \times (0, \infty)$ the Schrödinger equation with the potential $u(x) = \frac{1}{2}x \cdot \Omega^2 x$ as a partial differential equation for each $\varphi \in \mathscr{F}(\mathbb{R}^d)$. The rest of such a proof, however, would be difficult; the comparison favours the method used here.

(b) Instead of $I^{\mathrm{P}}(\,.\,)$, another F-integral may be used to evaluate the propagator $\{V_t\}$; one excepts, of course, to get formula (2.5) again. Some results are known – cf. Problem 52 and the notes.

In the remainder of this section, we are going to discuss some properties of the solution (2.5) in the simplest case $d = 1$, where $\Omega = \omega - i\nu$, $0 < \nu \leqslant \omega$. First we shall treat the *non-damped limit*. The propagator kernel of the non-damped harmonic oscillator is well known to be

$$K_t(x, y) = K_t^F(x, y)M(t), \tag{2.20}$$

where

$$K_t^F(x, y) = (2\pi i)^{-1/2}\left(\frac{\omega}{|\sin \omega t|}\right)^{1/2} \times$$

$$\times \exp\left\{\frac{i\omega}{2 \sin \omega t}\,[(x^2 + y^2)\cos \omega t - 2xy]\right\} \tag{2.20a}$$

$$M(t) = \exp\left\{-\frac{\pi i}{2} \operatorname{Ent} \frac{\omega t}{\pi}\right\}, \tag{2.20b}$$

if t is not equal to a multiple of the half-period, $\omega t \neq k\pi$. Here $\operatorname{Ent} z \equiv [z]$ means the entire part of a real number z. If $\omega t = k\pi$, the kernel may be written as

$$K_t(x, y) = \exp\left\{-\frac{\pi i}{2} k\right\} \delta(x - (-1)^k y). \tag{2.21}$$

The expression (2.20) bears close similarity to (2.5b), but it has peculiar jumps in phase at every half-period given by the term (2.20b), the so-called *Maslov correction*. The points at which the jumps occur are just the turning points of the classical harmonic oscillator. There are various ways in which the phase-changing factor can be derived. Let us show that it emerges naturally in the non-damped limit of the propagator (2.5).

PROPOSITION 6.2.12. *Let $d = 1$ and $\Omega = \omega - i\nu$ with $0 < \nu \leqslant \omega$. Suppose that $\omega t \neq k\pi$, $k = 0, 1, \ldots$, and $\varphi \in L^2(\mathrm{IR})$ has a compact support, then*

$$\lim_{\nu \to 0+} (V_t\varphi)(x) = \int_{\mathrm{IR}} K_t(x,y)\varphi(y)\,\mathrm{d}y, \tag{2.22a}$$

where $K_t(x,y)$ is given by (2.20). *On the other hand, we have*

$$\lim_{\nu \to 0+} (V_t\psi)(x) = \exp\left\{-\frac{\pi i}{2}k\right\}\psi((-1)^k x) \tag{2.22b}$$

for $t = k\pi/\omega$ and $\psi \in \mathscr{S}(\mathrm{IR})$.

Proof. Suppose first $\omega t \neq k\pi$ and denote

$$h_x(y) = \exp\left\{\frac{i\Omega}{2\sin\Omega t}(y^2\cos\Omega t - 2xy)\right\}.$$

It is not difficult to see that

$$|h_x(y)| = \exp\left\{\frac{\omega\nu t}{2\,|\sin\Omega t|^2}\left[(y^2\cos\omega t - 2xy\,\mathrm{ch}\,\nu t)\frac{\sin\omega t}{t} - \right.\right.$$

$$\left.\left. -(y^2\,\mathrm{ch}\,\nu t - 2xy\cos\omega t)\frac{\mathrm{sh}\,\nu t}{t}\right]\right\}$$

so simple estimates lead to the inequality

$$|h_x(y)| \leqslant \exp\{\omega\,|y|\,(|y| + 2\,|x|)\,\mathrm{sh}\,\nu t\,\sin^{-2}\omega t\}$$

showing that the dominated-convergence theorem may be applied to (2.5) if φ has a compact support. In this way, we obtain

$$\lim_{\nu \to 0+} (V_t\varphi)(x) = \lim_{\nu \to 0+} \exp\{\tfrac{1}{2}g_\nu(t)\}\int_{\mathrm{IR}} K_t^F(x,y)\varphi(y)\,\mathrm{d}y,$$

where the function g_ν is defined by $g_\nu(t) = \arg(\Omega/\sin\Omega t)$. One can express it more explicitly: it holds that

$$g_\nu(t) = \mathrm{arctg}(\mathrm{th}\,\nu t\,\mathrm{ctg}\,\omega t) - \mathrm{arctg}(\nu/\omega) - k\pi \tag{2.23a}$$

for $k\pi < \omega t < (k+1)\pi$, where the last term is chosen so that the r.h.s. is continuous at the points $t = k\pi/\omega$ and tends to zero as $t \to 0+$ which certainly must be true for g_ν. For a fixed t, the relation (2.23a) gives

$$\lim_{\nu \to 0+} g_\nu(t) = -k\pi \qquad \text{if} \quad k\pi < \omega t < (k+1)\pi; \tag{2.23b}$$

combining this result with (2.20), we get (2.22a). Figure 1 illustrates how the jumps appear in the limit.

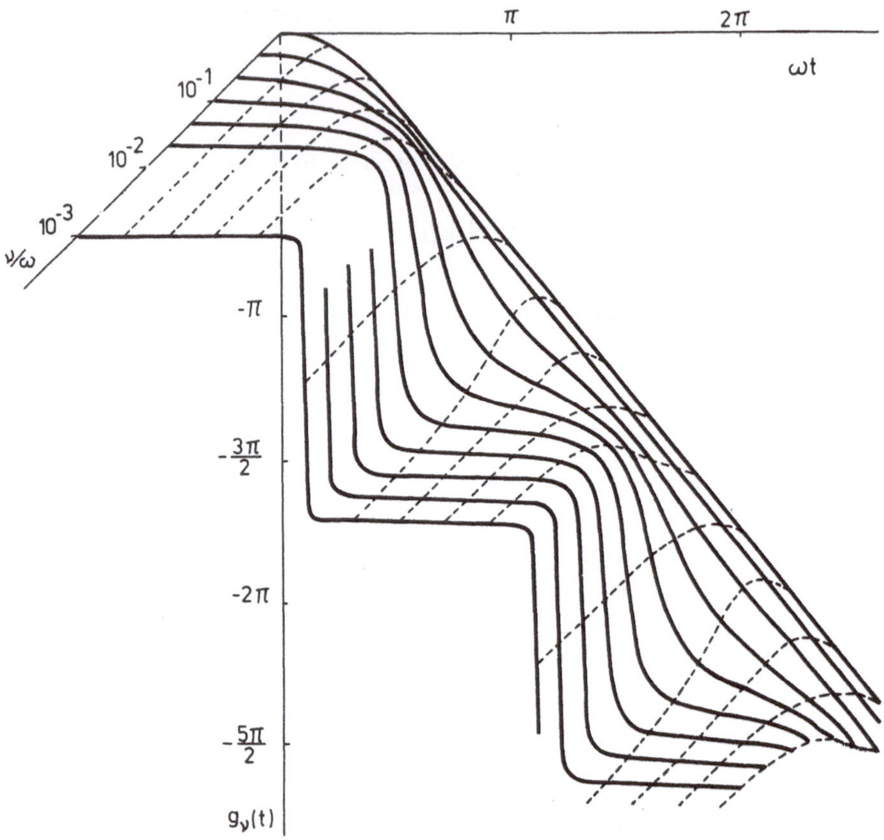

6.1. Emergence of the Maslov correction in the non-damped limit.

Let now, in turn, $\omega t = k\pi$ and express $V_t\psi$ from Proposition 2.9. Since

$$\left| \exp\left\{ -\frac{y^2 t \,\mathrm{th}\, \nu t}{2(k\pi - i\nu t)} - \frac{i(-1)^k xy}{\mathrm{ch}\, \nu t} \right\} \right| \leqslant 1,$$

the dominated-convergence theorem may be applied if only $F_1\psi \in L(\mathbb{R})$; it gives

$$\lim_{\nu \to 0+} (V_t\psi)(x) = (2\pi)^{-1/2} \exp\left\{ -\frac{\pi i}{2} k \right\} \int_{\mathbb{R}} e^{i(-1)^k xy} (F_1\psi)(y)\, dy.$$

The desired result then follows from simple properties of the Fourier–Plancherel operator F_1. ∎

Now we shall discuss the *classical limit* of the solution (2.5). It should be stressed that there is no *a priori* reason to expect that the limit will yield the motion of the classical (linearly) damped harmonic oscillator, because the pseudo-Hamiltonian under consideration has *not* been obtained by quantization of the latter. Nevertheless, the limit is worth examining; it is presumably better to check it on an example than to attempt a general formulation.

EXAMPLE 6.2.13. We restrict our attention to the cases when the initial wave-packets are Gaussian, especially those that are obtained by shifting the 'ground state': we consider the vectors $\varphi = \varphi_{L,\alpha,\kappa} \in L^2(\mathbb{R})$ of the form

$$\varphi(x) = (\pi l^2)^{-1/4} \exp\left\{-\frac{1}{2L^2}(x-\alpha)^2 + \frac{i}{\hbar}\kappa x\right\} \tag{2.24a}$$

with L complex, $\operatorname{Re} L^2 > 0$, $l^{-2} = |L|^{-4} \operatorname{Re} L^2$, and α, κ real. The expectations and dispersions of position and momentum in such a state are

$$\begin{aligned} &\langle Q\rangle_\varphi = \alpha, && \langle P\rangle_\varphi = \kappa, \\ &(\Delta Q)_\varphi = 2^{-1/2}\, l, && (\Delta P)_\varphi = 2^{-1/2}\,\hbar l\,|L|^{-2}; \end{aligned} \tag{2.25}$$

notably φ is not a minimum-uncertainty state unless L^2 is real. Of course, we need the propagator referring to arbitrary values of the mass m and Planck's constant $\hbar$. It is easy to see that one can obtain it from (2.5) by the substitutions $t \to (\hbar/m)t$ and $\Omega \to (m/\hbar)\Omega$. After this modification, we may apply Theorem 2.5 and perform some simple calculations (which are left to the reader) to obtain

$$\begin{aligned} (V_t\varphi)(x) = (\pi l^2)^{-1/4} &\left[\cos\Omega t + i\left(\frac{\Lambda}{L}\right)^2 \sin\Omega t\right]^{-1/2} \times \\ &\times \exp\left\{-\frac{i}{2\Lambda^2}\,\frac{\sin\Omega t - i(\Lambda/L)^2\cos\Omega t}{\cos\Omega t + i(\Lambda/L)^2 \sin\Omega t} \times\right. \\ &\left.\times \left[x^2 - \frac{2xz\Lambda^2 + \Lambda^4 z^2 \sin\Omega t}{\sin\Omega t - i(\Lambda/L)^2\cos\Omega t}\right] - \tfrac{1}{2}\left(\frac{\alpha}{L}\right)^2\right\}, \end{aligned} \tag{2.26a}$$

where $\Lambda^2 = \hbar/m\Omega$ and $z = \kappa\hbar^{-1} - i\alpha L^{-2}$. This expression simplifies considerably if we choose the particular value of L that corresponds to the 'ground state',

$$L^2 = \Lambda^2 = \frac{\hbar}{m\Omega} \tag{2.24b}$$

(cf. (2.31) below). In that case, we have

$$\begin{aligned} (V_t\varphi)(x) = (\pi\lambda^2)^{-1/4} \exp&\left\{-\frac{i}{2}\Omega t - \frac{1}{2\Lambda^2}\left[x - \left(\alpha + \frac{i}{\hbar}\kappa\Lambda^2\right)e^{-i\Omega t}\right]^2 + \right. \\ &\left. + \frac{1}{2\Lambda^2}\left(\alpha + \frac{i}{\hbar}\kappa\Lambda^2\right)^2 e^{-i\Omega t}\cos\Omega t - \tfrac{1}{2}\left(\frac{\alpha}{\Lambda}\right)^2\right\}, \end{aligned} \tag{2.26b}$$

where $\lambda^2 = \hbar/m\omega$. The probability density of finding the particle at x is then given by

$$|(V_t\varphi)(x)|^2 = (\pi\lambda^2)^{-1/2} \exp\left\{-\nu t - \frac{1}{\lambda^2}(x - x_0(t))^2 + y(t)\right\}, \qquad (2.27)$$

where

$$x_0(t) = (\alpha \cos \omega t + \beta \sin \omega t)\, e^{-\nu t} \qquad (2.27a)$$

and

$$y(t) = \frac{1}{2}\left(\frac{\gamma}{\lambda}\right)^2 -$$

$$-\frac{1}{2\lambda^2}\left[(\beta^2 - \gamma^2)\cos 2\omega t + \frac{\nu}{\omega}(\alpha^2 - \gamma^2)\sin 2\omega t - (\alpha^2 + \beta^2)\right] e^{-2\nu t} \qquad (2.27b)$$

with

$$\beta = \frac{\kappa - m\alpha\nu}{m\omega}, \qquad \gamma = \frac{\kappa}{m\,|\Omega|}. \qquad (2.27c)$$

Thus the function (2.27) is Gaussian-shaped and has the following properties:

(i) The peak diminishes with time, for large t approximately as $e^{-\nu t}$.

(ii) Its width λ does not change; it is negligible in the classical limit when $\alpha^2 + \beta^2 \gg \lambda^2$.

(iii) The peak travels along $x = x_0(t)$ which is the trajectory of the classical damped harmonic oscillator (described by complex frequency Ω) with the initial position $x_0(0) = \alpha$. The corresponding initial momentum, however, assumes the value $m\dot{x}_0(0) = \kappa - 2m\alpha\nu$ instead of κ suggested by (2.25), but this is not surprising in view of the remark preceding this example. If $x_c(\,.\,)$ denotes the trajectory of the classical damped oscillator referring to the initial data (α, κ), then $x_c(t) - x_0(t) = 2\alpha(\nu/\omega)\, e^{-\nu t} \sin \omega t$. In particular, the difference is negligible in the case of weak damping, $\nu \ll \omega$ (see also Problem 55).

Finally, let us show what the *spectrum* of the pseudo-Hamiltonian under consideration looks like.

PROPOSITION 6.2.14. *Let H be the generalized Schrödinger operator on $L^2(\mathbb{R})$ referring to the potential $u : u(x) = \frac{1}{2}\Omega x^2$ with $\Omega = \omega - i\nu, 0 < \nu \leqslant \omega$. The spectrum of this operator is purely discrete and consists of the eigenvectors*

$$\psi_n : \psi_n(x) = N_{nn}^{-1/2} H_n(\Omega^{1/2}x) \exp\{-\tfrac{1}{2}\Omega x^2\},\; n = 0, 1, 2, \ldots \qquad (2.28a)$$

where H_n are Hermite polynomials and

$$H\psi_n = \Omega(n + \tfrac{1}{2})\psi_n. \qquad (2.28b)$$

In general, the eigenvectors (2.28a) *are not orthonormal: it holds that* $(\psi_n, \psi_m) = N_{nn}^{-1/2} N_{mm}^{-1/2} N_{nm}$, *where*

$$N_{n,n+2s+1} = 0$$

$$N_{n,n+2s} = \left(\frac{\pi}{\omega}\right)^{1/2} \frac{n!(n+2s)!}{(n+s)!} \, \omega^{-(n+s)} \, |\Omega|^n \Omega^s \times$$

$$\times \sum_{k=0}^{[n/2]} \sum_{l=0}^{[n/2]+s} (-1)^{k+l} \binom{2(n+s-k-l)}{n-2k} \times$$

$$\times \binom{k+l}{k} \binom{n+s}{k+l} \omega^{k+l} (\bar{\Omega})^{-k} \, \Omega^{-l} \tag{2.29}$$

for $s = 0, 1, 2, \ldots$, $[\,.\,]$ *being the entire part.*

Proof. In view of Theorem 4.3.15, $\sigma(H)$ is purely discrete. The relations (2.28), (2.29) are verified by straightforward computation. Let us show that H has no other eigenvectors. For an arbitrary complex λ, the equation

$$\psi'' + (2\lambda - \Omega^2 x^2)\psi = 0 \tag{2.30a}$$

has the following fundamental solution

$$\psi_\lambda(x) = [\alpha\Phi(\tfrac{1}{4} - \gamma, \tfrac{1}{2}; \Omega x^2) + \beta x \Phi(\tfrac{3}{4} - \gamma, \tfrac{3}{2}; \Omega x^2)] \exp(-\tfrac{1}{2}\Omega x^2), \tag{2.30b}$$

where $2\Omega\gamma = \lambda$ and Φ is the degenerate hypergeometric function. We have $\mathrm{Re}\, \Omega x^2 = \omega x^2 \geqslant 0$ so the asymptotic behaviour of (2.30b) for large $|x|$ is given by

$$\psi_\lambda(x) = C(\alpha, \beta, \lambda, \Omega) x^{1/4 - \gamma} \exp(\tfrac{1}{2}\Omega x^2) \, [1 + O(|x|^{-1})]$$

(Abramowitz and Stegun, 1964, 13.1.4), except for the cases when one of the functions in (2.30b) reduces to some of (2.28a) and the other is absent. Since the coefficient $C(\alpha, \beta, \lambda, \Omega)$ is non-zero unless $\alpha = \beta = 0$, the functions (2.28a) represent the only solutions to (2.30a) contained in $L^2(\mathbb{R})$. ■

The fact that the residual spectrum of H is void does not need reference to Theorem 4.3.15: it is easy to see that the set $M = \{\psi_n\}_{n=0}^{\infty}$ is complete in $L^2(\mathbb{R})$ so $(H - \lambda)M_{\mathrm{lin}} = M_{\mathrm{lin}}$ is dense if $\lambda \neq \Omega(n + \tfrac{1}{2})$, $n = 0, 1, 2, \ldots$. Proposition 2.14 can also be used to determine the spectrum of H for $d > 1$ if only $\Omega^2 = 2(A - iB)$ can be diagonalized; it is clear that the spectrum is not necessarily simple in that case. Moreover, some results remain true even if A, B are not simultaneously diagonalizable. For instance, one can check readily that the 'ground state' vector

$$\psi_0(x) = [\det(\mathrm{Re}\, \Omega)/\pi]^{d/4} \exp(-\tfrac{1}{2} x \cdot \Omega x) \tag{2.31}$$

corresponds to the eigenvalue $\tfrac{1}{2}\,\mathrm{Tr}\,\Omega$ for any A, B which obey the assumptions of Theorem 2.3.

6.3. The 'Feynman Paths'

"The pronounced stupidity is like a slope. I go on rolling downhill."

R. MERLE, *Malevil*

Considerations of the preceding two sections show how one can use Feynman path integrals to express the semigroup time evolution governed by a Schrödinger pseudo-Hamiltonian, despite the fact that the standard way of motivation mentioned in the opening to Chapter 5 is generally no longer applicable. It is therefore worthy to look at the motivating argument more closely.

To this end, we return to the limit of continual observation studied in Section 2.4. Recall that it represents rather a mathematically useful concept than a physically reasonable way of describing a microsystem subjected to a permanent measuring process. With this reservation in mind, we may generalize now the evolution operator (2.3.4). Let the system under consideration be described by a pseudo-Hamiltonian H on some state Hilbert space $\mathscr{H}$. Suppose, further, that a projection-valued function $E : J^b \to \mathscr{B}(\mathscr{H})$ is given for some positive b. We call it the *apparatus function* and interpret its value $E(\tau)$ as the appropriate 'yes–no' experiment performed on the system at the instant τ. Now we consider some $t \in J^b$ and a partition $\sigma \in \mathscr{P}(J^t)$, and set

$$U(t,0;E,\sigma) = E(t)V_{\delta_{n-1}}E(\tau_{n-1})V_{\delta_{n-2}}\cdots V_{\delta_0}E(0), \tag{3.1a}$$

where $V_t = e^{-iHt}$. This is clearly the evolution operator referring to the situation when the system has suffered measurements at the instants $0, \tau_1, \tau_2, \ldots, t$ provided that only the positive outcomes are taken into account. Consequently, the limit

$$U(t,0;E) = \operatorname*{s-lim}_{\delta(\sigma)\to 0} U(t,0;E,\sigma) \tag{3.1b}$$

(if it exists) may be deemed as the evolution operator in the case when the system is under continual observation described by the apparatus function E. For each $t \in J^b$, we have

$$\|U(t,0;E)\| \leqslant 1, \tag{3.2a}$$

$$(\operatorname{Ker} U(t,0;E))^\perp \subset E(0)\mathscr{H}, \qquad \operatorname{Ran} U(t,0;E) \subset E(t)\mathscr{H}. \tag{3.2b}$$

Furthermore, the strong sequential continuity of operator multiplication shows that

$$U(t,\tau;E)U(\tau,0;E) = U(t,0;E) \tag{3.3}$$

holds for $0 \leqslant \tau \leqslant t \leqslant b$ if the respective operators make sense.

In Section 2.4, we have established various other properties of the evolution operator (3.1b) in the particular situation when E is time-independent. If we are going now to treat the case of a non-constant apparatus function, we should first explain why it is needed. Recall Feynman's thought experiment with a particle that passes through a sequence of many-slit screens, and suppose that each slit is watched by a counter. If the system of counters is not functioning, the particle motion can be described in terms of the coherent sum over all possible 'trajectories' through the sandwiched screens. In order to select a particular 'trajectory', one has to switch the monitor on and watch which counters will click. Knowledge of the evolution operator of such a process allows us (at least, in principle) to draw conclusions about the weight factor which the particular 'trajectory' contributes to the coherent sum.

This provides a natural motivation to investige the limit (3.1b) that is expected to describe the motion of the system along the 'path' in $\mathscr{H}$ determined by the apparatus function. One is particularly interested in whether some 'Zeno-type' effect would occur again, i.e. whether the continual observation is able to force the system to stay within the moving subspace $E(\,.\,)\mathscr{H}$. If it is so, then the notion of 'path' in $\mathscr{H}$ associated with the apparatus function makes good sense. One should stress, however, that these 'paths' are likely to have *no operational meaning*. The arguments presented in Section 2.4 are easily adapted to the more general situation; they show that we cannot realize such a 'Feynman path' experimentally, and that it is even difficult to come close to the 'Zeno's limit'.

After this introduction, let us examine the 'Feynman paths' in the simplest case when $\dim E(t) = 1$ within J^b. The sought result is then obtained if some suitable smoothness assumptions are imposed on $E(\,.\,)$.

THEOREM 6.3.1. *Let H be maximal dissipative and assume that $E(t)$ is a one-dimensional projection for each $t \in J^b$ containing a unit vector $\psi_t \in D(H)$ in its range. Suppose that the function $t \mapsto H\psi_t$ is continuous on J^b and $t \mapsto \psi_t$ is continuously differentiable with $t \mapsto \|\dot{\psi}_t\|$ bounded on* $(0, b)$, *then the corresponding operator* (3.1b) *exists and*

$$\varphi_t = U(t, 0; E)\varphi = \exp\left\{ i \int_0^t L(\tau)\, \mathrm{d}\tau \right\} (\psi_0, \varphi)\psi_t, \tag{3.4a}$$

$$L(t) = -i(\dot{\psi}_t, \psi_t) - (\psi_t, H\psi_t) \tag{3.4b}$$

holds for all $t \in [0, b)$ and $\varphi \in \mathscr{H}$. Moreover, the function $U(\,.\,, 0; E)\varphi$ obeys the equation

$$i\,\frac{\mathrm{d}\varphi_t}{\mathrm{d}t} = [i\dot{E}(t)E(t) + E(t)HE(t)]\varphi_t. \tag{3.5}$$

REMARK 6.3.2. The operator $U(t, 0; E)$ depends actually on $E(\,.\,)$ only, and not on the representing vector function. If $\psi_t = \tilde{\psi}_t\, e^{i\alpha(t)}$ with α absolutely continuous, then $L(t) = L(t) - \dot{\alpha}(t)$ so the substitution to (3.4a) gives $\tilde{\varphi}_t = \varphi_t$.

Proof. We take an arbitrary partition $\sigma \in \mathscr{P}(J^t)$ and $\varphi \in \mathscr{H}$, then $U(t, 0; E, \sigma)\varphi = f(t, 0; \sigma)(\psi_0, \varphi)\psi_t$, where

$$f(t, 0; \sigma) = \prod_{j=0}^{n-1} g(\tau_{j+1}, \tau_j),$$

$$g(s, t) = (\psi_s, \mathrm{e}^{-iH(s-t)}\,\psi_t).$$

Using the fact that $t \mapsto \mathrm{e}^{-iHt}$ is a continuous contractive semigroup and the smoothness assumptions of the theorem, one can check that $g(\,.\,, t)$ is continuously differentiable for each $t \in (0, b)$ with

$$\left.\frac{\partial g(s, t)}{\partial s}\right|_{s=t} = iL(t)$$

and that $L(\,.\,)$ is continuous on $(0, b)$. Since $g(t, t) = 1$, the last inequality yields the relation

$$L(t) = -i\left.\frac{\partial \ln g(s, t)}{\partial s}\right|_{s=t}.$$

Further, we want to express $\ln f(t, 0; \sigma)$ with the help of the Lagrange remainder theorem. To this end, one must consider the real and imaginary parts separately, because the theorem does not apply to complex functions. We have

$$\ln f(t, 0; \sigma) = \sum_{j=0}^{n-1} \ln g(\tau_{j+1}, \tau_j) = i\sum_{j=0}^{n-1} [\mathrm{Re}\, L(\xi_j) + i\, \mathrm{Im}\, L(\eta_j)]\,\delta_j$$

for some numbers $\xi_j, \eta_j \in (\tau_j, \tau_{j+1}), j = 0, 1, \ldots, n-1$, but $\mathrm{Re}\, L(\,.\,)$ and $\mathrm{Im}\, L(\,.\,)$ are continuous and bounded on $(0, b)$ due to the assumptions, and are therefore Riemann integrable. In this way, we obtain the relation

$$\lim_{\delta(\sigma)\to 0} \ln f(t, 0; \sigma) = i\int_0^t \mathrm{Re}\, L(\tau)\, \mathrm{d}\tau - \int_0^t \mathrm{Im}\, L(\tau)\, \mathrm{d}\tau$$

$$= i\int_0^t L(\tau)\, \mathrm{d}\tau,$$

which proves (3.4). The check of (3.5) is straightforward. ■

We see that the above-formulated question is answered positively under the assumptions of the theorem: if the initial state is not disjoint with ψ_0, it is driven by the operator (3.1b) so that it cannot leave the 'path' specified by the chosen apparatus function. By that, it acquires a phase factor. In order to clarify its nature, let us consider a particular case with $\mathscr{H} = L^2(\mathbb{R}^d)$.

COROLLARY 6.3.3. *Let ψ be a unit vector from $\mathscr{S}(\mathbb{R}^d)$ and $\gamma, \pi \in C^1[J^t; \mathbb{R}^d]$, i.e. we assume that $\gamma(\,.\,)$ is continuously differentiable with $|\dot{\gamma}(\,.\,)|$ bounded on $(0, b)$, and the same for $\pi(\,.\,)$. Let, further, H be a maximal dissipative operator on $L^2(\mathbb{R}^d)$ whose domain $D(H) \supset \mathscr{S}(\mathbb{R}^d)$. We denote*

$$\psi_t : \psi_t(x) = \psi(x - \gamma(t))\, \mathrm{e}^{ix \cdot \pi(t)} \tag{3.6a}$$

and assume the function $t \mapsto H\psi_t$ to be continuous on J^b, then the corresponding evolution operator (3.1b) *is given by* (3.4a) *with*

$$L(t) = \sum_{k=1}^{d} \dot{Q}_k(t)P_k(t) - E(t) - i\,\frac{\mathrm{d}}{\mathrm{d}t} \sum_{k=1}^{d} Q_k(t)P_k(t), \tag{3.7}$$

where $E(t) = (\psi_t, H\psi_t)$ and similarly $Q_k(t)$, $P_k(t)$ are expectations of the standard coordinate and momentum operators on $L^2(\mathbb{R}^d)$, respectively, in the state ψ_t.

Proof. One has just to specify the first term on the r.h.s. of (3.4b). Since $\psi \in \mathscr{S}(\mathbb{R}^d)$, we have

$$\frac{\mathrm{d}}{\mathrm{d}t}\psi_t(x)$$

$$= \sum_{k=1}^{d} \left\{ -\dot{\gamma}_k(t) \left.\frac{\partial \psi(y)}{\partial y_k}\right|_{y = x - \gamma(t)} + ix_k\dot{\pi}_k(t)\psi(x - \gamma(t)) \right\} \mathrm{e}^{ix \cdot \pi(t)}. \tag{3.8a}$$

We use these relations to prove that the derivative $\dot{\gamma}_t$ in the sense of Hilbert-space norm is given by

$$\dot{\gamma}_t = -i \sum_{k=1}^{d} \dot{\gamma}_k(t)P_k\psi_t + i[\dot{\gamma}(t) \cdot \pi(t) + x \cdot \dot{\pi}(t)]\,\psi_t. \tag{3.8b}$$

To this end, we denote $\psi_k = \partial\psi/\partial y_k$ and employ the Lagrange remainder theorem:

$$\left\| \frac{\psi_{t+\delta} - \psi_t}{\delta} + \sum_{k=1}^{d} \{\dot{\gamma}(t)\psi_k(x - \gamma(t)) - ix_k\dot{\pi}_k(t)\psi(x - \gamma(t))\}\, \mathrm{e}^{ix \cdot \pi(t)} \right\|^2$$

$$= \int_{\mathbb{R}^d} \left| \left.\frac{\mathrm{d}}{\mathrm{d}\tau} \operatorname{Re} \psi_\tau(x)\right|_{\tau = \xi_1} - \operatorname{Re} \frac{\mathrm{d}}{\mathrm{d}t}\psi_t(x) \right|^2 \mathrm{d}x \;+$$

$$+ \int_{\mathbb{R}^d} \left| \left.\frac{\mathrm{d}}{\mathrm{d}\tau} \operatorname{Im} \psi_\tau(x)\right|_{\tau = \xi_2} - \operatorname{Im} \frac{\mathrm{d}}{\mathrm{d}t}\psi_t(x) \right|^2 \mathrm{d}x$$

for some $\xi_1, \xi_2 \in (t, t+\delta)$. We express the derivatives with respect to τ from (3.8a), then the squared norm can be estimated by

$$\sum_{l=1}^{2} \sum_{k=1}^{d} \int_{\mathbb{R}^d} |\dot{\gamma}_k(\xi_l)\psi_k(x-\gamma(\xi_l)) - \dot{\gamma}_k(t)\psi_k(x-\gamma(t))|^2 \, dx +$$

$$+ \sum_{l=1}^{2} \sum_{k=1}^{d} \int_{\mathbb{R}^d} |\dot{\pi}_k(\xi_l)x_k\psi(x-\gamma(\xi_l)) - \dot{\pi}_k(t)x_k\psi(x-\gamma(t))|^2 \, dx$$

$$\leqslant 2 \sum_{l=1}^{2} \sum_{k=1}^{d} |\dot{\gamma}_k(\xi_l) - \dot{\gamma}_k(t)|^2 \int_{\mathbb{R}^d} |\psi_k(x-\gamma(\xi_l))|^2 \, dx +$$

$$+ 2 \sum_{l=1}^{2} \sum_{k=1}^{d} |\dot{\gamma}_k(t)|^2 \int_{\mathbb{R}^d} |\psi_k(x-\gamma(\xi_l)) - \psi_k(x-\gamma(t))|^2 \, dx +$$

$$+ 2 \sum_{l=1}^{2} \sum_{k=1}^{d} |\dot{\pi}_k(\xi_l) - \dot{\pi}_k(t)|^2 \int_{\mathbb{R}^d} |x_k\psi(x-\gamma(\xi_l))|^2 \, dx +$$

$$+ 2 \sum_{l=1}^{2} \sum_{k=1}^{d} |\dot{\pi}_k(t)|^2 \int_{\mathbb{R}^d} |x_k\psi(x-\gamma(\xi_l) - x_k\psi(x-\gamma(t))|^2 \, dx.$$

According to the assumptions, the functions $\dot{\gamma}_k$ are bounded on $(0, b)$ and $\psi_k \in \mathscr{S}(\mathbb{R}^d)$, thus the first sum on the r.h.s. tends to zero as $\delta \to 0+$. So does the second one, because the group of translations on $L^2(\mathbb{R}^d)$ is strongly continuous. A similar argument combined with the boundedness of γ_k applies to the remaining terms. Hence we obtain the relation (3.8b) which yields readily an expression for $(\dot{\psi}_t, \psi_t)$. It holds that

$$Q_k(t) = \gamma_k(t) + (\psi, Q_k\psi), \tag{3.9a}$$

$$P_k(t) = \pi_k(t) + (\psi, P_k\psi) \tag{3.9b}$$

for $k = 1, \ldots, d$; using the derivative of the first relation, we arrive at (3.7). ■

The first two terms on the r.h.s. of (3.7) call to mind the standard expression for action. This correspondence becomes even more apparent if we specify further the 'path' (3.6a) and consider the particular case when H is (the closure of) a Schrödinger pseudo-Hamiltonian $\frac{1}{2}P^2 + U$. Notice first that the phase of ψ can be fixed (cf. Remark 3.2) in such a way that the last term in (3.9b) is zero. Then we choose

$$\dot{\gamma}_k = \pi_k, \quad k = 1, \ldots, d \tag{3.6b}$$

so $\dot{Q}_k(t) = P_k(t)$ holds within $(0, b)$. Using further $(\Delta P_k)^2_{\psi_t} = (\psi_t, P_k^2 \psi_t) - P_k(t)^2$, we can write

$$L(t) = \frac{1}{2} \sum_{k=1}^{d} P_k(t)^2 - U(t) - \frac{1}{2} \sum_{k=1}^{d} (\Delta P_k)^2_{\psi_t} - i \frac{\mathrm{d}}{\mathrm{d}t} \sum_{k=1}^{d} Q_k(t) P_k(t) \tag{3.10}$$

with $U(t) = (\psi_t, U\psi_t)$.

Consider now the quasiclassical situation in a sense similar to Example 2.13, i.e. assume that the size of the wavepacket (3.6a) is negligible in both the coordinate and momentum scales given by the ranges of the functions γ, π. In that case, the total phase accumulated along the 'path' (3.6) consists of three parts. The first of them can be given naturally with the meaning of action,

$$S(\psi_t) = \int_0^t \left[\frac{1}{2} \sum_{k=1}^{d} P_k(t)^2 - U(t) \right] \mathrm{d}t. \tag{3.11}$$

This part is the only essential. The second one is negligible due to the assumption; notice that the evolution of ψ_t is controlled from outside and not governed by the dynamics of the system, and therefore the wavepacket does not spread with time. Finally, the last part contributes with a factor which is obviously path-independent.

It is important to note that these considerations *are not bound to the case of real-valued potentials*. The summary phase along the chosen 'path' is also given essentially by (3.11) for a complex u. Of course, the corresponding classical evolution equation cannot, in general, be derived from the complex action, but it does not matter for the present purpose. It is particularly encouraging that no detailed knowledge of U was needed; the argument would work even if the latter was not a multiplication operator.

On the other hand, this way of motivation has its drawbacks. In particular, there is a taste of phase-space reasoning in it: in order to get the expression (3.10), we had to insert the classical relation (3.6b) between the coordinate and momenta into a definition of the 'path'. It would be more natural in the configuration-space approach adopted here to perform the analogous considerations with the one-dimensional $E(t)$ replaced by the projection to $L^2(M_t)$, where M_t is some 'moving' region of $\mathbb{R}^d$. Aside from the motivational aspect discussed here, the solution to such a problem would provide interesting comparisons with the W-measure theory (cf., e.g. Problem 48). Unfortunately, infinite dimensionality of the projections involved makes the problem difficult. For some particular results, we can refer to Example 2.4.11 and to the notes to this section, but the full answer is far from known.

Notes to Chapter 6

SECTION 6.1. For the methods of describing the classical dissipative systems see, e.g. Goldstein (1951, §§1.4, 10.5) or Landau and Lifshitz (1965, §V.26). The relations (1.1) are frequently named Feynman–Kac formula in the physical literature. We prefer the name used here, because it allows us to distinguish between the real- and imaginary-time cases, with the substantially different functional integrals involved. Cameron (1960) was the first who obtained a rigorous expression of the solution to a Schrödinger-type equation (under rather restrictive analycity hypotheses imposed on the potential) by means of the limiting F-integral. A little later, Itô (1961) used independently a modified Gaussian-regularized F-integral to prove the analogous formula for the free-particle case and for the linear potential.

• There are many modifications of the FCI-formula for various real-valued potentials; e.g. those by Babbitt (1963), Nelson (1964), Faris (1967b), Itô (1967), Albeverio and Høegh-Krohn (1976), Truman (1976), Combe *et al.* (1978) and others mentioned in the notes to Section 5.1. There are also analogous results for (sufficiently nice) complex-valued potentials (Cameron, 1960; Albeverio and Høegh-Krohn, 1979; Johnson and Skoug, 1981c), eventually formulated for the integral equation (4.2.13) (Cameron and Storvick, 1968; Johnson and Skoug, 1973). The dissipativity condition is not required, however, in these papers, because their results are not given with a physical interpretation (of course, with the exception of the 'real' case).

• Theorem 1.1 was first proved for real-valued potentials (and with $\alpha = g$) by Itô (1967). The idea of the present proof is due to Albeverio and Høegh-Krohn (1976, chap. 3); see also Truman (1976), Exner and Kolerov (1981c). The assertions (b) and (c) of Theorem 1.3 for real-valued potentials and the regular product F-integral are due to Nelson (1964); for the generalization to the complex case and the uniform limit see Exner and Kolerov (1982b) or Exner (1982b). As for Theorem 1.4, a slightly more general 'real' result is due to Faris (1967b); the complex version has been proved in Exner and Kolerov (1982b) under the additional assumptions that $d \leqslant 3$ and U_1 is time-independent. Notice also that the methods discussed here are restricted to the cases when the Schrödinger operators under consideration are (at least, essentially) maximal dissipative. This does not mean, however, that the FCI-formula could not hold in other cases too. This is illustrated, e.g., by an example given by Nelson (1964) which concerns the pseudo-Hamiltonian related to the operator $\widetilde{T}_\alpha$ from Example 4.3.8; the corresponding contractive semigroup is expressed by means of the analytic F-integral. Another formula applicable when $H_0 + V$ referring to a real-valued potential v need not be e.s.a., is due to Ichinose (1980).

• Various F-integral expressions have been derived from the wave operators and the S-matrix. The results differ in the level of mathematical rigour. Generally speaking, the existence problem for the 'improper' F-integrals which appear in such expressions is more difficult than in the finite-time case. Using a method similar to the proof of Theorem 1.1, Albeverio and Høegh-Krohn (1976, chap. 3) solved the problem for the real-valued potentials which are bounded, smooth and decay fast enough at infinity. An expression of the S-matrix via a phase-space F-integral was given by Campbell *et al.* (1975), whose results were further extended by Vasiliev and Kuzmenko (1977, 1979) and Gerry and Singh (1980); see also DeWitt-Morette *et al.* (1979, §2.4). Yet another approach is due to B. Nelson (1983).

• Evaluation of the decay probability from the semiclassical barrier-penetration formula was first used by Gamow (1928) in his theory of α-radioactivity. A vast amount of literature is devoted to this subject, and we limit ourselves to quoting recent papers by Drukarev *et al.* (1979), Levit *et al.* (1980), Campbell (1980) and Patrascioiu (1981). For the influence-functional method see, e.g. Wells (1961), Feynman and Vernon (1963), or Feynman and Hibbs (1965, chap. 12).

SECTION 6.2. Propagator of the (non-damped) harmonic oscillator can be obtained by the well-known path-integral argument (Feynman and Hibbs, 1965, chap. 3), which is, however, only formal (Hjorth, 1982). Itô (1967) was the first to obtain this result rigorously using the Gaussian-regularized F-integral. Other existing path-integral derivations employ the regular cylindrical F-integral (Truman, 1977a) and the product F-integral (Combe *et al.*, 1978; Hjorth, 1982).

• Description of the damped harmonic oscillator in a quantum-theoretical framework has been considered by many authors. Most of them, however, approach the problem via quantization of the classical (linearly) damped harmonic oscillator, eventually in combination with some *ad hoc* assumptions. This usually leads to time-dependent Hamiltonians (in which even the kinetic part varies with time) or to non-linear evolution equations – see, e.g., Haase (1975), Brinati and Mizrahi (1980), Gisin (1981), and a large review by Dekker (1981). In some cases, the corresponding propagators may be evaluated by path integration (Khandekar and Lawande, 1978, 1979; Jannussis *et al.*, 1979). As we have stressed in Section 1.2, such methods can yield a physically acceptable model if there is reasonable similarity between the classical and quantum mechanisms of damping in a particular case of interest. There is also another type of damped-oscillator models that are physically more specified: they assume a non-damped harmonic oscillator coupled to some other quantum system; e.g. to a massless scalar field (Arai, 1981) or to a certain heat bath (Streater, 1982).

• The presentation of this section follows Exner (1983a). The detailed discussion of the example has mainly methodical and illustrative purposes; we are not going to examine the physical situations which the pseudo-Hamiltonian under consideration could suit. Notice that applicability of such models is somewhat limited because H has no stable ground state, but this property seems to be common for many Schrödinger pseudo-Hamiltonians (cf. Corollary 4.3.12 and Theorem 4.3.13). Notice further that multidimensionality is substantial in our example: the problem cannot be generally reduced to the direct sum of one-dimensional damped oscillators, because A, B need not be simultaneously diagonalizable. The estimates used in Lemmas 2.1 and 2.2 are well known from some self-adjointness proofs. The trick employed to prove Theorem 2.3 calls to mind the way in which the symmetric version of the Kato–Rellich theorem is obtained (cf. Reed and Simon (1975, theorem X.13) and Problem 33). The propagator (2.5) can also be calculated by means of other F-integrals, but in distinction to the presented proof, one does not know then *a priori* that the result is a continuous contractive semigroup generated by iH. Evaluation of the corresponding regular cylindrical F-integral is proposed as Problem 52. In the case when $A = 0$ but B is possibly time-dependent, existence of the analytic (Johnson and Skoug, 1981b) and Fresnelian (Johnson, 1982) F-integrals has been recently established.

• The fact that the phase can change discontinuously at caustics was established by Maslov (1961, 1965); related results are due to Morse and other people. A differential-geometric method of evaluating the Maslov correction that yields explicit result for the harmonic oscillator is due to Souriau (1976); Horváthy (1979) obtained this correction by modification of Feynman's formal argument. The presented derivation comes from Exner and Kolerov (1981c) and Exner (1983a). A link between the Maslov correction and the Cameron–Martin formula has been pointed out by Elworthy and Truman (1982).

• An alternative way of evaluating the harmonic oscillator propagator relies on the Cameron–Martin formula; it was used first by Itô (1967). A somewhat more general problem concerns an oscillator with bounded anharmonicity, say, $v(x) = \frac{1}{2}x \cdot \Omega^2 x + v_1(x)$, where $v_1 \in \mathscr{F}(\mathbb{R}^d)$. We have mentioned in Section 5.6 that the solution can be obtained by means of the generalized Fresnelian F-integral relative to the quadratic form $\Delta : \Delta(\gamma) = \frac{1}{2}\int_0^t [|\dot{\gamma}(\tau)|^2 - \gamma(\tau) \cdot \Omega^2\gamma(\tau)]\, d\tau$ (Albeverio and Høegh-Krohn, 1976, chap. 5; Rezende, 1983); see also Albeverio *et al.* (1982). We have remarked also that this F-integral might by obtained from some 'standard' one by means of the CM-formula (Elworthy and Truman, 1982). Hence one could avoid use of the generalized F-integral by reducing the original problem via the CM-formula with a suitably chosen operator K to the evaluation of Fresnelian F-integral of the function $\gamma \mapsto \exp\{-i\int_0^t v_1(\gamma(\tau) + x)\, d\tau\}\varphi(\gamma(0) + x)$. A sketch of such a construction was given by Truman (1979). Notice that this approach is attractive, because such an evaluation of the F-integral employs, at least in part, the calculus (change-of-variable transformation) rather than a direct computation.

SECTION 6.3. The idea of observing the path of a micro-object by a sequence of successive position measurements goes back to Heisenberg (1927). Feynman in his original paper (1948) noted that these measurements might be made infinitely dense (see also Feynman and Hibbs (1965, §1.4)). Further references to the papers dealing with the limit of continual observation can be found in the notes to Section 2.4 – particularly, Bloch and Burba (1974) treated the case of permanent localization to a region moving in space; see also Mensky (1979, 1983). The term 'Feynman path' was invented by Aharonov and Vardi (1980) who applied this concept to situations not necessarily connected with position monitoring. They insisted, however, that such 'paths' can be given with an operational meaning, which is improbable in view of the considerations presented in Section 2.4. In their paper, a heuristic argument leading to a particular case of the relation (3.10) was formulated. The presentation of this section essentially follows Exner (1982a).

• The problem of finding $U(t, 0; E)$ in the case when $E(t)$ is a projection to $L^2(M_t)$ for some space region $M_t \subset \mathbb{R}^d$ and H is a Schrödinger pseudo-Hamiltonian on $L^2(\mathbb{R}^d)$ with a potential u, is difficult. Notice that even for the constant $E(t) = E_M$ considered in Example 2.4.11, the existence of $U(t, 0; E_M)$ was not established. It would be useful to be able to obtain the propagator $U(\,.\,, 0; E)$, say, by solving a differential equation instead of calculating the appropriate limit. This problem, too, remains open for a non-constant $E(\,.\,)$. One possibility is represented by equation (3.5) which was derived formally by Bloch and Burba (1974); however, $\dot{E}(t)$ need not exist, especially if the projections $E(t)$ are infinite-dimensional. Friedman (1976) proposed the equation $i(d\varphi_t/dt) = H_{E(t)}\varphi_t$, where $H_{E(t)}$ is a suitable extension of $E(t)HE(t)$, presumably with the Dirichlet Laplacian relative to M_t as the free part. He also established the existence of the solution in the case when M_t is a smoothly varying interval in $\mathbb{R}$; however, it is not known whether it would obey $\varphi_t = U(t, 0; E)\varphi$.

• Various results are known concerning the W-measure of such 'paths', i.e. sets of γ that do not leave some moving space region (from obvious reasons, they are often called *Wiener sausages*). In particular, one can obtain in this way a more rigorous formulation of the commonplace statement about 'two-dimensionality' of a typical Wiener path – cf. Simon (1979, §22).

Selected Problems

1. Prove Proposition 1.4.3. Generalize its assertion to the orthogonal sums of a larger (not necessarily finite) family of Hilbert spaces.
2. Suppose we have Hilbert space $\mathscr{G}_j \subset \mathscr{H}_j, j = 1, 2$, a topological group G, and weakly continuous positive-definite functions $V_j : G \to \mathscr{B}(\mathscr{G}_j)$ such that $V_j(e) = I_{\mathscr{G}_j}$. Let P_j, $P = P_1 \otimes P_2$ be projections with the ranges $\mathscr{G}_j$ and $\mathscr{G} = \mathscr{G}_1 \otimes \mathscr{G}_2$, respectively. If the triples $\{\mathscr{H}_j, U_j, P_j\}, j = 1, 2$, are minimal unitary dilations of V_j, then $\{\mathscr{H}, U, P\}$ with $U : U(g) = U_1(g) \otimes U_2(g)$ is a unitary dilation of $V : V(g) = V_1(g) \otimes V_2(g)$ which, in general, is *not* minimal. Prove that this dilation is minimal if at least one of the V_j's is a unitary group. Generalize the assertion to the case of an n-fold tensor product.

Hint: In order to show that the dilation need not be minimal, consider $\mathscr{H}_1 = \mathscr{H}_2 = L^2(\mathbb{R})$ with both the $U_j(\,.\,)$ being translation groups and $\mathscr{G}_j$ one-dimensional and spanned by the same vector from $L^2(\mathbb{R})$; then the minimal dilation refers to the symmetric subspace in $\mathscr{H} = L^2(\mathbb{R}^2)$.

3. Theorem 1.3.4 may be used to check the existence of s-$\lim_{n\to\infty} (EU_{t/n}E)^n$, where $\{U_t : t \in \mathbb{R}\}$ is a one-parameter unitary group and E a projection on some Hilbert space $\mathscr{H}$. If (1.3.7b) with some $\alpha > 1$ is valid, then $\{\mathrm{pr}_E\, U_t : t \geqslant 0\}$ is a semigroup and the limit trivially exists. The argument extends to the case $\alpha = 1$: prove that if $\|EU_{t+s}E - EU_tEU_sE\| \leqslant Cts$ for all $t, s \geqslant 0$, then the sequence $\{(EU_{t/n}E)^n\}$ converges in the operator norm (actually, it remains true when $\{U_t\}$ is a continuous contractive semigroup on $\mathscr{H}$).

Hint: As in the proof of Theorem 1.3.4, we get $\|EU_tE - (EU_{t/n}E)^n\| \leqslant Kt^2$. This inequality further implies $\|(EU_{t/m}E)^m - (EU_{t/mn}E)^{mn}\| \leqslant Kt^2/m$ so the sequence under consideration is Cauchy (cf. Misra and Sinha (1977)).

4. Prove the assertion mentioned in the notes to Section 1.5: if there are some vectors $\varphi, \psi \in \mathscr{H}_{\mathrm{u}}$, $(\varphi, \psi) \neq 0$, and a positive constant γ such that $|(\varphi, V_t\psi)| \leqslant \exp(-\gamma t)$, then $\sigma(H) = \mathbb{R}$.

Hint: One cannot apply the procedure used to prove Theorem 1.5.7, because we have no information about the behaviour of $f(t) = (\varphi, V_t\psi)$ for $t < 0$. Instead, we write $f(t) = \int_{\mathbb{R}} e^{-i\lambda t}\, d\omega(\lambda)$ and define $\omega_\pm(\lambda) = (2\pi)^{-1} \int_{\mathbb{R}} f(t)\, e^{i\lambda t}\, \Theta(\pm t)\, dt$. The functions $\omega_\pm$ are analytic in the upper (lower) halfplane, they are also defined for real λ in the sense of tempered distributions and $\omega(\lambda) = \omega_+(\lambda) + \omega_-(\lambda)$. If there is an open set $M \subset \mathbb{R}$ which does not belong to $\sigma(H)$, then $\omega_+(\lambda) = -\omega_-(\lambda)$ for $\lambda \in M$ and ω_+, ω_- are different pieces of one analytic function, further values of ω coincide with the discontinuity of this function on $\mathbb{R}$. The exponential bound implies that ω_+ is analytic for $\operatorname{Im} \lambda > -\gamma$, so we come to the contradictary conclusion $\omega = 0$. For details see Williams (1971).

5. The conclusion of Theorem 2.2.3 is preserved if the chamber structure is *discrete* and periodic, i.e. it obeys (c2) and there is a natural k_p such that $W[m+k_p, n+k_p] = W[m,n]$ for all m, n.

Hint: For $P(t) = \exp(-\Gamma t)$, equation (2.1.7c) is solved by $E_0(t) = P(t)$ so one can express $E(t)$ explicitly and perform the needed estimates (cf. Exner (1977b)).

6. Proposition 2.4.1 can be strengthened: actually $\lim_{\lambda \to \infty} \gamma(\lambda) = \Gamma_0$ holds under given assumptions.

Hint: Find a suitable upper bound on $P(t)$.

7. Consider an unstable system which suffers the repeated 'non-decay' measurements described by a function W (cf. the relations (2.1.1), (2.1.2)). Suppose that the system is initially described by some ρ and that each measurement yields a positive result. Write down operator integral equations which govern the time evolution of the density matrix. Under appropriate restrictions on W, check the existence of the solution and find its properties. How does the time evolution differ when the measurements are performed but their outcomes are not registered?

Hint: Cf. the papers quoted in the notes to Section 2.1.

8. Let H_g be the Friedrichs Hamiltonian from Section 3.2. How are its eventual eigenvalues related to the function v?

Hint: Cf. Horwitz and Marchand (1971) for the particular case when v has no zeros.

9. Prove in a straightforward way that the convergence in (3.2.7a) is uniform in each compact interval $\subset (0, \infty)$ of λ.

Hint: Use the same trick as in the continuity proof of $I(\,.\,, v)$.

10. Prove the formulae (3.2.8). Find the fourth-order terms.

11. For sufficiently small real g, equation (3.2.12a) has a unique solution.

Hint: Apply the implicit-function theorem to a suitably modified equation, say, $z^2(z - w_\alpha(z,g))\varphi(\arg z) = 0$, where $\varphi \in C_0^\infty[0, 2\pi]$ such that $\varphi(\beta) = 0$ iff $\beta = 3\pi/2$. The relations (3.2.12b, c) ensure that the sought solution does not fall to the negative imaginary axis.

12. Consider Friedrichs model with the particular choice $|v(\lambda)|^2 = \sqrt{\lambda}\, e^{-a\lambda}$. For which values of the parameters a, g and λ_0 the pole-and-real-axis approximation can be justified?

Hint: Cf. Demuth (1976).

13. Prove Proposition 3.2.7. Show that a stronger assertion is valid: if the intervals are centred at $\lambda_0 + g\lambda_u$, then we obtain spectral concentration with $\Delta_g = (\lambda_0 + g\lambda_u - \alpha g^p, \lambda_0 + g\lambda_u + \alpha g^p)$ for each $\alpha > 0$ and $p < 2$.

Hint: Using the relations (3.1.1) and (3.2.5b) together with $v(\lambda_0) \neq 0$, one obtains

$$\langle \psi_u, E_{H_g}((\lambda, \mu))\psi_u \rangle$$

$$= \int_\lambda^\mu \frac{g^2 |v(\xi)|^2}{(-\xi + \lambda_0 + g\lambda_u + g^2 I(\xi, v))^2 + (\pi g^2 |v(\xi)|^2)^2} \, d\xi;$$

then the limit is performed by means of the dominated-convergence theorem.

14. Prove Theorem 3.3.1.

Hint: The reduced resolvent is given by Proposition 3.2.1 to be $R_u(z, H_g) = h(z, g)^{-1}$, where the matrix elements of $h(z, g)$ with respect to the eigenvectors of H_0 are

$$h(z, g)_{jk} = (-z + \lambda_j)\delta_{jk} + gM_{jk} + g^2 \int_{\mathbb{R}} \frac{(v_j(\lambda), v_k(\lambda))_{\mathcal{G}}}{z - \lambda} \, d\lambda.$$

According to the analycity assumption, the matrix function $h(\,.\,, g)$ may be continued across $\mathbb{R}$ into Ω_-, where its (jk)th element acquires the additional term $-2\pi i g^2 (v_j(z), v_k(z))_{\mathcal{G}}$; similarly $h^{\mathrm{II}}(\,.\,, g)$ on $\mathbb{R}$ is defined in analogy with (3.2.6a). Now one has to apply the implicit-function theorem to $f : f(g, z) = \det h^{\mathrm{II}}(z, g)$ at the points $(0, \lambda_j)$, which is possible because $(\partial f/\partial z)(0, \lambda_j) = -\Pi_{k \neq j}(\lambda_k - \lambda_j) \neq 0$. Evaluation of the derivatives needed for the proof of the formulae (3.3.3) is slightly more involved than in the one-dimensional case, but up to the second order in g it yields one additional term in (3.3.3a) only.

15. Construct an example of the embedded-eigenvalue perturbation, in which positions of the poles depend non-analytically on the coupling constant.

Hint: Consider the two-dimensional Friedrichs model as described in Theorem 3.3.1 with $\mathcal{G} = \mathbb{C}^2$, $M = \kappa I$, $\kappa > 0$, but assume now the eigenvalue to be degenerate, say $\lambda_1 = \lambda_2 = 0$. The pole position is given by the equation $z = \kappa g + \frac{1}{2} g^2 [\mathrm{Tr}\, F(z) \pm H(z)^{1/2}]$, where $H(z) = (\mathrm{Tr}\, F(z))^2 - 4 \det F(z))$ and

$$F(z)_{jk} = \int_{\mathbb{R}} \frac{\overline{v_j(\lambda)} v_k(\lambda)}{z - \lambda} \, d\lambda, \qquad j, k = 1, 2.$$

If $H(\,.\,)$ has simple zero at $z = 0$, then both solutions to the above equation are expressed by a Puiseux series containing $g^{5/2}$ with a non-zero coefficient. Thus one has to choose suitable v_1, v_2, say $v_1(\lambda) = (2/\pi)^{1/2}(1 + \lambda^2)^{-1}$ and

$$v_2(\lambda) = \begin{cases} (2 - 2\eta)^{-1/2} \, \mathrm{sg}\, \lambda & \ldots \quad 0 < \eta < |\lambda| < 1 \\ 0 & \ldots \quad \text{otherwise} \end{cases}$$

with an appropriate η (cf. Howland (1974)).

16. Prove the equality (3.4.3b) using the explicit form of $\rho_1(\rho)$.

Hint: Let $\rho = \Sigma_k w_k E_{\chi_k}$ with the unit vectors $\chi_k = \Sigma_{ij} \alpha_{ij}^k \varphi_i \otimes \psi_j$, where $\{\varphi_i\}$, $\{\psi_j\}$ are orthonormal bases in $\mathscr{H}_u^{(1)}, \mathscr{H}^{(2)}$, respectively. The reduction formulae yield

$$\rho_1(\rho)\varphi_i = \sum_{klj} w_k \alpha_{lj}^k \bar{\alpha}_{ij}^k \varphi_l$$

so both the decay laws are expressed by $\Sigma_{ijkl} w_k \alpha_{lj}^k \bar{\alpha}_{ij}^k (V_t^{(1)}\varphi_i, V_t^{(1)}\varphi_l)_1$ – cf. Blank and Exner (1977; 1980, §7.6).

17. Consider the Hilbert space (3.5.8a) and the representation $\tilde{U}$ which acts on it according to (3.5.9) and (3.5.8b). As pointed out in the notes to Section 3.5, the subspace

$$\tilde{\mathscr{H}}_u' = \{f\}_{\text{lin}} \otimes L^2(B_\epsilon, d^3k/2k_0) \otimes \mathbb{C}^{2s+1}$$

is not invariant under space translations. Estimate the distance on which this violation of translational invariance becomes essential for the mass distributions $|f(\,.\,)|^2$ referring to real unstable particles.

Hint: The mass cut-off (3.5.13a) may be chosen so that $\eta \approx 10^2\Gamma$.

18. Let H be a dissipative operator on $\mathscr{H}$ and $\lambda > 0$. Without reference to Theorem 4.2.5, prove that H is maximal dissipative if $\text{Ran}(H - i\lambda) = \mathscr{H}$, and that H is closed iff $\text{Ran}(H - i\lambda)$ is closed.

Hint: Use (4.2.2a) and the fact that $\{\psi, H\psi\} \mapsto (H - i\lambda)\psi$ maps bijectively the graph $\mathscr{G}(H)$ onto $\text{Ran}(H - i\lambda)$ – cf. Phillips (1959).

19. If H is maximal symmetric, then either iH or $-iH$ generates a continuous semigroup of isometries.

Hint: The Cayley transform defines a bijective correspondence between symmetric operators on $\mathscr{H}$ and isometries of $\mathscr{H}$. If V is maximal isometric, then either $D(V)$ or Ran V equals $\mathscr{H}$ – cf. Phillips (1959) or Davies (1980a, §6.1).

20. In connection with the previous problem, find example of an operator H on $\mathscr{H}$ such that both $\pm iH$ generate continuous contractive semigroups, which however do not extend to a continuous one-parameter unitary group on $\mathscr{H}$.

Hint: Consider translations on a halfline (cf. Example 1.2.5). Notice the substantially different character of the semigroups generated by $\pm iH$.

21. Let H be a dissipative operator on a finite-dimensional $\mathscr{H}$. The relation $\lim_{t\to\infty} e^{-iHt}\psi = 0$ holds for each $\psi \in \mathscr{H}$ unless H has some real eigenvalue.

Hint: Use the fact that any matrix representation of H can be brought to Jordan form. Prove that $e^{-iHt} = S\tilde{V}_t S^{-1}$, where S is an invertible operator and $\tilde{V}_t$ may be represented by a quasidiagonal matrix which contains blocks of the form

$$e^{-i\lambda t}\begin{pmatrix} 1 & -it & (-it)^2/2! & \dots & (-it)^{n-1}/(n-1)! \\ 0 & 1 & -it & \dots & (-it)^{n-2}/(n-2)! \\ \vdots & & & & \\ 0 & 0 & 0 & \dots & 1 \end{pmatrix}$$

where the numbers λ are eigenvalues of H.

22. Let an operator H on $\mathscr{H}$ be maximal dissipative and $V_t = e^{-iHt}$. The following conditions are sufficient for H to be completely dissipative:

    (i) $\text{s-lim}_{t\to\infty} V_t = 0$;
    (ii) $\bigcap_{t\geqslant 0} V_t\mathscr{H} = \{0\}$;
    (iii) to each non-zero $\psi \in \mathscr{H}$, there is $t > 0$ such that $\|V_t\psi\| < \|\psi\|$.

Hint: Cf. proof of Theorem 4.2.10.

23. Find an example of a completely dissipative operator H such that iH generates an isometric semigroup.

Hint: Cf. Problem 20.

24. Sum of two completely dissipative operators need not be completely dissipative.

Hint: Cf. Problem 20.

25. Let H be a closed dissipative operator whose resolvent set $\rho(H)$ contains some real number μ, then H is maximal dissipative.

Hint: Without loss of generality, one may set $\mu = 0$. Since H is closed, $\rho(H)$ is an open set in $\mathbb{C}$ so Theorem 4.2.1 can be used.

26. Consider Theorem 4.3.5 with the following additional hypothesis: there is a positive $\delta < \pi/2$ such that $0 \leqslant \arg[u(x) - \gamma] \leqslant -\delta$ holds a.e. in $\mathbb{R}^d$. Then the semiopen sector $\{z : z \neq 0, 0 < \arg(z - \gamma) \leqslant 3\pi/2 - \delta\}$ belongs to the resolvent set $\rho(\bar{H})$.

Hint: The only place in the proof, where the replacement of $-\mu$ by $\xi = -\mu + i\lambda$ leads to an additional restriction, concerns the inequality $\|(\bar{H}_c - \xi)\psi\| \geqslant |\xi|\,\|\psi\|$ whose validity requires $\mu V + \lambda W \geqslant 0$.

27. Let $\mathscr{H} = \Sigma_{l=0}^{\infty}{}^{\oplus}\mathscr{H}_l$ and $H = \Sigma_{l=0}^{\infty}{}^{\oplus}H_l$, where H_l is an operator on $\mathscr{H}_l$ and $D(H)$ consists of all $\psi = \{\psi_l\}_{l=0}^{\infty}$ with $\psi_l \in D(H_l)$. The operator H is (maximal, essentially maximal) dissipative iff all the H_l are (maximal, essentially maximal) dissipative.

28. Let $\mathscr{H} = \mathscr{H}_1 \otimes \mathscr{H}_2$ and assume an operator H_1 to be e.m.d. on $D(H_1) \subset \subset \mathscr{H}_1$. Then the operator $H_1 \otimes I_2$ with the domain $D(H_1) \odot \mathscr{H}_2$, which consists of all linear combination of the vectors $\varphi_1 \otimes \varphi_2$ with $\varphi_1 \in D(H_1)$ and $\varphi_2 \in \mathscr{H}_2$, is e.m.d.

Hint: The dissipativity verifies directly; further, $\text{Ran}(H_1 \otimes I_2 + i) = \text{Ran}(H + i) \odot \mathscr{H}_2$, so Proposition 4.2.7 may be applied.

29. Let $\mathscr{H}_1$, $\mathscr{H}_2$ be Hilbert spaces and $S : \mathscr{H}_1 \to \mathscr{H}_2$ a unitary operator. For an operator H_1 on $\mathscr{H}_1$, we define $H_2 = SH_1S^{-1}$. Show that H_2 is (maximal, essentially maximal) dissipative iff the same is true for H_1.

30. Consider the operator T_α from Example 4.3.8. Construct the one-parameter set of its self-adjoint extensions, and relate to them the operator $\tilde{T}_\alpha$ with $\alpha \in [-1/4, 3/4)$.

Hint: The construction is standard using the von Neumann's theory of symmetric extensions (see also Radin (1975)).

31. Consider again Example 4.3.8 but without the assumption $m = \hbar = 1$. Find the conditions under which the fall to the centre of attraction is possible.

Compare the result with the motion of a classical particle of mass m and angular momentum J under influence of the potential (4.3.21).

Hint: Cf. Nelson (1964).

32. Theorem 4.2.8 remains valid if one requires $H - iB$ to be dissipative instead of accretivity of B. Furthermore, B may be only closable, then $H - iB$ is again e.m.d. but $D(\bar{H}) \subset D(\bar{B})$ and $\overline{H - iB} = \bar{H} - i\bar{B}$.

33. The following symmetric version of Theorem 4.2.8 is valid. Let the operators H, G be dissipative with $D(H) \cap D(G)$ containing a dense linear manifold D such that $B \equiv i(G - H) \upharpoonright D$ is closable. Further, let some constants $a \in [0, 1)$ and $b \geqslant 0$ exist such that $\|B\varphi\| \leqslant a(\|H\varphi\| + \|G\varphi\|) + b\,\|\varphi\|$ holds for all $\varphi \in D$. Then $G \upharpoonright D$ is e.m.d. iff the same is true for $H \upharpoonright D$, and in that case the domains of their closures coincide.

Hint: Consider the operators $H(\alpha) = H - i\alpha B$. For $2a\beta/(1 - a) < 1$, use the preceding problem to show that if $H(\alpha) \upharpoonright D$ is e.m.d., then the same is true for $H(\alpha + \beta) \upharpoonright D$ and the domains of the closures coincide. Starting from $H(0) = H \upharpoonright D$ and applying this argument a finite number of times, we arrive at $H(1) = G \upharpoonright D$.

34. Theorem 4.3.3 remains valid when $d \geqslant 5$ and $p = \frac{1}{2}d$.

Hint: According to the Strichartz theorem – see Reed and Simon (1975, theorem X.21 and its corollary) – a potential $u \in L^{d/2}(\mathbb{R}^d)$ is H_0-bounded with arbitrarily small relative bound if $d \geqslant 5$.

35. Let H be a dissipative Schrödinger operator related to a potential $u \in L^p_{\mathrm{loc}}(\mathbb{R}^d)$, where $p = \max\{d/2, 2\}$ if $d \neq 4$, and $p > 2$ for $d = 4$. Further, let a finite limit $\gamma = \lim_{|x| \to \infty} u(x)$ exist, with the possible exception of a zero subset in $\mathbb{R}^d$. Then H is maximal dissipative and $\sigma_{\mathrm{ess}}(H) = \{z : z = \gamma + a, a \geqslant 0\}$.

Hint: Check that u is L^p-integrable on each compact $M \subset \mathbb{R}^d$. Maximal dissipativity then follows from Theorem 4.3.3 and Problem 34; further, one has to apply Theorem 4.3.15 to the operator $H - \gamma$.

36. Without reference to (A.14), show that the constant 2 on the r.h.s. of (4.4.17) is the best possible.

Hint: Choose H_0 to be the generator of translations in $L^2(\mathbb{R})$ and G as a suitable one-dimensional projection – cf. Exner and Úlehla (1983)).

37. Assume the hypotheses of Proposition 4.4.9 to be valid, then $\|S_t\varphi - S'_t\varphi\| \leqslant 2g\epsilon^{\alpha}(1 + t^{\alpha})\,|\varphi|_{\alpha}$.

Hint: Express $S_t\varphi$ and $S'_t\varphi$ with the help of (4.4.9) and use an estimate analogous to (4.4.37) – cf. Davies (1978).

38. Let C, H be closed operators on $\mathscr{H}$ such that C is relatively compact with respect to H, then C is H-bounded with the relative bound zero.

Hint: Suppose the statement is false, then there are $\epsilon > 0$ and $\{\varphi_n\} \subset D(H)$ such that $\|C\varphi_n\|^2 > \epsilon^2\|\varphi_n\|^2 + n^2\|\varphi_n\|^2$. The sequence $\{\psi_n\}$, $\psi_n = \varphi_n/\|C\varphi_n\|$, is $\|\,.\,\|_H$-bounded so there is a subsequence $\{\psi_{n_k}\}$ which converges weakly in $\langle D(H), \|\,.\,\|_H\rangle$ to some ψ. The sequence $\{C\psi_{n_k}\}$ then converges to $C\psi$ so $\|C\psi\| = 1$, but $\psi_n \to 0$ together with the closedness of C give $\|C\psi\| = 0$ – cf. also Weidmann (1980, theorem 9.7).

39. For a self-adjoint operator H, the set $\mathscr{M}(H)$ given by (4.5.1) is dense in $\mathscr{H}_{ac}(H)$ defined in the standard way.

Hint: First use unitarity of e^{-iHt} and the monotone-convergence theorem to show that the integration in (4.5.1) may be performed by the same right over $\mathbb{R}$. Then apply lemma 1 from Reed and Simon (1979, §XI.3). See also Davies (1980a, theorem 6.24).

40. Let H be a pseudo-Hamiltonian on a finite-dimensional $\mathscr{H}$. Show that the corresponding subspaces introduced in Section 4.5 fulfil $\mathscr{H}_{nre} = \mathscr{H}_{dec} = \mathscr{H}_{ac} = \mathscr{H}^{\perp}_{bound}$.

Hint: Use Corollary 4.2.11.

41. The *generalized wave operators* corresponding to a bounded operator J on $\mathscr{H}$ are defined by the relations

$$\Omega_-(H, H_0, J)\varphi = \lim_{t\to\infty} V_t J V^0_{-t}\varphi,$$

$$\Omega^*_+(H, H_0, J)\varphi = \lim_{t\to\infty} V^0_{-t} J V_t\varphi$$

(compare to (4.5.4)). Show that Theorem 4.5.4 and 4.5.6 generalize to this case, when $U = H - H_0$ is replaced by $HJ - JH_0$ and $JH - H_0J$, respectively.

42. Under the assumptions of Theorem 4.5.8, $S = \Omega^*_+\Omega_-$ exists.

Hint: One has to check $\operatorname{Ran}\Omega_- \subset \mathscr{H}^{\perp}_{bound}$. Since the spectrum of H_0 is absolutely continuous, $V^0_t \underset{w}{\to} 0$ as $t\to\infty$. For a bound state φ, the intertwining relation between V_t and V^0_t gives $|(\Omega^*_-\varphi, V^0_t\Omega^*_-\varphi)| = \|\Omega^*_-\varphi\|^2$ so $\mathscr{H}_{bound} \subset \operatorname{Ker}\Omega^*_-$ – cf. Simon (1979b).

43. Let $\mathscr{H}$ be a real Hilbert space. A non-trivial measure μ with the following properties:

(i) μ assumes finite values on all bounded sets $\subset \operatorname{Bor}(\mathscr{H})$,
(ii) μ is invariant under Euclidean transformations of $\mathscr{H}$,

exists iff $\dim\mathscr{H} < \infty$.

Hint: In an infinite-dimensional $\mathscr{H}$, the unit ball contains infinitely many disjoint balls of the radius 1/4.

44. For a real non-zero s, the conclusions of Example 5.2.10 can be derived using a one-dimensional $\mathscr{H} = \mathbb{R}$ only.

Hint: Define $\mu_n : \mu_n(M) = \int_{M_n} \exp((is/2)x^2)\,dx$, where $M_n = M \cap [-n, n]$. The corresponding f_n obviously fulfil $I^f_s(f_n) = 1$; use Lemma 5.2.6 to check (5.2.22) – cf. Exner and Kolerov (1981b).

45. Consider the self-adjoint operator $A : (A\varphi)(\tau) = -\varphi''(\tau)$ on $L^2(J^t; \mathbb{R}^d)$ defined on all $\varphi \in AC^2[J^t; \mathbb{R}^d]$ which fulfil the boundary conditions $\varphi(t) = \varphi'(0) = 0$. Show that A is the integral operator whose kernel G is given by (5.3.9).

46. Let $f \in \mathscr{F}^{uc}_s(\mathscr{H})$ be a cylindrical function on $\mathscr{H} = AC_0[J^t; \mathbb{R}^d]$ based in some $P(\tilde\sigma)\mathscr{H}$ corresponding to a fixed $\tilde\sigma \in \mathscr{P}(J^t)$. Then the integrals

(5.4.20a) fulfil $\lim_{\delta(\sigma)\to 0} I_s(f;\sigma) = I_s(f;\bar\sigma)$ for each $s \in \mathbb{C}_F^0$ provided the limit is restricted to those partitions σ which commute with $\bar\sigma$. In the same sense, the improper integrals (5.4.21a) fulfil $\lim_{\delta(\sigma)\to 0} I_s'(f;\sigma) = I_s'(f;\bar\sigma)$ for each $s \in \mathbb{C}_F$.

Hint: If $\delta(\sigma)$ is small enough, then $\sigma \supset \bar\sigma$.

47. Let $f \in \mathscr{F}(\mathscr{H})$ and $s = \rho e^{-i\pi\beta/2}$ with $\rho > 0$ and $0 \leqslant \beta \leqslant 1$, then the Fresnelian F_s-map fulfils $|I_s^f(f)| \leqslant \|f\|_0^{1-\beta}\,\|f\|_\infty^\beta$.

Hint: According to Propositions 5.2.3 and 5.4.9, the function $g : g(s) = |I_s^f(f)|\,\|f\|_0^{\beta-1}\|f\|^{-\beta}$ fulfils $|g(s)| = 1$ on the edges of the fourth quadrant so an easy modification of Hadamard's interpolation theorem (Reed and Simon, 1975, appendix to §IX.4) yields the result.

48. Prove the relation

$$w_c\{\gamma \in C_0[J^t;\mathbb{R}] : \|\gamma\|_\infty \leqslant \alpha\} = \frac{4}{\pi}\sum_{m=0}^{\infty} \frac{(-1)^m}{2m+1} \exp\left\{-\frac{\pi^2 ct}{8\alpha^2}(2m+1)^2\right\}.$$

Find the asymptotic behaviour of $\omega_1(\,.\,, ct)$ for large α.

Hint: According to Simon (1979a, lemma 7.10), the sought expression equals $\int_{-\alpha}^{\alpha} P_D(x, 0; ct)\,dx$, where $P_D(\,.\,,\,.\,;s)$ is the kernel of $\exp(-H_\alpha s)$ and H_α is the self-adjoint operator on $L^2([-\alpha,\alpha])$, $H_\alpha\varphi = -\frac{1}{2}\varphi''$, specified by the boundary conditions $\varphi(-\alpha) = \varphi(\alpha) = 0$.

49. Proposition 5.2.9 extends to the uniform (or regular) cylindrical F_s-maps. Consider a function $f \in \mathscr{F}_s^{uc}(\mathscr{H}_x)$ on $\mathscr{H}_x = AC_0[J^t;\mathbb{R}^d]$ and a regular isometric operator R on $\mathscr{H}_0$, then

$$\int_{\mathscr{H}_x}^{uc} f(x + R(\gamma - x))\,D\phi_s(\gamma) = \int_{\mathscr{H}_x}^{uc} f(\gamma)\,D\phi_s(\gamma),$$

and the analogous formula holds for I_s^{rc}.

50. Show that $\int_{\mathbb{R}} f(x)\,e^{ix^2/2s}\,dx$, understood as the oscillatory integral (cf. (5.6.1)), makes sense if f is an arbitrary polynomial.

Hint: Denote $I_k^\epsilon(\varphi) = \int_{\mathbb{R}} x^k e^{ix^2/2s}\varphi(\epsilon x)\,dx$, where $\varphi \in \mathscr{S}(\mathbb{R})$ with $\varphi(0) = 1$. Use (5.1.11) to show that $\lim_{\epsilon\to 0} I_0^\epsilon(\varphi) = \varphi(0)$. The integration by parts yields the recursive relations

$$I_k^\epsilon(\varphi) = is(k-1)I_{k-2}^\epsilon(\varphi) + is\epsilon I_{k-1}^\epsilon(\varphi'), \quad k \geqslant 2,$$

$$I_1^\epsilon(\varphi) = is\epsilon I_0^\epsilon(\varphi'),$$

which give $\lim_{\epsilon\to 0} I_{2k-1}^\epsilon(\varphi) = 0$ and $\lim_{\epsilon\to 0} I_{2k}^\epsilon(\varphi) = (2k-1)!!(is)^k$.

51. Consider $X = C_0[J^t;\mathbb{R}^d]$ and $\mathscr{H} = AC_0[J^t;\mathbb{R}^d]$ equipped with the natural Banach- and Hilbert-space norm, respectively. It holds $\|\gamma\|_X \leqslant t^{1/2}\|\gamma\|_{\mathscr{H}}$ for each $\gamma \in \mathscr{H}$, but the norms are not equivalent on $\mathscr{H}$.

52. Prove that the FCI-formula with α = rc holds for the damped harmonic oscillator described by Schrödinger pseudo-Hamiltonian with the potential (6.2.1).

Hint: One has to show that the corresponding F-integral yields the expression appearing on the r.h.s. of (6.2.5b) – cf. Exner and Kolerov (1981c).

53. Prove the relations (6.2.27).

Hint: Use Lemma 6.2.6 to derive the expression (6.2.26a).

54. Let A be an observable of a system whose time evolution is described by a pseudo-Hamiltonian H. For a given $\psi \in \mathscr{H}$, we denote $\psi_t = e^{-iHt}\psi$, and assume that $\psi_t \in D(H^*A - AH) \cap D(H^* - H)$ and the function $t \mapsto \|A\psi_t\|$ is bounded on $(0, b)$, then the relation

$$\frac{d}{dt}\langle A\rangle_{\psi_t} = \langle i(H^*A - AH)\rangle_{\psi_t} - \langle A\rangle_{\psi_t}\langle i(H^* - H)\rangle_{\psi_t}$$

holds in this interval.

Hint: Since $\langle A\rangle_{\psi_t} = (\psi_t, A\psi_t)\|\psi_t\|^{-2}$, one has only to check that the differentiation may be carried out.

55. Let H be a Schrödinger pseudo-Hamiltonian on $L^2(\mathbb{R}^d)$ referring to a mass m and some potential $u = v - iw$. Let further $u \in C^\infty(\mathbb{R}^d)$ such that $U\mathscr{S}(\mathbb{R}^d) \subset \mathscr{S}(\mathbb{R}^d)$, and denote by F_k^g the operator of multiplication by $f_k^g(x) = -(\partial g/\partial x_k)(x)$, $g = u, v, k = 1, \dots, d$. Suppose that H is e.m.d. on $\mathscr{S}(\mathbb{R}^d)$, the vectors $\psi_t = \exp(-i\bar{H}t)\psi$ for a given $\psi \in \mathscr{S}(\mathbb{R}^d)$ and $t \in (0, b)$ belong to $\mathscr{S}(\mathbb{R}^d)$, and the functions $t \mapsto \|Q_k\psi_t\|$, $t \mapsto \|P_k\psi_t\|$ are bounded in $(0, b)$. Use the preceding problem to derive the relations

$$\frac{d}{dt}\langle Q_k\rangle_{\psi_t} = \frac{1}{m}\langle P_k\rangle_{\psi_t} - \frac{2}{\hbar}\langle WQ_k\rangle_{\psi_t} + \frac{2}{\hbar}\langle W\rangle_{\psi_t}\langle Q_k\rangle_{\psi_t},$$

$$\frac{d}{dt}\langle P_k\rangle_{\psi_t} = \langle F_k^v\rangle_{\psi_t} - \frac{1}{\hbar}\langle WP_k + P_kW\rangle_{\psi_t} + \frac{2}{\hbar}\langle W\rangle_{\psi_t}\langle P_k\rangle_{\psi_t}$$

$$= \langle F_k^u\rangle_{\psi_t} - \frac{2}{\hbar}\langle WP_k\rangle_{\psi_t} + \frac{2}{\hbar}\langle W\rangle_{\psi_t}\langle P_k\rangle_{\psi_t}.$$

Find some conditions under which the $\hbar$-dependent terms are negligible so a sort of Ehrenfest theorem holds.

56. Let the operators H, B on $\mathscr{H}$ be self-adjoint and such that $H - iB$ is maximal dissipative. Let $\psi_t = e^{-i(H-iB)t}\psi$ for a given $\psi \in \mathscr{H}$, then $\varphi_t := \psi_t/\|\psi_t\|$ obeys the Gisin equation

$$i\dot{\varphi}_t = H\varphi_t + i(\langle B\rangle_{\varphi_t} - B)\varphi_t.$$

Appendix

(A.1) *The uniform boundedness principle for convex functionals* was first formulated by Gel'fand (1936); it can be found, e.g. in Akhiezer and Glazman (1966, §II.21) or Havlíček *et al.* (1975, theorem 2.10). Even more general formulation in which the assumption of convexity is slightly weakened is due to Kato (1966, theorem III.1.29): *Let* $\{p_\alpha\}$ *be a family of positive continuous functionals on a Banach space* X, *which obey* $p_\alpha(x+y) \leqslant p_\alpha(x) + p_\alpha(y)$ *and* $p_\alpha(-x) = p_\alpha(x)$ *for all* α *and each* $x, y \in X$. *If the set* $\{p_\alpha(x)\} \subset \mathbb{R}_+$ *is bounded for any fixed* $x \in X$, *then the family* $\{p_\alpha\}$ *is uniformly bounded in the unit ball* $\|x\| \leqslant 1$.

(A.2) *Equivalence of strong and weak continuity of a one-parameter semigroup* is easily verified for contractive semigroups on a Hilbert space; however, neither the contractivity nor the Hilbert-space structure of the underlying space is necessary (Yosida, 1966, theorem IX.1.1): *Let* $\{T_t : t \geqslant 0\}$ *be a semigroup of bounded operators on a Banach space* X, *i.e.* $T_0 = I$ *and* $T_t T_s = T_{t+s}$ *for all* $t, s \geqslant 0$, *then* $\text{s-lim}_{t \to s} T_t = T_s$ *holds for each* $s \geqslant 0$ *iff* $\text{w-lim}_{t \to 0+} T_t = I$.

(A.3) *The Fourier–Stieltjes inversion formulae.* Let $f(t) = \int_{\mathbb{R}} e^{-i\lambda t}\, d\omega(\lambda)$, where ω is a Lebesgue–Stieltjes measure (generated by a right-continuous function of bounded variation denoted again by ω). The following inversion formulae hold (Bochner, 1959, theorems 17 and 24)):

$$\tfrac{1}{2}[\omega(\lambda+0) + \omega(\lambda-0)] = \text{const} + \frac{1}{2\pi} \lim_{C \to \infty} \int_{-C}^{C} f(t) \frac{e^{i\lambda t} - 1}{it}\, dt,$$

$$\omega(\lambda+0) - \omega(\lambda-0) = \lim_{T \to \infty} \frac{1}{2T} \int_{-T}^{T} f(t)\, e^{i\lambda t}\, dt.$$

The cut-off function $g_C = \chi_{[-C,C]}$ is the first relation can be eventually replaced, e.g., by $g_\eta : g_\eta(t) = e^{-\eta|t|}$, which is then removed by $\eta \to 0+$. If ω is real, we have $\overline{f(t)} = f(-t)$ and an easy rearrangement gives (Akhiezer and Glazman, 1966, §VI.69):

$$\tfrac{1}{2}[\omega(\lambda+0) + \omega(\lambda-0)] = \text{const} + \frac{1}{\pi} \lim_{\eta \to 0+} \int_0^{\lambda} \operatorname{Re} \varphi(\xi + i\eta)\, d\xi,$$

$$\varphi(\xi + i\eta) = \int_0^{\infty} f(t)\, e^{it(\xi + i\eta)}\, dt, \qquad \eta > 0.$$

(A.4) *The image-measure theorem* is the common name for many different assertions dealing with substitution or change of variables in an integral. As a sufficiently general representative, we choose the following formulation due to Dunford and Schwartz (1958, lemma III.10.8): *Consider sets* S_1, S_2 *and a mapping* $\varphi : S_1 \to S_2$. *If* $\mathscr{A}_2$ *is a* (σ-)*algebra of subsets in* S_2, *then* $\mathscr{A}_1 = \{\varphi^{(-1)}(M) : M \in \mathscr{A}_2\}$ is a (σ-)*algebra in* S_1. *If* $\mu_2 : \mathscr{A}_2 \to \mathbb{C}$ *is an additive set function, then* $\mu_1 : \mu_1(\varphi^{(-1)}(M)) = \mu_2(M)$ *is an additive set function on* $\mathscr{A}_1$. *Moreover,*

(a) *if* μ_2 *is* σ*-additive so is* μ_1;
(b) *if* μ_2 *is finite so is* μ_1;
(c) *the total variations obey* $|\mu_1|(\varphi^{(-1)}(M)) = |\mu_2|(M)$ *for each* $M \in \mathscr{A}_2$;
(d) *if* $f : S_2 \to \mathbb{C}$ *is* μ_2*-measurable, then* $f(\varphi(\,.\,))$ *is* μ_1*-measurable*;
(e) *if* μ_2 *is non-negative and* σ*-additive and* $f : S_2 \to \mathbb{R}_+$ *is* μ_2*-measurable, then* $\int_M f(x_2)\,\mathrm{d}\mu_2(x_2) = \int_{\varphi^{(-1)}(M)} f(\varphi(x_1))\,\mathrm{d}\mu_1(x_1)$;
(f) *if* $f \in L(S_2, \mu_2)$, *then* $f(\varphi(\,.\,))$ *belongs to* $L(S_1, \mu_1)$ *and the above equality between the integrals holds; this assertion as well as* (d) *are also valid for vector-valued functions.*

(A.5) *Radon–Nikodým and related theorems.* It is not difficult to verify that a measure ν constructed from another measure μ on $(S, \mathscr{A})$ and some function $f \in L(S, \mu)$ by $\nu(M) = \int_M f(x)\,\mathrm{d}\mu(x)$ is absolutely continuous with respect to μ. A partial converse to this assertion is known as Radon–Nikodým theorem (see, e.g. Dunford and Schwartz (1958, theorem III.10.2)): *Let* μ *be a positive* σ*-finite measure on* $(S, \mathscr{A})$. *If a finite measure* ν *on* $(S, \mathscr{A})$ *is absolutely continuous with respect to* μ, *then there is a unique function* $f \in L(S, \mu)$ *such that* $\nu(M) = \int_M f(x)\,\mathrm{d}\mu(x)$ *for each* $M \in \mathscr{A}$. Moreover, the total variation of ν equals the L-norm of f. The function f is usually called the Radon–Nikodým derivative of ν with respect to μ. Further, μ is allowed to be complex, if only it is finite (*ibid.*, theorem III.10.7). Integrals corresponding to the measures ν, μ are related by the following assertion (*ibid.*, cor. III.10.6): *Let* ν, μ, f *be as above, then a function* g *on* S, *which assumes values from some Banach space* X, *belongs to* $L(S, \nu; X)$ *iff* $gf \in L(S, \mu; X)$, *and in this case* $\int_M g(x)\,\mathrm{d}\nu(x) = \int_M g(x)f(x)\,\mathrm{d}\mu(x)$ *holds for each* $M \in \mathscr{A}$.

(A.6) *The Fubini theorem* also holds for complex measures (Dunford and Schwartz, 1958, theorem III.11.13): *if* $f \in L(S \times T, \mu \otimes \nu)$, *then*

$$\int_{S\times T} f(x,y)\,\mathrm{d}(\mu\otimes\nu)(x,y) = \int_S \mathrm{d}\mu(x) \int_T \mathrm{d}\nu(y)\, f(x,y)$$

$$= \int_T \mathrm{d}\nu(y) \int_S \mathrm{d}\mu(x)\, f(x,y);$$

the only difference from the positive-measure case is that μ, ν *are allowed to be finite, not* σ*-finite.*

(A.7) *Product formulae* is the common name for various generalizations to the classical Lie formula $\lim_{n\to\infty} (\exp(tA/n)\exp(tB/n))^n = \exp(t(A + B))$, where

A, B are arbitrary square matrices. The central result is that of Trotter (1959): *Let A, B and $\overline{A+B}$ be generators of continuous contractive semigroups on a Banach space X, then*

$$\operatorname*{s-lim}_{n\to\infty} (\exp(-tA/n)\exp(-tB/n))^n = \exp(-t(\overline{A+B}))$$

holds for all $t \geqslant 0$. In particular, if A, B and $\overline{A+B}$ are self-adjoint operators on a Hilbert space $\mathscr{H}$, then the corresponding unitary groups are related in the following way:

$$\operatorname*{s-lim}_{n\to\infty} (\exp(itA/n)\exp(itB/n))^n = \exp(it(\overline{A+B})).$$

The proof is complicated but simplifies considerably in the particular case when $A + B$ itself is the generator – cf. Nelson (1964) or Reed and Simon (1972, theorem VIII.30; 1975, §X.8).

(A.8) *Further generalizations of the Trotter formula* (among many others, some of which are mentioned below and in the main text) belong to Chernoff (1968, 1970, 1974). We have already used one of his results (1974, theorem 2.5.3) in a substantially simplified form in our Lemma 1.3.3 (see also notes to Section 1.3). Here we present two others. Assume that a strongly continuous (linear-)contraction-valued function $F : \mathbb{R}_+ \to \mathscr{B}(X)$ on a Banach space X is given, which obeys $F(0) = I$. The first assertion to be mentioned (1970, theorem 2.1; 1974, theorem 3.1) is: *Let the strong derivative $F'(0)$ be densely defined. If for all $t \geqslant 0$ the limit $G(t) := \text{s-lim}_{n\to\infty} F(t/n)^n$ exists, then $\{G(t) : t \geqslant 0\}$ is a continuous contractive semigroup, $G(t) = e^{-tA}$, and its generator fulfils $A \supset -F'(0)$. Moreover, the convergence of $F(t/n)^n$ to $G(t)$ is uniform with respect to t in each compact interval $\subset \mathbb{R}_+$.* Notice further that if $\{G(t) : t \geqslant 0\}$ is known *a priori* to form a semigroup and the convergence is uniform on compact intervals of t, then $A \supset -F'(0)$ whether the latter is densely defined or not (1974, theorem 3.7). On the other hand, imposing a natural assumption on $F'(0)$, one can ensure existence of the limit (1968, theorem): *Assume that the closure $C = -\overline{F'(0)}$ is the generator of a continuous contractive semigroup, then $\text{s-lim}_{n\to\infty} F(t/n)^n = e^{-tC}$ holds for all $t \geqslant 0$* (with the particular choice $F(t) = e^{-tA}e^{-tB}$, this result gives the Trotter formula).

(A.9) Particular interest concerns the *product formulae* in the case of *self-adjoint contractive semigroups on a Hilbert space $\mathscr{H}$*. It is useful to consider the generators which are in general not densely defined, i.e. corresponding to semigroups which act on some subspaces of $\mathscr{H}$ only (and are assumed *ex definitio* to be zero on the orthogonal complements). The following general result was proved by Kato (1978) (see also Simon (1979a)): *Let A, B be positive self-adjoint operators and denote by E, F the projections on $\overline{D(A)}$, $\overline{D(B)} \subset \mathscr{H}$, respectively. Let further P be the projection referring to $\overline{D}$, where $D = D(A^{1/2}) \cap D(B^{1/2})$, and C the 'form sum' of A and B (i.e. the self-adjoint operator on $P\mathscr{H}$ associated with the closed* *densely defined quadratic form q on D* : $q(\varphi) = \|A^{1/2}\varphi\|^2 + \|B^{1/2}\varphi\|^2$). *Then the relation*

$$\operatorname*{s-lim}_{n\to\infty} [e^{-tA/n} E\, e^{-tB/n} F]^n = e^{-tC} P$$

holds for all $t \geqslant 0$. This assertion generalizes further to the case of non-linear operators and to arbitrary finite number of semigroups (Kato and Masuda, 1978). If the operators involved are densely defined and $A + B$ is e.s.a., then we arrive back at the Trotter formula. On the other hand, D may even contain the zero vector only. An explicit example, where A is a positive Schrödinger operator, B is the multiplication operator by a (positive, but highly pathological) function, and the limit in the last formula is zero, was given by Chernoff (1974, prop. 7.9). For other particular cases of the considered formula, see Faris (1967a), Friedman (1972), Chernoff (1974, §8). Notice that the familiar expression for the projection corresponding to the intersection of ranges of two (non-commuting) projections also turns out to be a consequence of the above result.

(A.10) *Singular values of a compact operator C* are defined as eigenvalues of $|C| = (C^*C)^{1/2}$ ordered decreasingly with repetition according to multiplicity. Among their properties (cf. Dunford and Schwartz (1963, §IX.9)), we shall mention the relations

$$\mu_j(BC) \leqslant \|B\|\mu_j(C), \qquad \mu_j(CB) \leqslant \|B\|\mu_j(C)$$

valid for any $B \in \mathscr{B}(\mathscr{H})$ and all $j = 1, 2, \ldots$. It is further known that for the trace-class operators, $C \in \mathscr{J}_1$, the singular values may be used to the estimate $|\det(I + \alpha C)| \leqslant \Pi_j (1 + |\alpha|\mu_j(C))$. This assertion has the following non-trivial generalization (Krein, 1964): *For* $C \in \mathscr{J}_1$ *and* $|\alpha| < \|C\|^{-1}$, *we have*

$$\|(I + \alpha C)^{-1}\| \leqslant \frac{\Pi_j(1 + |\alpha|\mu_j(C))}{|\det(I + \alpha C)|}.$$

(A.11) The so-called *preparation theorem* (Vorbereitungsatz) of Weierstrass generalizes the well-known property of analytic functions of one complex variable, namely that they behave around each zero point z_0 as some (entire) power of $z - z_0$. It may be formulated for analytic functions of n complex variables (cf., e.g. Shabat (1976, §II.8)), but we shall need the case $n = 2$ only: *Let f be an analytic function in some neighbourhood* $Z \times W$ *of the point* $(z_0, 0) \in \mathbb{C}^2$. *Assume that it is not identically zero and* $f(z, 0) = (z - z_0)^m$. *Then there is a neighbourhood* $Z' \times W' \subset Z \times W$ *of* $(z_0, 0)$ *and unique analytic functions* $p_1, \ldots, p_m$, q *on* W' *such that*

$$f(z, w) = [(z - z_0)^m + p_1(w)(z - z_0)^{m-1} + \ldots + p_m(w)]\, q(w)$$

for all $(z, w) \in Z' \times W'$, *and* $p_1(0) = \cdots = p_m(0) = 0$ *and* $q(w)$ *is non-zero within* W'. Roots of the polynomial may be grouped into so-called *Puiseux cycles*, $\{z_{j1}(w), \ldots, z_{jp_j}(w)\}$, $j = 1, \ldots, s$, and $\Sigma_{j=1}^{s} \mu_j p_j = m$, where μ_j denotes multiplicity of the jth cycle. Members of each cycle are branches of one analytic function of w : they undergo permutations between themselves when g is moved around the origin to its original position. It is not difficult to see that roots from the jth cycle are expressed within W' by the *Puiseux series* (e.g. Kato (1966, §2.1.2):

$$z_{jk}(w) = z_0 + \sum_{n=1}^{\infty} \beta_{jn} \exp(2\pi ikn/p_j)\, w^{n/p_j}, \quad k = 1, \ldots, p_j.$$

(A.12) *Solution to the evolution equation* (4.1.10), or more generally, to the equation

$$\frac{\mathrm{d}}{\mathrm{d}t}\psi_t = -A(t)\psi_t \tag{$*$}$$

on a Banach space X, may be constructed in the following way. We divide the considered interval $J = (r, v)$ into n subintervals of lengths $\delta = (v - r)/n$; then we set

$$V_n(t, s) = \exp(-A(j\delta/n)(t - s)) \quad \text{for } j\delta/n \leqslant s \leqslant t \leqslant (j+1)\delta/n,\ j = 0, 1, \ldots, n-1,$$

$$V_n(u, s) = V_n(u, t)V_n(t, s) \quad \text{for } r \leqslant s \leqslant t \leqslant u \leqslant v$$

and define the operators

$$V(t, s) = \operatorname*{s\text{-}lim}_{n\to\infty} V_n(t, s), \tag{$**$}$$

which are intended to play the role of propagator for equation (∗). The idea has its roots in the classical method elaborated for ordinary differential equations by Cauchy; its first successful application to the case of unbounded operators $A(t)$ is due to Kato (1953). For a more complete exposition we refer to Yosida (1966, §XIV.4) or Reed and Simon (1975, §X.12), where the following theorem may be found: *Let X be a Banach space and J an open interval in* $\mathbb{R}$. *For each $t \in J$, let $A(t)$ be a generator of a continuous contractive semigroup on X such that* 0 *belongs to the resolvent set $\rho(A(t))$ and*

(a) *all the operators $A(t)$ have a common domain D,*
(b) *let $C(t, s) \equiv A(t)A(s)^{-1} - I$, then the function $(s, t) \mapsto (t - s)^{-1}C(t, s)$ is uniformly continuous in the strong operator topology and uniformly bounded in s and t, if s and t are non-equal and belong to an arbitrary fixed compact subinterval of J,*
(c) *the limit $C(t) \equiv \text{s-lim}_{s\to t}(t - s)^{-1}C(t, s)$ exists uniformly with respect to t in any compact subinterval; further $C(\,.\,)$ is strongly continuous and bounded.*

Then for all $s \leqslant t$ in any compact subinterval of J, the limit (∗∗) *exists uniformly with respect to s, t. Furthermore, if $\varphi \in D$, then $\psi_t := V(t, s)\varphi$ belongs to D for all $t \geqslant s$, fulfils $\|\psi_t\| \leqslant \|\varphi\|$, and solves the equation* (∗) *with the initial datum $\psi_s = \varphi$.*

(A.13) For *perturbations of continuous one-parameter semigroups* we have the following useful result (Davies, 1976, theorem 1.9.1): *Let A be the generator of a strongly continuous one-parameter semigroup $\{V_t\}$ on a Banach space X. If B is a bounded operator on X, then $A + B$ generates a strongly continuous one-parameter semigroup $\{\tilde{V}_t\}$ related to the original one by the relation*

$$\tilde{V}_t\psi = V_t\psi - \int_0^t \tilde{V}_{t-s}BV_s\psi\, ds,$$

valid for all $\psi \in X$ and $t \geqslant 0$.

(A.14) *The smooth-perturbation technique* was introduced by Kato (1966b). We shall quote here few basic results referring to the original paper for a detailed exposition; see also Reed and Simon (1978, §XIII.7): *Let H, C be closed densely defined operators on $\mathscr{H}$ with $\sigma(H) \subset \mathbb{R}$ and $D(C) \supset D(H)$. The operator C is called 'H-smooth' if there is a positive M such that*

$$\int_{\mathbb{R}} (\|CR(\lambda + i\epsilon, H)\varphi\|^2 + \|CR(\lambda - i\epsilon, H)\varphi\|^2)\, d\lambda \leqslant 4\pi^2 M^2\, \|\varphi\|^2$$

holds for all $\varphi \in \mathscr{H}$ and $\epsilon > 0$. The smallest of the constants M for which it is true is called $\|C\|_H$. If $\{e^{-iHt} : t \geqslant 0\}$ is a continuous contractive semigroup, then each H-smooth operator C fulfils the inequality

$$\int_{\mathbb{R}} \|C\, e^{-iHt}\varphi\|^2\, dt \leqslant 2\pi\, \|C\|_H^2\, \|\varphi\|^2 \qquad (*)$$

for all $\varphi \in \mathscr{H}$, where $\|C\|_H$ cannot be replaced by any smaller number (cf. lemma 3.6 of Kato's paper). In addition, if H is self-adjoint, we have many equivalent expressions of the norm $\|C\|_H$ (*ibid.*, theorem 5.1). For this purpose, assume that J varies over all finite intervals $(\alpha, \beta]$ with $|J| = \beta - \alpha$; further, φ, ψ goes over the set of unit vectors in $\mathscr{H}$ and $D(C^*)$, respectively, and $z \in \mathbb{C}\backslash\mathbb{R}$. Then $\|C\|_H^2$ is equal to the following numbers:

$$a_1 = \sup_{J,\varphi} \|CE_H(J)\varphi\|^2\, |J|^{-1},$$

$$a_2 = \sup_{J,\psi} \|E_H(J)C^*\psi\|^2\, |J|^{-1},$$

$$a_3 = \frac{1}{2\pi} \sup_{z,\psi} |([R(z, H) - R(\bar{z}, H)]\, C^*\psi, C^*\psi)|,$$

$$a_4 = \frac{1}{\pi} \sup_{z,\psi} |\mathrm{Im}\, z|\, \|R(z, H)C^*\psi\|^2,$$

$$a_5 = \frac{1}{\pi} \sup_{z,\varphi} |\mathrm{Im}\, z|\, \|CR(z, H)\varphi\|^2,$$

$$a_6 = \frac{1}{\pi} \sup_{z,\varphi} |\mathrm{Im}\, z|\, (\|CR(z, H)\varphi\|^2 + \|CR(\bar{z}, H)\varphi\|^2).$$

Moreover, C is H-smooth iff the common value of $a_j, j = 1, \ldots, 6$, is finite. Notice that there is one more equivalent expression related to (∗), namely

$$a_7 = \frac{1}{2\pi} \sup_{\varphi} \int_{\mathbb{R}} \|C\, e^{-iHt} \varphi\|^2 \, dt.$$

(A.15) *Let* $\{V_t : t \geqslant 0\}$ be a continuous semigroup of *isometries*, which is *completely non-unitary*. The existence of its minimal unitary dilation is established by Theorems 1.4.1 and 1.4.7. In this particular case, however, construction of the dilation is simplified by the following assertion (Sz.-Nagy and Foias, 1970, theorem III.9.3), Davies (1980a, theorem 6.18)): *The semigroup* $\{V_t\}$ *is isomorphic to a unilateral shift, i.e. a semigroup* $\{\tilde{V}_t : t \geqslant 0\}$ *acting on* $\tilde{\mathscr{H}} = L^2(\mathbb{R}_+; \mathscr{G})$, *where* $\mathscr{G}$ *is some auxiliary Hilbert space, as* $(\tilde{V}_t\varphi)(x) = \varphi(x - t)\Theta(x - t)$.

(A.16) The complete non-unitarity may be used as an alternative *sufficient condition for absolute continuity* (Davies, 1980a, theorem 6.21): *Let a continuous contractive semigroup* $\{V_t : t \geqslant 0\}$ *be completely non-unitary, then the spectrum of* H, *the generator of its minimal unitary dilation, is absolutely continuous (and equal to* $\mathbb{R}$ *– cf. Theorem* 1.5.1).

(A.17) *The Paley–Wiener theorem.* Various PW-theorems are known (cf., e.g., Yosida (1966, §VI.4) or Rudin (1973, §7.2)). We are particularly interested in the following result of Paley and Wiener (1934, theorem XII): *Let* φ *be a non-zero non-negative function from* $L^2(\mathbb{R})$. *The integral* $\int_{\mathbb{R}} |\ln \varphi(x)| (1 + x^2)^{-1} dx$ *converges iff there is a function* $g \in L^2(\mathbb{R})$ *with a semibounded support (i.e.* $g(y) = 0$ *for almost all* y *less than some* y_0) *such that* $\varphi(x) = |f(x)|$ *a.e. in* $\mathbb{R}$, *where* $f = Fg$ *and* F *is the Fourier–Plancherel operator.*

(A.18) *The Dirichlet–Jordan theorem* is the standard result concerning Fourier series (cf., e.g., Jarník (1955, theorem 185)): *Let* J *be a finite interval in* $\mathbb{R}$ *and assume that a function* $f \in L(J; \mathbb{R})$ *has a bounded variation in some* $[a, b] \subset J$. *Then*

(a) *for each* $x \in [a, b]$, *the Fourier series of* f *(relative to* J*) converges to* $\frac{1}{2}[f(x + 0) + f(x - 0)]$;

(b) *if* f *is continuous, the convergence is uniform in each closed subinterval of* (a, b).

Bibliography

Abbot, L. F. and Wise, M. B.: 1981, 'Dimension of a Quantum-Mechanical Path', *Amer. J. Phys.* **49**, 37–39.

Abramowitz, M. and Stegun, I. (eds): 1964, *Handbook of Mathematical Functions*, National Bureau of Standards.

Aguilar, J. and Combes, J. M.: 1971, 'A Class of Analytic Perturbations for One-Body Schrödinger Hamiltonians', *Commun. Math. Phys.* **22**, 269–279.

Aharonov, Y. and Vardi, M: 1980, 'Meaning of an Individual "Feynman Path"', *Phys. Rev.* **D21**, 2235–2240.

Akhiezer, N. I. and Glazman, I. M.: 1966, *The Theory of Linear Operators in Hilbert Space* (2nd edn), Nauka, Moscow (in Russian; English trans.: F. Ungar, New York. 1961, 1963).

Albeverio, S., Blanchard Ph. and Høegh-Krohn, R.: 1982, 'Feynman Path Integrals and the Trace Formula for the Schrödinger Operators', *Commun. Math. Phys.* **83**, 49–76.

Albeverio, S. A. and Høegh-Krohn, R. J.: 1976, *Mathematical Theory of Feynman Path Integrals*, Lecture Notes in Mathematics, Vol. 523, Springer, Berlin.

Albeverio, S. A. and Høegh-Krohn, R. J.: 1977, 'Oscillatory Integrals and Method of Stationary Phase, with Applications to the Classical Limit of Quantum Mechanics', *Invent. Math.* **40**, 59–106.

Albeverio, S. A. and Høegh-Krohn, R. J.: 1979, 'Feynman Path Integrals and the Corresponding Method of Stationary Phase', in S. Albeverio *et al.* (eds.), *Feynman Path Integrals*, Lecture Notes in Physics, Vol. 106, Springer, Berlin, pp. 3–57.

Albeverio S. A. and Høegh-Krohn, R. J.: 1982, 'Perturbation of Resonances in Quantum Mechanics', Preprint, Universität Bochum; *J. Math. Anal. Appl.* (to appear).

Alda, V., Kundrát, V. and Lokajíček, M. V.: 1974, 'Exponential Decay Law and Irreversibility of Decay and Collision Processes', *Aplikace mat.* **19**, 307–315.

Alguard, M. J. and Drake, C. W.: 1973, 'Stark Beats in Lyman-Series Emission', *Phys. Rev.* **A8**, 27–36.

Ali, S. T., Fonda, L. and Ghirardi, G. C.: 1975, 'Pertinence of the Semigroup Law in the Theory of the Decay of an Unstable Elementary Particle', *Il Nuovo Cimento* **25A**, 134–148.

Ali, S. T. and Ghirardi, G. C.: 1974, 'Unstable Systems and Measurement Processes', *Il Nuovo Cimento* **24A**, 220–238.

Alicki, R.: 1978, 'The Theory of Open Systems in Application to Unstable Particles', *Rep. Math. Phys.* **14**, 27–42.

Alicki, R.: 1981, 'On the Scattering Theory for Quantum Dynamical Semigroups', *Ann. Inst. H. Poincaré* **A35**, 97–103.

Alicki, R. and Messer, J.: 1983, 'Nonlinear Quantum Dynamical Semigroups for Many-Body Open Systems', *J. Stat. Phys.* **32**, 299–312.

Allcock, G. R.: 1969, 'The Time of Arrival in Quantum Mechanics. I, Formal Considerations;

II, The Individual Measurement; III, The Measurement Ensemble', *Ann. Phys.* **53**, 253–285, 286–310, 311–348.

Alzetta, R. and d'Ambrogio, E.: 1966, 'Evolution of a Resonant State', *Nucl. Phys.* **82**, 683–9.

Amrein, W. O.: 1981, *Non-relativistic Quantum Dynamics*, Mathematical Physics Studies, Vol. 2, D. Reidel, Dordrecht.

Amrein, W. O., Jauch, J. M. and Sinha, K. B.: 1977, *Scattering Theory in Quantum Mechanics*, Lecture Notes and Supplements in Physics, Vol. 16, W. A. Benjamin, Reading.

Andrä, H. J.: 1974, 'Fine Structure, Hyperfine Structure and Lamb Shift Measurements by the Beam-Foil Technique', *Physica Scripta* **9**, 257–280.

Arai, A.: 1981, 'On a Model of a Harmonic Oscillator Coupled to a Quantized, Massless, Scalar Field I, II', *J. Math. Phys.* **22**, 2539–2548, 2549–2552.

Araki, H., Munakata, Y., Kawaguchi, M. and Gotō, T.: 1957, 'Quantum Field Theory of Unstable Particles', *Progr. Theor. Phys.* **17**, 419–442.

Ashbaugh, M. S. and Harrell, E., II: 1982, 'Perturbation Theory for Shape Resonances and Large Barrier Potentials', *Commun. Math. Phys.* **83**, 151–170.

Ashbaugh, M. S. and Sundberg, C.: 1984, 'An Improved Stability Result for Resonances', *Trans. Amer. Math. Soc.* **281**, 347–360.

Avron, J. E.: 1982, 'The Lifetime of Wannier Ladder States', *Ann. Phys.* **143**, 33–53.

Babbitt, D. G.: 1963, 'A Summation Procedure for Certain Feynman Integrals', *J. Math. Phys.* **4**, 36–41.

Babbitt, D. G.: 1965, 'The Wiener Integral and the Schrödinger Equation', *Trans. Amer. Math. Soc.* **116**, 66–78; correction **121** (1966), 549–552.

Babbitt, D. and Balslev, E.: 1976, 'Local Distortion Techniques and Unitarity of the S-Matrix for the 2-Body Problem', *J. Math. Anal. Appl.* **54**, 316–349.

Bailey, T. K. and Schieve, W. C.: 1978, 'Complex Energy Eigenstates in Quantum Decay Models', *Il Nuovo Cimento* **47A**, 231–250.

Balslev, E.: 1978, 'Analytic Scattering Theory of Two-Body Schrödinger Operators', *J. Funct. Anal.* **29**, 375–396.

Barchielli, A., Lanz, L. and Prosperi, G. M.: 1982, 'A Model for Macroscopic Description and Continuous Observations in Quantum Mechanics', *Il Nuovo Cimento* **72B**, 79–121.

Barut, A. O. and Rączka, R.: 1977, *Theory of Group Representations and Applications*, PWN, Warsaw.

Baumgärtel, H.: 1974, "Kompakte Störungen nichtisolierter Eigenwerte endlicher Vielfachheit', *Math. Nachr.* **59**, 265–273.

Baumgärtel, H.: 1975, 'Partial Resolvent and Spectral Concentration', *Math. Nachr.* **69**, 107–121.

Baumgärtel, H. and Demuth, M.: 1976, 'Perturbation of Unstable Eigenvalues of Finite Multiplicity', *J. Funct. Anal.* **22**, 187–203.

Baumgärtel, H., Demuth, M. and Wollenberg, M.: 1978, 'On the Equality of Resonances (Poles of the Scattering Amplitude) and Virtual Poles', *Math. Nachr.* **86**, 167–174.

Baz', A. I., Zel'dovich, Ya. B. and Perelomov, A. M.: 1966, *Scattering, Reactions and Decays in Nonrelativistic Quantum Mechanics*, Nauka, Moscow (in Russian; English version: Israel Program for Scientific Translations, Jerusalem 1966).

Beekman, J. A.: 1965, 'Gaussian Processes and Generalized Schrödinger Equation', *J. Math. Mech.* **14**, 789–806.

Bell, J. S. and Goebel, C. J.: 1965, 'Double Poles and Nonexponential Decays', *Phys. Rev.* **138B**, 1198–1201.

Beltrametti, E. G. and Luzzatto, G.: 1965, 'Representations of the Poincaré Group Associated to Complex Energy-Momentum', *Il Nuovo Cimento* **36**, 1217–1229.

Bencze, G. and Chandler, C.: 1982, 'The Energy Independent Optical Potential and its Physical Interpretation', Preprint, Albuquerque.

Berezin, F. A.: 1971, 'Non-Wiener Functional Integrals', *Teor. mat. fiz.* **6**, 194–212 (in Russian).

Berezin, F. A.: 1980, 'Functional Integral over Paths in Phase Space', *Usp. Fiz. Nauk* **132**, 497–548 (in Russian).

Berg, H. P. and Tarski, J.: 1981, 'Fourier Transforms of Distributions and Associated Feynman Integrals', *J. Phys.* **A14**, 2207–2213.

Bernido, Ch. C. and Inomata, A.: 1981, 'Path Integrals with a Periodic Constraint: the Aharonov–Bohm Effect', *J. Math. Phys.* **22**, 715–718.

Berry, M. V. and Mount, K. E.: 1972, 'Semiclassical Approximations in Wave Mechanics', *Rep. Progr. Phys.* **35**, 315–398.

Beskow, A. and Nilsson, J.: 1967, 'The Concept of Wave Function and the Irreducible Representations of the Poincaré Group, II. Unstable Systems and the Exponential Decay Law', *Arkiv Fys.* **34**, 561–569.

Bethe, H. A.: 1940, 'A Continuum Theory of the Compound Nucleus', *Phys. Rev.* 57, 1125–1144.

Blank, J. and Exner, P.: 1976, 1977, 'Remarks on Tensor Products and their Applications in Quantum Theory; I, General Considerations; II, Spectral Properties', *Acta Univ. Carolinae, Math. Phys.* **17**, 75–98; **18**, 3–35.

Blank, J. and Exner, P.: 1978, 1980, *Selected Topics of Mathematical Physics. Linear Operators on Hilbert Space*, Vols. II, III, SPN, Prague (in Czech).

Blank, J., Exner, P. and Havlíček, M.: 1975, *Selected Topics of Mathematical Physics. Linear Operators on Hilbert Space*, Vol. I, SPN, Prague (in Czech).

Blank, J., Exner, P. and Havlíček, M.: 1979, 'Quantum-Mechanical Pseudo-Hamiltonians', *Czech. J. Phys.* **B29**, 1325–1341.

Blankenbecler, R., Goldberger, M. L. and Simon, B.: 1977, 'The Bound States of Weakly-Coupled Long-Range One-Dimensional Quantum Hamiltonians', *Ann. Phys.* **108**, 69–78.

Bloch, I. and Burba, D. A.: 1974, 'Presence of a Particle in a Given Space-Time Region and the Continuous Action of a Particle Detector', *Phys. Rev.* **D10**, 2306–2318.

Blokhintsev, D. I. and Barbashov, B. M.: 1972, 'Application of Functional Integrals in Quantum Mechanics and Field Theory', *Usp. fiz. nauk* **106**, 593–616 (in Russian).

Bochner, S.: 1959, *Lectures on Fourier Integrals*, Princeton Univ. Press, Princeton.

Bogoliubov, N. N., Logunov, A. A. and Todorov, I. T.: 1969, *Foundations of Axiomatic Approach in Quantum Field Theory*, Nauka, Moscow (in Russian; English trans.: Benjamin, New York 1975).

Bogoliubov, N. N. and Shirkov, D. V.: 1976, *Introduction to the Theory of Quantized Fields* (3rd edn), Nauka, Moscow (in Russian; English trans.: Wiley-Interscience, New York 1980).

Böhm, A.: 1980, 'Decaying States in a Rigged Hilbert Space Formalism of Quantum Mechanics', *J. Math. Phys.* **21**, 1040–1043.

Böhm, A.: 1981, 'Resonance Poles and Gamow Vectors in the Rigged Hilbert Space Formulation of Quantum Mechanics', *J. Math. Phys.* **22**, 2813–2823.

Bollé, D., Gesztesy, F. and Grosse, H.: 1983, 'Time Delay for Long-Range Interactions', *J. Math. Phys.* **24**, 1529–1541.

Bollé, D. and Osborn, T. A.: 1979, 'Time Delay in *N*-Body Scattering', *J. Math. Phys.* **20**, 1121–1134.

Bratelli, O. and Robinson, D. W.: 1979, *Operator Algebras and Quantum Statistical Mechanics*, Vol. I, Springer, Berlin.

Bricman, C. *et al.* (Particle Data Group): 1978, 'Review of Particle Properties', *Phys. Letters B*, Vol. 75.

Brinati, J. R. and Mizrahi, S. S.: 1980, 'Quantum Friction in the *c*-Number Picture: the Damped Harmonic Oscillator', *J. Math. Phys.* **21**, 2154–2158.

Brodsky, I. S. and Lifshitz, M. S.: 1958, 'Spectral Analysis of Non-Selfadjoint Operators and the Intermediate Systems', *Usp. mat. nauk* **13** (1), 1–85.

Burzlaff, J.: 1979, 'Canonical Quantization of Dissipative Systems', *Rept. Math. Phys.* **16**, 101–110.

Caldirola, P.: 1983, 'On the Schrödinger Equation for a Class of Open Systems', *Lett. Nuovo Cimento* **36**, 385–388.

Calucci, G. and Ghirardi, G. C.: 1968, 'Correspondence between Unstable Particles and Poles in *S*-Matrix Theory: the Exponential Decay Law', *Phys. Rev.* **169**, 1339–1342.

Cameron, R. H.: 1960, 'A Family of Integrals Serving to Connect the Wiener and Feynman Integrals', *J. Math. and Phys.* **39**, 126–140.

Cameron, R. H.: 1962–63, 'The Ilstow and Feynman Integrals', *J. d'Anal. Math.* **10**, 287–361.

Cameron, R. H.: 1968, 'Approximation to Certain Feynman Integrals', *J. d'Anal. Math.* **21**, 337–371.

Cameron, R. H. and Martin, T. W.: 1945, 'Transformations of Wiener Integrals under a General Class of Linear Transformations', *Trans. Amer. Math. Soc.* **58**, 184–219.

Cameron, R. H. and Storvick, D. A.: 1966, 'A Translation Theorem for Analytic Feynman Integrals', *Trans. Amer. Math. Soc.* **125**, 1–6.

Cameron, R. H. and Storvick, D. A.: 1968, 'An Operator-Valued Function Space Integral and a Related Integral Equation, *J. Math. Mech.* **18**, 517–552.

Cameron, R. H. and Storvick, D. A.: 1973, 'An Operator-Valued Function Space Integral Applied to Multiple Integrals of Functions of Class L_1', *Nagoya Math. J.* **51**, 91–122.

Cameron, R. H. and Storvick, D. A.: 1980, 'Some Banach Algebras of Analytic Feynman Integrable Functions', in J. Lawrynowicz (ed.), *Analytic Functions, Kozubnik 1979*, Lecture Notes in Mathematics, Vol. 798, Springer, Berlin, pp. 18–67.

Campbell, D. K.: 1980, 'Tunneling Decay of Unstable Particles in a Quantum Field Theoretical Model', *Ann. Phys.* **129**, 249–272.

Campbell, W. B., Finkler, P., Jones, C. E. and Mischeloff, N. N.: 1975, 'Path-Integral Formulation of Scattering Theory', *Phys. Rev.* **D12**, 2363–2369.

Campesino-Romeo, E., d'Olivo, J. C. and Socolovsky, M.: 1982, 'Haussdorf Dimension for the Quantum Harmonic Oscillator', *Phys. Letters* **89A**, 321–324.

Carleson, L.: 1966, 'On Convergence and Growth of Partial Sums of Fourier Series', *Acta Math.* **116**, 135–157.

Chernoff, P. R.: 1968, 'Note on Product Formulas for Operator Semigroups', *J. Funct. Anal.* **2**, 238–242.

Chernoff, P. R.: 1970, 'Semigroup Product Formulas and Addition of Unbounded Operators', *Bull. Amer. Math. Soc.* **76**, 395–398.

Chernoff, P. R.: 1972, 'Perturbations of Dissipative Operators of Relative Bound One', *Proc. Amer. Math. Soc.* **33**, 72–74.

Chernoff, P. R.: 1974, 'Product Formulas, Nonlinear Semigroups and Addition of Unbounded Operators', *Mem. Amer. Math. Soc.* No. 140, Providence, Rhode Island.

Chiu, C. B., Misra, B. and Sudarshan, E. C. G.: 1982, 'The Time Scale for Quantum Zeno Paradox and Proton Decay', *Phys. Letters* **117B**, 34–40.

Chiu, C. B., Sudarshan, E. C. G. and Misra, B.: 1977, 'Time Evolution of Unstable Quantum States and a Resolution to Zeno's Paradox', *Phys. Rev.* **D16**, 520–529.

Clark, T. E., Menikoff, R. and Sharp, D. H.: 1980, 'Quantum Mechanics on the Half-Line Using Path Integrals', *Phys. Rev.* **D22**, 3012–3016.

Coester, F.: 1954, 'Influence of Extranuclear Fields on Angular Correlations', *Phys. Rev.* **93**, 1304–1308.

Combe, Ph., Høegh-Krohn, R., Rodriguez, R., Sirugue, M. and Sirugue-Collin, M.: 1981, 'Generalized Poisson Processes in Quantum Mechanics and Field Theory', *Phys. Rep.* **77**, 221–234.

Combe, Ph., Rideau, G., Rodriguez, R., Sirugue-Collin, M.: 1978, 'On the Cylindrical Approximation of the Feynman Path Integrals', *Rep. Math. Phys.* **13**, 279–294.

Combe, Ph., Rodriguez, R., Sirugue, M. and Sirugue-Collin, M.: 1981, 'Applications of the Stochastic Jump Processes to the Definition of Feynman Path Integrals', Preprint CNRS 81/P.1341, Marseille.

Crandall, M. G. and Phillips, R. S.: 1968, 'On the Extension Problem for Dissipative Operators', *J. Funct. Anal.* **2**, 147–176.

Creutz, M. and Freedman, B.: 1981, 'A Statistical Approach to Quantum Mechanics', *Ann. Phys.* **132**, 427–462.
Daletskii, Yu. L.: 1961, 'Functional Integrals Connected with Certain Differential Equations and Systems', *Dokl. Akad. Nauk USSR* **137**, 268–271 (in Russian).
Daletskii, Yu. L.: 1962, 'Functional Integrals Connected with Operator Evolution Equations', *Usp. mat. nauk* **17** (5), 3–115.
Davies, E. B.: 1969, 'Quantum Stochastic Processes', *Commun. Math. Phys.* **15**, 277–304.
Davies, E. B.: 1975a, 'Resonances, Spectral Concentration and Exponential Decay', *Lett. Math. Phys.* **1**, 31–35.
Davies, E. B.: 1975b, 'A Model for Absorption or Decay', *Helv. Phys. Acta* **48**, 365–382.
Davies, E. B.: 1976, *Quantum Theory of Open Systems*, Academic Press, London.
Davies, E. B.: 1978, 'Two-Channel Hamiltonians and the Optical Model of Nuclear Scattering', *Ann. Inst. H. Poincaré* **A29**, 395–413.
Davies, E. B.: 1980a, *One-parameter Semigroups*, London Mathematical Society Monographs, Vol. 15, Academic Press, London.
Davies, E. B.: 1980b, 'Non-unitary Scattering and Capture; I, Hilbert Space Theory', *Commun. Math. Phys.* **71**, 299–309.
Davies, E. B. and Eckmann, J.-P.: 1975, 'Time Decay for Fermion Systems with Persistent Vacuum', *Helv. Phys. Acta* **48**, 731–742.
Degasperis, A., Fonda, L. and Ghirardi, G. C.: 1974, 'Does the Lifetime of an Unstable System Depend on the Measuring Apparatus?', *Il Nuovo Cimento* **21A**, 471–484.
Dekker, H.: 1981, 'Classical and Quantum Mechanics of the Damped Harmonic Oscillator', *Phys. Repts* **80**, 1–112.
Demuth, M.: 1974, 'On the Perturbation Theory of Unstable Isolated Eigenvalues', *Math. Nachr.* **64**, 345–356.
Demuth, M.: 1976, 'Pole Approximation and Spectral Concentration', *Math. Nachr.* **73**, 65–72.
DeWitt-Morette, C.: 1972, 'Feynman Path Integrals, Definition without Limiting Procedure', *Commun. Math. Phys.* **28**, 47–67.
DeWitt-Morette, C.: 1974, 'Feynman Path Integrals: I, Linear and Affine Techniques; II, The Feynman-Green Function', *Commun. Math. Phys.* **37**, 68–81.
DeWitt-Morette, C.: 1976, 'The Semiclassical Expansion', *Ann. Phys.* **97**, 367–399; erratum **101**, 682–683.
DeWitt-Morette, C.: 1979, 'A Reasonable Method for Computing Path Integrals on Curved Space', in S. Albeverio *et al.* (eds), *Feynman Path Integrals*, Lecture Notes in Physics, Vol. 106, Springer, Berlin, pp. 227–233.
DeWitt-Morette, C., Maheswari, A. and Nelson, B.: 1979, 'Path Integration in Non-relativistic Quantum Mechanics', *Phys. Repts* **50**, 255–372.
Dirac, P. A. M.: 1933, 'The Lagrangian in Quantum Mechanics', *Phys. Zeit. Sowjetunion* **3**, 64–72.
Dirac, P. A. M.: 1945, 'On the Analogy between Classical and Quantum Mechanics', *Rev. Mod. Phys.* **17**, 195–199.
Dirac, P. A. M.: 1959, *The Principles of Quantum Mechanics* (4th edn), Clarendon, Oxford.
Dixmier, J.: 1969, *Les algèbres des opérateurs dans l'espace Hilbertien (algèbres de von Neumann)*, (2nd edn), Dunod, Paris.
Dolejší, J. and Exner, P.: 1977, 'Corrections to the Exponential Decay Law: Are They Observable?', *Czech. J. Phys.* **B27**, 855–864.
Dolph, C. L.: 1961, 'Recent Developments in Some Non-selfadjoint Problems of Mathematical Physics', *Bull. Amer. Math. Soc.* **67**, 1–69.
Doyle, S. D., Eck, J. S., Thompson, W. J. and Weaver, O. L.: 1975, 'Optical-Model and Coupled-Channel Calculations in Quantum Mechanical Scattering', *Amer. J. Phys.* **43**, 677–682.
Drukarev, G., Fröman, N. and Fröman, P. O.: 1979, 'The Jost Function Treated by the F-Matrix Phase Integral Method', *J. Phys.* **A12**, 171–186.
Duerinckx, G.: 1983, 'On the Point Spectrum of the N-Point Friedrichs Model', *J. Phys.* **A16**, L289–293.

Dunford, N. and Schwartz, J. T.: 1958–71, *Linear Operators: I, General Theory; II, Spectral Theory (Selfadjoint Operators in Hilbert Space); III, Spectral Operators*, Interscience Publishers, New York.

Dyson, F. J.: 1949a, 'The Radiation Theories of Tomonaga, Schwinger and Feynman', *Phys. Rev.* **75**, 486–502.

Dyson, F. J.: 1949b, 'The *S* Matrix in Quantum Electrodynamics', *Phys. Rev.* **75**, 1736–1755.

Dyson, F. J.: 1972, 'Missed Opportunities', *Bull. Amer. Math Soc.* **78**, 635–652.

Ekstein, H. and Siegert, A. J. F.: 1971, 'On a Reinterpretation of Decay Experiments', *Ann. Phys.* **68**, 509–520.

Elworthy, K. D. and Truman, A.: 1981, 'Classical Mechanics, the Diffusion (Heat) Equation and the Schrödinger Equation on Riemannian Manifolds', *J. Math. Phys.* **22**, 2144–66.

Elworthy, K. D. and Truman, A.: 1982, 'A Cameron-Martin Formula for Feynman Integrals (the Origin of Maslov Indices)', in R. Schrader *et al.* (eds.), *Mathematical Problems of Theoretical Physics*, Lecture Notes in Physics, Vol. 153, Springer, Berlin, pp. 288–294.

Elworthy, K. D. and Truman, A.: 1983, 'Feynman Maps, Cameron-Martin Formulae and Anharmonic Oscillators', Preprint, Univ. of Warwick, Coventry.

Emch, G. G. and Sinha, K. B.: 1979, 'Weak Quantization in a Non-perturbative Model', *J. Math. Phys.* **20**, 1336–1340.

Enss, V.: 1978, 'Asymptotic Completeness for Quantum Mechanical Potential Scattering', *Commun. Math. Phys.* **61**, 258–291.

Erdös, P. and Kac, M.: 1946, 'On Certain Limit Theorems of the Theory of Probability', *Bull. Amer. Math. Soc.* **52**, 292–302.

d'Espagnat, B. (ed.): 1971, 'Foundations of Quantum Mechanics', *Proc. Int. School 'Enrico Fermi'*, Vol. 49, Academic Press, New York.

Exner, P.: 1976a, 'Remark on the Decay of a Mixed State', *Czech. J. Phys.* **B26**, 976–982.

Exner, P.: 1976b, 'Remark on the Energy Spectrum of a Decaying System', *Commun. Math. Phys.* **50**, 1–10.

Exner, P.: 1977a–c, 'Unstable Systems and Repeated Measurements: I, General Considerations; II, Examples (Exponential Primary Decay Law, Idealized Spark Chamber); III, Example (Homogeneous Chamber), Conjecture for the General Case and Discussion', *Czech. J. Phys.* **B27**, 117–126, 233–246, 361–372.

Exner, P.: 1977d, *Unstable Quantum Systems*, Charles Univ. Prague (in Czech; unpublished).

Exner, P.: 1980, 'Bounded Energy Approximation to an Unstable Quantum System', *Rep. Math. Phys.* **17**, 275–285.

Exner, P.: 1982a, 'On the "Feynman Paths"', *Lett. Math. Phys.* **6**, 215–220.

Exner, P.: 1982b, 'Complex Potentials and Rigorous Feynman Integrals', *Czech. J. Phys.* **B32**, 628–632.

Exner, P.: 1983a, 'Complex-Potential Description of the Damped Harmonic Oscillator', *J. Math. Phys.* **24**, 1129–1135.

Exner, P.: 1983b, 'On the Representations of Poincaré Group Associated with Unstable Particles', *Phys. Rev.* **D28**, 2621–2627.

Exner, P. and Kolerov, G. I.: 1981a, 'On Hilbert Spaces of Paths', *Czech. J. Phys.* **B31**, 470–4.

Exner, P. and Kolerov, G. I.: 1981b, 'Feynman Maps without Improper Integrals', *Czech. J. Phys.* **B31**, 1207–1224.

Exner, P. and Kolerov, G. I.: 1981c, 'Path-Integral Expression of Dissipative Dynamics', *Phys. Letters* **83A**, 203–206.

Exner, P. and Kolerov, G. I.: 1982a, 'Polygonal-Path Approximations on Path Spaces of Quantum-Mechanical Systems', *Int. J. Theor. Phys.* **21**, 397–417.

Exner, P. and Kolerov, G. I.: 1982b, 'Uniform Product Formulae with Application to the Feynman–Nelson Integral for Open Systems', *Lett. Math. Phys.* **6**, 153–159.

Exner, P. and Úlehla, I.: 1983, 'On the Optical Approximation in Two-Channel Systems', *J. Math. Phys.* **24**, 1542–1547.

Faddeev, L. D.: 1969, 'Feynman Integral for Singular Lagrangians', *Teor. mat. fiz.* **1**, 3–18 (in Russian).

Faris, W. G.: 1967a, 'The Product Formula for Semigroups Defined by Friedrichs Extensions', *Pacific J. Math.* **22**, 47–70.

Faris, W. G.: 1967b, 'Product Formulas for Perturbation of Linear Operators', *J. Funct. Anal.* **1**, 93–107.

Feldman, J.: 1963, 'On the Schrödinger and Heat Equations', *Trans. Amer. Math. Soc.* **110**, 251–264.

Fermi, E.: 1960, *Notes on Quantum Mechanics. A Course Given at the University of Chicago in 1954, edited by E. Segrè*, The Univ. of Chicago Press.

Feshbach, H.: 1958, 1962, 'Unified Theory of Nuclear Reactions I, II', *Ann. Phys.* **5**, 357–390, **19**, 287–313.

Feshbach, H., Porter, C. E. and Weisskopf, V. F.: 1954, 'Model for Nuclear Reactions with Neutrons', *Phys. Rev.* **96**, 448–464.

Feynman, R. P.: 1948, 'Space-Time Approach to Non-relativistic Quantum Mechanics', *Rev. Mod. Phys.* **20**, 367–387.

Feynman, R. P.: 1949a, 'The Theory of Positrons', *Phys. Rev.* **76**, 749–759.

Feynman, R. P.: 1949b, 'Space-Time Approach to Quantum Electrodynamics', *Phys. Rev.* **76**, 769–789.

Feynman, R. P.: 1950, 'Mathematical Formulation of the Quantum Theory of Electromagnetic Interaction', *Phys. Rev.* **80**, 440–457.

Feynman, R. P.: 1951, 'An Operator Calculus Having Applications in Quantum Electrodynamics', *Phys. Rev.* **84**, 108–128.

Feynman, R. P. and Hibbs, A. R.: 1965, *Quantum Mechanics and Path Integrals*, McGraw-Hill, New York.

Feynman, R. P. and Vernon, F. L., Jr: 1963, 'The Theory of a General Quantum System Interacting with a Linear Dissipative System', *Ann. Phys.* **24**, 118–173.

Fiziev, P. P.: 1983, 'Approximation of Phase-Space Paths in Feynman Integral', Preprint JINR P2-82-528, Dubna; *Bulg. J. Phys.* (to appear).

Fleming, G. N.: 1973, 'A Unitary Bound on the Evolution of Nonstationary States', *Il Nuovo Cimento* **16A**, 232–240.

Fonda, L.: 1977, 'A Critical Discussion on the Decay of Quantum Unstable Systems', *Fortsch. Phys.* **25**, 101–121.

Fonda, L. and Ghirardi, G. C.: 1972, 'Some Remarks on the Origin of the Deviations from the Exponential Decay Law of an Unstable Particle', *Il Nuovo Cimento* **7A**, 180–184; addendum **10A**, 850.

Fonda, L., Ghirardi, G. C. and Rimini, A.: 1974, 'Evolution of Quantum Systems Subject to Random Measurements', *Il Nuovo Cimento* **18B**, 1–10.

Fonda, L., Ghirardi, G. C. and Rimini, A.: 1978, 'Decay Theory of Unstable Quantum Systems', *Repts Progr. Phys.* **41**, 587–631.

Fonda, L., Ghirardi, G. C., Rimini, A. and Weber, T.: 1973, 'On the Quantum Foundations of the Exponential Decay Law', *Il Nuovo Cimento* **15A**, 689–704; erratum **18A**, 805.

Frey, R. B. and Thiele, E.: 1968, 'Expansion of the Time-Dependent Wavefunctions in Quasi-stationary Energy Eigenfuctions', *J. Chem. Phys.* **48**, 3240–3245.

Friedman, Ch. N.: 1972, 'Semigroup Product Formulas, Compressions and Continuous Observation in Quantum Mechanics', *Indiana Univ. Math. J.* **21**, 1001–1011.

Friedman, Ch. N.: 1976, 'Continual Measurements in Space-Time Formation of Nonrelativistic Quantum Mechanics', *Ann. Phys.* **98**, 87–97.

Friedrichs, K. O.: 1948, 'On the Perturbation of Continuous Spectra', *Commun. (Pure and) Appl. Math.* **1**, 361–406.

Fuks, B. A.: 1962, *Introduction to the Theory of Analytic Functions of Many Complex Variables*, Fizmatgiz, Moscow (in Russian).

Gadella, M.: 1983, 'A Rigged Hilbert Space of Hardy Class Functions: Applications to Resonances', *J. Math. Phys.* **24**, 1462–1469.

Gal, A., Toker, G. and Alexander, Y.: 1981, 'Optical Potential Study of Σ Nuclear States', *Ann. Phys.* **137**, 341–377.

Gamow, G.: 1928, 'Zur Quantentheorie des Atomkernes', *Zeit. Phys.* **51**, 204–212.

Gantmakher, F. R.: 1966, *The Theory of Matrices* (2nd edn), Nauka, Moscow (in Russian; English trans.: Chelsea Publishing Company, New York 1959).

Garrod, C.: 1966, 'Hamiltonian Path-Integral Methods', *Rev. Mod. Phys.* **38**, 483–494.

Gel'fand, I. M.: 1936, 'On a Lemma from the Theory of Linear Spaces', *Soobsh. Khark. Mat. Obsh.* **13**, 35–40 (in Russian).

Gel'fand, I. M. and Raikov, D. A.: 1943, 'Irreducible Unitary Representations of Locally Bicompact Groups', *Sbornik Mosk. mat. obsh.* **13**, 301–316 (in Russian).

Gel'fand, I. M. and Yaglom, A. M.: 1956, 'Integration in Function Spaces', *Usp. mat. nauk* **11** (1), 77–114 (in Russian; English trans. in *J. Math. Phys.* **1** (1960), 48–69).

Gerry, Ch. C. and Singh, V. A.: 1980, 'Path-Integral Formulation of Scattering Theory: Central Potentials', *Phys. Rev.* **D21**, 2979–2985.

Gesztesy, F., Grosse, H. and Thaller, B.: 1982a, 'Spectral Concentration in the Non-relativistic Limit', *Phys. Letters* **116B**, 155–157.

Gesztesy, F., Grosse, H. and Thaler, B.: 1982b, 'First Order Relativistic Corrections and Spectral Concentration', Preprint CERN TH-3344, Geneva; *Adv. Appl. Math.* (to appear).

Gesztesy, F., Grosse, H. and Thaler, B.: 1984, 'A Rigorous Approach to Relativistic Corrections of Bound State Energies for Spin 1/2 Particles', *Ann. Inst. H. Poincaré* **40**, 159–174.

Gisin, N.: 1981, 'A Simple Nonlinear Dissipative Quantum Evolution Equation, *J. Phys.* **A14**, 2259–2267.

Gisin, N.: 1982a, 'Microscopic Derivation of a Class of Non-linear Dissipative Schrödinger-Like Equations', *Physica* **A111**, 364–370.

Gisin, N.: 1982b, 'A Model of Dissipative Quantum Dynamics', (PhD. dissertation) Preprint UGVA-DPT 1982/07-356, Geneva.

Glaser, V. and Källén, G.: 1957, 'A Model of an Unstable Particle, *Nucl. Phys.* **2**, 706–722.

Glazman, I. M.: 1957, 'On an Analogue of the Extension Theory of Hermitean Operators and the Non-symmetric One-Dimensional Problem on the Semiaxis', *Dokl. Akad. Nauk USSR* **115**, 214–216 (in Russian).

Glazman, I. M.: 1963, *Direct Methods of Qualitative Spectral Analysis of Singular Differential Operators*, Fizmatgiz, Moscow (in Russian; English version: Israel Program for Scientific Translations, Jerusalem 1965).

Glimm, J. and Jaffe, A.: 1981, *Quantum Physics: a Functional Integral Point of View*, Springer, New York.

Goldberger, M. L. and Watson, K. M.: 1964, 'Lifetime and Decay of Unstable Particles in *S*-Matrix Theory', *Phys. Rev.* **136B**, 1472–1480.

Goldstein, Ch.: 1970, 1971, 'Perturbation of Non-selfadjoint Operators I, II', *Arch. Rat. Mech. Anal.* **37**, 268–296; **42**, 380–402.

Goldstein, H.: 1951, *Classical Mechanics*, Addison-Wesley, Reading.

Goodman, M.: 1981, 'Path Integral Solution to the Infinite Square Well', *Amer. J. Phys.* **49**, 843–847.

Gorini, V., Frigerio, A., Verri, M., Kossakowski, A. and Sudarshan, E. C. G.: 1978, 'Properties of Quantum Markovian Master Equations', *Rep. Math. Phys.* **13**, 149–173.

Gradshtein, I. S. and Ryzhik, I. M.: 1971, *Tables of Integrals, Sums, Series and Products* (5th edn), Nauka, Moscow (in Russian; English trans.: Academic Press, New York 1965).

Graffi, S. and Grecchi, V.: 1981, 'Resonances in Stark Effect of Atomic Systems', *Commun. Math. Phys.* **79**, 91–110.

Grossmann, A.: 1964, 'Nested Hilbert Spaces in Quantum Mechanics', *J. Math. Phys.* **5**, 1025–37.

Gudder, S. and Naroditsky, V.: 1981, 'Finite-Dimensional Quantum Mechanics', *Int. J. Theor. Phys.* **20**, 619–643.

Gustafson, K.: 1966, 'A Perturbation Lemma', *Bull. Amer. Math. Soc.* **72**, 334–338.

Gustafson, K. and Sinha, K.: 1980, 'On the Eisenbud–Wigner Formula for Time-Delay', *Lett. Math. Phys.* **4**, 381–385.

Gutzwiller, M. C.: 1967, 1969, 'Phase-Integral Approximation in Momentum Space and the Bound States of an Atom I, II', *J. Math. Phys.* **8**, 1979–2000; **10**, 1004–1020.

Gutzwiller, M. C.: 1970, 'Energy Spectrum According to Classical Mechanics', *J. Math. Phys.* **11**, 1791–1806.

Gutzwiller, M. C.: 1971, Periodic Orbits and Classical Quantization Conditions', *J. Math. Phys.* **12**, 343–359.

Gutzwiller, M. C.: 1978, 'Path Integrals and the Relation between Classical and Quantum Mechanics', in G. J. Papadopoulos and J. T. Devreese (eds), *Path Integrals and their Applications in Quantum, Statistical and Solid State Physics*, Plenum Press, New York, pp. 163–200.

Haag, R.: 1972, 'Quantum Field Theory', in R. F. Streater (ed.), *Mathematics of Contemporary Physics*, Academic Press, New York, pp. 1–16.

Haag, R. and Kastler, D.: 1964, 'An Algebraic Approach to Quantum Field Theory', *J. Math. Phys.* **5**, 848–861.

Haase, R. W.: 1975, 'On the Quantum Mechanical Treatment of Dissipative Systems', *J. Math. Phys.* **16**, 2005–2011.

Haase, R. W.: 1978, 'Approaches to Nuclear Friction', *Repts Progr. Phys.* **41**, 1027–1101.

Hack, M. N.: 1982, 'Long Time Tails in Decay Theory', *Phys. Letters* **90A**, 220–221, addendum **95A** (1983), 477.

Hagedorn, G. A.: 1979, 'A Link between Scattering Resonances and Dilation Analytic Resonances in Few Body Quantum Mechanics', *Commun. Math. Phys.* **65**, 181–188.

Harrell, E. M., II: 1982, 'General Lower Bounds for Resonances in One Dimension', *Commun. Math. Phys.* **86**, 221–225.

Havlíček, M. and Exner, P.: 1973, 'Note on the Description of an Unstable System', *Czech. J. Phys.* **B23**, 594–600.

Heisenberg, W.: 1927, 'Über den anschaulichen Inhalt der quantentheoretischen Kinematik und Mechanik', *Zeit. Phys.* **43**, 172–198.

Heitler, W.: 1954, *The Quantum Theory of Radiation* (3rd edn), Clarendon Press, Oxford.

Hellund, E. J.: 1953, 'The Decay of Resonance Radiation by Spontaneous Emission', *Phys. Rev.* **89**, 919–922.

Helton, J. W.: 1972, 'The Characteristic Function of Operator Theory and Electrical Network Realization', *Indiana Univ. Math. J.* **22**, 403–414.

Herbst, I. W.: 1979, 'Dilation Analycity in Constant Electric Field: I, The Two-Body Problem', *Commun. Math. Phys.* **64**, 279–298.

Hille, E. and Phillips, R. S.: 1957, *Functional Analysis and Semigroups*, American Mathematical Society Colloquium Publications, Vol. 31, Providence.

Hjorth, P. G.: 1982, 'On the Trotter Formula Computation of the Kernel for the Harmonic Oscillator', Preprint, La Jolla.

Hodgson, P. E.: 1978, *Nuclear Heavy-Ion Reactions*, Clarendon Press, Oxford.

Hodgson, P. E.: 1980–81, *Growth Points in Nuclear Physics*, Vols 2 and 3, Pergamon Press, Oxford.

Höhler, G.: 1958, 'Über die Exponentialnäherung beim Teilchenzerfall', *Zeit. Phys.* **152**, 546–565.

Horváthy, P. A.: 1979, 'Extended Feynman Formula for Harmonic Oscillator', *Int. J. Theor. Phys.* **18**, 245–250.

Horwitz, L. P. and Katznelson, E.: 1983a, 'Is the Proton Decay Measurable?', *Phys. Rev. Letters* **50**, 1184–1186; **51**, 1602.

Horwitz, L. P. and Katznelson, E.: 1983b, 'A Partial Inner Product Space of Analytic Functions for Resonances', *J. Math. Phys.* **24**, 848–859.

Horwitz, L. P., LaVita, J. A. and Marchand, J.-P.: 1971, 'The Inverse Decay Problem', *J. Math. Phys.* **12**, 2537–2543.

Horwitz, L. P. and Marchand, J.-P.: 1969a, 'Unitary Sum Rule and the Time Evolution of Neutral *K*-Mesons', *Helv. Phys. Acta* **42**, 801–807.

Horwitz, L. P. and Marchand, J.-P.: 1969b, 'Formal Scattering Treatment of the Neutral *K* Meson System', *Helv. Phys. Acta* **42**, 1039–1054.

Horwitz, L. P. and Marchand, J.-P.: 1971, 'The Decay Scattering System', *Rocky Mountain J. Math.* **1**, 225–253.

Horwitz, L. P. and Sigal, I. M.: 1978, 'On a Mathematical Model for Non-stationary Physical Systems', *Helv. Phys. Acta* **51**, 685–715.

Howland, J. S.: 1968, 'Perturbation of Embedded Eigenvalues by Operators of Finite Rank', *J. Math. Anal. Appl.* **23**, 575–584.

Howland, J. S.: 1970, 'On the Weinstein–Aronszajn Formula', *Arch. Rat. Mech. Anal.* **39**, 323–339.

Howland, J. S.: 1971, 'Spectral Concentration and Virtual Poles, II', *Trans. Amer. Math. Soc.* **162**, 141–156.

Howland, J. S.: 1972, 'Perturbation of Embedded Eigenvalues', *Bull. Amer. Math. Soc.* **78**, 280–283.

Howland, J. S.: 1974, 'Puiseux Series for Resonances at an Embedded Eigenvalue', *Pacific J. Math.* **55**, 157–176.

Howland, J. S.: 1975, 'The Livsic Matrix in Perturbation Theory', *J. Math. Anal. Appl.* **50**, 415–437.

Ichinose, T.: 1980, 'A Product Formula and its Application to the Schrödinger Equation', *Pub. Res. Inst. Math. Sci., Kyoto Univ.* **16**, 585–600.

Ichinose, T.: 1982, 'Path Integral for the Dirac Equation in Two Space-Time Dimensions', *Proc. Japan Acad.* **58**, 290–293.

Ichinose, T.: 1984, 'Path Integral for a Hyperbolic System of the First Order', *Duke Math. J.* **51**, 1–36.

Itô, K.: 1961, 'Wiener Integral and Feynman Integral', in *Proc. Fourth Berkeley Symp. on Mathematical Statistics and Probability*, Vol. 2, Univ. of California Press, pp. 227–238.

Itô, K.: 1967, 'Generalized Uniform Complex Measures in the Hilbertian Metric Space with their Application to the Feynman Integral', in *Proc. Fifth Berkeley Symp. on Mathematical Statistics and Probability*, Vol. 2, Part 1, Univ. of California Press, pp. 145–161.

Jacob, R. and Sachs, R. G.: 1961, 'Mass and Lifetime of Unstable Particles', *Phys. Rev.* **121**, 350–356.

Jannussis, A. D., Brodinas, G. N. and Streclas, A.: 1979, 'Propagator with Friction in Quantum Mechanics', *Phys. Letters* **74A**, 6–10.

Jarník, V.: 1955, *Integral Calculus II*, Czechoslovak Academy of Sciences Publishing House, Prague (in Czech).

Jarník, V.: 1956, *Differential Calculus II* (2nd edn), Czechoslovak Academy of Sciences Publishing House, Prague (in Czech).

Jauch, J. M.: 1968, *Foundations of Quantum Mechanics*, Addison-Wesley, Reading.

Jauch, J. M., Misra, B. and Sinha, K. B.: 1972, 'Time-Delay in Scattering Processes', *Helv. Phys. Acta* **45**, 398–426.

Jensen, A.: 1977, 'Local Distortion Technique, Resonances and Poles of the *S*-Matrix', *J. Math. Anal. Appl.* **59**, 505–513.

Jensen, A.: 1980, 'Resonances in an Abstract Scattering Theory', *Ann. Inst. H. Poincaré* **A33**, 209–223.

Jensen, A.: 1981, 'Time-Delay in Potential Scattering: Some 'Geometric' Results', *Commun. Math. Phys.* **82**, 435–456.

Jersák, J.: 1969a, 'Number of Wavefunctions of an Unstable Particle', *Yadernaya fiz.* **9**, 458–461 (in Russian).

Jersák, J.: 1969b, 'Wave Functions of Unstable Particles', *Czech. J. Phys.* **B19**, 1523–1532.

Jersák, J.: 1970, 'On Properties of Unstable Particles', *Helv. Phys. Acta* **43**, 93–98.

Johnson, G. W.: 1982, 'The Equivalence of Two Approaches to the Feynman Integral', *J. Math. Phys.* **23**, 2090–2096.

Johnson, G. W.: 1983, 'A Bounded Convergence Theorem for the Feynman Integral', Preprint, Univ. of Nebraska.

Johnson, G. W. and Skoug, D. L.: 1973, 'A Banach Algebra of Feynman Integrable Functions with Applications to an Integral Equation Formally Equivalent to Schrödinger Equation', *J. Funct. Anal.* **12**, 129–152.

Johnson, G. W. and Skoug, D. L.: 1975, 'The Cameron–Storvick Function Space Integral: the L_1 Theory', *J. Math. Anal. Appl.* **50**, 647–667.

Johnson, G. W. and Skoug, D. L.: 1976, 'The Cameron–Storvick Function Space Integral: an $L(L_p, L_{p'})$ Theory', *Nagoya Math. J.* **60**, 93–137.

Johnson, G. W. and Skoug, D. L.: 1979, 'Scale-Invariant Measurability in Wiener Space', *Pacific J. Math.* **83**, 157–176.

Johnson, G. W. and Skoug, D. L.: 1981, 'Notes on Feynman Integral, I–III', *Pacific J. Math.* **93**, 313–324; *J. Funct. Anal.* **41**, 277–289; Preprint, Univ. of Nebraska.

Jona-Lasinio, G., Martinelli, F. and Scoppola, E.: 1982, 'Decaying Quantum-Mechanical States: an Informal Discussion within Stochastic Mechanics', *Lett. Nuovo Cimento* **34**, 13–17.

Kac, M.: 1949, 'On Distributions of Certain Wiener Functionals', *Trans. Amer. Math. Soc.* **65**, 1–13.

Kato, T.: 1951, 'Fundamental Properties of Hamiltonians of Schrödinger Type', *Trans. Amer. Math. Soc.* **70**, 195–211.

Kato, T.: 1953, 'Integration of the Equation of Evolution in a Banach Space', *J. Math. Soc. Japan* **5**, 208–234.

Kato, T.: 1966a, *Perturbation Theory for Linear Operators*, Die Grundlagen der Mathematischen Wissenschaften, Band 132, Springer, Berlin.

Kato, T.: 1966b, 'Wave Operators and Similarity for Some Non-self-adjoint Operators', *Math. Annal.* **162**, 258–279.

Kato, T.: 1973, 'Schrödinger Operators with Singular Potentials', *Israel J. Math.* **13**, 135–148.

Kato, T.: 1978, 'Trotter's Product Formula for an Arbitrary Pair of Self-adjoint Contraction Semigroups', in I. Gohberg and M. Kac (eds), *Topics in Functional Analysis*, Advances in Mathematics Supplementary Studies, Vol. 3, Academic Press, New York, pp. 185–195.

Kato, T. and Masuda, K.: 1978, 'Trotter's Product Formula for Non-linear Semigroups Generated by Subdifferentials of Convex Functionals', *J. Math. Soc. Japan* **30**, 169–178.

Kawai, T. and Gotō, M.: 1969, 'Nonunitary Representations of the Poincaré Group and Field Theory of Unstable Particles', *Il Nuovo Cimento* **60B**, 21–39.

Kerler, W. and Petzold, J.: 1965, 'Über einige Probleme bei instabilen quantenmechanischen Systemen', *Zeit. Phys.* **186**, 168–189.

Khalfin, L. A.: 1957a, 'To the Decay Theory of a Quasistationary State', *Dokl. Akad. Nauk USSR* **115**, 277–280 (in Russian).

Khalfin, L. A.: 1957b, 'To the Decay Theory of a Quasistationary State', *Zh. eksper. teor. fiz.* **33**, 1371–1382 (in Russian).

Khalfin, L. A.: 1968, 'A Phenomenological Theory of K^0-Mesons and the Decay-Law Nonexponentiality', *Piśma zh. eksper. teor. fiz.* **8**, 106–108 (in Russian).

Khalfin, L. A.: 1982, 'The Proton Nonstability and Nonexponentiality of the Decay Law', *Phys. Letters* **112B**, 223–226.

Khandekar, D. C. and Lavande, S. W.: 1978, 'Exact Propagator for a Quadratic Lagrangian', *Phys. Letters* **67A**, 175–176.

Khandekar, D. C. and Lavande, S. W.: 1979, 'Exact Solution of a Time-Dependent Quantal Harmonic Oscillator with Damping and a Perturbative Force', *J. Math. Phys.* **20**, 1870–1877.

Kholevo, A. S.: 1980, *Probabilistic and Statistical Aspects of the Quantum Theory*, Nauka, Moscow (in Russian).

Klaus, M. and Simon, B.: 1980, 'Coupling Constant Thresholds in Non-relativistic Quantum Mechanics: I, Short-Range Two-Body Case; II, Two-Cluster Thresholds in N-Body Systems', *Ann. Phys.* **130**, 251–281; *Commun. Math. Phys.* **78**, 153–168.

Kolmogorov, A. N. and Fomin, S. V.: 1976, *Elements of the Theory of Functions and Functional Analysis* (4th edn), Nauka, Moscow (in Russian; English trans.: Graylock Press, Rochester 1957).

Kovalchik, M. I.: 1963, 'Wiener Integral', *Usp. mat. nauk* **18** (1), 97–134 (in Russian).

Koyama, T. and Ichinose, T.: 1981, 'On the Trotter Product Formula', *Proc. Japan Acad.* **57**, 95–100.

Kraus, K.: 1981, 'Measuring Process in Quantum Mechanics: I, Continuous Observation and the Watchdog Effect', *Found. Phys.* **11**, 547–576.

Kree, P.: 1979, 'Feynman Path Integrals and Theory of Forms', in S. Albeverio *et al.* (eds), *Feynman Path Integrals*, Lecture Notes in Physics, Vol. 106, Springer, Berlin, pp. 120–136.

Krein, M. G.: 1964, 'On Some New Investigations in the Perturbation Theory of Self-adjoint Operators', in *Proc. First Mathematical Summer School, Kanev 1963*, Naukova Dumka, Kiev, pp. 103–187 (in Russian).

Krylov, N. S. and Fock, V. A.: 1947, 'On Two Basic Interpretations of the Energy-Time Uncertainty Relation', *Zh. eksper. teor. fiz.* **17**, 93–107 (in Russian).

Kuo, H.-H.: 1975, *Gaussian Measures in Banach Spaces*, Lecture Notes in Mathematics, Vol. 463, Springer, Berlin.

Kyselka, A.: 1981, 'Complex Mass in Nonrelativistic Quantum Mechanics', *Int. J. Theor. Phys.* **20**, 13–17.

Landau, L. D. and Lifshitz, E. M.: 1965, *Mechanics, A Course of Theoretical Physics*, Vol. 1 (2nd edn), Nauka, Moscow (in Russian; English trans.: Pergamon Press, Oxford 1960).

Landau, L. D. and Lifshitz, E. M.: 1974, *Quantum Mechanics. Non-relativistic Theory. A Course of Theoretical Physics*, Vol. 3 (3rd edn, revised), Nauka, Moscow (in Russian; English trans.: Pergamon Press, New York 1974).

Lapidus, M. L.: 1981, 'Generalization of the Lie–Trotter Formula', *Integral Equations and Operator Theory* **4**, 365–415.

Lapidus, M. L.: 1983, 'Product Formula for Imaginary Resolvents with Application to a Modified Feynman Integral', Preprint, Univ. of South California, Los Angeles.

von Laue, M.: 1958, *Geschichte der Physik*, Ullstein, Frankfurt a/M.

Lax, P. D. and Phillips, R. S.: 1967, *Scattering Theory*, Academic Press, New York.

Lee, T. D.: 1954, 'Some Special Examples in Renormalizable Field Theory', *Phys. Rev.* **95**, 1329–1334.

Lemos, N. A.: 1981, 'The Heisenberg Picture is not Privileged for the Canonical Quantization of Dissipative Systems', *Phys. Rev.* **D24**, 2338–2340.

Levit, S., Negele, J. W. and Paltiel, A.: 1980, 'Barrier Penetration and Spontaneous Fission in the Time-Dependent Mean-Field Approximation', *Phys. Rev.* **C22**, 1979–1995.

Lévy, M.: 1959a, 'On the Description of Unstable Particles in Quantum Field Theory', *Il Nuovo Cimento* **13**, 115–143.

Lévy, M.: 1959b, 'On the Validity of the Exponential Law for the Decay of an Unstable Particle', *Il Nuovo Cimento* **14**, 612–624.

Liance, V. E.: 1969, 'Non-selfadjoint Second-Order Differential Operator on the Semiaxis', appendix to second edition of the monograph by M. A. Naimark, *Linear Differential Operators*, Nauka, Moscow (in Russian).

Lidskii, V. B.: 1957, 'Conditions of Complete Continuity for the Resolvent of a Non-selfadjoint Differential Operator', *Dokl. Akad. Nauk USSR* **113**, 28–31 (in Russian).

Lidskii, V. B.: 1960, 'Non-selfadjoint Operator of Sturm–Liouville Type with a Discrete Spectrum', *Trudy Mosk. mat. obsh.* **9**, 45–79.

Lumer, G. and Phillips, R. S.: 1961, 'Dissipative Operators in a Banach Space', *Pacific J. Math.* **11**, 679–698.

Lupieri, G.: 1983, 'Generalized Stochastic Processes and Continual Observations in Quantum Mechanics', *J. Math. Phys.* **24**, 2329–2339.

Lurçat, F.: 1968, 'Strongly Decaying Particle and Relativistic Invariance', *Phys. Rev.* **173**, 1461–1473.

Macek, J.: 1970, 'Theory of Atomic Lifetime Measurements', *Phys. Rev.* **A1**, 618–627.

Maikov, E. V.: 1958, 'On Non-equivalence of Two Definitions of the Functional Integral', *Nauch. dokl. vys. shk. Fiz. mat. nauk.* **3**, 85–87 (in Russian).

Malik, F. B. and Sabatier, P. C.: 1973, 'Decay Formula and Fission Barrier', *Helv. Phys. Acta* **46**, 303–315.

Mandel'shtam, L. I. and Tamm, I. E.: 1945, 'The Energy–Time Uncertainty Relation in Nonrelativistic Quantum Mechanics', *Izv. Akad. Nauk USSR, Ser. fiz.* **9**, 122–128 (in Russian).

Marinov, M. S.: 1980, 'Path Integrals in Quantum Theory: an Outlook of Basic Concepts', *Phys. Rep.* **60**, 1–57.

Martin, Ph. A.: 1975, 'Scattering Theory with Dissipative Interaction and Time Delay', *Il Nuovo Cimento* **30B**, 217–238.

Martin, Ph. A.: 1976, 'On the Time-Delay of Simple Scattering Systems', *Commun. Math. Phys.* **47**, 221–227.

Martin, Ph. A.: 1981, 'Time Delay in Quantum Scattering Processes', in H. Mitter and L. Pittner (eds), *New Developments in Mathematical Physics, Acta Phys. Austriaca*, Suppl. XXIII, pp. 157–208.

Martinson, I.: 1974, 'Recent Progress in Studies of Atomic Spectra and Transition Probabilities by Beam-Foil Spectroscopy', *Physica Scripta* **9**, 281–296.

Maslov, V. P.: 1961, 'Quasiclassical Asymptotics to Solutions of some Problems of Mathematical Physics, I, II', *Zh. vych. mat. mat. fiz.* **1**, 113–128, 638–663 (in Russian).

Maslov, V. P.: 1965, *Perturbation Theory and Asymptotical Methods*, Moscow State Univ. Press (in Russian; French trans.: Dunod, Paris 1972).

Maslov, V. P. and Chebotarev, A. M.: 1976, 'Generalized Measure in the Feynman Functional Integral', *Teor. mat. fiz.* **28**, 291–307 (in Russian).

Matthews, P. T. and Salam, A.: 1958, 1959, 'Relativistic Field Theory of Unstable Particles, I, II', *Phys. Rev.* **112**, 283–287; **115**, 1079–1084.

Mensky, M. B.: 1979, 'Quantum Restrictions for Continuous Observation of an Oscillator', *Phys. Rev.* **D20**, 384–387.

Mensky, M. B.: 1983, *The Path Group. Measurements, Fields, Particles*, Nauka, Moscow (in Russian).

Messer, J.: 1979, 'Friction in Quantum Mechanics', *Acta Phys. Austriaca* **50**, 75–91.

Messiah, A.: 1959, *Mécanique Quantique, I. II*, Dunod, Paris.

Misra, B. and Sinha, K. B.: 1977, 'A Remark on the Rate of Regeneration in Decay Processes', *Helv. Phys. Acta* **50**, 99–104.

Misra, B. and Sudarshan, E. C. G.: 1977, 'The Zeno's Paradox in Quantum Theory', *J. Math. Phys.* **18**, 756–763.

Mizrahi, M. M.: 1975, 'The Weyl Correspondence and Path Integrals', *J. Math. Phys.* **16**, 2201–2206.

Mizrahi, M. M.: 1977, 'On the Semiclassical Expansion in Quantum Mechanics, for Arbitrary Hamiltonians', *J. Math. Phys.* **18**, 786–790.

Mizrahi, M. M.: 1979, 'Correspondence Rules and Path Integrals', in S. Albeverio *et al.* (eds), *Feynman Path Integrals*, Lecture Notes in Physics, Vol. 106, Springer, Berlin, pp. 234–253.

Mizrahi, M. M.: 1980, 'Phase Space Path Integrals, without Limiting Procedure', *J. Math. Phys.* **21**, 298–307; erratum **21**, 1965.

Mizrahi, M. M.: 1981a, 'Correspondence Rules and Path Integrals', *Il Nuovo Cimento* **61B**, 81–98.

Mizrahi, M. M.: 1981b, 'On the WKB Approximation to the Propagator for Arbitrary Hamiltonians', *J. Math. Phys.* **22**, 102–107.

Molchanov, A. M.: 1953, 'On Conditions for Discreteness of the Spectrum of Self-adjoint Second-Order Differential Equations', *Trudy Mosk. mat. obsh.* **2**, 169–200 (in Russian).

Mozzochi, Ch. J.: 1971, *On the Pointwise Convergence of Fourier Series*, Lecture Notes in Mathematics, Vol. 199, Springer, Berlin.

Naimark, M. A.: 1943, 'Positive Definite Operator-Valued Functions on a Commutative Group', *Izv. Akad. Nauk USSR, Ser. mat.* **7**, 237–244.

Naimark, M. A.: 1954, 'Investigation of the Spectrum and Eigenfunction Expansion for a Non-selfadjoint Second-Order Differential Operator on the Semiaxis', *Trudy Mosk. mat. obsh.* **3**, 181–270 (in Russian).

Namiki, M. and Mugibayashi, N.: 1953, 'On the Radiation Damping and the Decay of an Excited State', *Progr. Theor. Phys.* **10**, 474–476.

Narnhofer, H.: 1980, 'Another Definition for Time Delay', *Phys. Rev.* **D22**, 2387–2390.

Narnhofer, H.: 1984, 'Time Delay and Dilation Properties in Scattering Theory', *J. Math. Phys.* **25**, 987–991.

Nelson, B.: 1983, 'Path Integral for Scattering Theory', *Phys. Rev.* **D27**, 841–852.

Nelson, B. and Sheeks, B.: 1982, 'Path Integration for Velocity Dependent Potentials', *Commun. Math. Phys.* **84**, 515–530.

Nelson, E.: 1964, 'Feynman Integrals and the Schrödinger Equation', *J. Math. Phys.* **5**, 332–343.

Newton, R. G.: 1961, 'The Exponential Decay Law of Unstable Systems', *Ann. Phys.* **14**, 333–345.

Newton, R. G.: 1966, *Scattering Theory of Waves and Particles*, McGraw-Hill, New York.

Nussenzweig, H. M.: 1961, 'On the Physical Interpretation of Complex Poles of the *S*-Matrix, II', *Il Nuovo Cimento* **20**, 694–714.

Okabayashi, T. and Sato, S.: 1957, 'A Method of Renormalization for Unstable Particles', *Progr. Theor. Phys.* **17**, 30–42.

Paley, R. E. A. C. and Wiener, N.: 1934, *Fourier Transforms in the Complex Domain*, American Mathematical Society Colloquium Publications, Vol. 19, New York.

Papadopoulos, G. J. and Devreese, J. T. (eds): 1978, *Path Integrals and their Applications in Quantum, Statistical and Solid State Physics*, NATO Advanced Study Institutes, Vol. B34, Plenum Press, New York.

Parravicini, G., Gorini, V. and Sudarshan, E. C. G.: 1980, 'Resonances, Scattering Theory, and Rigged Hilbert Spaces', *J. Math. Phys.* **21**, 2203–2216.

Parthasarathy, K. R. and Sinha, K. B.: 1982, 'Feynman Path Integrals of Operator-Valued Maps', *J. Math. Phys.* **23**, 1459–1462.

Patrascioiu, A.: 1981, 'Complex Time and Gaussian Approximation', *Phys. Rev.* **D24**, 496–504.

Peres, A.: 1980, 'Nonexponential Decay Law', *Ann. Phys.* **129**, 33–46.

Perkins, D.: 1972, *Introduction to High-Energy Physics*, Addison-Wesley, Reading.

Petzold, J.: 1959a, 'Wie gut gilt das Exponentialgesetz beim α-Zerfall?', *Zeit. Phys.* **155**, 422–432.

Petzold, J.: 1959b, 'Zum Anfangwertproblem zerfallender Zustände', *Zeit. Phys.* **157**, 122–129.

Phillips, R. S.: 1959, 'Dissipative Operators and Hyperbolic Systems of Partial Differential Equations', *Trans. Amer. Math. Soc.* **90**, 193–254.

Piron, C.: 1969, 'Les règles de supersélection continues', *Helv. Phys. Acta* **42**, 330–338.

Popov, V. N.: 1976, *Functional Integrals in Quantum Field Theory and Statistical Physics*, Atomizdat, Moscow (in Russian).

Prokhorov, L. V.: 1982, 'Hamiltonian Functional Integrals', *Prob. fiz. elem. chas. atom. yadra* **13**, 1094–1156 (in Russian).

Rączka, R.: 1973, 'A Theory of Relativistic Unstable Particles', *Ann. Inst. H. Poincaré* **A19**, 341–356.

Radin, Ch.: 1975, 'Some Remarks on the Evolution of a Schrödinger Particle in an Attractive $1/r^2$ Potential', *J. Math. Phys.* **16**, 544–547.

Rau, J.: 1963, 'Relaxation Phenomena in Spin and Harmonic Oscillator Systems', *Phys. Rev.* **129**, 1880–1888.

Rauch, J.: 1980, 'Perturbation Theory for Eigenvalues and Resonances for Schrödinger Hamiltonians', *J. Funct. Anal.* **35**, 304–315.

Reed, M. and Simon, B.: 1972–79, *Methods of Modern Mathematical Physics: I, Functional Analysis; II, Fourier Analysis. Self-adjointness; III, Scattering Theory; IV, Analysis of Operators*, Academic Press, New York.

Rezende, J.: 1983, 'Remark on the Solution of the Schrödinger Equation for Anharmonic Oscillators via Feynman Path Integral', *Lett. Math. Phys.* **7**, 75–84.

Richtmyer, D. R.: 1978, *Principles of Advanced Mathematical Physics*, I, Springer, New York.

Roffman, E. H.: 1967, 'Unitary and Non-unitary Representations of the Complex Inhomogeneous Lorentz Group', *Commun. Math. Phys.* **4**, 237–257.

Romo, W. J.: 1980, 'A Study of Completeness Properties of the Resonant States', *J. Math. Phys.* **21**, 311–326.

Rudin, W.: 1973, *Functional Analysis*, McGraw-Hill, New York.
Santhanam, T. S. and Sinha, K. B.: 1978, 'Quantum Mechanics in Finite Dimensions', *Australian J. Phys.* **31**, 233–238.
Scher, G., Smith, M. and Baranger, M.: 1980, 'Numerical Calculations in Elementary Quantum Mechanics using Feynman Path Integrals', *Ann. Phys.* **130**, 290–306.
Schulman, L. S.: 1970, 'Unstable Particles and Poincaré Semigroup', *Ann. Phys.* **59**, 201–218.
Schulman, L. S.: 1981, *Techniques and Applications of Path Integration*, Wiley-Interscience, New York.
Schwartz, L.: 1967, *Analyse mathématique*, I, Hermann, Paris.
Schweber, S. S.: 1961, *An Introduction to Relativistic Quantum Field Theory*, Row, Peterson and Co., Evanston.
Schwinger, J.: 1960, 'Field Theory of Unstable Particles', *Ann. Phys.* **9**, 169–193.
Shabat, B. V.: 1976, *Introduction to Complex Analysis: II, Functions of Several Variables* (2nd edn), Nauka, Moscow (in Russian).
Sherry, T. N. and Sudarshan, E. C. G.: 1978, 1979, 'Interaction between Classical and Quantum Systems: a New Approach to Quantum Measurement, I, II', *Phys. Rev.* **D18**, 4580–4589; **D20**, 857–868.
Sherry, T. N., Sudarshan, E. C. G. and Gantham, S. R.: 1979, 'Interaction between Classical and Quantum Systems: a New Approach to Quantum Measurement: III, Illustration', *Phys. Rev.* **D20**, 3081–3094.
Shilov, G. E.: 1963, 'Integration in Infinite-Dimensional Spaces and the Wiener Integral', *Usp. mat. nauk* **18** (2), 99–120 (in Russian).
Shirokov, M. I.: 1975, 'On the Decay Law Exponentiality', *Yadernaya fiz.* **21**, 674–687.
Simon, B.: 1971, *Quantum Mechanics for Hamiltonians Defined as Quadratic Forms*, Princeton Univ. Press, Princeton.
Simon, B.: 1973, 'Resonances in *N*-Body Quantum Systems with Dilation Analytic Potentials and the Foundation of Time-Dependent Perturbation Theory', *Ann. Math.* **97**, 247–272.
Simon, B.: 1974, *The $P(\varphi)_2$ Euclidian (Quantum) Field Theory*, Princeton Univ. Press, Princeton.
Simon, B.: 1979a, *Functional Integration and Quantum Physics*, Academic Press, New York.
Simon, B.: 1979b, 'Phase Space Analysis of Simple Scattering Systems: Extensions of Some Work of Enss', *Duke Math. J.* **46**, 119–168.
Simonius, M.: 1970, 'Description of Unstable Particles by Nonunitary Representations of the Poincaré Group', *Helv. Phys. Acta* **43**, 223–234.
Sinha, K.: 1972, 'On the Decay of an Unstable Particle', *Helv. Phys. Acta* **45**, 619–628.
Sinha, K.: 1978, 'Relative Scattering and Time Delay in Presence of a Uniform Electric Field', *Rep. Math. Phys.* **14**, 65–73.
Sinha, K.: 1979, 'Time Observable in Quantum Mechanics', A talk on Einstein Centenary Symposium, Calcutta.
Siroid, I.-P. P.: 1981, 'Sufficient Conditions in Terms of the Potential for Absence of Spectral Singularities of a Non-selfadjoint Schrödinger Operator', *Sibirskii mat. zh.* **22**, 151–157 (in Russian).
Skorokhod, A. V.: 1975, *Integration in Hilbert Space*, Nauka, Moscow (in Russian).
Slavnov, A. A. and Faddeev, L. D.: 1978, *Introduction to the Quantum Theory of Gauge Fields*, Nauka, Moscow (in Russian; English trans.: Benjamin, Reading 1980).
Smolianov, O. G.: 1982, 'Infinite-Dimensional Pseudodifferential Operators and Schrödinger Quantization', *Dokl. Akad. Nauk USSR* **263**, 558–562 (in Russian).
Souriau, J.-M.: 1976, 'Construction explicite de l'indice de Maslov', in A. Janner, T. Janssen and M. Boon (eds.), *Group Theoretical Methods in Physics*, Lecture Notes in Physics, Vol. 50, Springer, Berlin, pp. 117–148.
Šťovíček, P.: 1980, 'Quantum Mechanics on a Finite Abelian Group', Czech Technical Univ., Prague (in Czech; unpublished).
Šťovíček, P. and Tolar, J.: 1979, 'Quantum Mechanics in a Discrete Space-Time', Preprint IC/79/147, Trieste; *Rep. Math. Phys.* (to appear).

Streater, R. F.: 1982, 'Damped Oscillator with Quantum Noise', *J. Phys.* **A15**, 1477–1485.

Streater, R. F. and Wightman, A. S.: 1964, *PCT, Spin, Statistics and All That*, W. A. Benjamin. New York.

Suárez, P.: 1984, 'Feynman Path Integrals for the Dirac Equation', *J. Phys.* **A17**, 1003–1009.

Sudarshan, E. C. G., Chiu, C. B. and Gorini, V.: 1978, 'Decaying States as Complex Energy Eigenvectors in Generalized Quantum Mechanics', *Phys. Rev.* **D18**, 2914–2929.

Sveshnikov, A. G. and Tikhonov, A. N.: 1971, *The Theory of Functions of a Complex Variable*, Mir, Moscow.

Sz.-Nagy, B.: 1953a, 'Sur les contractions de l'espace Hilbert', *Acta Scient. Math., Szeged* **15**, 87–92.

Sz.-Nagy, B.: 1953b, 'Transformations de l'espace Hilbert, fonctions de type positif sur une groupe', *Acta Scient. Math., Szeged* **15**, 104–114.

Sz.-Nagy, B.: 1955, 'Prolongements des transformations de l'espace de Hilbert qui sortent de cet espace', Budapest, also as an appendix to the monograph by F. Riesz and B. Sz.-Nagy, *Leçons d'analyse fonctionelle*, since the 1960 English edition (Ungar, New York).

Sz.-Nagy, B. and Foias, C.: 1970, *Harmonic Analysis of Operators on Hilbert Space*, North-Holland, Amsterdam.

Tarski, J.: 1979, 'Feynman-Type Integrals Defined in Terms of General Cylindrical Approximations', in S. Albeverio *et al.* (eds), *Feynman Path Integrals*, Lecture Notes in Physics, Vol. 106, Springer, Berlin, pp. 254–279.

Tarski, J.: 1981, 'Path Integrals over Phase Space, their Definition and Simple Properties', Preprint IC/81/192, Trieste.

Taylor, J. R.: 1972, *Scattering Theory. The Quantum Theory of Non-relativistic Collisions*, Wiley, New York.

Terentiev, M. V.: 1965, 'The Decay $K_2 \to 2\pi$ and a Possible CP-Parity Non-conservation', *Usp. fiz. nauk* 86, 231–262 (in Russian).

Terentiev, M. V.: 1972, 'On the Exponential Decay Law of Nonstable Particle', *Ann. Phys.* **74**, 1–42.

Titchmarsh, E. C.: 1951, 'Some Theorems on Perturbation Theory, III, IV', *Proc. Roy. Soc.* **A207**, 321–328, **A210**, 30–47.

Trotter, H.: 1959, 'On the Product of Semigroups of Operators', *Proc. Amer. Math. Soc.* **10**, 545–551.

Truman, A.: 1976, 'Feynman Path Integrals and Quantum Mechanics as $\hbar \to 0$', *J. Math. Phys.* **17**, 1852–1862.

Truman, A.: 1977a, 'The Classical Action in Nonrelativistic Quantum Mechanics', *J. Math. Phys.* **18**, 1499–1509.

Truman, A.: 1977b, 'Classical Mechanics, the Diffusion (Heat) Equation and the Schrödinger Equation, *J. Math. Phys.* **18**, 2308–2315.

Truman, A.: 1978a, 'The Feynman Maps and the Wiener Integral', *J. Math. Phys.* **19**, 1742–1750; erratum **20**, 1832–1833.

Truman, A.: 1978b, 'Some Applications of Vector Space Measures to Non-relativistic Quantum Mechanics', in R. Aron and S. Dinceu (eds), *Vector Space Measures and Applications*, Lecture Notes in Mathematics, Vol. 644, Springer, Berlin 1978, pp. 418–441.

Truman, A.: 1979, 'The Polygonal Path Formulation of the Feynman Path Integrals', in S. Albeverio *et al.* (eds), *Feynman Path Integrals*, Lecture Notes in Physics, Vol. 106, Springer, Berlin, pp. 73–102.

Uglanov, A. V.: 1978, 'On a Construction of the Feynman Integral', *Dokl. Akad. Nauk USSR* **243**, 1406–1409.

Úlehla, I., Gomolčák, L. and Pluhař, Z.: 1964, *Optical Model of the Atomic Nucleus*, Czechoslovak Academy of Sciences Publishing House, Prague.

Vasiliev, A. N.: 1976, *Functional Methods in Quantum Field Theory and Statistics*, Leningrad State Univ. Press, Leningrad (in Russian).

Vasiliev, A. N. and Kuzmenko, A. V.: 1977, 'Functional-Integral Representation of the Scattering Amplitude and Quasiclassical Asymptotics in Quantum Mechanics, *Teor. mat. fiz.* **31**, 313–326 (in Russian).

Vasiliev, A. N. and Kuzmenko, A. V.: 1979, 'Functional Integral for the Scattering Amplitude in Presence of Long-Range Forces', *Teor. mat. fiz.* **41**, 12–25 (in Russian).

Vladimirov, V. S.: 1971, *Equations of Mathematical Physics* (2nd edn), Nauka, Moscow (in Russian).

Vock, E. and Hunziker, W.: 1982, 'Stability of Schrödinger Eigenvalue Problems', *Commun. Math. Phys.* **83**, 281–302.

Votruba, V.: 1969, *Foundations of the Special Theory of Relativity*, Academia, Prague (in Czech).

Weidmann, J.: 1980, *Linear Operators in Hilbert Space*, Graduate Texts in Mathematics, Vol. 68, Springer, New York.

Weisskopf, V. F. and Wigner, E. P.: 1930a, 'Berechnung der natürlichen Linienbreite auf Grund der Diracschen Lichttheorie', *Zeit. Phys.* **63**, 54–73.

Weisskopf, V. F. and Wigner, E. P.: 1930b, 'Über die natürliche Linienbreite in der Strahlung des harmonischen Oszillators, *Zeit. Phys.* **65**, 18–27.

Weldon, H. A.: 1976, 'Description of Unstable Particles in Quantum Field Theory', *Phys. Rev.* **D14**, 2030–2063.

Wells, W. H.: 1961, 'Quantum Formalism Adapted to Radiation in a Coherent Field', *Ann. Phys.* **12**, 1–40.

Wigner, E. P.: 1939, 'On Unitary Representations of the Inhomogeneous Lorentz Group', *Ann. Math.* **40**, 149–204.

Wigner, E. P.: 1955, 'Lower Limit for the Energy Derivatives of the Scattering Phase Shift', *Phys. Rev.* **98**, 145–147.

Wigner, E. P.: 1970, *Symmetries and Reflections*, Indiana Univ. Press, Bloomington.

Wildermuth, K. and Tang, Y. C.: 1977, *A Unified Theory of the Nucleus*, Vieweg, Braunschweig.

Williams, D. N.: 1971, 'Difficulty with a Kinematic Concept of Unstable Particles: the Sz.-Nagy Extension and the Matthews–Salam–Zwanziger Representation', *Commun. Math. Phys.* **21**, 314–333.

Winter, R. G.: 1961, 'Evolution of a Quasistationary State', *Phys. Rev.* **123**, 1503–1507.

Wollenberg, M.: 1977, 'Levinsontheorem, instabile Eigenwerte und Streuquerschnittmaxima', *Math. Nachr.* **78**, 223–230.

Yasue, K.: 1978, 'Quantum Mechanics of Non-conservative Systems', *Ann. Phys.* **114**, 479–496.

Yosida, K.: 1966, *Functional Analysis*, Die Grundlehren der Mathematischen Wissenschaften, Band 123, Springer, Berlin.

Yourgrau, W.: 1968, 'A Budget of Paradoxes in Physics', in I. Lakatos and A. Musgrave (eds), *Problems in Philosophy in Science*, North-Holland, Amsterdam, pp. 178–199.

Zwanziger, D.: 1963, 'Representations of the Lorentz Group Corresponding to Unstable Particles', *Phys. Rev.* **131**, 2818–2819.

Subject Index